HYDROGEN ION CONCENTRATION

New Concepts in a Systematic Treatment

HYDROGEN ION CONCENTRATION

New Concepts in a Systematic Treatment

By JOHN E. RICCI

PRINCETON, NEW JERSEY
PRINCETON UNIVERSITY PRESS
1952

Published, 1952, *by Princeton University Press*
London: Geoffrey Cumberlege, Oxford University Press

Printed in the United States of America

Preface

THE quantitative relations determining the hydrogen ion concentration of aqueous solutions constitute a purely mathematical problem subject to exact and systematic treatment and capable of simple and uniform presentation. On looking into the best available discussions, however, one finds that except for some isolated problems the subject has not been treated in any truly general way. The usual procedure is to approach the derivation of formulas involving the hydrogen ion concentration not on the basis of strict mathematical definitions but through simplifying assumptions according to some special theory of the process of ionization. As a consequence the problem of the precise calculation in all the possible variety of cases appears to be very complicated, and certain simple, interesting, and important conclusions have been missed.

The purpose of this study is to consider the problem without special assumptions and without dependence on theories of the mechanism of the process of ionization, but purely on the basis of mathematical definitions, and to present a systematic derivation of interrelated formulas for calculations involving the hydrogen ion concentration in various aqueous solutions of almost any degree of complexity.

The classical and still familiar approach is based on the theory of dissociation as a mechanism for the process of ionization, and although this theory led to the postulation of "dissociating compounds" such as "H_2SO_3" and "NH_4OH," which are otherwise both unnecessary and unjustifiable, it did succeed in giving us immediately various useful formulas of first approximation, such as the Ostwald dilution law. Accordingly, the procedure for the derivation of better or second approximations, for difficult cases or for extreme conditions, has generally been that of introducing further special assumptions into the original, already approximate, derivations. But unless all the approximations (first, second, third, etc.), although of some undeniable practical usefulness, are related to truly exact equations, their applicability in any given case is always somewhat in question, and the error may sometimes be enormous. One of the examples to be developed later concerns the important first approximation, the "common ion" equation, also known as Henderson's equation, for a mixture of a weak acid (or base) and its salt, for which, we shall find, it is possible to define explicit conditions of applicability. In the case of this particular approximation it will even be possible by means of the systematic procedure to be followed to predict the error in its use so that the first approximation formula, which is so easy to handle, may then be used to obtain a much more accurate value.

As for "second approximations," we find that because of the method commonly used for their derivation, they do not even seem to be the same in various texts, nor is it always apparent which refinement is to be used in any given case. The reason for this is that the manner in which the simplifying assumptions are made usually hides or confuses the actual number and relation of the variables involved, and that the conditions of applicability are usually stated in terms of the very quantity being calculated—the hydrogen ion concentration itself, or the degree of ionization, etc. It is therefore not always evident, at the end, whether the derived equation applies for certain ranges of values of the concentrations, of the ionization constants, or of both. Indeed, the various statements made are sometimes even contradictory.

If second approximations, or more accurate formulas, are at all desirable and useful, their relation to general and exact equations should be shown so that the nature of the assumptions may be readily appreciated and the effect of concomitant variables estimated, if possible. It is important, moreover, that in a subject which is so purely mathematical, the word "exact," unqualified, should never be used for even an excellent approximation. But even this fault may be noted in some of the best work in the field. The word "exact" must involve no question of the degree of accuracy in the final calculated value, for it is otherwise misleading and may leave the true solution of many problems unnecessarily obscured.

As an example of the way in which the usual procedure of derivation does not allow any appreciation of either the sign or the extent of the effect of certain neglected variables, we may cite the familiar first approximation formula for the hydrogen ion concentration in a solution of the salt of a weak acid and a weak base. Despite the length and apparent complexity of the usual derivation of this formula, making use of the nondefinable expression "degree of hydrolysis" of the salt, we have merely to combine the three definitions, $W = [H^+][OH^-]$ for the ion-product constant of water, $A = [H^+][X^-]/[X^0]$ for the ionization constant of the acid, and $B = [M^+][OH^-]/[M^0]$ for that of the base, *assuming* that $[X^-]/[X^0] = [M^+]/[M^0]$, to obtain the result, $[H^+] = \sqrt{WA/B}$. The effect of the concentration, c, of the salt, then, can not possibly be estimated, although we know that $[H^+]$ must equal $\sqrt{W}$ when c is zero for any value of A/B. It has been pointed out (by Griffith) that the formula "requires slight modification in extremely dilute solution" because of the failure, then, of the "assumption" that the weak acid and weak base are practically completely unionized. This, however, is not a necessary assumption in the derivation of the formula at all, as we have just seen. (Nor, as will be shown later, does the applicability of the formula depend on c alone.) The only necessary assumption seems to be that $[X^-]/[X^0] = [M^+]/[M^0]$, which is possible for the salt solution strictly only when $A = B$, in which case $[H^+]$ remains equal to $\sqrt{W}$, as in pure water, for all values of c.

Clark has derived the following relation for this case: $[H^+] = \sqrt{WA(B + [M^+])/B(A + [X^-])}$. According to Clark this leads to various familiar expressions, including the formula $[H^+] = \sqrt{WA/B}$, "if B and A are small in relation to $[M^+]$ and $[X^-]$ and if the solution is sufficiently dilute, so that $[M^+]$ and $[X^-]$ each approximate the salt concentration," c. Actually, however, the simple formula $[H^+] = \sqrt{WA/B}$ increases in accuracy not as c decreases but as c increases. We shall find, in fact, contrary to what seems generally to be expected, that when c approaches zero, one of the ratios, $[M^+]/c$ and $[X^-]/c$, increases while the other decreases, unless $A = B$, in which case a change in concentration has no effect at all on these ratios. Clark's equation therefore, though exact, is not useful since we do not know how, in general, to introduce c into it. If the unknowns, $[M^+]$ and $[X^-]$, are eliminated, the equation is of the fourth degree in $[H^+]$.

This equation, we may note, is typical of many of the expressions in the literature in which $[H^+]$ is shown as a function of quantities themselves unknown, and to be estimated in particular cases. When these related unknowns are eliminated, the resulting equations are generally not directly useful since they are usually of high degree in $[H^+]$. One of the purposes of this study is to simplify these general exact expressions into usable forms of various degrees of approximation, and with limits or conditions of applicability stated, as far as possible, in terms not of the unknowns involved but of the known parameters, the analytical concentrations, and the equilibrium constants.

The very nature of certain of the special assumptions made in familiar derivations is not clear at all except from the point of view of a special theory of ionization. The most frequent of these is "the neglect of the hydrogen ion coming from the water" in an aqueous acid. But the hydrogen ion concentration or activity in an aqueous acid, unless the solute itself contains hydrogen exchangeable with that of the water, is altogether that of the water made more active by the acidic solute; the above expression therefore can have absolutely no meaning in the case of aq. CO_2, for example, or aq. SO_3, etc. We must obviously concentrate throughout on the hydrogen ion concentration of the medium as a whole, and in most cases the hydrogen ion must "come" completely from the water itself. In other words, the expression "to neglect the ionization of the water" is absurd, for the very thing being calculated, in the calculation of the "hydrogen ion concentration" of an aqueous solution, is the effect of the solute or solutes on the relation between $[H^+]$ and $[OH^-]$ (for example, the difference $[H^+] - [OH^-]$) of the medium.

Certain problems of high apparent difficulty can be solved with surprising ease on the basis of the general exact equations, whereas they appear sometimes too formidable even to be attempted with our usual method of approach. The reason for this is that the ordinary method is that of starting with an "obvious" first approximation such as the Henderson equation for titration, and modifying this with further special assumptions to arrive at greater accuracy. This leads always to relatively involved expressions which are nevertheless only approximate, requiring much labor and time in their application and still giving answers which may sometimes be far from the truth. The full equations defining $[H^+]$ in mixtures of two weak electrolytes are at least of the fourth degree in $[H^+]$, and it seems to be felt that a simpler method of thinking will solve the problem so long as it does not lead to equations of such high degree. But the idea is false for we can never then know how to solve the problem in the case of certain extreme possible values of the parameters involved. We usually neglect $[OH^-]$ relative to $[H^+]$ in acid solution, but we find cases in which $pH \cong 7$ to be often those of the greatest importance, expecially in certain indicator problems; and then the usual methods actually involve laborious processes which, it will be shown, become unnecessary so long as we never make any assumptions in our thinking and in our derivations, and leave all approximations to the final numerical solution of the truly exact equation. This is after all the only correct and ultimately easy method in any purely mathematical problem. To derive approximate practical formulas by simplification of exact mathematical equations is a legitimate and safe procedure, but to derive practical formulas on the basis of approximate thinking is unsafe, unsound, and in this case unnecessary.

A few examples will be mentioned here of the usefulness of this general point of view: (1) The problem of finding the ratio of the two forms of an indicator to give an indicator solution of specified p*H*, particularly for $[H^+] = \sqrt{W}$, will be given a very simple solution; (2) The problem of estimating the "feasibility" of various titrations, including that of one acid in the presence of another, will be easily handled; (3) The theorem of isohydric solutions, also, has always appeared so difficult either to prove or to disprove with the usual equations that it is not quite clear when it may or may not be applied. If "isohydric indicators" are to be used, however, their applicability depends on the validity of the theorem. The general conditions for the validity of this theorem, so fundamental to the application of many of the formulas to be derived, will also be explained by the systematic treatment; (4) The value of $[H^+]$ in solutions of salts such as NH_4HSO_3 and $NaNH_4HPO_4$, even as function of the concentration, seems to be a problem hardly approachable, in view of the number of variable parameters involved, with the usual methods. Yet even for this general type of salt solution, relatively simple formulas will be clearly and systematically derivable with the exact approach; (5) In fact it seems quite difficult to answer the simple question whether a solution of a salt of the type $NaNH_4X$ (given the two ionization constants of the dibasic acid and that of the weak base) will be acid or alkaline with the usual "formulas" and treatment; such a question, however, turns out to have an extremely simple answer.

The degree of the exact algebraic equation for $[H^+]$ in solutions not involving complex ions and species is found to be equal to one plus the number of equilibrium constants involved, so that even for the case of a single ionization constant, in addition to the ion-product constant of water, the equation is already cubic. For these solutions, however, of the type which we shall later define as "noncomplex" (limited to complete dissociation of all possible salts and to the absence of any complex-ion formation), the full exact equation is always linear with respect to the analytical concentrations. This relation makes possible the direct exact calculation of the composition (the relations and values of the analytical concentrations) for any specified value of $[H^+]$, thus solving in surprisingly simple fashion many problems otherwise very difficult on the basis of the same equations considered as definitions of $[H^+]$ as the dependent variable.

At the same time, of course, a problem of equal practical importance, especially in analytical chemistry, is the calculation of $[H^+]$ as function of constants and concentrations, and unless we are to revert to approximate derivations leading to simple equations of degree not higher than two, we must find a way of dealing with the exact equation, of any degree, for the calculation of the numerical value of $[H^+]$. Because of the apparent variety and complexity and the high degree of most of the equations to be encountered, it may easily be felt that, although "interesting," such expressions can not in general be of practical use; the usual numerical solution by means of successive trial and error approximations certainly does seem to leave the equations little more than academic interest. Two principles or observations will be offered, however, which should remove this difficulty and enable us to use these general exact equations, which are after all algebraically simple to derive and which, if properly used, then assure us that no relatively large effect of any possible variable has been overlooked in the particular problem.

The first point is a systematic relatively simple procedure for the numerical solution of the equations. The method allows us easily to choose the best approximation for any specific problem purely on mathematical grounds, and even to estimate roughly the accuracy of the approximate answer. Because of the nature of the physical relations to which the otherwise cumbersome equations actually apply, this method turns out to give accurate numerical answers in almost any particular problem. In addition, of course, the applicability of the various possible approximations of the general equation applying to a problem will be stated, as far as possible, in terms of the actual data of the problem, the explicit parameters, or the known analytical concentrations and the ionization constants, so that the work of selecting the correct approximate solution is necessary essentially only in the original investigation of the particular type of problem.

The second point has to do with the general interrelation of the equations. It is found, for example, that all the equations dealing with $[H^+]$ as function of one ionization constant (besides W) are mathematically identical, being variants or special cases of a single mathematical equation. A second fundamental equation describes all solutions involving two ionization constants pertaining to two independent solutes. A third covers all cases in which two constants pertain to a single solute. This fundamental identity, possible through the mathematical transformation of ionization constants—nothing more than redefinition—obviously reduces the mathematical operations greatly. The problem of the hydrogen ion concentration of the solution of a weak acid and the problem of the hydrogen ion concentration of the solution of the salt of a strong acid and a weak base, for example, are not merely similar but mathematically—exactly—*identical*. That for the salt of weak acid and weak base is merely the case of a special equivalent point in the titration, with strong base, of a mixture of two weak acids (or with strong acid, of a mixture of two weak bases) at equal concentrations.

A few complexities such as the ionization of a "dimeric" acid and the association of the various species, charged or uncharged, of an acid in solution are considered in Chapter xv, which includes hydrofluoric acid, chromic acid, and boric acid in its examples. Complex ions have otherwise not been treated, the main concern being with the general relations of acids and bases as such in solution.

A systematic development of the relations between hydrogen ion concentration and the solubility of electrolytes has been attempted in the last five chapters. The relations are applied to certain problems of separation and to the question of the conditions and errors of certain volumetric methods involving precipitation. The aim throughout is to express the dependence of the solubility of a substance not simply upon the hydrogen ion concentration of the solution but explicitly upon the analytical concentrations of reagents used to control the hydrogen ion concentration. The treatment is again in general exact, and it is distinguished by the use of graphical representations related to ordinary phase rule diagrams. In this way it is felt that it becomes possible to deal with solubility relations involving a considerable number of equilibrium constants without too great difficulty. In twofold saturation with respect to both a simple ampholyte and its salt of weak monobasic acid, for example, we have to consider simultaneously two solubility products, three independent ionization constants, and the ion-product constant of the solvent. Such a condition may arise if an excess of the solid ampholyte is suspended in a solution of the alkali salt of the weak acid, resulting in "anion exchange" upon the suspended solid. The relations in such situations are more easily sorted and appreciated with the aid of certain "phase diagrams" as background than on the basis of algebraic expressions alone.

This systematic elaboration of interrelated "p*H*" problems constituted an activity both stimulating and satisfying to the writer, who hopes that the perusal of it may prove equally interesting and suggestive to others. He also hopes that there are not too many errors in the almost innumerable equations and formulas presented.

I take this opportunity to express my appreciation to Professor H. Austin Taylor of New York University for his kindness in reading the manuscript in its early stages and for his guidance and criticism in many discussions, and to Dr. Nathaniel Thon of Princeton University for valuable comments upon his reading the manuscript. For the drawings I am indebted to Mr. William P. Millé, lithographer, of Yonkers, N.Y.

Finally, I am deeply indebted to the Princeton University Press for its willingness to accept this undertaking, and in particular to Dean Hugh S. Taylor for his help and encouragement in the appraisal of the manuscript.

J. E. Ricci

New York University
February 1, 1951

Contents

Symbols and Abbreviations

1. Capitals, Roman

C^0, $C^{-(+)}$: complex species

M: a specific base

M^0, M^+, M^{++} · · · $M^{+z'}$: its species in solution

MOH, $M(OH)_2$, $M(OH)_3$, NOH, $N(OH)_2$, etc.: various specific bases, in problems involving more than one

M_sOH: a specific strong monoacid (or univalent) base

X: a specific acid

X^0, X^-, $X^=$ · · · X^{-z}: its species in solution

HX^0, X^-, HY^0, Y^- }
H_2X^0, HX^-, $X^=$ } species of acids of various basicities
H_3X^0, H_2X^-, $HX^=$, $X^{\equiv}$ }

HX_s: a specific strong monobasic (or univalent) acid

MX, M_sX, MX_s, M_sX_s: various salts

X_2: dimer of X

U, U^0, U^+, U^-, $U^\pm$: an ampholyte and its species

Z, Z^0, Z^-, Z^+, etc.: a solute, in general, and its species

All chemical symbols and formulas.

2. Capitals, Italic and Bold

Note: Mass constants (equilibrium constants in terms of concentrations) are printed in italic; the corresponding thermodynamic constants (equilibrium constants in terms of activities) are printed in bold face.

A ($\mathbf{A}$): ionization constant of a monobasic acid

A_1, A_2 · · · A_z: ionization constants of various monobasic acids

B ($\mathbf{B}$), B_1, B_2 · · ·: same, for monoacid bases

$\mathbf{C}_0$, $\mathbf{C}_1$, $\mathbf{C}_2$ · · ·: coefficients of a polynomial

D: $H - OH$

$\mathbf{D}$: $\mathbf{H} - \mathbf{OH}$

D': $-D = OH - H$

E: relative titration error; $100E$ = percentage error

E: electromotive force

F, F_-, F_+: feasibility functions

F: the faraday

G: feasibility function in complex case

H: concentration of hydrogen ion; $[H^+]$

H: activity of hydrogen ion; a_{H^+}

$K(\mathbf{K})$: an equilibrium constant, in general

K^*: conjugate value of the ionization constant K; W/K

$K_1, K_2 \cdots K_z$: successive ionization constants of a polybasic acid

K_{12}, K_{123}, etc.: the products $K_1K_2, K_1K_2K_3$, etc.

$K'_1, K'_2 \cdots K'_{z'}$: successive ionization constants of a polyacid base

$K_a(\mathbf{K}_a)$: the acid ionization constant of a 1 : 1 ampholyte

$K_b(\mathbf{K}_b)$: the base ionization constant of a 1 : 1 ampholyte

$K_{a_1} \cdots K_{a_{z_-}}$: successive acid ionization constants of a general ampholyte

$K_{b_1} \cdots K_{b_{z_+}}$: successive base ionization constants of a general ampholyte

K_w: ionization constant of water

L: the liquid phase

M: molarity

OH: concentration of hydroxyl ion; $[OH^-]$

OH: activity of hydroxyl ion; a_{OH^-}

$P(\mathbf{P})$: solubility product

P_A: solubility product of acid salt

P_B: solubility product of basic salt

P_a, P_b, P_s: forms of the solubility product of an ampholyte

Q: solution of a quadratic equation

R_n: the "symmetrical ratio" of hydrogen ion concentrations in titration, or $H_n/H_{(1-n)}$, in which n is the titration ratio b_s/a for titration of an acid, a, with strong base, b_s; for dibasic acid, $R_n = H_n/H_{(2-n)}$

R: ratio of salt concentrations

S: solubility

T: temperature

V: volume

$W(\mathbf{W})$: ion-product constant of water

Note: The letters **J, X, Y** are used as temporary arbitrary abbreviations for various functions.

3. Small Capitals

A, B: functions in the Debye-Hückel limiting law

K, K′: special equilibrium constants for boric acid, Chapter xv

K_1, K_2, etc: special equilibrium constants for boric acid and for phosphoric acid, Chapter xv

K: the distance parameter in the Debye-Hückel second approximation

P: pressure

P: a point in a diagram

R: the gas constant

4. Small Letters, Italic

a: concentration of an acid

$a_1, a_2, a_3 \cdots; a_X, a_Y$: concentrations of various acids, distinguished either by numbers or by the symbols X, Y, etc.

a_s: concentration of strong monobasic acid

b: concentration of a base

$b_1, b_2, b_3 \cdots; b_M, b_N$: concentrations of various bases

b_s: concentration of strong monoacid base

c: concentration in general; usually of a salt or of an ampholyte

c_A, c_B: concentration of a particular acid (or base) solution used in titration

d: sign of differentiation

e: an error

$f(\ )$: function of ( )

g: $a_s - b_s$

g': $-g$

h, h': degree of hydration of a solute species

i: ρ/α $(= 1/r)$

j: the number of acids (or of bases, or of solutes in general in a solution)

k, k': specific equilibrium constants

k_i: specific equilibrium constant for ionization

k_T: specific equilibrium constant for tautomerization

m: valence of the acid radical of a salt

m': valence of the base radical of a salt

n: b/a; titration ratio; usually b_s/a

n': a/b; titration ratio; usually a_s/b

n^*: $2b_s/a$; titration ratio for pseudodibasic acid, Chapter XV

p: percentage

q, q': special abbreviations in Chapter XIV

$r_1 \cdots r_z$: $[X^-]/[X^0]$, $[X^=]/[X^-]$ $\cdots$ $[X^{-z}]/[X^{-(z-1)}]$

r: $[X^-]/[X^0]$ for monobasic acid; α/ρ

$s_1 \cdots s_z$: $[X^-]/[X^0]$, $[X^=]/[X^0]$ $\cdots$ $[X^{-z}]/[X^0]$

s: same as r

t: total concentration

u, v, w: coefficients of a quadratic equation

x: a temporary symbol

y: K_2/K_1 for a polybasic acid; also A_2/A_1 for two monobasic acids

y_1: K_3/K_2

z: basicity of an acid; number of ionization constants of an acid; maximum negative charge of the acid

z': corresponding symbol for a base

z_-: number of acid ionization constants (maximum negative charge) of an ampholyte

z_+: number of base ionization constants (maximum positive charge) of an ampholyte

5. Small Letters, Bold

$\mathbf{a}$: moles of an acid per liter of system

$\mathbf{b}$: moles of a base per liter of system

$\mathbf{c}$: moles of a solute per liter of system

$\mathbf{c}_e$: $c_A c_B/(c_A + c_B)$

$\mathbf{g}$: value of g in original solution

$\mathbf{h}$: $\mathbf{a} - \mathbf{b}$

$\mathbf{h}^*$: $2\mathbf{b} - \mathbf{a}$, in Chapter XIX

$\mathbf{h}'$: $\mathbf{a} - (\mathbf{b}_M + \mathbf{b}_N)$, in Chapter XIX

$\mathbf{k}$: $S\rho$

$\mathbf{p}$: K_1K_2

$\mathbf{p}_1$: K_2K_3

$\mathbf{q}$: moles of a precipitate per liter of system

r: $[K_2CrO_4]/[K_2Cr_2O_7]$, in Chapter xv

v, **w**: *coefficients of a quadratic equation*

y: $K_2/\sqrt{K}$, for chromic acid, in Chapter xv

z: the number of monobasic acids in a solution

6. Small Letters, Roman

$a_{(\ )}$: activity of the species ( )

p( ): $-\log_{10}(\ )$

Points on diagrams

7. Greek Letters

α: degree of ionization of a monobasic acid

$(\alpha)_1, (\alpha)_2 \cdots (\alpha)_{\mathbf{z}}$: ionization fractions for a number (**z**) of monobasic acids

$\alpha_1, \alpha_2 \cdots \alpha_z$: successive ionization fractions of a polybasic acid

α_m: the m'th ionization fraction of a polybasic acid

α'
$(\alpha')_1, (\alpha')_2, (\alpha')_3 \cdots$
$\alpha'_1, \alpha'_2 \cdots \alpha'_{z'}$
$\alpha'_{m'}$
} corresponding symbols, for bases

α_0: unionized fraction of an ampholyte

$\alpha_0, \alpha_-, \alpha_+$: ionization fractions of a 1 : 1 ampholyte

$\alpha_{-1}, \alpha_{-2} \cdots \alpha_{-z_-}$: acidic (negative) ionization fractions of a higher order ampholyte

α_{-m}: the m'th negative ionization fraction of same

$\alpha_{+1}, \alpha_{+2} \cdots \alpha_{+z_+}$: basic (positive) ionization fractions of same

$\alpha_{+m'}$: (m')'th positive ionization fraction of same

α_t: total fraction ionized, for any solute, such as $1 - \rho$ for an acid

β: charge coefficient of a solute; net number of negative solute charges per mole of solute

β': net number of positive solute charges per mole of solute; used only for a base, for which $\beta' = -\beta$

β_C: charge coefficient for a complex solute

β_S: charge coefficient for a salt

$\gamma_{(\ )}$: activity coefficient of the species ( )

γ: activity coefficient of a univalent ion according to the Debye-Hückel limiting law

Δ: an increment

ε: a very small quantity

η: $n/(1-n)$, $= b/(a-b)$, during titration

θ_{MX}: P_{MX}/α_{MOH}

θ'_{M_2X}: $P_{M_2X}/(\alpha'_{MOH})^2$

θ''_{MX_2}: $P_{MX_2}/(\alpha'_2)_{M(OH)_2}$

λ: conductivity coefficient of a solute

Λ: equivalent conductivity

μ: ionic strength

ν: osmotic coefficient of a solute

ν_a: number of acid radicals in the formula of a salt

ν_b: number of base radicals in the formula of a salt

π: buffer capacity with respect to strong base or strong acid; $= | db_s/dpH | = | da_s/dpH | = | dg/dpH |$

π_a: buffer capacity with respect to weak acid; $= | da/dpH |$

π_b: buffer capacity with respect to weak base; $= | db/dpH |$

π_j: $| c_j d\beta_j/dpH |$

Π: product

ρ: unionized fraction of an acid

ρ': unionized fraction of a base

σ: a special summation, Chapter XII

Σ: summation

Σ_z: summation for all values to z

Σ_i^z: summation from i to z

τ: temporary constant in Chapter XV

ϕ_{MX}: $P_{MX}/\alpha'_{MOH}\,\alpha_{HX}$

ψ: the unknown in a quadratic equation

8. Special Subscripts and Abbreviations

$(\ )_a$: value of $(\ )$ at point a in a diagram

$(\ )_{col}$: colorimetrically determined value

$(\ )_{crit}$: critical value

$(\ )_e$: value at equivalent point

$(\ )_{end}$: value at end-point of titration

$(\ )_f$: final value

$(\ )_i$: value at inflection point of titration curve

$(\ )_{ie}$: value at iso-electric point

$(\ )_{inc}$: value pertaining to incongruence of solubility

$(\ )_{ind}$: pertaining to an indicator

$(\ )_j$: pertaining to the j'th solute

$(\ )_{K_1}$: value at $H = K_1$

$(\ )_{max}$: value at a maximum

$(\ )_{min}$: value at a minimum

$(\ )_{mix}$: pertaining to a mixture

$(\ )_n$: value at the titration ratio n

$(\ )_{\mathbf{n}}$: value for neutrality

$(\ )_0$: value in pure water solution

$(\ )_{orig}$: original value

$(\ )_{\mathrm{P}}$: value at pressure P

$(\ )_{-p/2}$: value at $p/2$ per cent of titration before equivalent point

$(\ )_{+p/2}$: value at $p/2$ per cent of titration beyond equivalent point

$(\ )_{req}$: required value of ( )

$(\ )_\infty$: value at infinite dilution, or at $c = 0$

$(\ )_*$: value at infinite concentration, or at $c = \infty$

$(\ )_-$: the lower value of ( ) from a quadratic equation

$(\ )_+$: the higher value of ( ) from a quadratic equation

$(\ )_I$, $(\ )_{II}$: values at the first and second inflection points respectively of a titration

b'_{M}: $pb_{\mathrm{M}}/100$, or p per cent of b_{M}

Δ_p: change in pH from $-p/2$ to $+p/2$ per cent of titration, through the equivalent point

$H_{\mathrm{XVII}(92)}$: value of H given by "Eq. XVII(92)," etc.

$H_{b_{\mathrm{M}}}$: value of H for saturation with respect to the base M at concentration b

$H_{46b_{\mathrm{M}}}$: value of H given by "Eq. (46)" with $b = b_{\mathrm{M}}$

$H_{\mathrm{MOH}\downarrow}$: value of H_{req} to cause precipitation of MOH

$H_{\mathrm{MOH}\uparrow}$: value of H_{req} to cause dissolving of MOH

H^{n}: the sequence of three terms of a polynomial in H beginning with the term in H^{n}

Q^n_J: the quadratic solution of the sequence H^n of "Eq. (J)"

$(\ )^2$, in $X \pm \sqrt{(\ )^2 + Z}$, stands for the square of the quantity, here X, outside the root

"Eq. $H(46 = 80)$" means the expression obtained when the expressions for H from "Eqs. (46) and (80)" are equated

Equilibrium Constants

Numerical values, some in round numbers, as used in illustrative examples. With the chief exceptions noted, the general reference is Latimer, Ref. III–1.

SUBSTANCE	EQUILIBRIUM RELATION	CONSTANT
Ag_2CO_3	$[Ag^+]^2[CO_3^=]$	$P = 8.2 \times 10^{-12}$
$AgCl$	$[Ag^+][Cl^-]$	$P = 2 \times 10^{-10}$
Ag_2CrO_4	$[Ag^+]^2[CrO_4^=]$	$P = 10^{-12}$
$AgOH$	$[Ag^+][OH^-]$	$P = 2 \times 10^{-8}$
$Al(OH)_3$	$[Al^{+++}][OH^-]^3$	$P_b = 1.9 \times 10^{-33}$
	$[H^+][H_2AlO_3^-]/[Al(OH)_3]$	$K_a = 4 \times 10^{-13}$
	$[Al^{+++}][OH^-]/[AlOH^{++}]$	$K_{b_3} = 7.1 \times 10^{-10}$
$BaCO_3$	$[Ba^{++}][CO_3^=]$	$P = 5 \times 10^{-9}$
$BaCrO_4$	$[Ba^{++}][CrO_4^=]$	$P = 2 \times 10^{-10}$
$BaSO_4$	$[Ba^{++}][SO_4^=]$	$P = 10^{-10}$
CH_3COOH	$[H^+][CH_3COO^-]/[CH_3COOH]$	$A = 1.75 \times 10^{-5}$
C_6H_5COOH[a]	$[H^+][C_6H_5COO^-]/[C_6H_5COOH]$	$A = 6.3 \times 10^{-5}$
$CaCO_3$	$[Ca^{++}][CO_3^=]$	$P = 5 \times 10^{-9}$
$Ca(OH)_2$(†)	$[Ca^{++}][OH^-]^2$	$P = 8 \times 10^{-6}$
$FeCO_3$	$[Fe^{++}][CO_3^=]$	$P = 2.1 \times 10^{-11}$
$Fe(OH)_2$(‡)	$[Fe^{++}][OH^-]^2$	$P = 1.7 \times 10^{-15}$
$Fe(OH)_3$[b]	$[Fe^{+++}][OH^-]^3$	$P = 10^{-37}$
	$[Fe(OH)_2^+][OH^-]/[Fe(OH)_3]$	$K_1' = 2.5 \times 10^{-8}$
	$[FeOH^{++}][OH^-]/[Fe(OH)_2^+]$	$K_2' = 5 \times 10^{-10}$
	$[Fe^{+++}][OH^-]/[FeOH^{++}]$	$K_3' = 2.9 \times 10^{-12}$
FeS	$[Fe^{++}][S^=]$	$P = 10^{-19}$
H_3BO_3	$[H^+][H_2BO_3^-]/[H_3BO_3]$	$A = 6 \times 10^{-10}$
	$[H_2B_4O_7]/[H_3BO_3]^4$	$K = 3.6 \times 10^{-3}$
	$[H^+][HB_4O_7^-]/[H_2B_4O_7]$	$K_1 = 10^{-4}$
	$[H^+][B_4O_7^=]/[HB_4O_7^-]$	$K_2 = 10^{-9}$

(†) $Ca(OH)_2$ and $Ba(OH)_2$ have been assumed strong in all numerical problems; for estimates of K_2' (as 0.05 and 0.23 respectively), see R. P. Bell and J. E. Prue, *J. Chem. Soc.*, 362 (1949).

(‡) T. V. Arden, *J. Chem. Soc.*, 882 (1950), reports $P = 2.4 \times 10^{-14}$.

Substance	Equilibrium Relation	Constant
HCN	$[H^+][CN^-]/[HCN]$	$A = 4 \times 10^{-10}$
H_2CO_3	$[H^+][HCO_3^-]/[H_2CO_3]$	$K_1 = 4 \times 10^{-7}$
	$[H^+][CO_3^=]/[HCO_3^-]$	$K_2 = 5 \times 10^{-11}$
$H_2CrO_4^c$	$[H^+][HCrO_4^-]/[H_2CrO_4]$	$K_1 = 0.2$
	$[H^+][CrO_4^=]/[HCrO_4^-]$	$K_2 = 3 \times 10^{-7}$
	$[Cr_2O_7^=]/[HCrO_4^-]^2$	$K = 40$
H_2O	$[H^+][OH^-]$	$W = 10^{-14}$
$H_3PO_4^d$	$[H^+][H_2PO_4^-]/[H_3PO_4]$	$K_1 = 7 \times 10^{-3}$
	$[H^+][HPO_4^=]/[H_2PO_4^-]$	$K_2 = 6 \times 10^{-8}$
	$[H^+][PO_4^{\equiv}]/[HPO_4^=]$	$K_3 = 5 \times 10^{-13}$
H_2S	$[H^+][HS^-]/[H_2S]$	$K_1 = 10^{-7}$
	$[H^+][S^=]/[HS^-]$	$K_2 = 10^{-15}$
H_2SO_4	$[H^+][SO_4^=]/[HSO_4^-]$	$K_2 = 10^{-2}$
$MgCO_3$	$[Mg^{++}][CO_3^=]$	$P = 5 \times 10^{-5}$
$Mg(OH)_2$	$[Mg^{++}][OH^-]^2$	$P = 10^{-11}$
$Mg(OH)_2^e$	$[Mg^{++}][OH^-]/[MgOH^+]$	$K_2' = 2.6 \times 10^{-3}$
NH_3	$[NH_4^+][OH^-]/[NH_3]$	$B = 1.8 \times 10^{-5}$
$SrCrO_4$	$[Sr^{++}][CrO_4^=]$	$P = 3.6 \times 10^{-5}$

References: *a*, v–3; *b*, xvii–2; *c*, xv–3 and xv–4; *d*, viii–3; *e*, xvii–3.

HYDROGEN ION CONCENTRATION

New Concepts in a Systematic Treatment

I

Definitions and Fundamental Relations

A. Neutrality and the Water Ions

The mathematical problem of the relations defining the hydrogen ion concentration of aqueous solutions is essentially that of the disturbance of the neutrality of water by various solutes. The "reaction" or condition of the pure solvent will be taken as the meaning of *neutrality*. Hence, on the basis of the principle of electroneutrality, that the sum of the positive ionic charges equals the sum of the negative charges in any solution, and on the assumption that the only ions in pure water are H^+ and OH^-, neutrality means the equality of the concentrations of "hydrogen ion" and "hydroxyl ion"

$$[H^+] = [OH^-], \qquad \text{or } H = OH. \tag{1}$$

A solution is then said to have an acid reaction, or to be acid, when $H > OH$, and basic if $H < OH$.

Pure water then contains finite and equal concentrations of these ions, and it is therefore conducting and electrolytic (decomposable by electric current). The ions H^+ and OH^- are the *water ions* and are always and already present in any aqueous solution. They pertain to the water and not to solutes; they are never "contributed" by solutes. By affecting the atmosphere (ionic and otherwise) surrounding these ions, all solutes whatever their nature affect the activities (absolute and relative) of these ions. But only those solutes which introduce ions other than H^+ and OH^- can affect what is defined as the neutrality of the solution, or the equality of the concentrations of the water ions. Such solutes, introducing foreign (solute) ions into the solution, increase the electrolytic properties of the solution and are customarily called "electrolytes" as distinguished from "nonelectrolytes" or those introducing no foreign ions.

B. Classes of Electrolytes

For our present purpose it is important to distinguish acids, bases, and ampholytes as one class and salts as another. The distinction is based on whether the solute (the physical "analytical" solute, not a "formula" or hypothetical substance) is or is not dissociated by reaction with water or otherwise in taking on charge and becoming ionic. On this basis the first class, or the acids, bases, and ampholytes, has this in common, that the solute ions, whether or not all of one sign, contain the solute molecule unbroken. Whatever may be the intermediate reactions (tautomerization, hydration, etc.) preliminary and necessary to ionization or to the formation of charged solute ions, these solute ions include the original solute composition in its entirety. This is obvious in the case of NH_3 and CO_2. The solute CO_2 may or may not have to rearrange itself from a "pseudo" to a "true" acid, then combine with water to form "H_2CO_3," which may then "dissociate" or "exchange a proton" with the "base" water, etc., etc., finally giving the ion HCO_3^- aq., or CO_2OH^- aq.; but this ion clearly still contains the solute itself, CO_2. The solute in "sulfuric acid" is SO_3, so that the same remarks apply to it too. The capture of hydroxyl ion from the water to form these negative ions can not be distinguished, in other words, either from dissociation of hydrogen ion from a (usually hypothetical) compound, or from "exchange of proton" between such a compound and water. The mathematical effect on the activity and concentration of hydrogen ion is in either case the same. This class of electrolytes then, the acids, bases, and ampholytes, suffers no analytical dissociation in aqueous solution. On the basis that H^+ and OH^- are ions of water itself not contributed or introduced by the solute, it does not seem possible to demonstrate analytically that a solute of this class is chemically or analytically dissociated into two new chemical species in aqueous solution.

These statements may appear to be contradicted in the case of aq. HCl and aq. NaOH in that the ions "Na^+" and "Cl^-" do not appear, as usually written, to contain the whole of the solute. But hydrated chloride ion can not be distinguished experimentally or mathematically from an ion formed by capture of hydroxyl ion by HCl; and hydrated sodium ion can not be distinguished from an ion formed by capture of hydrogen ion by the solute NaOH. The exchange (isotopic) of hydrogen or oxygen occurring between these solutes and the solvent can of course be explained from either point of view.

The same is true for ampholytes, whether of the amino-acid type or of the amphoteric hydroxide type. These solutes may appear in various forms: species with no net charge, whether molecular or dipolar-ionic, and both positive and negative ions. But the solute is nevertheless unbroken, or complete in composition, in any of these forms. All this simply means that we adhere to the position that the ions H^+ and OH^- are water ions, constituent parts if not merely aspects of the structure of the water or of the aqueous solution itself, and that it is only the difference in their concentrations ($D = H - OH$) that is affected by solutes.

Salts, on the other hand, are substances which are dissociated by reaction with water so as to introduce foreign species, both charged and uncharged, of different composition. There seems in fact to be no particle or species left with the composition of the salt itself in a (dilute) "salt solution." Electrolysis brings about a physical separation of the parts of a salt aside from the reactions occurring at the electrodes; whereas for the electrolytes called acids, bases, and ampholytes there is no separation of parts of the solute in electrolysis except for the reduction or oxidation which may take place at the electrodes.

The individuality of a salt as a solute, or as a component we may say, seems to be entirely lost in a salt solution. At least in dilute solution, then, we shall assume no particles or species of "undissociated salt." The individuality of a salt makes its appearance only at saturation or in connection with its solubility product. But in respect to the effect on the neutrality of a solution, a "salt solution" will be considered simply as an equivalent point between an acid and a base (in some cases between an ampholyte and either an acid or a base). If the concentration of a salt is represented by c, and the concentrations of acids and bases by a and b respectively, then for salts of a polybasic acid, $c = a$, while b may equal a, $2a$, $3a$, etc. The word "salt," then, has no implication about the "reaction" of its solution, i.e. whether $H \gtrless OH$. Nor does it even imply for our present purpose the existence of a crystallizable compound of the supposed composition. We shall use the word merely for convenience, but its meaning will be strictly an *equivalent point*. The use of the word "salt" in its other connection, signifying an ionic solid substance out of solution, has nothing to do with the "reaction" of that substance in solution, which is our present concern; and it must never be taken to imply "neutrality," or $H = OH$.

The distinction here drawn between salts as one class of electrolytes on the one hand, and acids, bases, and ampholytes on the other, involves distinction between the terms *dissociation* and *ionization*. Ionization may be said to be *observed* to take place in solutions of both classes of electrolytes; it is a descriptive term referring to the enhanced electrolytic properties of these solutions as compared to water. Dissociation, on the other hand, may be said to be observed in the case of salts in solution, but it is not *observed* for the real solutes constituting the other class of electrolytes. We shall, in other words, use the word dissociation in the actual sense of chemical dissociation, analytically proved. A strong acid in solution is completely ionized (observation) and may or may not be dissociated at all; the latter is a question of our theory about the mechanism of ionization. A salt in solution is completely dissociated, but may or may not be completely ionized; the latter is a question of the "strengths" of its constituent acid and base.

The mathematical relations defining the hydrogen ion concentration of aqueous solutions will not depend, of course, on the mechanism assumed for the process of ionization in solution. The same final results will be obtained if developed always in complete mathematical exactness, whether we assume the dissociation by proton exchange or otherwise of HCl so as to "contribute" both Cl^- and H^+, or whether we assume, as Werner suggested,[1] that the HCl as a solute coordinates with the negative part, hydroxyl, of water to produce the hydrated negative "chloride ion," at the same time increasing the concentration of the hydrogen ion in accordance with the principle of electroneutrality and subject to the equilibrium constants involved. But since all the reactions of so-called hydrogen ion are already observable in pure water, modified only in degree by the introduction of the acid, it appears simpler, expecially since the desired mathematical relations are thereby reached more directly and with fewer assumptions in general, to use the point of view that an electrolyte of this class (acids, bases, ampholytes) introduces only one foreign species, whatever its state of charge, into solution, the solute species itself remaining unbroken, undissociated, in this process of ionization. The mathematical convenience of this point of view will be apparent in the application.

C. Charge Coefficient of Solute

The neutrality of a solution is of necessity disturbed if the solute, as in the case of acids and bases, can furnish foreign ionic charges of only one sign; to repeat, ions other than H^+ and OH^-, or the water ions, will be called foreign or solute ions. Salts and ampholytes, on the other hand, furnishing foreign ionic charges of both signs, may or may not affect the reaction of the solution. In general, the neutrality of the solution is disturbed by electrolytes only when they furnish unequal numbers of foreign ionic charges of opposite sign. The effect upon the neutrality of the solution does not depend upon the total number of solute ions per mole of solute, nor upon the number of ions of one particular kind, but on the net number of solute ionic charges of one sign. It is this number which, through the requirement of electroneutrality, affects the equality of the concentrations of the water ions.

We shall now define the *charge coefficient* as the net number of foreign solute ionic charges of one sign produced per mole of dissolved solute. The charge coefficient will be represented by the symbol β when referring to the net number of *negative* solute charges per mole (the coefficient β itself may then be either positive or negative as a number), and by the symbol β' when referring to the net number of *positive* solute charges per mole. Although one such symbol is sufficient, the symbol β' is convenient in connection with the effect of bases and will be used only as the charge coefficient of a base.

Applying the condition of electroneutrality, then, and defining D as the algebraic difference $H - OH$, we have

$$D = H - OH = \Sigma c\beta, \tag{2}$$

(1) A. Werner, *New Ideas on Inorganic Chemistry* (E. P. Hedley, transl.), Longmans, London, 1911; esp. pp. 212–13.

in which c is the concentration of a solute with charge coefficient β; or, if we segregate the solutes classified as bases, with concentrations b,

$$D = \Sigma c\beta - \Sigma b\beta'. \tag{3}$$

Since a salt is merely the equivalent point between an acid and a base, the charge coefficient of a salt, β_s, may be either positive or negative, being, like that for an ampholyte, a compound coefficient consisting of at least two terms. Given a solution containing an acid and a base, we have

$$D = H - OH = a\beta - b\beta'; \tag{4}$$

and a salt solution is simply one in which a and b are integrally related, i.e. equivalent. In such an expression, a represents the concentration of an acid, alone or mixed with bases and other solutes, and "neutralized" or not. The "concentration of sulfuric acid," in other words, will be the analytical concentration of SO_3 as it would be determined by precipitation as $BaSO_4$ in a solution "containing" Na_2SO_4, H_2SO_4, and SO_3. Similarly, b represents the concentration of a base, in the same sense; all the ammonia available, by treatment with excess of KOH, represents the base, ammonia, in a solution containing various ammonium salts with or without some "free" ammonia.

For a 1 : 1 salt, then,

$$D = H - OH = c(\beta - \beta') = c\beta_s, \tag{5}$$

in which the charge coefficients β and β' refer to the constituent acid and base of the salt, each at concentration c, or in which β_s represents a charge coefficient referred to the salt. Neutrality is always disturbed unless $\beta_s = 0$, or unless the "strengths" of the acid and base are equal. If salts are considered completely dissociated, then, the relations in salt solutions require no special equilibrium constants not already involved in the relations between water and the constituent acids and bases. The equilibrium or ionization constants pertaining to acids, bases, and ampholytes, and to water, will suffice to describe all possible salt solutions merely by the restriction of stoichiometric or equivalent relations of the concentrations.

D. Ionization Constants

The equilibrium constants relate the activities of the various species, ionic and otherwise, in the solution. The activities are related to the concentrations through activity coefficients which are defined as having the limiting value of 1 in "infinitely dilute" solution—or more strictly, at zero ionic strength if the coefficients depend upon ionic strength.

The most fundamental equilibrium constant involved is that pertaining to the solvent itself:

$$\mathbf{K}_w = \mathrm{a}_{H^+}\mathrm{a}_{OH^-}/(\mathrm{a}_{H_2O})^h = \mathbf{H}(\mathbf{OH})/(\mathrm{a}_{H_2O})^h. \tag{6}$$

The equilibrium process for the ionization of a solute, Z, to form a univalent solute anion, and leading to an increase in the activity and concentration of hydrogen ion, is an "acidic process," and involves a constant of the form

$$\mathbf{K} = a_{H^+}a_{Z^-}/a_{Z^0}(a_{H_2O})^{h'}. \qquad (7)$$

The various ions are all "hydrated," and the exponents h and h' involve the unknown, possibly specific and variable, degrees of hydration. If the variation in the activity of "H_2O" in dilute aqueous solutions is considered negligible, then whether the activity of "H_2O" in pure water is called unity or is assigned some constant value (a molality) such as "55.35" at 25° C., we may write as the thermodynamic or activity constants,

$$\mathbf{W} = a_{H^+}a_{OH^-} = \mathbf{H(OH)}, \qquad (8)$$

which will be called the ion-product constant of the solvent, water, and

$$\mathbf{K}_1 = a_{H^+}a_{Z^-}/a_{Z^0}, \qquad (9)$$

the first "acid ionization constant" of the solute. If h and h' are equal, the ratio of these constants has not been changed by this simplification. By analogy with Eq. (9) the formation of a divalent anion of the solute involves the equilibrium constant

$$\mathbf{K}_2 = a_{H^+}a_{Z^=}/a_{Z^-}, \qquad (10)$$

while for the formation of solute cations we have

$$\mathbf{K}_1' = a_{Z^+}a_{OH^-}/a_{Z^0}, \qquad (11)$$

and

$$\mathbf{K}_2' = a_{Z^{++}}a_{OH^-}/a_{Z^+}. \qquad (12)$$

These constants represent *processes*. The first type, involving negative ionization of the solute, tends to make the solution "acidic" (or $H > OH$) and is called an acidic process, described by "acid ionization constants," or **K**'s. The second type, formation of positive solute ions, tends to make the solution "basic" ($H < OH$), and therefore involves "base ionization constants," or primed **K**'s. A solute may react with water in both kinds of processes, giving rise to species of the solute both with positive and with negative (net) charge. Such a solute is an ampholyte, and its relations with water require both acid and base ionization constants (at least one of each) for their mathematical description. If a solute takes part in only one of these two types of processes—or at least if one of the processes is not observable, being vanishingly insignificant—it is an acid or a base: an acid if the solute appears only in anions, with a charge coefficient (β) always positive, and a base if it appears only in cations, with β always negative. An acid is said to be monobasic, dibasic, etc., according to the highest valence (z) or charge of its anion, and hence according to the number (z) of ionization constants required for its mathematical relations; similarly for the monoacidity or diacidity, etc. (z') of a base. Obviously the mathematical relations for an ampholyte will reduce to those of an "acid" or a "base" if all its basic or all its acidic ionization constants are set equal to zero. Mathematically an acid or a base is from this point of view the limiting case of an ampholyte. There seems to be no *a priori* basis for expecting a particular solute of this class (ampholytes, acids, and bases) to act merely as an acid or merely as a base in water solution. Nor does there seem to be any necessary or limiting relation between the acid ionization constant of a simple ampholyte and its base ionization constant. Also, the ionization constant of a simple acid may be larger or smaller than either of the constants of a simple ampholyte.

Finally, since the ionization constants are constants only in terms of activities, while the equation of electroneutrality, such as Eqs. (2–5), can be used only in terms of concentrations, we must either introduce and use activity coefficients, for which we have no general or universally exact expression, or set limitations to our treatment for the sake of mathematically "exact" equations which represent at least the forms or models for limiting or ideal conditions. We shall therefore assume such conditions that all activity coefficients are sensibly equal to 1. We may then add and subtract activities in place of concentrations in the equation of electroneutrality; or what is the same thing, we may deal with the thermodynamic or activity constants as if they applied in terms of concentrations. Only such ideal relations between concentrations and equilibrium constants may be called "exact," but they apply correctly only as limiting relations and hence only to unreal cases. For real conditions, or finite concentrations, we must either find some valid extrapolation to infinite dilution to correct for the nonideality of the actual relations, or introduce activity coefficients, which are never exact since they are of the nature of specific deviations from ideality. The first procedure is usually available and to be preferred in the determination of some equilibrium constant. The second is required in the calculation of concentrations or activities from the constants and other data. But the activity coefficients then have to be estimated essentially as empirical corrections or as approximations on the basis of some limiting law (Debye-Hückel), as functions of the ionic strength.

We shall give examples later of the introduction of activity coefficients in modifying the ideal relations, but we shall be concerned essentially only with the ideal exact relations between the analytical concentrations of solutes, the equilibrium constants, and the hydrogen ion concentration of the solution.

We must now consider how the β coefficients, required in the application of electroneutrality, may be expressed first in terms of "ionization fractions" (degrees of ionization) and second in terms of H and ionization constants, so that we may finally relate H explicitly to analytical concentrations and constants.

Before proceeding, however, we must note that the words "acidity" and "acid" have here been used in a distinctly limited and special sense, purely for use in connection with the mathematical relations involving the hydrogen ion concentration of aqueous solutions. No generality, in fact no descriptive content, is intended in the definitions here adopted. "Acidic" and "basic" will here mean simply conditions of inequality of H and OH, and "acidity" simply a measure of this inequality. In this mathematical definition the reference point is the condition of the solvent itself. In a binary solution it is conceivable that we might distinguish between the difference D_1 for the ions of one component and D_2 for the ions of the other, one of which ions may even be common to those of the first component. Nevertheless, the "acidity," defined as $D = H - OH$ in an aqueous solution of SO_3, has no meaning when pure SO_3 is reached, in which there can be neither H^+ nor OH^- ions. The ionization constants here defined are strictly *aqueous* ionization constants, involving the water ions and the interaction between the solute and the specific solvent, water.

E. Ionization Fractions (of Acids, Bases, and Ampholytes)

(1) Acids will be represented as substances of formula X or, in order to indicate their basicity, as H_zX. Their ionization fractions will be represented by the symbol α:

$$\begin{aligned} &\alpha_1 \text{ (or simply } \alpha\text{, for monobasic acid)} = [X^-]/a, \\ &\alpha_2 = [X^=]/a, \cdots, \alpha_z = [X^{-z}]/a, \\ &\rho = [X^0]/a, \end{aligned} \tag{13}$$

in which the symbols X^0, X^-, etc., represent the solute, an acid, with the charges indicated. The fraction ρ is called the unionized residue or fraction. For bases, we shall use the same symbols, primed, for the fractions, and M as the symbol of the solute:

$$\begin{aligned} &\alpha'(= \alpha_1') = [M^+]/b, \cdots, \alpha_{z'}' = [M^{+z'}]/b, \\ &\rho' = [M^0]/b. \end{aligned} \tag{14}$$

(2) At the same time, the ionization constants are defined as follows: for acids,

$$\begin{aligned} A = K_1 &= H[X^-]/[X^0] = H\alpha_1/\rho, \\ K_2 &= H[X^=]/[X^-] = H\alpha_2/\alpha_1, \\ K_z &= H[X^{-z}]/[X^{-(z-1)}] = H\alpha_z/\alpha_{z-1}, \end{aligned} \tag{15}$$

while for bases,

$$\begin{aligned} B = K_1' &= OH[M^+]/[M^0] = OH\alpha_1'/\rho', \\ K_{z'}' &= OH[M^{+z'}]/[M^{+(z'-1)}] = OH\alpha_{z'}'/\alpha_{z'-1}'. \end{aligned} \tag{16}$$

For brevity, the symbols A and B alone will be used for the constants of monobasic acid and monoacid base, respectively. The symbols H, $[X^-]$, $[X^0]$, etc., represent concentrations, so that the mass "constants," or the K's, are equal to the true constants, the **K**'s, only if all activity coefficients equal 1. With this assumption we shall write "ideal" mathematical equations relating concentrations and ionization constants which are truly constants only in terms of activities.

(3) Various combinations of these constants are sometimes of value:

$$H^2[X^=]/[X^0] = K_1K_2 \doteq \mathbf{p};$$

$$H^2[X^{\equiv}]/[X^-] = K_2K_3 = \mathbf{p}_1;$$

$$H^2[X^{-z}]/[X^{-(z-2)}] = K_{(z-1)}K_z. \tag{17}$$

Also,

$$H^3[X^{\equiv}]/[X^0] = K_1K_2K_3, \text{ etc.} \tag{18}$$

and

$$[X^0][X^=]/[X^-]^2 = K_2/K_1 = y,$$

$$[X^-][X^{\equiv}]/[X^=]^2 = K_3/K_2 = y_1, \text{ etc.} \tag{19}$$

(4) By combining the definitions of the constants with those of the fractions, we may express all the fractions of a polybasic acid in terms of its first fraction.

$$\alpha_2 = \alpha_1K_2/H, \qquad \alpha_3 = \alpha_1K_2K_3/H^2, \cdots, \qquad \alpha_z = \alpha_1K_2 \cdots K_z/H^{(z-1)}, \tag{20}$$

while

$$\rho = \alpha_1H/K_1. \tag{21}$$

Similar expressions hold for bases, with OH in place of H.

(5) Now *if*

$$a = [X^0] + [X^-] + [X^=] + \cdots + [X^{-z}], \tag{22}$$

then

$$\rho = 1 - \Sigma_z\alpha_z, \tag{23}$$

and

$$\alpha_1 = 1/(1 + H/K_1 + K_2/H + K_{23}/H^2 + \cdots + K_{2\ldots z}/H^{z-1}). \tag{24}$$

[*Note:* The symbol K_{23} here represents the product K_2K_3. Such condensed symbols will be used frequently in subsequent expressions.] Then

$$\beta = \alpha_1 + 2\alpha_2 + \ldots + z\alpha_z = \Sigma_z\, z\alpha_z, \tag{25}$$

and

$$\beta = \frac{1 + 2K_2/H + 3K_{23}/H^2 + \cdots + zK_{2\ldots z}/H^{z-1}}{1 + H/K_1 + K_2/H + K_{23}/H^2 + \cdots + K_{2\ldots z}/H^{z-1}}, \tag{26}$$

for a polybasic acid; analogously for a polyacid base,

$$\beta' = \frac{1 + 2K'_2/OH + 3K'_{23}/(OH)^2 + \cdots + z'K'_{2\ldots z'}/(OH)^{z'-1}}{1 + OH/K_1 + K'_2/OH + K'_{23}/(OH)^2 + \cdots + K'_{2\ldots z'}/(OH)^{z'-1}}. \quad (27)$$

These expressions for the charge coefficient may then be used in the general equation of electroneutrality,

$$D = H - OH = \Sigma a\beta - \Sigma b\beta', \quad (28)$$

for a solution containing any number of acids and bases (and therefore including all possible "salts"), thus leading to explicit analytical equations for the quantity $H - OH$ in such solutions, as function of concentrations, H, and constants.

(6) For a simple ampholyte (1 : 1 type, with z_- and z_+, the number of acid and base ionization constants respectively, both equal to 1), using "U" as the symbol for the ampholyte at concentration c, we have

$$\alpha_- = [U^-]/c, \qquad \alpha_+ = [U^+]/c, \qquad \alpha_0 = [U^0]/c, \quad (29)$$

with

$$K_a = H[U^-]/[U^0] = H\alpha_-/\alpha_0,$$
$$K_b = OH[U^+]/[U^0] = OH\alpha_+/\alpha_0. \quad (30)$$

Now if the species U^0, the form with no net charge, exists entirely in one form, i.e. either all molecular or all dipolar (zwitter-ion), and if

$$c = [U^0] + [U^-] + [U^+], \quad (31)$$

so that

$$\alpha_0 = 1 - \alpha_- - \alpha_+, \quad (32)$$

then

$$\alpha_- = K_aOH/(K_aOH + K_bH + W), \quad (33)$$

$$\alpha_+ = K_bH/(K_aOH + K_bH + W), \quad (34)$$

$$\alpha_0 = W/(K_aOH + K_bH + W). \quad (35)$$

[The effect of assuming that U^0 consists of both a molecular and a dipolar-ion form, $U^{\pm}$, will be discussed later; Chapter VI.]

Hence

$$\beta = \alpha_- - \alpha_+ = (K_aOH - K_bH)/(K_aOH + K_bH + W). \quad (36)$$

Meaning of the Term "Strong" as Applied to Acid or Base. We shall define as a "strong acid," one for which ρ (the unionized fraction) is zero; for a "strong base," similarly, $\rho' = 0$.

For a monobasic acid this means $A = \infty$, and $\alpha = \beta = 1$ under all conditions. This is then only a limiting case, for whatever the nature or mechanism of the ionization process, the equilibrium constant A cannot be infinite. But mathematically, a solute such as HCl is to be assumed to be either always or never completely ionized, whatever the concentration and whatever the composition of the solution; α is either always strictly 1 or never 1. It is important to note, however, that this definition of a "strong acid" means that $[X^-] = a_s$ (a_s and b_s will represent concentrations of strong acid and strong base, respectively), but *not* that $H = a_s$ under any conditions, for this would be inconsistent with the most fundamental definition of all, that $H(OH) = W$. In other words, the fraction α equals $[X^-]/a$, not H/a.

For a polybasic acid, this definition of "strong" (that $\rho = 0$), means that $K_1 = \infty$, and that the total degree of ionization, $\alpha_t (= \Sigma_z \alpha_z) = 1$, since ρ always equals $(1 - \alpha_t)$. Since each "degree of ionization" of such an acid, "strong" in respect to K_1, is still, as in Eqs. (20–24), a function of all the ionization constants of the acid, it is important to note in this connection that although $K_1 = \infty$, $\alpha_1 < 1$. Or, for a dibasic acid "strong" in respect to K_1,

$$\alpha_1 = H/(K_2 + H), \qquad \alpha_2 = K_2/(K_2 + H), \tag{37}$$

and

$$\beta = \alpha_1 + 2\alpha_2 = 1 + \alpha_2 = (H + 2K_2)/(K_2 + H). \tag{38}$$

For an acid like sulfuric acid, for example, generally treated as "strong" in respect to K_1, α_1 approaches 0 while α_2 approaches 1 as pH increases.

For certain purposes it may be convenient to treat a triacid base such as $Fe(OH)_3$ as "strong" in respect to both its first and its second ionization constants. This may be necessary simply because of our lack of knowledge of any of the constants but K_3'. In such a case, both ρ and α_1' are zero, and

$$\alpha_2' = OH/(K_3' + OH), \qquad \alpha_3' = K_3'/(K_3' + OH), \tag{39}$$

while

$$\beta' = \alpha_1' + 2\alpha_2' + 3\alpha_3' = 2 + \alpha_3' = [2(OH) + 3K_3']/(K_3' + OH). \tag{40}$$

For mathematical purposes, then, the term "strong" acid or base will be an arbitrary one, not a question of degree. For mathematical simplicity we merely set $\rho = 0$ for a "strong" acid or base. If the equilibrium constants are ever evaluated, these solutes then become ordinary or "weak" acids, etc., and the general equations will be readily modifiable by the introduction of the appropriate value of A, K, etc. in each case, to express the α's and the β's. Finally, the word "strong," unqualified, will hereafter be used only for strong monobasic acid or base; and we shall otherwise say "strong in respect to K_1," etc.

F. Complex and Noncomplex Solutions

For Eqs. (22) and (31) to hold, or in general, for the condition

$$c = [Z^0] + \Sigma_{z-}[Z^{-z_-}] + \Sigma_{z+}[Z^{+z_+}], \tag{41}$$

it is necessary to assume the absence of interaction of solute species (ions or molecules of solutes, not the ions H^+ and OH^-) to form complexes which may or may not be charged. Such interactions are described by what may be called "complex" constants, such as the stability constant of a complex ion MX^-,

$$K_{MX^-} = [MX^-]/[M^0][X^-]; \tag{42}$$

that for the "incompletely dissociated salt," MX,

$$K_{MX} = [MX]/[M^+][X^-]; \tag{43}$$

or that for the "polyacid" type of ion, $[X_2^-]$,

$$K_{X_2^-} = [X_2^-]/[X^0][X^-]. \tag{44}$$

The number of such possible specific interaction constants between solute species has no limit even in the case of the very simplest solution, since we would have to consider the possibility of the association to any extent of the molecules and ions, together or separately, even of a single solute. If such "complexes" are involved, then, instead of Eq. (22), we have

$$a = [X^0] + [X^-] + [X^=] + \cdots + [X^{-z}] + [C^0] + [C^{-(+)}], \tag{45}$$

in which $[C^0]$ and $[C^{-(+)}]$ represent concentrations of complex species of various charges, involving the particular solute X. Hence although the interrelations of the fractions remain as in Eqs. (20), we now have

$$\alpha_1 = \frac{1 - ([C^0] + [C^{-(+)}])/a}{1 + H/K_1 + K_2/H + \cdots + K_{2\,\ldots\,z}/H^{z-1}}, \tag{46}$$

and

$$\beta_c = \Sigma_z z\alpha_z \overset{+}{(-)} [C^{-(+)}]/a. \tag{47}$$

This charge coefficient (now a "complex" one, written β_c) is not a function of H and K's alone, but also of the concentration, possibly even of the concentrations of more than one solute, if various solutes interact. It is only in the absence of such complex charge coefficients that the quantity $H - OH$, defined by Eq. (28), becomes the sum of terms of the form $c\,f(H,K)$; the second factor of this term is otherwise a function, possibly very complicated, of the concentration itself. The simple case, then, which assumes complete dissociation of all salts and absence of "complex ions," or in which the charge coefficients are independent of the concentrations, will be referred to as that of a "noncomplex solution," in contrast to a "complex solution." In the "noncomplex" case, then

$$D = H - OH = \Sigma[c\,f(H,K)]. \tag{48}$$

A special but unimportant exception will be noted later of the hypothetical case of a solution containing a strong acid and a strong base assumed to form an "incompletely dissociated salt," which is still described by Eq. (48).

G. Mathematical Transformation of Constants ("Conjugate" Constants)

An "acid process" is one in which a component of the solution (a real solute) produces an anion and hydrogen ion by reaction with the water, and thermodynamically it is equivalent to a process consuming hydroxyl ion. The reverse of such a specific process is a hypothetical process, since it is not the effect upon the solution of a component of the solution but of an ion, which is in mathematical strictness an abstraction, never a component, and not a substance in the sense that the actual solute is a substance.(2) This hypothetical reverse process (the "reaction" of the anion with the solution) may likewise be considered either as the consumption of hydrogen ion or as the production of hydroxyl ion. In the first case the equilibrium constant would be the reciprocal of the real acid ionization constant, or $1/K$. In the second case the equilibrium constant has the form of a basic ionization constant and the value W/K. Thus,

$$\text{if } K = H[\mathrm{X}^-]/[\mathrm{X}^0], \text{ then } K'^* = W/K = OH[\mathrm{X}^0]/[\mathrm{X}^-]; \qquad (49)$$

and

$$\text{if } K' = OH[\mathrm{M}^+]/[\mathrm{M}^0], \text{ then } K^* = W/K' = H[\mathrm{M}^0]/[\mathrm{M}^+]. \qquad (50)$$

The acidic ionization process of a real solute, such as HCN for example, may always be described either by an acid ionization constant or by a *conjugate* base ionization constant. Similarly the basic ionization process of a *substance*, such as ammonia, may always be described either by a "real" base ionization constant, K', or by a "conjugate" (relatively hypothetical) acid ionization constant, $K^* = W/K'$. The actual or real constants pertaining to reactions of real solutes or components with the water will be represented as simple constants, such as K, K', etc.; the constants of the hypothetical conjugate acids and bases, defined as in Eqs. (49) and (50), will always be distinguished by asterisks.

(2) A *component* is not simply one of the molecular or ionic species (supposed or otherwise) in a system, but one of the substances in a system capable of independent variation in respect to concentration. Water is a component of an aqueous solution, but neither H^+ nor OH^- is a component, since the concentration of one of these ions is automatically determined by the concentration of the other. It is clear that no ion is a component of an aqueous solution in this real and only consistent sense of the word. Although it may appear that we can vary the concentration of sodium ion, for example, arbitrarily by addition or subtraction of a soluble salt of sodium, we can not do so without at the same time changing the concentrations of other ions in the solution. There is no such substance as "sodium ion" that can be added to a solution. Furthermore, a *substance* is a species which can constitute a component of a system in the sense just defined. Hence, in this strict and mathematical sense at least, an "ion" in an aqueous solution is not a substance; it remains part of our mathematical operations, a mathematical concept convenient or required in explaining and describing the relations in the system. The logical difference between ions and substances as categories in this connection may be appreciated by noting that in the application of the phase rule, a pair of reciprocal salts, although consisting of four ions, consists of only three components, or three substances.

Such transformation is purely mathematical and hence always possible. It is of some use in connection with certain theories (the Brønsted theory, for example). But it must be remembered that the fact that the transformation is possible, or even that it is useful, has nothing to do with supporting any particular theory.

The chief point is that it is useful in that such transformations can reduce apparently diverse mathematical problems to cases of one problem. It will be possible to treat any ampholyte (of any "order," i.e. "basicity" as an acid or "acidity" as a base) as the mathematical equivalent of a particular salt of a polybasic acid (or polyacid base). It is in fact possible to treat any acid (or base) as though it were the salt (mathematically) of a corresponding ("conjugate" in terms of the Brønsted theory) base (or acid). What is even more important practically, however, is that it is possible then, in reverse, to treat an actual salt as though it were (mathematically) an acid or a base. These transformations are convenient as simplifications of the apparent variety of the mathematical problems; but as such they prove nothing about the actual constitution of the substances involved or about the mechanism of their interaction with water or with each other.

H. Examples of Explicit Analytical Equations for *H*

The electroneutrality equation, Eq. (28), may now be expanded by means of the various expressions for the charge coefficient into equations relating H explicitly with the concentrations and the ionization constants. The examples here considered are limited to "noncomplex" cases.

(1) Strong acid (a_s) and strong base (b_s).

$$D = H - OH = a_s - b_s = g. \tag{51}$$

We note here that in all equations defining H, the effect of a_s is the negative of that of b_s. The quantity b_s may be substituted at any time for $-a_s$, and for generality we shall sometimes write g for $(a_s - b_s)$, which is the effective quantity in a solution containing both a_s and b_s; g may of course be either positive or negative.

(2) Solution containing j acids and j' bases (any number of weak monobasic acids and bases).

$$D = \Sigma_j a\alpha - \Sigma_{j'} b\alpha' = \Sigma_j[aA/(A + H)] - \Sigma_{j'}[bB/(B + OH)]. \tag{52}$$

(3) Salt, MX, of weak (monobasic) acid and base. Here $a = b = c$, the concentration of the salt, and

$$D = c(\alpha - \alpha') = c[A/(A + H) - B/(B + OH)]. \tag{53}$$

(4) Pure acid (monobasic).

$$D = a\alpha = aA/(A + H). \tag{54}$$

(5) Salt, MX_s, of strong acid and weak base.

$$D = c(1 - \alpha') = c\rho' = cOH/(B + OH), \tag{55}$$

$$= c(W/B)/(W/B + H) = cA^*/(A^* + H), \tag{56}$$

through the transformation of Eq. (50). We note here the identity, mathematically, of Eqs. (54) and (56). Whatever approximations we shall establish later for the practical application of Eq. (54) will hold, by mere substitution of W/B for A, for the numerical solution of Eq. (55) or (56). The weak acid is mathematically therefore "like" a salt of strong acid and weak base with $B = W/A$; or the salt MX_s is like a weak acid with $A = W/B$. The analogy used is simply that which happens to be more convenient. Ideally, or except for the effect of ionic strength, the two solutions are identical in respect to "buffer properties"—effect of dilution upon H, and behavior in titrations —provided $A = W/B$.

(6) Polybasic acid.

$$D = a(\alpha_1 + 2\alpha_2 + \cdots + z\alpha_z), \tag{57}$$

$$= a\,\frac{1 + 2K_2/H + 3K_{23}/H^2 + \cdots + zK_{2\ldots z}/H^{z-1}}{1 + H/K_1 + K_2/H + K_{23}/H^2 + \cdots + K_{2\ldots z}/H^{z-1}}. \tag{58}$$

(7) Dibasic acid, strong in K_1 ($\alpha_1 + \alpha_2 = 1$).

$$D = a(1 + \alpha_2) = a(H + 2K_2)/(H + K_2). \tag{59}$$

(8) Polybasic acid and strong base at concentration $b_s = na$ (where $n =$ the "titration ratio," or b_s/a). From Eq. (25),

$$D = a(\beta - n) = a[\Sigma_z(z - n)\alpha_z - n\rho]. \tag{60}$$

When n is integral ($= 1, 2, \cdots, z$), we have the various alkali salts of the acid (primary, secondary, $\cdots$, normal) at concentration $c = a$.

(9) Polyacid base.

$$D' = -D = OH - H = b(\alpha'_1 + 2\alpha'_2 + \cdots + z'\alpha'_{z'}) \tag{61}$$

$$= b\,\frac{1 + 2K'_2/OH + \cdots + z'K'_{2\ldots z'}/(OH)^{z'-1}}{1 + OH/K'_1 + K'_2/OH + \cdots + K'_{2\ldots z'}/(OH)^{z'-1}}. \tag{62}$$

(10) Polyacid base and strong acid.

$$D' = -D = OH - H = b(\beta' - n') = b[\Sigma_{z'}(z' - n')\alpha'_{z'} - n'\rho'], \tag{63}$$

in which n' now represents the titration ratio a_s/b. For the individual salts, at concentration $c = b$, n' is some integer from 1 to z'; when $n' < z'$, the salt is a "basic salt."

Eqs. (60) and (63) can be transformed into each other. If the ionization constants of a polyacid base are transformed, according to Eq. (50), so that

$$K_1^* = W/K'_{z'},\ K_2^* = W/K'_{(z'-1)},\ \cdots,\ K_{z'}^* = W/K'_1, \tag{64}$$

and these "conjugate acid constants" are used as those of a polybasic acid, then Eq. (63), for the system polyacid base and strong acid, becomes Eq. (60), with the condition that $n = z' - n'$. The primary strong acid salt of a tetracid base ($n' = 1$) is, in other words, mathematically identical in relations defining D, or $H - OH$, with the tertiary strong base salt of a tetrabasic acid, its constants being defined as in Eq. (64). The pure base itself ($n' = 0$) corresponds to the normal salt ($n = z$) of a polybasic acid.

(11) Salts of polybasic acid and polyacid base (or bases). For a salt of polybasic acid and polyacid base,

$$D = c\beta_s = c(\nu_a\beta - \nu_b\beta'). \tag{65}$$

Here ν_a and ν_b are integers representing the number of "acid radicals" and of "base radicals," respectively, in the formula weight or "mole" of the salt. The salt is an acid salt if m, the valence of the acid radical of the salt, is less than z, the basicity of the acid, and it is a basic salt if m', the valence of the base radical, is less than z'; it may even possibly be both as, at least hypothetically, in a "uranyl acid phosphate."

For a mixed salt of a polybasic acid and several bases,

$$D = c[\nu_a\beta - \Sigma(\nu_b\beta')]. \tag{66}$$

(12) Simple ampholyte.

$$D = c(\alpha_- - \alpha_+) = c[K_a(OH) - K_bH]/[K_a(OH) + K_bH + W]. \tag{67}$$

This reduces to simple monobasic or monoacid base if one of the constants is zero. It is furthermore identical with the equation for the alkali acid salt of a dibasic acid ($n = 1$, $z = 2$) in Eq. (60) if

$$K_1^* = W/K_b \text{ and } K_2^* = K_a, \tag{68}$$

when

$$D = c(K_2^*/H - H/K_1^*)/(1 + H/K_1^* + K_2^*/H); \tag{69}$$

or, referring now to the hypothetical dibasic acid at its first equivalent point with strong base,

$$D = c(\alpha_2 - \rho), \tag{70}$$

identical with Eq. (60). This means that the ampholyte is mathematically equivalent to the acid salt NaHX, and that the hydrochloride of the ampholyte is mathematically equivalent to a dibasic acid. On comparing Eqs. (67) and (70), we have identified α_+, α_0, and α_- of the ampholyte with ρ, α_1, and α_2 respectively of this hypothetical dibasic acid.

But Eq. (67) is similarly identical with the equation for a diacid base at its first equivalent point with strong acid, if

$$K_1'^* = W/K_a \text{ and } K_2'^* = K_b, \tag{71}$$

when

$$D' = -D = c(\alpha_2' - \rho') = c\,\frac{K_2'^*/OH - OH/K_1'^*}{1 + OH/K_1'^* + K_2'^*/OH}, \tag{72}$$

in accordance with Eq. (63) for the case $z' = 2$ and $n' = 1$. The identification now is between α_-, α_0, α_+ of the ampholyte and ρ', α_1' and α_2' of the hypothetical diacid base.

(13) Ampholyte of higher order, or with several acid and several base ionization constants. By extension of the above considerations, an ampholyte with z_+ basic constants and z_- acid constants may be treated mathematically as the equivalent of the acid salt (with strong base and of "degree" $m = z_+$), of a polybasic acid of basicity $z = z_+ + z_-$; or as the basic salt (with strong acid) of a polyacid base with $z' = z_+ + z_-$ and $m' = z_-$. The constants are transformed as follows, with transformation to polybasic acid as the example:

$$K_1^* = W/K_{b_{z_+}}, \qquad K_2^* = W/K_{b_{(z_+-1)}}, \;\cdots,\; K_{z_+}^* = W/K_{b_1};$$
$$K_{(z_++1)}^* = K_{a_1}, \qquad K_{(z_++2)}^* = K_{a_2}, \;\cdots,\; K_{(z_++z_-)}^* = K_{a_{z_-}}. \tag{73}$$

The fractions are identified as follows (with the ampholyte concentration equal to c):

$$\begin{aligned}
\alpha_{+z_+} &= [X^{+z_+}]/c = \rho \text{ (of the "polybasic acid")};\\
\alpha_{+(z_+-1)} &= [X^{+(z_+-1)}]/c = \alpha_1;\\
\alpha_{+1} &= [X^+]/c = \alpha_{(z_+-1)};\\
\alpha_0 &= [X^0]/c = \alpha_{z_+};\\
\alpha_{-1} &= [X^-]/c = \alpha_{(z_++1)};\\
\alpha_{-2} &= [X^{-2}]/c = \alpha_{(z_++2)};\\
\alpha_{-z_-} &= [X^{-z_-}]/c = \alpha_{(z_++z_-)}.
\end{aligned} \tag{74}$$

Hence, for such an ampholyte,

$$D = c(\Sigma_{z_-} z_-\alpha_{-z_-} - \Sigma_{z_+} z_+\alpha_{+z_+}) = c(\mathbf{J} - \mathbf{Y})/(\mathbf{J}' + \mathbf{Y}' + W), \tag{75}$$

in which

$$\mathbf{J} = K_{a_1}OH[1 + 2K_{a_2}/H + 3K_{a_2}K_{a_3}/H^2 + \cdots], \tag{76}$$

$$\mathbf{Y} = K_{b_1}H[1 + 2K_{b_2}/OH + 3K_{b_2}K_{b_3}/(OH)^2 + \cdots], \tag{77}$$

and $\mathbf{J}'$ and $\mathbf{Y}'$ are the same functions but with all coefficients equal to 1. This reduces to Eq. (67) if $z_+ = z_- = 1$, to Eq. (58) if all the K_b's $= 0$, and to Eq. (62) if all K_a's $= 0$. In terms of the normal hydrochloride (z_+ moles HCl per mole of ampholyte) considered as a polybasic acid, the ampholyte is (mathematically) the acid salt with $b_s = z_+c$; hence

$$D = c(\beta - z_+) = c[(1 - z_+)\alpha_1 + (2 - z_+)\alpha_2 + \cdots + z_-\alpha_{(z_++z_-)} - z_+\rho], \tag{78}$$

in which the fractions refer to the hypothetical acid defined as in Eqs. (73) and (74).

I. Derivatives of the Fractions and of the Charge Coefficient, With Respect to pH

Since the derivative of β (noncomplex charge coefficient) with respect to pH (p$H = -\log_{10}H$) will be important in connection with the buffer capacity of solutions, we shall write here for future reference the derivatives of the various fractions (α's and ρ) and of the charge coefficient, β, with respect to pH.

1. The Derivative With Respect to pH. We note, first, that

$$d/d\mathrm{p}H = -(\ln 10)H(d/dH) = -2.3H(d/dH)^{(3)}. \tag{79}$$

Since α and β are in terms of concentrations, pH here means $-\log[\mathrm{H}^+]$· Referring to hydrogen ion activity, we would have, since H or $[H^+] = a_{\mathrm{H}^+}/\gamma_{\mathrm{H}^+} = \mathbf{H}/\gamma_{\mathrm{H}^+}$,

$$d/d\mathrm{p}\mathbf{H} = (1 + d\log\gamma_{\mathrm{H}^+}/d\mathrm{p}\mathbf{H})d/d\mathrm{p}H; \tag{80}$$

but the second term of the parenthesis can not be evaluated without regard to the general composition of the solution.

From Eq. (79),

$$d/dH = -(d/d\mathrm{p}H)/2.3H, \tag{81}$$

$$d/d\mathrm{p}(OH) = -d/d\mathrm{p}H, \tag{82}$$

$$d/d(OH) = (d/d\mathrm{p}H)/2.3(OH) = -(d/dH)H^2/W. \tag{83}$$

2. Derivatives Involving H and OH Alone.

(a) $$dH/d\mathrm{p}H = -2.3H; \qquad d^2H/d\mathrm{p}H^2 = (2.3)^2H. \tag{84}$$

$$d(OH)/d\mathrm{p}H = 2.3(OH); \qquad d^2(OH)/d\mathrm{p}H^2 = (2.3)^2(OH). \tag{85}$$

(b) For the difference D ($= H - OH$):

$$dD/d\mathrm{p}H = -2.3(H + OH); \qquad d^2D/d\mathrm{p}H^2 = (2.3)^2D. \tag{86}$$

(c) For the sum, $H + OH$:

$$d(H + OH)/d\mathrm{p}H = -2.3D; \qquad d^2(H + OH)/d\mathrm{p}H^2 = (2.3)^2(H + OH). \tag{87}$$

3. Polybasic Acid. The derivatives of fractions and of charge coefficients here given will apply only to "noncomplex" cases, where $\rho = 1 - \Sigma_z\alpha_z$

(3) The quantity $\log_e 10$ will always be abbreviated as 2.3.

and $\beta = \Sigma_z z\alpha_z$. Hence, for the m'th fraction, or α_m, where $\alpha_m = [X^{-m}]/a$, of an acid of basicity z,

$$d\alpha_m/d\text{p}H = 2.3\alpha_m[m\rho + (m-1)\alpha_1 + (m-2)\alpha_2 + \cdots + (m-z)\alpha_z] = 2.3\alpha_m(m-\beta). \tag{88}$$

Here m is an integer from 1 to z; by definition $z > \beta$. Hence $d\alpha_z/d\text{p}H$ is always positive, while the other fractions, α_1 to α_{z-1}, all have maxima in respect to pH. Also,

$$d\rho/d\text{p}H = -2.3\rho\Sigma_z z\alpha_z = -2.3\rho\beta, \tag{89}$$

and hence always negative. Finally,

$$d\beta/d\text{p}H = 2.3\Sigma_z[z\alpha_z(z-\beta)], \tag{90}$$

also always positive, as expected by definition of β. This may also be written

$$d\beta/d\text{p}H = 2.3(\Sigma_z z^2\alpha_z - \beta^2). \tag{91}$$

Expanded, Eq. (90) becomes

$$d\beta/d\text{p}H = 2.3\Sigma_z\{z\alpha_z[z\rho + (z-1)\alpha_1 + (z-2)\alpha_2 + (z-3)\alpha_3 + \cdots]\}. \tag{92}$$

For a dibasic acid "strong in K_1," $\rho = 0$ by definition, and

$$d\alpha_1/d\text{p}H = -d\alpha_2/d\text{p}H = -2.3\alpha_1\alpha_2; \tag{93}$$

$$d\beta/d\text{p}H = 2.3[(\alpha_1 + 4\alpha_2) - (\alpha_1 + 2\alpha_2)^2]. \tag{94}$$

4. Polyacid Base.

$$d\alpha'_{m'}/d\text{p}H = -2.3\alpha'_{m'}(m' - \beta'); \tag{95}$$

$$d\rho'/d\text{p}H = 2.3\rho'\beta', \tag{96}$$

which is always positive.

$$d\beta'/d\text{p}H = -2.3\Sigma_{z'}[z'\alpha'_{z'}(z'-\beta')] = -2.3[\Sigma_{z'}(z')^2\alpha'_{z'} - (\beta')^2], \tag{97}$$

so that $d\beta'/d\text{p}H$ is always negative, again as expected by definition.

5. Simple Ampholyte.

$$d\alpha_-/d\text{p}H = 2.3\alpha_-(\alpha_0 + 2\alpha_+), \tag{98}$$

$$d\alpha_+/d\text{p}H = -2.3\alpha_+(\alpha_0 + 2\alpha_-), \tag{99}$$

$$d\alpha_0/d\text{p}H = 2.3\alpha_0(\alpha_+ - \alpha_-), \tag{100}$$

$$d\beta/d\text{p}H = 2.3[\alpha_0(\alpha_- + \alpha_+) + 4\alpha_-\alpha_+]; \tag{101}$$

the last is always positive.

6. General Ampholyte. If

$$\alpha_0 = 1 - (\Sigma_{z_-}\alpha_- + \Sigma_{z_+}\alpha_+), \tag{102}$$

and

$$\beta = \Sigma_{z_-} z_-\alpha_- - \Sigma_{z_+} z_+\alpha_+, \tag{103}$$

then for the (m')'th positive ionization fraction,

$$d\alpha_{+m'}/d\text{p}H = -\ 2.3\alpha_{+m'}(m' + \beta), \tag{104}$$

and for the m'th negative ionization fraction,

$$d\alpha_{-m}/d\text{p}H = +\ 2.3\alpha_{-m}(m - \beta), \tag{105}$$

while

$$d\alpha_0/d\text{p}H = -\ 2.3\alpha_0\beta. \tag{106}$$

Also,

$$d\beta/d\text{p}H = 2.3\{\Sigma_{z_-}[z_-\alpha_-(z_- - \beta)] + \Sigma_{z_+}[z_+\alpha_+(z_+ + \beta)]\}, \tag{107}$$

which is always positive since, algebraically, $z_- > \beta$. Eqs. (98–101) are the simplest examples of Eqs. (104–107). Writing out Eq. (105) for α_{-2}, for an ampholyte with z_- acid constants and z_+ base constants, we have

$$d\alpha_{-2}/d\text{p}H = 2.3\alpha_{-2}[(z_+ + 2)\alpha_{+z_+} + (z_+ + 1)\alpha_{+(z_+-1)} + z_+\alpha_{+(z_+-2)} + (z_+ - 1)\alpha_{+(z_+-3)} + \cdot\cdot\cdot + 2\alpha_0 + \alpha_{-1} - \alpha_{-3} - 2\alpha_{-4} - \cdot\cdot\cdot - (z_- - 2)\alpha_{-z_-}]. \tag{108}$$

J. Buffer Capacity

The buffer capacity of a solution, with respect to a particular solute, is the change in concentration of that solute required for unit change in pH of the solution. This derivative is therefore positive with respect to addition of base and negative for addition of acid, but only the absolute value will be significant in our considerations. Although, then, $db_s/d\text{p}H = -\ da_s/d\text{p}H$, since mathematically $b_s = -\ a_s$, we shall speak of "π" in either case as the buffer capacity with respect to either strong base or strong acid; π is therefore defined as

$$\pi = |\ dg/d\text{p}H\ | = |\ d(a_s - b_s)/d\text{p}H\ |. \tag{109}$$

The buffer capacity with respect to weak base and that with respect to weak acid will be distinguished as π_b and π_a respectively. But the symbols π, π_b, and π_a will in each case be understood to be the absolute value of the buffer capacity in question, regardless of the sign of the actual effect.

The most important buffer capacity is π, that with respect to strong base or strong acid. Since

$$D = \Sigma c\beta - b_s, \tag{110}$$

$$\pi = d(\Sigma c\beta - D)/d\text{p}H = \Sigma_j[d(c\beta)_j/d\text{p}H] + 2.3(H + OH). \quad (111)$$

In "noncomplex" solutions $\Sigma(c\beta) = \Sigma[cf(H,K)]$ and the charge coefficients, β_j, are independent of each other, being functions of H and constants only. We may therefore write

$$\pi = \Sigma_j\pi_j + 2.3(H + OH), \quad (112)$$

in which

$$\pi_j = c_j(d\beta_j/d\text{p}H). \quad (113)$$

Now if the quantity $2.3(H + OH)$ is neglected in Eq. (112), the buffer capacity of the solution becomes the sum of terms attributable to the individual solutes, the characteristic term of each solute being simply the derivative of β by pH, as given in the preceding section, multiplied by c.

The reciprocal of π gives the slope of the titration curve with strong base, or $d\text{p}H/db_s$. But π, as defined by any of the above equations, even with neglect of the quantity $(H + OH)$, is an implicit function of H. The value of π for a solution at any specified value of H is directly and exactly calculable by Eq. (112); but the value of π, and hence of $d\text{p}H/db_s$, for a specified value of b_s can not so directly be calculated, for H is itself in general a complicated function of b_s in such a solution. Expressions for π obtained from Eq. (112), in terms of c, H, K, are therefore exact; but being implicit functions of H, they can not be used to find π as a function of concentrations and constants alone, unless a formula is first available to define H as $f(c,K)$. Despite this limitation, the exact expressions for π corresponding to Eq. (112) are extremely useful in discussing the general nature of titration curves.

K. Relation Between β and Other Coefficients of Solutes

We have seen that as a result of the "conjugate" transformation of constants, physically different solutions may be considered identical in respect to the mathematical relations defining H. Given $D = c\beta$, in which c is the analytical concentration of a solute whatever its nature, then if β for two different solutes is the same mathematical function of some constant (or constants) and H, the two solutions, however different in other respects, have the same "buffer properties." "Ideally," they behave in exactly the same way in titration (reaction with acids and bases) and in the effect of dilution upon H, as in Eqs. (54) and (56).

Despite the differences in conductivities and in osmotic (colligative) properties, we may even write a "dilution law" for the solution of the salt MX_s (of weak base and strong acid) identical in form with that for the solution of weak acid alone. Expanding Eq. (54) and neglecting terms in W, we may write the result as

$$A \cong a\alpha^2/\rho, \quad (114)$$

which is the Ostwald dilution law. Similarly if Eq. (56) is expanded and if the corresponding terms in W are neglected, the result is

$$W/B \text{ or } A^* \cong c(\rho')^2/\alpha'. \tag{115}$$

With reference to Eq. (114) this is as if we had an acid, with ionization constant W/B, and with a "degree of ionization," ρ', increasing with dilution. Eq. (114) is usually applied through conductivities because in the case of a pure acid the charge coefficient is very nearly equal to what may be called the conductivity coefficient λ, as shown below. But for a salt the two effects, or the two coefficients, are quite different, and the conductivity is not a sensitive measurement of the relation in Eq. (115) as an effect of dilution.

The rather complex relation of the buffer properties of a solution to its conductivity and to its osmotic properties may be appreciated by defining the various coefficients involved on the basis of the fundamental definitions of W and α given in preceding sections. We shall distinguish three coefficients for an ideal solute: ν, the osmotic coefficient, is the number of particles (ions or molecules) per mole of solute in solution; λ, the conductivity coefficient, is the number of positive (or of negative, the number being of necessity equal) charges per mole; and β, the charge coefficient, is the *net* number of negative *solute* ion charges per mole. The exact interrelation of these coefficients varies with the type of solute. For example:

(1) Salt, M_sX_s, of strong acid and strong base.

$$\beta = 0;$$
$$\lambda = 1 + \sqrt{W/c};$$
$$\nu = 2\lambda. \tag{116}$$

This last follows if H^+ ion is a "particle."

(2) Strong acid.

$$\beta = 1;$$
$$\lambda = H/a_s = 1/2 + \sqrt{1/4 + W/a_s^2};$$
$$\nu = 2\lambda. \tag{117}$$

(3) Weak acid (ionization constant A).

$$\beta = \alpha;$$
$$\lambda = H/a;$$
$$\nu = 1 + (H + OH)/a = 1 + \lambda + W/Ha. \tag{118}$$

We shall see later, however, that H is now a cubic function of a and A. Therefore only approximations are practical at any concentration in expressing λ. It should be remembered that they are always approximations, and that only the implicit expression, $\lambda = H/a$, is exact.

(4) Salt, MX_s, of strong acid and weak base (ionization constant B).

$$\beta = 1 - \alpha' = \rho';$$
$$\lambda = \alpha' + H/c = 1 + W/Hc;$$
$$\nu = 2 + (H + OH)/c = 1 + \lambda + H/c. \tag{119}$$

As above, H is a cubic function of c and B.

(5) Salt, MX, of weak acid and weak base.

$$\beta = \alpha - \alpha';$$
$$\lambda = \alpha + OH/c = \alpha' + H/c;$$
$$\nu = 2 + (H + OH)/c. \tag{120}$$

If $A = B$,

$$\beta = 0;$$
$$\lambda = \alpha + \sqrt{W}/c;$$
$$\nu = 2\lambda. \tag{121}$$

It is seen that even in the simplest case, of a 1 : 1 strong-strong salt, the important coefficients λ and ν are not independent of c, and especially so in extrapolation to "infinite dilution." Ideally, the conductivity is proportional to λ (other things being equal, such as the ion mobilities, etc.), the colligative properties depend on ν, and the "acidity," or D $(= H - OH)$, or the "buffer properties" in general, depend on β; and the three coefficients are not very simply interrelated because of the mathematical definitions of W and α. Hence the only exact (not necessarily the only accurate) determination of the ionization constants, A and B, and of the ion-product constant W, is a method such as the E.M.F. method involving the direct determination of H, and hence operating through the coefficient β directly, and not methods which involve λ or ν, coefficients which are themselves functions of the concentrations and which require the use of approximations for their application.

On the basis of the definitions here given, the coefficients λ and ν are readily seen to increase with dilution, the theoretical limit in both cases being infinity at infinite dilution for any solute. The effect of dilution on the coefficient β (or in simple cases on α) will be discussed later (Chapters III and IV).

II

The Theorem of Isohydric Solutions

A. Conditions for the Validity of the Theorem

We shall here examine the conditions for the validity of the theorem of isohydric solutions, that isohydric solutions (solutions of the same H) mix without change of H. The original statement was that of Arrhenius[1] in terms of conductivity, that when solutions of different acids are mixed, having the same conductivity (and therefore "isohydric" according to Arrhenius) the conductivity is not changed on mixing. The proof given by Arrhenius is only an approximate one based on the assumption that $H = a\alpha$ and amounting practically to an application of the Ostwald dilution law; on mixing, according to Arrhenius, H would equal $(a\alpha)_1 + (a\alpha)_2 + \cdots$. This is still the usual proof used.[2] Michaelis's[3] proof of the theorem (in terms explicitly of H rather than of conductivity) is also an approximation, assuming that H may be represented by $H = \sqrt{(aA)_1 + (aA)_2}$ in the mixture (only a rough approximation actually, as will be discussed later under the problem of two acids in Chapter XI). Since these proofs are both limited and approximate, it may strictly be concluded on the basis of such proofs only that the theorem is approximately valid in the cases considered. Michaelis moreover adds, "But this law is only valid for mixtures of free acids alone in the absence of their salts. It is *not* applicable to two isohydric solutions of *any* composition."

If the problem is attempted through equations which are explicit in terms of H, it certainly seems hopeless of proof (or disproof), because the exact equations, when solved for H as the dependent variable in solutions of unlimited complexity, could not be handled for the purpose. Michaelis's statement simply means that neither the proof nor the disproof is apparent with the ordinary "pH" equations. The problem is possible and the general conditions for the validity of the theorem can be established only if we drop all approximations, use the general equation already developed, namely Eq. I(2), and look at it as an expression for D, or $H - OH$, which may or

(1) S. Arrhenius, *Z. phys. Chem.*, **2**, 284 (1888).

(2) Thus, C. A. Kraus, *The Properties of Electrolytically Conducting Systems*, Chemical Catalog Co., N.Y., 1922, p. 219. The "proof," applied to acids, assumes that $H = [X^-]$, which is the same as $H = a\alpha$.

(3) L. Michaelis, *Hydrogen Ion Concentration*, Williams and Wilkins, Baltimore, 1926; pp. 42–43.

may not be linear in respect to the analytical concentrations of solutes. It results that for "noncomplex" solutions (those not involving the "complex" class of equilibrium constants) whose mixture is still "noncomplex," D is a linear function of the concentrations, and isohydric solutions can be proved to mix without change of H; the theorem is otherwise not valid. [It is furthermore understood, of course, that the solutions mix without precipitation unless the precipitate is that of the normal salt of acid and base of equal strength.]

If the solution is noncomplex, the expression for D, Eq. 1(2),

$$D = H - OH = \Sigma c\beta, \tag{1}$$

contains no terms involving β_c (a complex charge coefficient, which is a function of one or more solute concentrations). The noncomplex charge coefficients are independent of the concentrations, being, as shown in Chapter 1, functions of H and ionization constants only. Every such β is a constant at constant H. Hence if the mixture of two isohydric solutions itself does not involve any terms in β_c arising from interaction of solutes of the two solutions, the process of mixing necessitates no readjustment of either H or the β's; or, it can involve no reaction of solute species, and both H and the β's remain unchanged.

A more explicit analytical proof will also be stated. Let us consider two noncomplex solutions of the same value of D, so that in Solution (1),

$$D = (\Sigma c\beta)_1, \tag{2}$$

and in Solution (2)

$$D = (\Sigma c\beta)_2. \tag{3}$$

The final values of these quantities after mixing will be distinguished by the subscript f. Some of the solutes in Solution (1) may be the same as those in Solution (2), etc. These solutions are to be mixed in the proportion, by volume, of x of (1) to $(1 - x)$ of (2), involving no change in total volume (inherent in the assumption that concentrations may be used in place of activities).

We shall now assume, temporarily, that the new values of D and of the charge coefficients are

$$D_f = D + \Delta D, \tag{4}$$

$$\beta_f = \beta + \Delta\beta. \tag{5}$$

Since the concentrations of Solution (1) are diluted to $x/1$ of the original values, and those of Solution (2) to $(1 - x)/1$ of the original values, the equation for the mixture (if it is assumed that the resulting solution is also noncomplex) is

$$D_f = x(\Sigma c\beta)_{1_f} + (1 - x)(\Sigma c\beta)_{2_f}. \tag{6}$$

Introducing Eqs. (4) and (5), we have

$$D + \Delta D = x(\Sigma c\beta)_1 + (1 - x)(\Sigma c\beta)_2 + x(\Sigma c\Delta\beta)_1 + (1 - x)(\Sigma c\Delta\beta)_2. \quad (7)$$

By Eqs. (2) and (3), then,

$$D + \Delta D = xD + (1 - x)D + \mathbf{Y}, \quad (8)$$

or

$$\Delta D = \mathbf{Y}, \quad (9)$$

in which

$$\mathbf{Y} = x(\Sigma c\Delta\beta)_1 + (1 - x)(\Sigma c\Delta\beta)_2. \quad (10)$$

But while $dD/dH > 0$, $d\beta/dH < 0$, by definition. Hence if H is assumed to have increased on mixing the two solutions, ΔD would be positive while $\mathbf{Y}$ would be negative; and vice versa if H is assumed to have decreased. Hence Eq. (9) is satisfied only if ΔD and therefore $\Delta H = 0$; and consequently all the $\Delta\beta$'s must also equal 0.

The theorem of isohydric solutions is therefore proved explicitly for the mixing of isohydric noncomplex solutions of any composition, which remain noncomplex on mixing. The condition is that all the β's are functions of H and K's only, or that there are no terms in the fundamental equation, Eq. (1), involving "complex" coefficients. The operation in writing Eq. (6), depending on linearity with respect to the concentrations, is otherwise impossible.

[Like all the other theorems and relations to be established, this also is an "ideal" theorem in that it assumes either that all the activity coefficients are unity, so that the "mass constants" remain unchanged after the mixing of the solutions, or that the ionic strength is high enough to make the change in γ negligible or zero.]

B. Complex Cases

We shall now consider some examples of equations resulting from the introduction of an equilibrium constant other than acid and base ionization constants:

(a) Strong acid and weak base, forming the complex ion MX_s^-, with K defined as in Eq. 1(42):

$$D = a_s + \tfrac{1}{2}\left\{\left[\frac{B(a_s - b)}{B + OH} + \frac{B}{K(OH)}\right] - \sqrt{(\quad)^2 + \frac{4bB^2}{K(B + OH)(OH)}}\right\}. \quad (11)$$

The symbol $(\quad)^2$ under the root will regularly indicate the square of the expression outside the root. Here a_s is the concentration of the strong acid "HX_s," b that of the weak base "MOH," with ionization constant B.

(b) Strong acid and weak base, forming the uncharged complex (or incompletely dissociated salt), MX_s^0, with K as in Eq. 1(43):

$$D = a_s - [M^+][1 + a_s K/(1 + K[M^+])], \tag{12}$$

in which

$$[M^+] = -\left[\frac{KB(a_s - b) + B + OH}{2K(B + OH)}\right] + \sqrt{(\quad)^2 + \frac{bB}{K(B + OH)}}. \tag{13}$$

(c) The formation of certain "acid salts" such as KHF_2 is ascribed not to the dibasicity of the acid but to the formation of a complex ion, here HF_2^- from F^- and HF, with $K_{HF_2^-}$ defined as $K_{X_2^-}$ of Eq. 1(44). In this case, if the acid is monobasic and weak, so that

$$A = H[F^-]/[HF^0], \tag{14}$$

and

$$K = [HF_2^-]/[F^-][HF^0], \tag{15}$$

while

$$a = [HF^0] + [F^-] + 2[HF_2^-], \tag{16}$$

and, for the pure acid solution,

$$D = H - OH = [F^-] + [HF_2^-], \tag{17}$$

then

$$[F^-] = -(H + A)/4HK + \sqrt{(\quad)^2 + aA/2HK}, \tag{18}$$

and

$$D = \frac{a}{2} + \frac{H^2 - A^2}{8HAK} + \left(\frac{A - H}{2A}\right)\sqrt{\left(\frac{H + A}{4HK}\right)^2 + \frac{aA}{2HK}}. \tag{19}$$

The effect of diluting a solution described by relations such as Eqs. (11), (12), and (19) does not involve mere multiplication of terms by the proportional change in dilution, as in writing Eq. (6). This of course merely makes the analytical proof of the theorem of isohydric solutions used above impossible. But it may be shown that the mixing of isohydric solutions involving these mathematical relations does involve a change of H, and hence that the theorem is valid in the absence of complex equilibrium constants and definitely not valid when they are involved.

Taking Eq. (11) as an example, we may find the neutralization ratio, $(a_s/b)_n$, or the value of a_s for a given value of b, making the solution neutral. Setting $H = OH = \sqrt{W}$,

$$\left(\frac{a_s}{b}\right)_n = \frac{B(bK\sqrt{W} - B - \sqrt{W})}{2bK\sqrt{W}(2B + \sqrt{W})} + \sqrt{\left(\quad\right)^2 + \frac{B^2}{bK\sqrt{W}(2B + \sqrt{W})}}. \tag{20}$$

Since this neutralization ratio is not independent of the concentration b itself, a neutral solution of this type does not remain neutral on dilution, i.e. on mixing with water, which is isohydric with it.

This is very different from the case of "noncomplex" solutions, which will be considered in detail in the following chapters. In the simple noncomplex case a neutral solution remains neutral on dilution, and if it is not neutral, $|D|$, or $|H - OH|$, continually decreases on dilution, approaching zero. But in the complex case, the originally neutral solution becomes acid or basic on dilution and finally, of course, approaches neutrality again on continued dilution to infinity. If such a solution is originally not neutral but, let us assume, acid, with (a_s/b_{orig}) greater than $(a_s/b_{\text{orig}})_n$—greater, that is, than the neutralization ratio at $b = b_{\text{orig}}$—the behavior on dilution depends on the effect of dilution on the neutralization ratio, $(a_s/b)_n$. The derivative, $(da_s/b)_n/db$, may be positive or negative. If it is negative and remains negative during dilution, dilution may take the originally acid solution down through neutrality into alkalinity and again ultimately to neutrality. If the derivative is positive, the solution may become more acid on dilution, and then approaches neutrality on further dilution. An example of these effects will be considered in Chapter XV.

Finally, if the same complex possibilities are considered in connection with strong acids and strong bases, or the limiting (strictly unreal) case in which the acids and bases are all strong, then the interaction of solute ions in any possible way to form charged or uncharged complexes leaves the equation for $H - OH$ linear in respect to the concentrations, being merely

$$D = H - OH = \Sigma a_s - \Sigma b_s. \tag{21}$$

The formation of an incompletely ionized salt of strong acid and strong base may be taken as an example. On the basis of the definition of "strong" already explained (Chapter I), ρ and ρ' here are both zero, whence $[X^-] = a_s - [M_sX_s]$ and $[M^+] = b_s - [M_sX_s]$. Hence $H - OH = [X^-] - [M^+] = a_s - b_s$, as in Eq. (21). Thus in this special (probably purely hypothetical and unreal) case, the introduction of "complex" equilibrium constants does not disturb the relations connected with the theorem of isohydric solutions or the special relations in noncomplex solutions.

C. Special Relations in Noncomplex Solutions

In solutions involving no "complex" equilibria, the following special relations hold:

(a) The neutralization ratio of any two materials is independent of the concentration.

(b) Dilution does not change the "reaction" of a neutral solution.

(c) For any nonneutral solution the quantity $|H - OH|$ decreases continuously upon dilution.

(d) Isohydric solutions mix without change in H.

It is only when ions or molecules of weak acids and bases take part in "complex reactions" that the theorem of isohydric solutions, with its consequences or implications, fails to apply. In some very exceptional cases in which two complex solutions to be mixed are not only isohydric but also isoplethic (of the same concentration) in respect to all the species involved in the complex equilibrium constant, the theorem again holds.

All further discussion with the exception of Chapter xv and some special parts of Chapter xx will now be limited to "noncomplex" solutions, which therefore behave according to the law of isohydric solutions. This means that the discussion will assume complete dissociation of salts and absence of complex ions.

III

Special Values and Limits of the Ionization Fractions for Acids and Bases

A. Relations Between α's and pH

1. Simple Acid or Base. For monobasic acid

$$\alpha = A/(A + H) = A/(A + 10^{-\mathrm{p}H}). \quad (1)$$

$$\rho = 1 - \alpha = H/(A + H). \quad (2)$$

$$H = A\rho/\alpha. \quad (3)$$

$$d\beta/d\mathrm{p}H = d\alpha/d\mathrm{p}H, (= \pi_j/c_j), = 2.3\alpha\rho = 2.3\alpha(1 - \alpha). \quad (4)$$

$$d\rho/d\mathrm{p}H = -2.3\alpha\rho. \quad (5)$$

$$d^2\alpha/d\mathrm{p}H^2 = (2.3)^2\alpha\rho(\rho - \alpha) = (2.3)^2\alpha(1 - \alpha)(1 - 2\alpha). \quad (6)$$

There is therefore an inflection (with maximum rate of change) in α, as $f(\mathrm{p}H)$, at $\alpha = \rho = 1/2$ and $H = A$.

For monoacid base

$$\alpha' = B/(B + OH). \quad (7)$$

$$\rho' = 1 - \alpha' = OH/(B + OH). \quad (8)$$

$$OH = B\rho'/\alpha'. \quad (9)$$

$$d\beta'/d\mathrm{p}H = d\alpha'/d\mathrm{p}H\ (= -\pi_{j'}/c_{j'}), = -2.3\alpha'\rho'. \quad (10)$$

$$d\rho'/d\mathrm{p}H = 2.3\alpha'\rho'. \quad (11)$$

$$d^2\alpha'/d\mathrm{p}H^2 = (2.3)^2\alpha'\rho'(\rho' - \alpha'). \quad (12)$$

Hence an inflection occurs at $\alpha' = \rho' = 1/2$, when $OH = B$ or $H = W/B$.
Graphically, the relations are shown in Figs. 3–1 and 3–2.

2. Dibasic Acid.

$$\alpha_1 = 1/(1 + H/K_1 + K_2/H). \quad (13)$$

$$\alpha_2 = (K_2/H)\alpha_1. \quad (14)$$

$$\rho = (H/K_1)\alpha_1. \quad (15)$$

We note that the ratio $\alpha_2/\rho = K_1K_2/H^2$.

For the derivatives,

$$d\alpha_1/dpH = 2.3\alpha_1(\rho - \alpha_2) \tag{16}$$

$$= 2.3\alpha_1^2(H/K_1 - K_2/H); \tag{17}$$

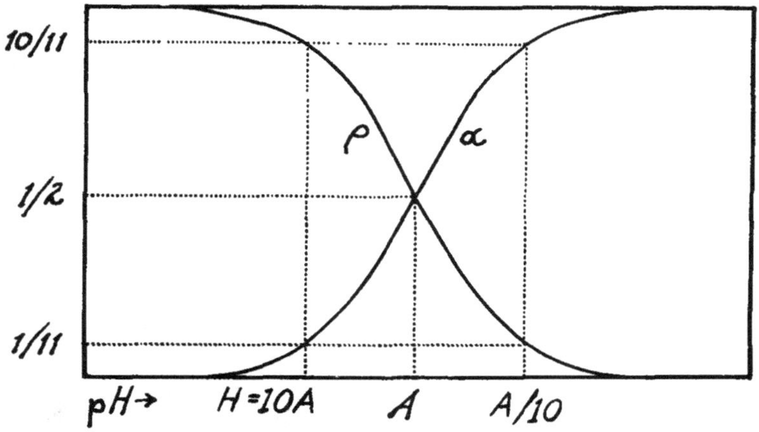

Fig. 3–1. Ionization fractions of monobasic acid.

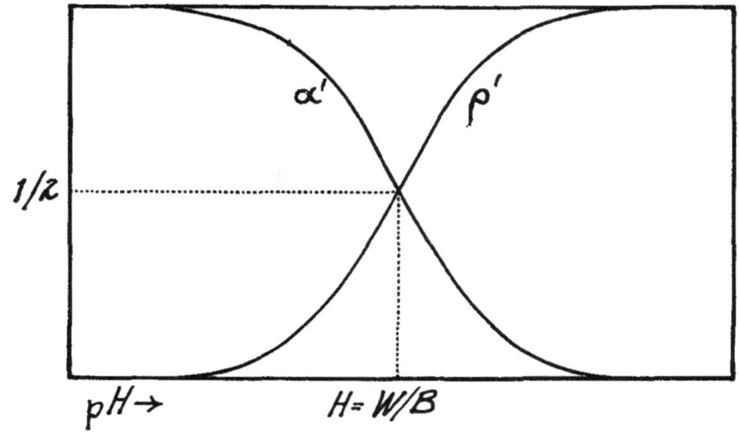

Fig. 3–2. Ionization fractions of monoacid base.

$$d\alpha_2/dpH = 2.3\alpha_2(\alpha_1 + 2\rho);\ > 0; \tag{18}$$

$$d\rho/dpH = -\ 2.3\rho(\alpha_1 + 2\alpha_2);\ < 0; \tag{19}$$

$$d\beta/dpH = 2.3[\alpha_1(\rho + \alpha_2) + 4\alpha_2\rho]; \tag{20}$$

$$d^2\alpha_1/dpH^2 = -\ (2.3)^2\alpha_1[\alpha_1\rho + \alpha_1\alpha_2 + 6\alpha_2\rho - \alpha_2^2 - \rho^2]; \tag{21}$$

$$d^2\alpha_2/dpH^2 = (2.3)^2\alpha_2[3\alpha_1\rho - \alpha_1\alpha_2 - 4\alpha_2\rho + \alpha_1^2 + 4\rho^2]; \tag{22}$$

$$d^2\rho/dpH^2 = (2.3)^2\rho[3\alpha_1\alpha_2 - \alpha_1\rho - 4\alpha_2\rho + \alpha_1^2 + 4\alpha_2^2]. \tag{23}$$

Inflection points of the fractions occur at values of H satisfying the following equations:

For α_1 (2 values): $H^4 - H^3K_1 - 6H^2K_{12} - HK_1^2K_2 + K_1^2K_2^2 = 0.$ (24)

For α_2 : $4H^3 + 3H^2K_1 + HK_1(K_1 - 4K_2) - K_1^2K_2 = 0.$ (25)

For ρ : $H^3 - H^2(K_1 - 4K_2) - 3HK_{12} - 4K_1K_2^2 = 0.$ (26)

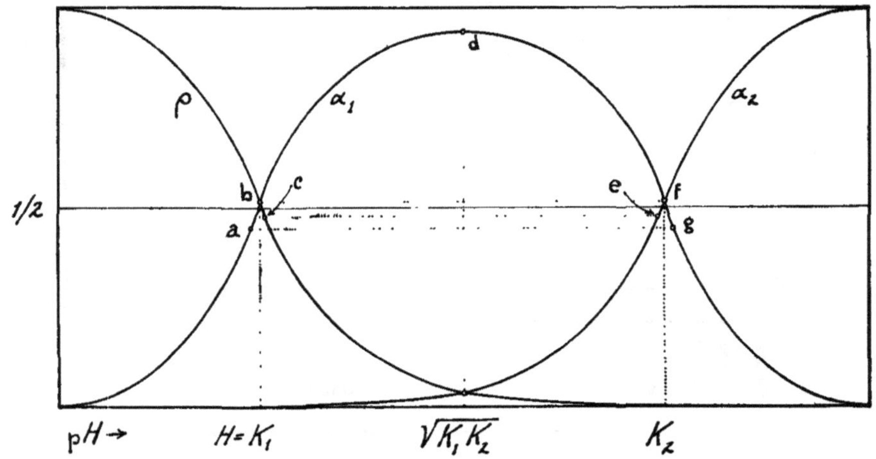

Fig. 3–3. Ionization fractions of dibasic acid with $K_1 \gg K_2$.

For the general case in which $K_1 \gg K_2$, the relations are shown graphically in Fig. 3–3. With increasing pH we have the following special values as lettered in Fig. 3–3, writing y for K_2/K_1:

Point a: first inflection in α_1, at $H \cong K_1(1 + 7y)$, and $\alpha_1 \cong (1 - 4y)/2$.
Point b: when $H = K_1$, $\alpha_1 = \rho = 1/(2 + y)$; $\alpha_2 = y/(2 + y) \cong y/2$.
Point c: inflection in ρ, at $H \cong K_1(1 - y)$, $\rho \cong (1 - y)/2$.
Point d: maximum of α_1, at $H = \sqrt{K_1K_2}$; $\alpha_1 = 1/(1 + 2\sqrt{y})$ and $\rho = \alpha_2 = 1/(2 + 1/\sqrt{y}) \cong \sqrt{y}$.

By symmetry,

Point e: inflection in α_2, at $H \cong K_2(1 + y)$; $\alpha_2 =$ value of ρ at point c.
Point f: when $H = K_2$, $\alpha_1 = \alpha_2 =$ value of α_1 and ρ at point b; $\rho =$ value of α_2 at point b.
Point g: second inflection in α_1, at $H \cong K_2(1 - 7y)$, and $\alpha_1 =$ value of α_1 at point a.

The relations for cases in which K_1 is not large compared to K_2 are discussed in Chapter XIV, Section B–5.

If the dibasic acid is strong in K_1, we have

$$d\alpha_1/dpH = -\,d\alpha_2/dpH = -\,2.3\alpha_1\alpha_2 = -\,2.3\alpha_2(1 - \alpha_2), \quad (27)$$

$$d^2\alpha_1/dpH^2 = (2.3)^2\alpha_1\alpha_2(\alpha_2 - \alpha_1) = -\,(2.3)^2\alpha_2(1 - \alpha_2)(1 - 2\alpha_2), \quad (28)$$

with an inflection in α_1 (and α_2) at $\alpha_1 = \alpha_2 = 1/2$ and $H = K_2$. Graphically, this case is shown in Fig. 3–4.

The relations for a diacid base, with $K_1' \gg K_2'$, correspond to Fig. 3–3 with the fractions reversed, or ρ', α_1', α_2' for α_2, α_1, ρ respectively, and with W/K_1' and W/K_2' in place of K_2 and K_1 respectively.

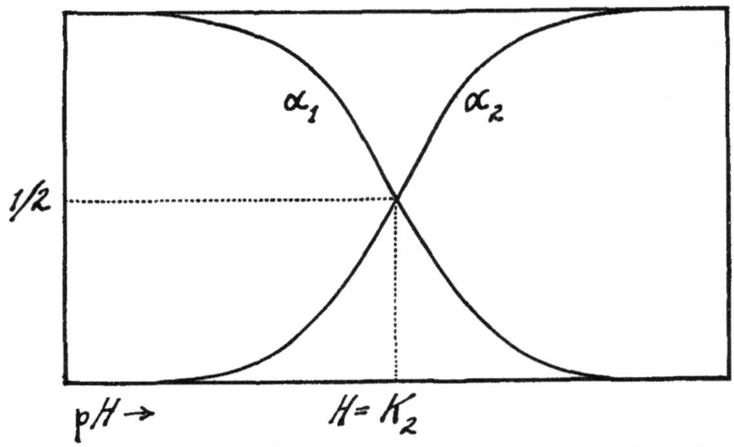

Fig. 3–4. Ionization fractions of dibasic acid strong in respect to K_1.

3. Tribasic Acid. For the derivatives,

$$d\alpha_1/dpH = 2.3\alpha_1(\rho - \alpha_2 - 2\alpha_3), \tag{29}$$

$$d\alpha_2/dpH = 2.3\alpha_2(2\rho + \alpha_1 - \alpha_3), \tag{30}$$

$$d\alpha_3/dpH = 2.3\alpha_3(3\rho + 2\alpha_1 + \alpha_2), \tag{31}$$

$$d\rho/dpH = -2.3\rho(\alpha_1 + 2\alpha_2 + 3\alpha_3); \tag{32}$$

$$d^2\alpha_1/dpH^2 = (2.3)^2\alpha_1(\rho^2 + \alpha_2^2 + 4\alpha_3^2 - \rho\alpha_1 - 6\rho\alpha_2 - 13\rho\alpha_3 - \alpha_1\alpha_2 - 4\alpha_1\alpha_3 + 3\alpha_2\alpha_3). \tag{33}$$

When $d^2\alpha_1/dpH^2 = 0$,

$$H^6 - H^5K_1 - 6H^4K_{12} - H^3K_{12}(K_1 + 13K_3) + H^2K_1^2K_2(K_2 - 4K_3) + 3HK_1^2K_2^2K_3 + 4K_1^2K_2^2K_3^2 = 0. \tag{34}$$

Graphically, for the case in which $K_1 \gg K_2 \gg K_3$, we have Fig. 3–5.

Besides maxima, α_1 and α_2 each have two inflections, point a and point d for α_1, point c and point f for α_2. These inflections, and those for ρ and α_3 (point b and point e) are estimated from equations like Eq. (34) for α_1. Here if $K_1 \gg K_2 \gg K_3$, then at high values of H, the first four terms of Eq. (34) give

$$H \cong K_1(1 + 7K_2/K_1), \tag{34a}$$

for inflection a. At low H, the five middle terms of Eq. (34) give

$$H \cong K_2(1 - 7K_2/K_1 - K_3/K_2), \tag{34b}$$

for inflection d. These values of course are similar to the approximations for a dibasic acid given above.

The maxima for α_1 and α_2 are found by setting the first derivatives equal to zero. For α_1, the maximum occurs when

$$H^3 - HK_1K_2 - 2K_{123} = 0, \tag{35}$$

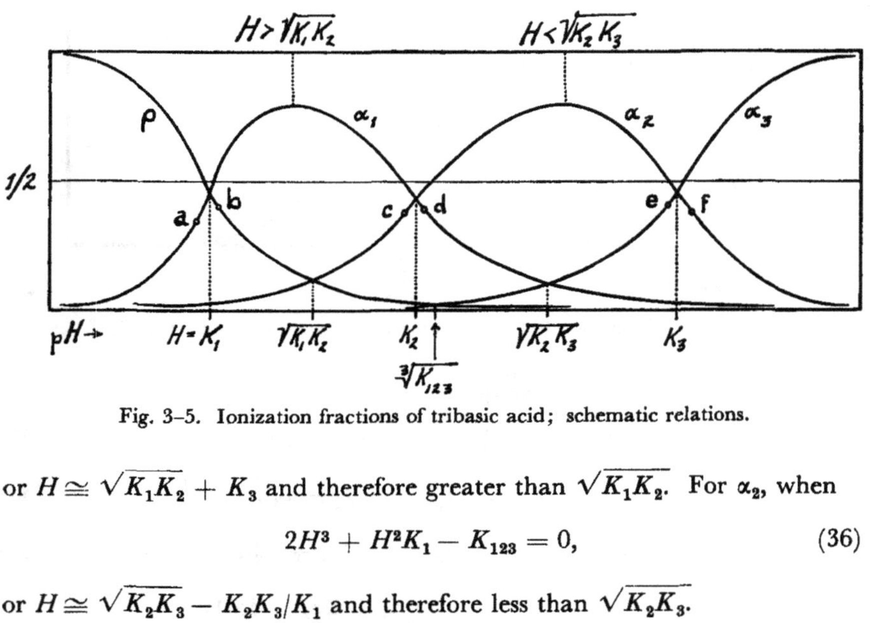

Fig. 3–5. Ionization fractions of tribasic acid; schematic relations.

or $H \cong \sqrt{K_1K_2} + K_3$ and therefore greater than $\sqrt{K_1K_2}$. For α_2, when

$$2H^3 + H^2K_1 - K_{123} = 0, \tag{36}$$

or $H \cong \sqrt{K_2K_3} - K_2K_3/K_1$ and therefore less than $\sqrt{K_2K_3}$.

For the crossings of the curves: $\rho = \alpha_1$, $\alpha_1 = \alpha_2$, $\alpha_2 = \alpha_3$, when $H = K_1$, K_2, K_3 respectively; $\rho = \alpha_2$, $\alpha_1 = \alpha_3$, when $H = \sqrt{K_1K_2}$, $\sqrt{K_2K_3}$, respectively; $\rho = \alpha_3$, when $H = \sqrt[3]{K_1K_2K_3}$, which may be greater or less than K_2, where $\alpha_1 = \alpha_2$.

The value of H at the special values of 1/3, 1/2, and 2/3 for any fraction may be found from the equation

$$\mathbf{C}_0H^3 + \mathbf{C}_1H^2K_1 + \mathbf{C}_2HK_1K_2 + \mathbf{C}_3K_1K_2K_3 = 0. \tag{37}$$

The values 2, 1, and 1/2 are given to the coefficient $\mathbf{C}_0$, for $\rho = 1/3$, 1/2, and 2/3 respectively, to the coefficient $\mathbf{C}_1$ for corresponding values of α_1, to the coefficient $\mathbf{C}_2$ for the corresponding values of α_2 and to the coefficient $\mathbf{C}_3$ for the corresponding values of α_3, all other coefficients being always -1. This applies also to a dibasic acid, where the term $\mathbf{C}_3K_{123} = 0$. Hence for dibasic or tribasic acid, the required (approximate) values of H to give these values of the fractions are shown in Table 3–1.

Table 3–1. Values of H for special values of the fractions.

Fraction	ρ	α_1	α_2	α_3
1/3	$K_1/2$	$2K_1$ or $K_2/2$	$2K_2$ or $K_3/2$	$2K_3$
1/2	$>K_1$	$< K_1$ or $> K_2$	$< K_2$ or $>K_3$	$< K_3$
2/3	$2K_1$	$K_1/2$ or $2K_2$	$K_2/2$ or $2K_3$	$K_3/2$

4. Polybasic Acid. The various derivatives may be written from the general formulas in Chapter I. The graphical relations would be similar, by extension, to those in Fig. 3–5. All the fractions except the terminal ones (ρ and α_z) have a maximum and two points of inflection; ρ and α_z have a single inflection point.

5. Special Note on Functions Replacing α. The fact that the function α does not, in general, vary in a simple manner with respect to pH, but shows a maximum for polybasic acids and an inflection point even in the simple case of a monobasic acid, suggests that some simpler function might be sought for the representation of these relations. Such simple functions (rising or falling with pH without change of sign of first and second derivatives) are, however, of the nature of *ratios* of the concentrations of species and not fractions of the total or analytical concentration. They do not possess the additivity, consequently, of the α's, in that $1 = \rho + \alpha_1 + \alpha_2 + \cdots$. The general equations can not, therefore, be so easily differentiated into simple useful forms for the representation of the buffer properties. As examples:

(a) If

$$s_1 = [X^-]/[X^0], \qquad s_2 = [X^=]/[X^0], \cdots, s_z = [X^{-z}]/[X^0], \quad (38)$$

then

$$\begin{aligned} s_1 &= K_1/H, \\ s_2 &= K_1K_2/H^2 = s_1K_2/H, \\ s_3 &= K_{123}/H^3 = s_1K_{23}/H^2, \text{ etc.} \end{aligned} \quad (39)$$

Also, $s_1 = \alpha_1/\rho$, and $\alpha_1 = s_1/(1 + s_1 + s_2 + s_3 + \cdots)$. Hence, for a polybasic acid and strong base,

$$D = a(s_1 + 2s_2 + 3s_3 + \cdots)/(1 + s_1 + s_2 + s_3 + \cdots) - b_s. \quad (40)$$

This is not as simple to handle as Eq. I(57) or I(58), although the individual derivatives of the s's are simple and there is never any inflection or maximum for s against pH. (Furthermore, however, while the α's are limited to values from 0 to 1, these ratios may approach infinity.) For the derivatives: $ds_1/dpH = 2.3s_1, \cdots, ds_z/dpH = 2.3zs_z$; $d^2s_1/dpH^2 = (2.3)^2s_1, \cdots, d^2s_z/dpH^2 = (2.3)^2z^2s_z$.

(b) If

$$r_1 = [X^-]/[X^0], \qquad r_2 = [X^=]/[X^-], \cdots, r_z = [X^{-z}]/[X^{-(z-1)}], \tag{41}$$

then

$$\begin{aligned} r_1 &= K_1/H, \\ r_2 &= K_2/H = r_1K_2/K_1, \\ r_3 &= K_3/H = r_1K_3/K_1, \text{ etc.} \end{aligned} \tag{42}$$

Also $r_1 = \alpha_1/\rho$, and $\alpha_1 = r_1/(1 + r_1 + r_1r_2 + r_1r_2r_3 + \cdots)$. Then for a polybasic acid and strong base,

$$D = a(r_1 + 2r_1r_2 + 3r_1r_2r_3 + \cdots)/(1 + r_1 + r_1r_2 + r_1r_2r_3 + \cdots) - b_s. \tag{43}$$

Again the individual derivatives are simple: $dr_1/dpH = 2.3r_1, \cdots, dr_z/dpH = 2.3r_z$; $d^2r_1/dpH^2 = (2.3)^2r_1, \cdots, d^2r_z/dpH^2 = (2.3)^2r_z$.

With either of these functions, the general equation for monobasic acid and strong base is quite simple (and furthermore identical for both functions and may be of use in connection with certain indicator problems in which the ratio of two forms of a substance is of direct significance:

$$D = H - OH = ar/(1 + r) - b_s, \tag{44}$$

in which r is the same as s, and represents α/ρ.

B. Limits of the Fractions in Respect to Concentration

1. Simple and Compound Charge Coefficient; H_*. Since the fractions are functions of H and K's, their limits in respect to concentration of solute depend on the limiting values of H in respect to concentration. The limit for $c = 0$ will be written H_∞, since it is customary to speak of "infinite dilution" at $c = 0$; $H_\infty = \sqrt{W}$ always, of course, for any possible solute. At the other extreme, or for the limit when $c = \infty$, we shall write H_*. Solutes in general may be divided into two classes mathematically, depending on whether H_* (the limiting value of H approached as the concentration increases indefinitely) is or is not independent of the concentration itself.

Class 1: Acids and bases, or solutes introducing solute ions of one sign only, so that all the terms of the charge coefficient β are of the same sign. For such solutes, β can never be zero at any pH and at any concentration, and we find that for this class, H_* and OH_* are $f(c)$. For acids, $H_* = \sqrt{aK_1}$, and for bases, $OH_* = \sqrt{bK_1'}$, (or $H_* = W/\sqrt{bK_1'}$), whatever the "basicity" of the acid or the "acidity" of the base.

Class 2: Solutes with a "compound" charge coefficient, or solutes introducing solute ions of both signs. For these solutes (salts and ampholytes) β is made up of both positive and negative terms, and hence it is possible for it to have the value zero at some particular value of H; in pure solution, moreover, β for this class approaches the value of zero as the concentration approaches infinity. Here H_*, as a limiting value, is independent of c and is a function of constants only.

In the present chapter we shall examine the limiting values of the fractions in the case of pure acids and pure bases alone, or for solutes of the first of these classes.

2. LIMITS AT $c = 0$; VALUES OF THE FRACTIONS AT "INFINITE DILUTION."

a. Monobasic case. For a monobasic acid, or for a simple base, it is seen from Eqs. (1) and (7) that the degree of ionization is never complete, since neither H nor OH can ever be zero. The limiting values at "infinite dilution" are

$$\alpha_\infty = A/(A + \sqrt{W}), \qquad \alpha'_\infty = B/(B + \sqrt{W}). \tag{45}$$

With $A = 10^{-5}$, α_∞ is only 100/101 (with $W = 10^{-14}$); for boric acid, with $A = 6 \times 10^{-10}$[1], α_∞ is only 3/503. The limiting value then of α_∞ is 1/2 if $A = \sqrt{W}$, less than 1/2 if $A < \sqrt{W}$, and greater than 1/2 (but always less than 1) if $A > \sqrt{W}$.

This limit, characteristic of the acid itself, is a mathematical consequence of the definitions of Chapter I. Any consistent theory leads to it. The "dissociation" theory, if correctly applied with regard to all the "dissociation constants" involved, or to the simultaneous "dissociation" of both solute and water, leads to it. So does the Brønsted theory.

Although thermodynamics, or mathematical definitions, will not decide between theories or mechanisms which can be made mathematically consistent with the definitions, and although the "dissociation" theory, at the expense of the invention of hypothetical "dissociating compounds," can be made consistent with the mathematics, the word has the connotation of dissociation in space and it will continue to imply and suggest dissociation in space with the false expectation that $\alpha_\infty = 1$. The expressions for α and ρ given in Eqs. (1), (2), (7), and (8) appear both in Clark's book[2] and in Michaelis's book (Ref. II–3), and are there used for the plotting of the relations between α and pH. But this particular point is not made at all, and the incorrect conclusion that $\alpha_\infty = 1$ is undeniably widespread simply because even in the most modern books and articles employing the most advanced theories of ionization of "acids and bases," α is currently still called a "degree of dissociation" and the constants are called "dissociation constants." Thus in the formula $\alpha = \Lambda/\Lambda_\infty$, "$\Lambda_\infty$" means strictly the sum

(1) W. M. Latimer, *Oxidation Potentials*, Prentice-Hall, N.Y., 1938.

(2) W. M. Clark, *The Determination of Hydrogen Ions*, Williams and Wilkins, Baltimore, 1922.

of the ionic mobilities, but the symbol *reads* "conductivity at infinite dilution," which is unfortunately mathematically false and is tantamount to the implication that $\alpha_\infty = 1$.

Values of the degrees of ionization greater than the characteristic limits of Eq. (45) are possible only upon the addition of base (or acid) to bring H (or OH) down to a value below $\sqrt{W}$. But then it seems still more artificial to call α any longer a degree of "dissociation," which must continue to

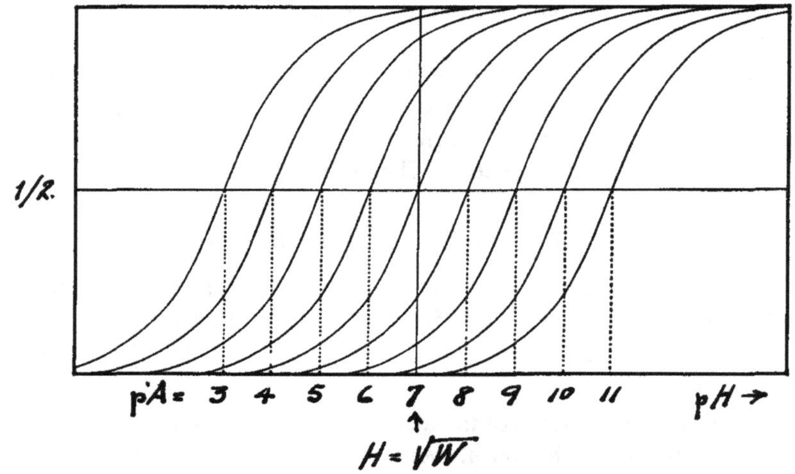

Fig. 3–6. Degree of ionization of monobasic acids of various strengths.

suggest approach to and attainment of complete dissociation by mere dilution. In no case, then, is α equal to 1 except by definition for a strong monobasic acid or strong (monoacid) base, for which $\alpha = 1$ at any pH and at any concentration.

The false conclusion that $\alpha_\infty = 1$ results then from identifying H (instead of D or $H - OH$) with $a\alpha$ and using the picture of dissociation rather than of an effect of solute upon the "reaction," $H - OH$, of water itself. It is in fact commonly "explained" that α increases with dilution because as the ions of the acid are brought farther apart their "recombination" becomes more difficult, and that $\alpha_\infty = 1$ because the ions are then too far apart to "recombine" at all. This reasoning is false. The ions H and OH are to begin with water ions, not ions "of" an acid or a base at all. The degree of ionization is only indirectly a function of the concentration, but directly $f(H,K)$. If we write $[X^-]/[X^0] = K/H$ from the definition of the ionization constant, and if we note that H is always greater than 0, it is immediately clear that the degree of ionization is never complete and that it is limited, in "infinite dilution," by the value $\sqrt{W}$ for H. This expression moreover also shows, clearly and correctly, why the degree of ionization increases with dilution, since by the very definition of an acid, H increases with the concentration of the solute.

Finally, the dependence of α_∞ upon the relation between A and $\sqrt{W}$ may be seen graphically in Fig. 3–6. The value of α_∞ is the intersection of the α curves with the vertical representing $H = \sqrt{W}$. Obviously $\alpha_\infty = 1/2$ if $A = \sqrt{W}$, less than 1/2 if $A < \sqrt{W}$, and greater than 1/2 if $A > \sqrt{W}$; but α is never 1 except for strong acid, a_s, for which the curve is by definition the horizontal at $\alpha = 1$.

b. Dibasic case. For a dibasic acid,

$$(\alpha_1)_\infty = 1/(1 + \sqrt{W}/K_1 + K_2/\sqrt{W}), \tag{46}$$

$$(\alpha_2)_\infty = (\alpha_1)_\infty K_2/\sqrt{W}, \tag{47}$$

$$\rho_\infty = (\alpha_1)_\infty \sqrt{W}/K_1. \tag{48}$$

If, for example, $K_1 = 10^{-6}$ and $K_2 = 10^{-10}$, $(\alpha_1)_\infty = 1/1.101$ and $(\alpha_2)_\infty = 0.001/1.101$, while ρ_∞ is not zero but $0.1/1.101$.

If the dibasic acid is "strong in K_1," $\rho = 0$ by definition, and

$$(\alpha_1)_\infty = 1 - (\alpha_2)_\infty = 1/(1 + K_2/\sqrt{W}). \tag{49}$$

If $K_2 = 10^{-5}$, $(\alpha_1)_\infty = 1/101$, and $(\alpha_2)_\infty = 100/101$.

3. Limits as the Concentration Increases Indefinitely ($c \overrightarrow{=} \infty$). The values of the fractions now depend upon H_*, the value of H approached as c increases indefinitely. From the equation for a polybasic acid, Eq. 1(58), it is seen that as the concentration a (and therefore as H) is made to increase indefinitely, we may neglect OH compared to H in D; then as H becomes very large compared to the constants, the expression simplifies to

$$H_* = a/(H/K_1) = \sqrt{aK_1}. \tag{50}$$

Hence the *lower limit* of α for a monobasic acid, in respect to concentration, or as $a \overrightarrow{=} \infty$, is

$$\alpha_* = 1/(1 + \sqrt{a/A}), \tag{51}$$

while for a dibasic acid,

$$(\alpha_1)_* = 1/(1 + \sqrt{a/K_1} + K_2/\sqrt{aK_1}), \tag{52}$$

$$(\alpha_2)_* = (\alpha_1)_*(K_2/\sqrt{aK_1}). \tag{53}$$

For a dibasic acid "strong in K_1," $H_* = a$, and

$$(\alpha_1)_* = 1/(1 + K_2/a), \tag{54}$$

with α_2 always equal to $(1 - \alpha_1)$.

C. Dependence of the Ion Concentrations Upon the Analytical Concentrations

By considering the general equation

$$D = a\beta_a + \Sigma c\beta, \tag{55}$$

for the derivatives $d\alpha_m/da$ and da/dH, we may write expressions for the variation of the concentration of a particular ion of the acid a with respect to the analytical concentration of the acid as it is being added to a solution containing the solutes represented in the sum $\Sigma c\beta$. For the cases here considered the c's are assumed to remain constant in the process. Variations in which increase of a in a saturated solution causes a change in the solubility of a salt involving one of the c's will be considered in Chapters XIX and XXI. For general use we note that $d/da = (d/dH)(dH/da) = (d/dH)/(da/dH)$ and that, from Eq. (55),

$$da/dH = \mathbf{Y}/\beta_a - (d\beta_a/dH)(D - \Sigma c\beta)/\beta_a^2, \tag{56}$$

$$= \mathbf{Y}/\beta_a - (d\beta_a/dH)a/\beta_a, \tag{56a}$$

in which we define $\mathbf{Y}$ as

$$\mathbf{Y} = d(D - \Sigma c\beta)/dH. \tag{57}$$

1. Monobasic Acid.

$$d[X^-]/da = d(a\alpha)/da = \mathbf{Y}dH/da. \tag{58}$$

Since by definition dD/dH is positive while $d\beta/dH$ is negative, then $\mathbf{Y}$ is positive; and then since dH/da is positive by definition, the expression in Eq. (58) is always positive, so that the concentration of the anion always increases with the analytical concentration, although α decreases whatever the composition of the solution. If the solution contains a weak base,

$$d(a\alpha)/da = \alpha/[1 + a\alpha\rho/(H + OH + b\alpha'\rho')]; \tag{59}$$

and if $b = 0$ or if the base is strong, this becomes simply

$$d(a\alpha)/da = \alpha/[1 + a\alpha\rho/(H + OH)]. \tag{60}$$

For the unionized form,

$$d[HX^0]/da = d(a\rho)/da = (dH/da)(\mathbf{Y}\rho/\alpha + a\rho/H), \tag{61}$$

which is always positive.

2. Dibasic Acid. For the univalent anion,

$$d[HX^-]/da = d(a\alpha_1)/da = (dH/da)(\mathbf{Y} - 2\alpha_2 da/dH - 2ad\alpha_2/dH). \tag{62}$$

When combined with Eq. (56a) this gives

$$d[HX^-]/da = (\alpha_1/\beta)(dH/da)(\mathbf{Y} + 2a\alpha_2/H), \tag{63}$$

which is always positive. When CO_2 is added to any solution, in other words, the concentration of the bicarbonate ion increases steadily with the analytical concentration of the CO_2, regardless of other acids and bases in the solution.

For the divalent anion, similar treatment leads to

$$d[X^=]/da = d(a\alpha_2)/da = (\alpha_2/\beta)(dH/da)(\mathbf{Y} - a\alpha_1/H). \tag{64}$$

This is always positive if $\Sigma(c\beta) = 0$, for if $a\alpha_1 = H - OH - 2a\alpha_2$ we then have

$$dD/dH - a\alpha_1/H = 2(OH + a\alpha_2)/H. \tag{65}$$

A little consideration shows that the derivative $d[X^=]/da$ is still always positive if the dibasic acid is being added to a solution already containing an additional acid. The carbonate ion concentration therefore increases steadily with the total CO_2 concentration if CO_2 is added to pure water, or in the presence of some acid. But if the additional solute is a base so that $\Sigma(c\beta) = -b\beta'$, the concentration of $X^=$ may pass through a maximum value as a is increased. The value of H at this point of maximum concentration of $X^=$ depends on the second ionization constant of the acid and on the concentration and nature of the base present. If this is a simple monoacid base at concentration b, then

$$D = a(\alpha_1 + 2\alpha_2) - b\alpha'. \tag{66}$$

If we substitute a from this equation into the last factor of Eq. (64) and set the result equal to zero, we obtain the condition for the value of H for the maximum value of $[X^=]$:

$$bH^2B(H^2B - 2K_2W) = 2(HB + W)^2(H^2K_2 + HW + K_2W). \tag{67}$$

Finally, the value of a at which the maximum occurs may then be obtained from Eq. (66) by substitution of the critical value of H, which will determine the values of D and the fractions.

If the base is strong and monoacid, Eq. (67) becomes

$$H = W/(b_s - 2K_2) + \sqrt{(\quad)^2 + 2K_2W/(b_s - 2K_2)}, \tag{68}$$

for H_{max}; for a divalent strong base such as $Ca(OH)_2$, "b_s" in this equation would represent its normality. It is seen that there will not be a maximum in the concentration of the ion $X^=$ unless the net normality of strong base (or $b_s - a_s$ if strong acid is also present) is greater than $2K_2$.

From these considerations it follows that the solubility of an acid salt of a dibasic acid, controlled by a solubility product constant of the form $[M^{++}][HX^-]^2$ or $[M^+][HX^-]$, is continually decreased by excess of the acid; this is so since the solubility is taken as the concentration of the base in presence of excess of the acid of the salt. On the other hand the solubility of the normal salt, measured again as the concentration of the base and with solubility product constant $[M^{++}][X^=]$ or $[M^+]^2[X^=]$, may pass through a minimum as excess of the acid is added. This effect, however, is possible only if the precipitating acid is being added to a solution containing excess of base over other acids present. If CO_2 is added to pure aq. $CaCl_2$, then, ideally, the precipitated $CaCO_3$ never redissolves in excess of CO_2; but if CO_2 is added to aq. $Ca(OH)_2$, the precipitate will redissolve in excess of CO_2.

For the uncharged form,

$$d[H_2X^0]/da = (\rho/\beta)(dH/da)[\mathbf{Y} + a(\alpha_1 + 4\alpha_2)/H], \qquad (69)$$

which is always positive.

If the dibasic acid is strong in respect to K_1, $[H_2X^0] = 0$ by definition. Otherwise, Eqs. (63) and (64) still hold. The concentration of HX^- increases steadily with the value of a whatever the composition of the solution, but the concentration of $X^=$ passes through a maximum in the presence of a base, provided that b_s, for example, is greater than $2K_2$. If sulfuric acid is strong in K_1 then, and if $K_2 \cong 10^{-2}$, the maximum concentration of $SO_4^=$, in the presence of alkali at the net concentration $b_s = 1$, occurs when according to Eq. (68) $H = 1.43 \times 10^{-8}$ and $a = 0.5$, with $[SO_4^=]_{max} = 0.5(1 - 1.4 \times 10^{-6}$; with $b_s = 1$ and $a = 1$, $H \cong [SO_4^=] \cong 0.0951$.

3. ALKALI ACID SALT OF DIBASIC ACID; AMPHOLYTE. With a now as the concentration of the acid salt NaHX, with charge coefficient $\beta_A = \alpha_1 + 2\alpha_2 - 1$,

$$D = a(\alpha_1 + 2\alpha_2 - 1) + \Sigma c\beta. \qquad (70)$$

Then

$$d[HX^-]/da = d(a\alpha_1)/da = (\alpha_1/\beta_A)(dH/da)[\mathbf{Y} + a(\rho + \alpha_2)/H]; \qquad (71)$$

$$d[X^=]/da = d(a\alpha_2)/da = (\alpha_2/\beta_A)(dH/da)(\mathbf{Y} + 2a\rho/H); \qquad (72)$$

$$d[H_2X^0]/da = d(a\rho)/da = (\rho/\beta_A)(dH/da)(\mathbf{Y} + 2a\alpha_2/H). \qquad (73)$$

These are all positive, since (dH/da) and β_A have the same algebraic sign.

From these results we deduce that the concentrations of the three species, U^+, U^0, and U^- of a 1 : 1 ampholyte, which in solution is mathematically equivalent to the acid salt NaHX, all increase with the concentration of the ampholyte whatever the composition of the solution. But the concentration of U^-, which corresponds to $X^=$ in the case of the acid salt NaHX, will pass through a maximum with respect to the concentration of the hydrochloride of the ampholyte, MCl, if this salt is added to a solution containing excess of foreign base. Similarly the concentration of U^+ passes through a maximum with respect to the concentration of the sodium salt of the ampholyte, if such salt is added to a solution containing excess of foreign acid.

IV

Limits of Ionization Fractions and the Iso-electric Point for Salts and Ampholytes

A. H_* (or H_{ie}) for Solutes with Compound β

1. H_* as H at $\beta = 0$. While again $H_\infty = \sqrt{W}$, as always, H_* is now estimated in a different but significant manner. Let us consider the equation for a simple ampholyte,

$$D = H - OH = c\beta = c(K_aOH - K_bH)/(K_aOH + K_bH + W), \tag{1}$$

which, on expansion, becomes

$$H^4K_b + H^3(cK_b + W) + H^2W(K_a - K_b) - HW(cK_a + W) - K_aW^2 = 0. \tag{2}$$

Unlike the case of simple acids and bases, the hydrogen ion concentration of the solution of an ampholyte or of a salt does not increase (or decrease) indefinitely as the concentration of the solute increases indefinitely. Or, if c approaches infinity for an acid, H also approaches infinity; but for an ampholyte or a salt, H (or OH) approaches some definite value characteristic of the equilibrium constants involved as the concentration approaches infinity. Hence the value of H_* follows from an equation like Eq. (2) by the neglecting of all terms not involving the concentration. But this is seen at once to be tantamount to setting $D = 0$ in Eq. (1), leaving the equation $0 = c\beta$; and for $c \neq 0$, the only solution is, of course, $\beta = 0$. The value of H_* then may be obtained directly merely by setting $\beta = 0$. Such a procedure is not to be followed in the case of acid or base alone, since β for such a solute can never equal zero at any value of c. But for salts and ampholytes, with β compounded of two terms, one positive and one negative, there can be a real and finite value of H corresponding to $\beta = 0$. The value so obtained may then be taken as the value of H approached as $c \rightrightarrows \infty$, or as H_*.

The first practical significance of H_*, determined by the equation $\beta = 0$, is that the resulting formula for H_*, a formula in terms of ionization constants only and independent of the concentration, is immediately the "first approximation" formula for H in the solution of the particular salt or ampholyte. It is clear that its applicability or accuracy increases with the concentration.

Furthermore, for these compound solutes (salts and ampholytes), this limiting value of H, designated as H_*, approached as c increases indefinitely, is that corresponding to the *iso-electric point* of the solute. When $\beta = 0$, and hence $H = H_*$, the total number of positive solute charges equals the total number of negative solute charges for the particular solute in question; this is by definition the iso-electric point of the solute.

2. Examples:

(a) As the first example we shall find H_* for a solution of a simple ampholyte. As just explained, we may either set $\beta = 0$ or, what is exactly equivalent, drop all terms in Eq. (2) not containing c and obtain

$$H_* = \sqrt{WK_a/K_b}. \tag{3}$$

(b) For a 1 : 1 salt, MX, of weak acid and weak base, with ionization constants A and B respectively,

$$\beta = A/(A + H) - B/(B + OH), \tag{4}$$

and when $\beta = 0$,

$$H_* = \sqrt{WA/B}. \tag{5}$$

(c) For the alkali acid salt of a dibasic acid, type $NaHSO_3$,

$$\beta = \alpha_2 - \rho. \tag{6}$$

Then when $\beta = 0$, this means $K_2/H - H/K_1 = 0$, or

$$H_* = \sqrt{K_1K_2}. \tag{7}$$

Formulas (3), (5), and (7) will be recognized as the familiar "first approximations" for H in each of the solutions, formulas independent of c and increasing in accuracy as c increases.

(d) For the normal salt of weak base and a dibasic acid "strong in K_1," type $(NH_4)_2SO_4$,

$$\beta = 1 + \alpha_2 - 2\alpha' = (H + 2K_2)/(H + K_2) - 2B/(B + OH). \tag{8}$$

When $\beta = 0$,

$$H_* = (W/2B) + \sqrt{(\quad)^2 + 2WK_2/B}. \tag{9}$$

(e) For the mixed salt of type $NaNH_4SO_4$, with the dibasic acid again strong in K_1,

$$\beta = \alpha_2 - \alpha' = K_2/(H + K_2) - B/(B + OH). \tag{10}$$

When $\beta = 0$,

$$H_* = \sqrt{WK_2/B}. \tag{11}$$

(f) For the acid salt of a dibasic acid and weak base, type NH_4HSO_3, $\beta = \alpha_1 + 2\alpha_2 - \alpha'$, or

$$\beta = (1 + 2K_2/H)/(1 + H/K_1 + K_2/H) - B/(B + OH). \tag{12}$$

Now H_* is defined by the cubic equation,

$$H_*^3B - H_*K_1(BK_2 + W) - 2K_1K_2W = 0. \tag{13}$$

(g) For the secondary alkali salt of a tribasic acid, Na_2HX,

$$\beta = \alpha_1 + 2\alpha_2 + 3\alpha_3 - 2. \tag{14}$$

Here

$$2H_*^3 + H_*^2K_1 - K_1K_2K_3 = 0, \tag{15}$$

and roughly,

$$H_* \cong \sqrt{K_2K_3}. \tag{16}$$

(h) An ampholyte with two basic ($K_{b_1} > K_{b_2}$) and one acid (K_a) ionization constants is mathematically the equivalent of the salt Na_2HX, in which the dihydrochloride of the ampholyte is considered as a tribasic acid. Hence H_* is defined by Eq. (15) with the condition that $K_1 = W/K_{b_2}$, $K_2 = W/K_{b_1}$, and $K_3 = K_a$ of the ampholyte; or roughly, then,

$$H_* \cong \sqrt{WK_a/K_{b_1}}. \tag{17}$$

In the last three cases and in many others which will be considered in later chapters, H_* is defined by a high degree equation, so that only approximate values suitable for certain relative values of the constants involved may be given. Nevertheless the fact is that such a value of H as H_*, independent of c, defined as above by setting $\beta = 0$, may always be expressed, if not evaluated, exactly, provided the nature of β is such as to make it possible for it to have the value of zero. And the value of H_* so obtained always represents the limit approached as c increases indefinitely. Since the effect of c falls off relatively as c increases, the formula for H_* becomes always a rough formula for the salt or ampholyte solution in which the effect of c is ignored and which increases in accuracy as c increases.

For certain types of salts (always involving "strong" acids or bases) it may appear impossible to write a formula for H_*, either because $\beta = 0$ for all values of c (as in the case of the salt of strong acid and strong base) or because β can not have the value zero for any real value ($0 < H < \infty$) of H. But this is so only because of the arbitrary nature of the mathematical definition of the term "strong"; in reality, the value of H_* is positive and finite in every such case. As examples of salts for which $\beta > 0$ for any value of H and of c we would have NH_4Cl, $NaHSO_4$, NH_4HSO_4; with $\beta < 0$ for all values of H, Na_2SO_4, Na_2SO_3, Na_3PO_4. Solutions of these salts are always acid or always alkaline respectively whatever the values of the constants involved (always on the assumption that the "strong" acids and bases have infinite constants). The limiting value of H_* for these salts is, as in the case of acids and bases themselves, a function of the concentration itself and never approaches independence of the concentration. Another way of expressing this is that in the case of such salts (for which β never equals zero) the expression obtained by neglecting all terms not involving c from the expanded equation in the form of Eq. (2) does not allow a positive finite value of H as its solution. Thus for NH_4HSO_4,

$$\beta = 1 + \alpha_2 - \alpha' = (H + 2K_2)/(H + K_2) - B/(B + OH), \tag{18}$$

and, on expanding $D = c\beta$, we have

$$H^4B + H^3(BK_2 + W) + H^2[K_2W - BW - c(BK_2 + W)] - HW[(BK_2 + W) + 2cK_2] - K_2W^2 = 0. \tag{19}$$

Inspection of Eq. (18) shows that $\beta > 0$ for all positive values of H, and the terms of Eq. (19) involving c give the equation $H = -2K_2W/(BK_2 + W)$, which is impossible. But it must be repeated that this is only the result of the arbitrary definition of the word "strong" for acids and bases.

3. H_* AS H_{ie}. The value of H_* for salts and ampholytes determined by $\beta = 0$ may also be written H_{ie}, the value of H for the iso-electric point of the salt or ampholyte. This value of H, being a function of ionization constants only, is in the ideal sense a constant, characteristic of the particular solute in water solution. Actually, however, both $[H^+]$ and a_{H^+} for the iso-electric point of a particular solute will depend upon the ionic strength and total composition of the particular *solution* involved. [It might be supposed that in terms of the true activity constants a definite value of the hydrogen ion activity would represent the iso-electric relations of a particular solute regardless of the total composition of the solution. But the idea of an iso-electric relation is defined by the equation of electroneutrality, which holds not in terms of activities but in terms of concentrations, so that in this sense the concentration of hydrogen ion corresponding to the iso-electric point is a more fundamental concept than is the activity of hydrogen ion.]

A pure solution of a salt or of an ampholyte, at any real finite concentration, is never at its iso-electric point. The solution *may be brought* to the iso-electric point, but then the final value of $[H^+]$ or of a_{H^+} required will depend upon the reagent used to bring about the necessary increase or decrease of H. The value of H_{ie} or of $\mathbf{H}_{ie}$ is a constant characteristic of the solute only in the ideal condition that all activity coefficients are unity. Thus for the salt NaHX (for which H_{ie} is given by Eq. (7) with symbols referring to concentrations and "mass-constants") we have, upon introducing activity coefficients and with the symbols now representing activity constants,

$$H_{ie} = [H^+]_{ie} = \sqrt{\mathbf{K}_1\mathbf{K}_2\gamma_{X^\circ}/\gamma^2_{H^+}\gamma_{X^=}}. \tag{20}$$

Now even if $\gamma_{X^\circ} = 1$,

$$\mathbf{H}_{ie} = (a_{H^+})_{ie} = \sqrt{\mathbf{K}_1\mathbf{K}_2/\gamma_{X^=}}, \tag{21}$$

with $\gamma_{X^=}$ depending on the total ionic strength. Even for a simple ampholyte,

$$H_{ie} = [H^+]_{ie} = \sqrt{\mathbf{W}\mathbf{K}_a\gamma_{U^+}/\mathbf{K}_b\gamma^2_{H^+}\gamma_{U^-}}, \tag{22}$$

and

$$\mathbf{H}_{ie} = (a_{H^+})_{ie} = \sqrt{(\mathbf{WK}_a/\mathbf{K}_b)\gamma_{U^+}/\gamma_{U^-}}, \tag{23}$$

so that $(a_{H^+})_{ie}$ is a constant and independent of the composition of the solution only if $\gamma_{U^+} = \gamma_{U^-}$, which at least is a very reasonable assumption.

B. Ampholytes

1. Simple Ampholyte.

a. Relations between fractions and p*H.* Since a simple ampholyte ($z_+ = z_- = 1$) is mathematically equivalent to the alkali acid salt of a dibasic acid,

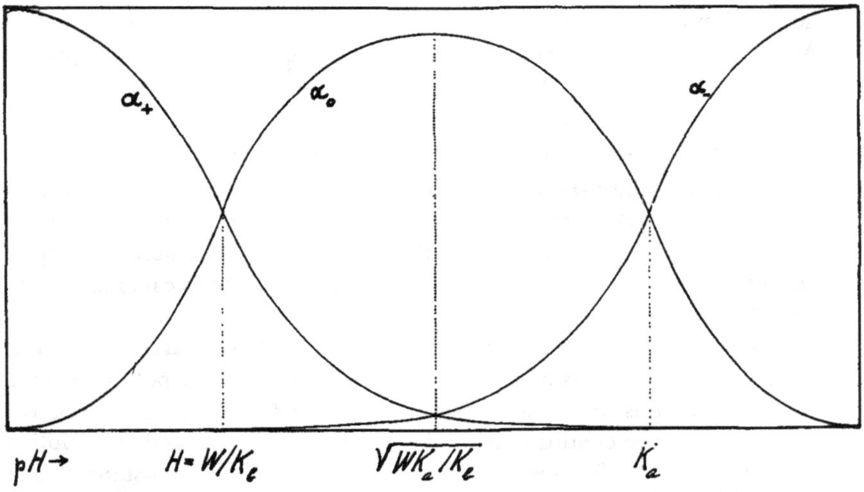

Fig. 4-1. Ionization fractions of simple ampholyte with $K_aK_b \ll W$.

we may identify α_+ with ρ, α_0 with α_1, and α_- with α_2, of such an acid. Then, graphically, provided $K_aK_b < W$, we have Fig. 4-1. At H_{ie}, α_0 is at its maximum value, with

$$(\alpha_0)_{max} = 1/(1 + 2\sqrt{K_aK_b/W}), \tag{24}$$

$$\alpha_- = \alpha_+ = 1/(2 + \sqrt{W/K_aK_b}); \tag{25}$$

at the same time, of course, since $1 = \alpha_0 + \alpha_- + \alpha_+$, $\alpha_t (= \alpha_- + \alpha_+)$ is a minimum, with

$$(\alpha_- + \alpha_+)_{min} = 1/(1 + \sqrt{W/4K_aK_b}). \tag{26}$$

For the ratio of the ionization fractions,

$$\alpha_-/\alpha_+ = K_aW/K_bH^2, \tag{27}$$

and

$$d(\alpha_-/\alpha_+)/dpH = 2(2.3)(\alpha_-/\alpha_+), \tag{28}$$

always greater than 0. This ratio equals K_a/K_b when the solution is neutral, and equals 1 at H_{ie}.

b. Limits of fractions in respect to concentration. When the solution is neutral, or when $c = 0$ and $H = \sqrt{W}$,

$$(\alpha_-)_\infty = K_a/(K_a + K_b + \sqrt{W}) \qquad (29a)$$

$$(\alpha_+)_\infty = K_b/(K_a + K_b + \sqrt{W}) \qquad (29b)$$

$$(\alpha_0)_\infty = \sqrt{W}/(K_a + K_b + \sqrt{W}). \qquad (29c)$$

As the concentration increases these values approach those at the iso-electric point. Such limits for the individual fractions are, of course, those obtainable in the solution of the pure ampholyte, or limits in respect to c only.

As c varies between 0 and ∞, α_- and α_+ undergo changes of opposite sign. If $K_a > K_b$, so that $H_{ie} > \sqrt{W}$, then α_- decreases (and α_+ increases) as c increases. At "infinite dilution," therefore:

(1) An ampholyte is first of all not "completely ionized," since the values of the two fractions are those corresponding to $H = \sqrt{W}$, Eq. (29 *a*, *b*).

(2) The two degrees at infinite dilution are not equal, but one (α_- if $K_a > K_b$) is at its maximum and the other (α_+) is at its minimum (in respect to concentration in pure water). They approach equality as c increases, and become equal only at $c = \infty$, when $H = H_{ie}$.

2. Significance of Iso-electric Point for General Ampholyte. The relations between fractions and pH are similar to those for a polybasic acid, with the correlations and transformations discussed in Chapter I. For the iso-electric point, the condition is by definition that $\beta = 0$, and the solution of the equation $\beta = 0$ gives H_{ie} in terms of ionization constants. The condition $\beta = 0$ means that

$$\Sigma_{z_-}(z_-\alpha_-) = \Sigma_{z_+}(z_+\alpha_+), \qquad (30)$$

not that $\Sigma_{z_-}\alpha_- = \Sigma_{z_+}\alpha_+$, which would correspond to "equal degrees of positive and negative ionization." The latter is true only in the special case of a simple 1 : 1 ampholyte, where $\beta = \alpha_- - \alpha_+$ or where $z_- = z_+ = 1$; then $\alpha_- = \alpha_+$ at the iso-electric point.

Also, the nonionized fraction, α_0, of an ampholyte is a maximum at H_{ie}, while α_t $(= \Sigma_{z_-}\alpha_- + \Sigma_{z_+}\alpha_+ = 1 - \alpha_0)$ or the total degree of ionization is a minimum. Thus, from Eq. I(106), $d\alpha_0/d\text{p}H = 0$ at H_{ie}, when $\beta = 0$. Moreover

$$d^2\alpha_0/d\text{p}H^2 = 2.3\alpha_0(2.3\beta^2 - d\beta/d\text{p}H), \qquad (31)$$

which is negative at H_{ie} when $\beta = 0$ since $d\beta/d\text{p}H > 0$. Hence α_0 is a maximum and α_t is a minimum, at H_{ie}.

Finally, the solubility is a minimum at H_{ie}. If P is used as the symbol for the solubility product constant, then

$$P_a = [\text{H}^+]^{z_-}[\text{U}^{-z_-}] = H^{z_-}S\alpha_{-z_-}, \qquad (32)$$

or

$$P_b = [U^{+z_+}][OH^-]^{z_+} = S\alpha_{+z_+}(OH)^{z_+}, \tag{33}$$

where S is the solubility. Then

$$S = P_a/H^{z_-}\alpha_{-z_-} = P_b/(OH)^{z_+}\alpha_{+z_+}. \tag{34}$$

In either case, or directly from $S\alpha_0 = \mathbf{k}$, a constant,

$$dS/dpH = 2.3S\beta = 0 \text{ at } H_{ie}, \tag{35}$$

and

$$d^2S/dpH^2 = 2.3S(2.3\beta^2 + d\beta/dpH), \tag{36}$$

which is always positive. Hence the solubility is a minimum at H_{ie}.

Again it must be pointed out that these are ideal relations, that a binary system ampholyte-water at finite concentrations is never at an iso-electric point by itself and that any *actual* value of the iso-electric point and the actual minimum solubility at this point will depend on the actual reagent used to bring the system to the "iso-electric point."

C. Salts

1. Significance of Iso-electric Point for Salts. The same mathematical meaning of the iso-electric point applies to salt solutions as to ampholytes. The value of H at the iso-electric point is H_* or H at $\beta = 0$, approached in a pure salt solution as $c \overset{\rightarrow}{=} \infty$. Again this means that the number of positive solute charges equals the number of negative solute charges per mole of salt, or as in Eq. (30) (with ν_a as the number of acid radicals and ν_b the number of base radicals in the formula of the salt),

$$\nu_a\Sigma_z z\alpha = \nu_b\Sigma_{z'} z'\alpha'; \tag{37}$$

but the iso-electric point, therefore, does not involve an equality of "degrees of ionization" of acid and base in the mere sense of ionization fractions. Since salts are completely dissociated in solution, it is impossible to speak of an "unionized fraction" ("ρ") for a salt, and hence there is no such condition as a maximum of "ρ" and a minimum of α_t for a salt at its iso-electric point. Various possible definitions of α_t will be discussed, but there is no consistent or useful meaning in any of them in any general sense, i.e. applicable to more than some special simple case. Finally, however, as for ampholytes, the solubility of a salt is a minimum at its iso-electric point.

This is easily seen for a simple 1 : 1 salt, for which

$$P = [M^+][X^-] = (\alpha' S)(\alpha S), \tag{38}$$

$$S = \sqrt{P/\alpha\alpha'}, \tag{39}$$

$$dS/dpH = (2.3/2)S\beta_s. \tag{40}$$

But β_s $(= \alpha - \alpha' = \rho' - \rho) = 0$ at the iso-electric point, by definition. Furthermore,

$$d^2S/dpH^2 = (2.3/2)S(2.3\beta_s^2/2 + d\beta_s/dpH), \tag{41}$$

which is always positive, so that S is a minimum at H_{le}.

For a more general case, we consider the salt $[M^{+z'}]_{\nu_b}[X^{-m}]_{\nu_b z'/m}$, in which z' is the "acidity" of the base (and also here the valence of the base radical of the salt), ν_b the number of base radicals in the empirical formula, and m the valence of the acid radical. Here

$$P = [M^{+z'}]^{\nu_b}[X^{-m}]^{\nu_b z'/m}, \tag{42}$$

$$S = [P/(\alpha_{z'}')^{\nu_b}(\alpha_m)^{\nu_b z'/m}]^{m/\nu_b(m+z')}, \tag{43}$$

and

$$dS/dpH = 2.3mS\beta_s/\nu_b(m + z'), \tag{44}$$

which is zero at H_{le} when $\beta_s = 0$. Also,

$$\frac{d^2S}{dpH^2} = \frac{2.3mS}{\nu_b(m + z')}\left[\frac{2.3m\beta_s^2}{\nu_b(m + z')} + \frac{d\beta_s}{dpH}\right]; \tag{45}$$

since this is always positive, S is a minimum at the iso-electric point. Incidentally, the formula of Eq. (43) applies generally to any salt consisting of only two radicals; m is the valence (as a positive number) of the anion, z' that of the cation, ν_b the number of cations in the empirical formula, and β_s is the charge coefficient of the salt as defined in Eqs. 1(65) and 1(66).

2. Ionization Fractions for a Simple 1 : 1 Salt.

a. The ionization fractions, as f(pH). The problem of the dependence of the ionization fractions upon pH in a salt solution is simply the superposition of the curves of the acid and those of the base, actually independent curves. Since there is no truly *general* case to consider we shall here discuss in detail only the most important case, that of the salt MX, of monobasic acid and base. Such a salt solution is of course merely a special point in a mixture of acid and base, where the concentrations a and b are equal. We shall therefore first consider the general interrelations of the fractions in a solution of weak acid and weak base, holding for *any* ratio of the concentrations, a and b. Graphically, these relations are shown in Figs. 4–2, 3, 4.

In each case the crossing of the α and α' curves (as also that of the ρ and ρ' curves) occurs at $H = \sqrt{WA/B}$; with respect to the salt this is the iso-electric point, where β_s $(= \alpha - \alpha') = 0$.

Algebraically, from the expressions $\alpha = A/(A + H)$ and $\alpha' = B/(B + OH)$, we have, for any values of a and b, relative and absolute:

$$\alpha = \rho'/[\rho' + (1 - \rho')W/AB], \tag{46}$$

$$\alpha' = \rho/[\rho + (1 - \rho)W/AB]. \tag{47}$$

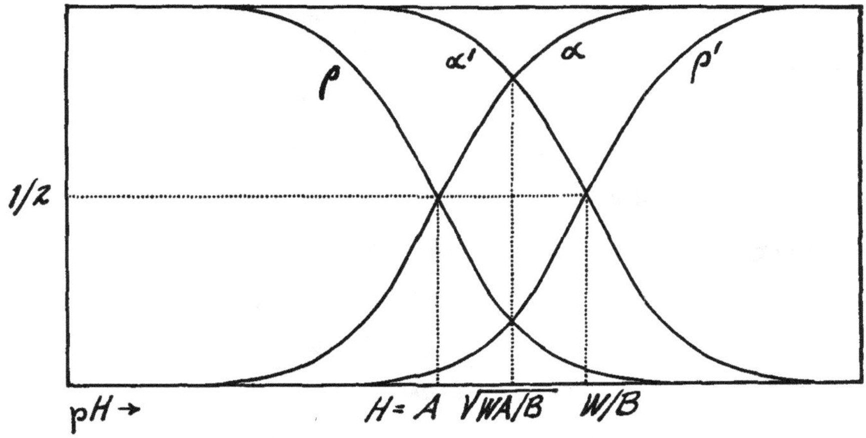

Fig. 4–2. Ionization fractions for simple 1 : 1 salt with $AB > W$.

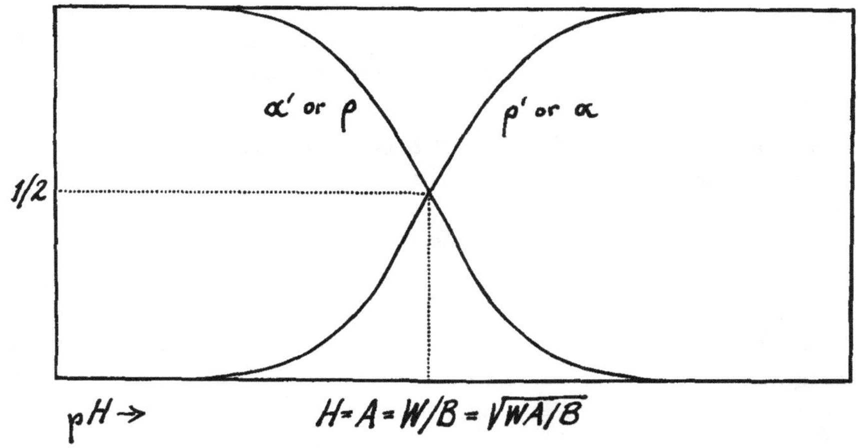

Fig. 4–3. Ionization fractions for simple 1 : 1 salt with $AB = W$.

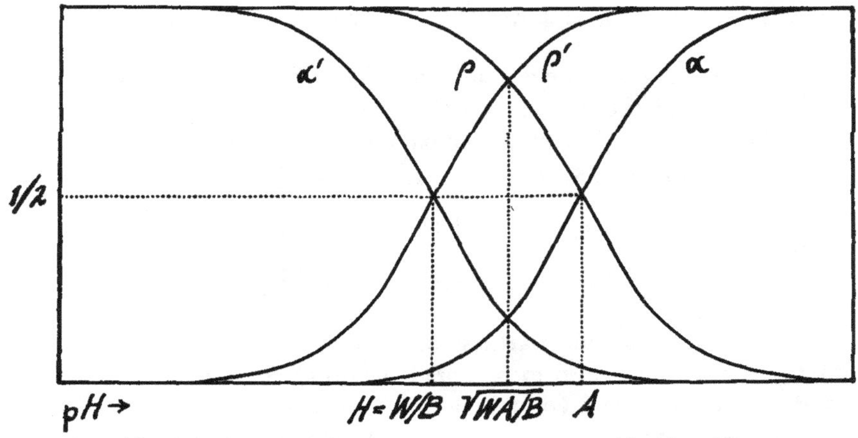

Fig. 4–4. Ionization fractions for simple 1 : 1 salt with $AB < W$.

Hence:

$$\text{if } AB > W, \text{ then } \alpha > \rho' \text{ and } \alpha' > \rho;$$
$$\text{if } AB = W, \text{ then } \alpha = \rho' \text{ and } \alpha' = \rho;$$
$$\text{if } AB < W, \text{ then } \alpha < \rho' \text{ and } \alpha' < \rho.$$

Also, in general, the ratios of the fractions are

$$\begin{aligned} \alpha/\alpha' &= A(B + OH)/B(A + H), \\ \rho/\rho' &= H(B + OH)/OH(A + H), \\ \alpha/\rho' &= A(B + OH)/OH(A + H), \\ \rho/\alpha' &= H(B + OH)/B(A + H). \end{aligned} \tag{48}$$

Roughly, then, for any ratio of a to b, these ratios become: in acid solution, $\sim A/H$, B/OH, AB/W, 1, respectively; and in alkaline solution, $\sim OH/B$, H/A, 1, W/AB, respectively. In neutral solution, their values depend on the relative values of A, B, $\sqrt{W}$.

We note also that at $H = \sqrt{WA/B}$ both $\alpha\alpha'$ and $\rho\rho'$ have maximum values, whether or not $a = b$ and whether or not $AB = W$; thus $d(\alpha\alpha')/dpH = \alpha\alpha'(\rho - \rho')$, while at $\rho = \rho'$, $d^2(\alpha\alpha')/dpH^2 = -\alpha\alpha'(\alpha\rho + \alpha'\rho')$. The maximum value of $\alpha\alpha'$ is, thereforef

$$(\alpha\alpha')_{\max} = 1/(1 + \sqrt{W/AB})^2. \tag{49}$$

b. Limits of the fractions, in respect to concentration, in pure salt solution. These depend on $H = \sqrt{W}$ at $c = 0$, and $H_* = H_{ie} = \sqrt{WA/B}$, at $c \overset{\rightarrow}{=} \infty$. At $c = 0$ (infinite dilution), $\alpha \neq \alpha'$ and $\rho \neq \rho'$; and

$$\alpha_\infty = A/(A + \sqrt{W}),\ \rho_\infty = \sqrt{W}/(A + \sqrt{W}); \tag{50}$$

$$\alpha'_\infty = B/(B + \sqrt{W}),\ \rho'_\infty = \sqrt{W}/(B + \sqrt{W}). \tag{51}$$

The α's are equal at infinite dilution only if $A = B$, but then they are equal at any concentration of the pure salt.

As c increases, α and α' approach equality, for they can be equal only at H_{ie} and hence at $c \overset{\rightarrow}{=} \infty$ for the pure salt solution. Thus their values become

$$\alpha_* = \alpha'_* = 1/(1 + \sqrt{W/AB}), \tag{52}$$

$$\rho_* = \rho'_* = 1/(1 + \sqrt{AB/W}). \tag{53}$$

The fractions are actually farthest apart (for the pure salt solution) at infinite dilution. It is also to be noted that (unless $A = B$, in which case α and α' are equal and constant at any concentration) the two degrees, α and α' (or ρ and ρ') change in opposite directions as the concentration changes. If $A > B$, $\alpha_\infty > \alpha'_\infty$; hence as such a salt solution is diluted, α increases while α' decreases; vice versa if $A < B$. Furthermore, the α's are high at high concentration if $AB \gg W$ (both $\cong 1$), but low at high concentration if $AB \ll W$ (both $\cong \sqrt{AB/W}$). But in any case they diverge as the solution is diluted. In the special case that $AB = W$ (whether or not $A = B$) it is seen that the limiting values at high concentration are 1/2 for all the fractions, and the manner in which they change as the solution is diluted then depends on whether $A \gtrless B$.

c. The "total degree of ionization" of a salt. The quantity, $\alpha + \alpha'$, for a simple 1 : 1 salt may be shown to be either a maximum or a minimum at the iso-electric point, depending on whether AB is greater or smaller than W:

$$d(\alpha + \alpha')/dpH = 2.3(\alpha\rho - \alpha'\rho'), \qquad = 0 \text{ at } H_{ie}; \tag{54}$$

$$d^2/dpH^2 = (2.3)^2[\alpha\rho(\rho - \alpha) + \alpha'\rho'(\rho' - \alpha')], \tag{55}$$

$$= 2(2.3)^2\alpha\rho(\rho - \alpha) \text{ at } H_{ie}. \tag{56}$$

Hence d^2/dpH^2 is negative at H_{ie} if $AB > W$ (Fig. 4–2) and positive if $AB < W$ (Fig. 4–4). The sum, $\alpha + \alpha'$, is therefore a maximum at H_{ie} and decreases with dilution if $AB > W$; it is a minimum at H_{ie} and increases with dilution if $AB < W$. If $AB = W$ (Fig. 4–3), $\alpha + \alpha' = 1$ at any pH and hence at any concentration of pure salt. Finally, if $A = B$, $H = \sqrt{W}$ at any concentration of the salt and the sum, $\alpha + \alpha'$, is a constant, independent of the concentration.

For salts in general, however (salts of acids and bases of basicity greater than 1 and which may or may not be "normal"), the iso-electric point, as defined by Eq. (37), does not correspond either to a maximum or to a minimum of any quantity which may consistently be defined as "α_t," or as a "total degree of ionization" of the salt. It is found, for example, that neither the sum $\nu_a\Sigma_z\alpha + \nu_b\Sigma_{z'}\alpha'$ nor the sum $\nu_a\Sigma_z z\alpha + \nu_b\Sigma_{z'}z'\alpha'$ is in general either a maximum or a minimum at such a point. This may be verified by differentiating such expressions for salts like MHX or M_2X with respect to pH and noting that the derivative is not in general equal to zero at H_{ie}.

3. ON THE SO-CALLED "DEGREE OF HYDROLYSIS" OF A SALT.

a. A term not mathematically definable. Despite the widespread use of the expression "degree of hydrolysis" for a salt, a general mathematical definition of such a quantity is impossible. The expression seems to be satisfactory in the case of simple 1 : 1 salts of strong-weak combinations, the "degree of hydrolysis" for the salt MX_s (weak base and strong acid) being taken as ρ', and as ρ for M_sX, the salt of strong base and weak acid. But we also read of the "degree of hydrolysis" of a salt in general. The rough approximation formula for the hydrogen ion concentration of a solution of the salt of weak acid and weak base, MX—namely $H \cong \sqrt{WA/B}$—is generally said to give the "degree of hydrolysis" of the salt; similarly the formula $H \cong \sqrt{K_1K_2}$ is called the "hydrolysis formula" for the salt $NaHSO_3$. But as it is impossible to define a "total degree of ionization" for a salt, it is similarly meaningless to speak of a "degree of hydrolysis" for a salt. Even for the simplest case of 1 : 1, there are two separate and equally significant "degrees" (ρ and ρ') interrelated according to Eqs. (46) and (47). The salts of strong-weak combinations are merely limiting cases in which one of these "degrees" is zero by definition. But it does not seem possible to define with mathematical distinctness and usefulness such a quantity as a single "degree of hydrolysis" for a salt in general, such as ammonium formate, ammonium bisulfite, magnesium ammonium phosphate, etc. Even if we arbitrarily choose either ρ or ρ' as the "degree of hydrolysis" for a simple salt MX, we find that given the concentration and one of these fractions for a salt, the second fraction can not be calculated unless at least one of the ionization constants is also known. In other words, whereas a and α are sufficient for the complete definition and description of a solution of an acid, c and ρ (or ρ') alone are not sufficient for the mathematical description of a salt solution. "Degree of hydrolysis" of a salt is not mathematically comparable to "degree of ionization" of an acid or a base.

We may question even the propriety of calling either of the fractions, ρ and ρ', a "degree of hydrolysis" on the basis of the meaning of the word "hydrolysis." Whereas it is said that the "degree of hydrolysis" of a salt increases with dilution (as one must expect if "hydrolysis" means decomposition of the salt by water, as the word *should* mean), we find instead that as c decreases, ρ increases (or decreases) while ρ' decreases (or increases) depending on whether A is smaller (or greater) than B. The limits obviously depend on the limits of H (or OH). With $A > B$, H ranges from $\sqrt{W}$ when $c = 0$ to an upper limit of $\sqrt{WA/B}$ at infinitely high concentration. Consequently, when $A > B$, ρ decreases from its value in Eq. (53) as an upper limit to that in Eq. (50), as $c \rightleftarrows 0$; while ρ' increases from Eq. (53) as a lower limit to its value in Eq. (51) as $c \rightleftarrows 0$.

Unless $A = B$, the two "degrees" (ρ and ρ') approach equality not at infinite dilution but at infinitely high concentration. For sodium acetate, with $A = 1.75 \times 10^{-5}$, ρ_∞ (when $c = 0$) is only 1/176 (not 1, as the concept of "hydrolysis," or decomposition of the salt by water, would lead one to expect) while ρ' is always zero.

The effect of concentration on what is called the "degree of hydrolysis" of a salt is usually said to be large for salts of strong-weak combinations and "practically zero" for weak-weak combinations. The effect is actually a function of the inequality of A and B; it is in opposite directions for ρ and ρ' when $A \neq B$, and it is exactly zero when $A = B$, for then $\rho = \rho' =$ a constant $= 1/(1 + A\sqrt{W})$ or $1/(1 + B/\sqrt{W})$ for all values of c. Again this result is not expected on the basis of the connotations of the word "hydrolysis," and it seems that it would be much clearer and sounder to explain the effects involved not as an interaction between salt and water but rather as one between aqueous acid and aqueous base at equal concentrations.

Some calculations for hypothetical "salt" solutions are shown in Table 4–1. The value of H, used then to calculate ρ and ρ', was obtained with the aid of the approximate equation VIII(15), which, as will be discussed in Chapter VIII, is capable of high accuracy for the data assumed. Such salts would ordinarily be said to be "very highly hydrolyzed" with the "degree of hydrolysis increasing (to an extent called 'practically complete') with dilution"; but with $A > B$, ρ decreases and ρ' increases as c decreases.(1)

Table 4–1. Unionized fractions in various salt solutions.

c	A	B	H	ρ	ρ'
10^{-2}	10^{-4}	10^{-10}	9.80×10^{-5}	0.495	0.505
10^{-5}	10^{-4}	10^{-10}	8.44×10^{-6}	0.0778	0.922
10^{-2}	10^{-4}	10^{-12}	6.70×10^{-4}	0.870	0.937
10^{-5}	10^{-4}	10^{-12}	9.15×10^{-6}	0.0838	0.99909
10^{-2}	10^{-2}	10^{-8}	7.05×10^{-5}	0.00700	0.01399
10^{-5}	10^{-2}	10^{-8}	2.70×10^{-6}	0.00027	0.2703

b. "Hydrolysis constant" for a salt mathematically unnecessary or not definable. Concerning the mathematical form of a "hydrolysis constant," we may consider the reaction of the substance MX with water according to the chemical equation

$$MX + HOH \rightleftarrows HX + MOH, \tag{57}$$

so that

$$K_h = [HX][MOH]/[MX][HOH]. \tag{58}$$

If all three of the substances MX, HX, and MOH are nonelectrolytes, then the hydrogen ion concentration, in ideal conditions, is not involved in the equilibrium. If MX is a salt, however, HX will be an acid and MOH a base. Now if the salt MX is completely dissociated so that $[MX] = 0$, "K_h" can have no mathematical meaning; the only equilibrium constants definable are the independent ionization constants of the acid and the base. But if the salt is incompletely dissociated so that $[MX] \neq 0$ and $K = [MX]/[M^+][X^-]$, as in Eq. I(43), then the solution is "complex" in the sense explained in Chapter I, section F. At the same time, however, we find that K_h defined as in Eq. (58) is not a new constant and that it represents nothing new. It is merely a function of K and the ionization constants of HX and MOH. This is true whatever the strengths of the "products" of hydrolysis, HX and MOH.

(1) Some similar calculations were made (not systematically) by R. O. Griffith, *Trans. Far. Soc.*, **17**, 525 (1922).

Thus, if both HX and MOH are weak, with ionization constants A and B respectively, Eq. (58) becomes

$$K_h = [H^+][X^-][M^+][OH^-]/ABK[M^+][X^-][HOH] = K_w/ABK. \quad (59)$$

If HX alone is strong, we have [HX] = 0, and reaction (57) must be written as

$$MX + HOH \rightleftarrows H^+ + X^- + MOH, \quad (60)$$

so that

$$K_h = \frac{[H^+][X^-][MOH]}{[MX][HOH]} = \frac{[H^+][X^-][M^+][OH^-]}{BK[M^+][X^-][HOH]} = \frac{K_w}{BK}. \quad (61)$$

Similarly, if MOH alone is strong, we obtain

$$K_h = [HX][M^+][OH^-]/[MX][HOH] = K_w/AK. \quad (62)$$

Finally, if both HX and MOH are strong,

$$K_h = [H^+][X^-][M^+][OH^-]/[MX][HOH] = K_w/K. \quad (63)$$

A "hydrolysis constant" for a salt, with correct mathematical definition, is therefore either impossible if the salt is completely dissociated or unnecessary if the salt is not completely dissociated since such a constant is then merely a function of the other fundamental constants of the salt solution. This remains true, furthermore, whether or not a "product" of the hydrolysis has a limited solubility, for the "constant" K_h would still have to be defined with respect to the concentrations of species in solution.

A true and significant hydrolysis constant applies only to those cases in which the reactant "MX" is not an electrolyte, or not a salt, therefore. Such reactions leading, for example, to two acids or to an acid and a nonelectrolyte, etc., as products of hydrolysis will be considered under the complex cases of Chapter xv.

D. Buffer Capacity With Respect to a Compound Solute (Salt or Ampholyte) at the Iso-electric Point

1. Adjustment of Solution to Iso-electric Point. As stated above, a pure solution of a salt or of an ampholyte at finite concentration is never at its iso-electric point, which requires $c = \infty$. It may, however, be brought to the iso-electric point (in respect to the particular solute in question) by addition of some reagent. The concentration of strong acid or strong base ($g = a_s - b_s$) required to bring the value of H to H_{ie} is a constant, characteristic of the compound solute (salt or ampholyte) and independent of the concentration of this solute. (This, as usual, assumes that all activity coefficients are unity.) Thus, if g_{ie} is the concentration required,

$$D_{ie} = (H - OH)_{ie} = c\beta_{ie} + g_{ie}. \quad (64)$$

But $\beta_{ie} = 0$; hence

$$g_{ie} = D_{ie} \tag{65}$$

for any value of c. For a simple ampholyte, therefore, for which $H_{ie} = \sqrt{WK_a/K_b}$, the concentration of strong acid required is

$$(a_s)_{ie} = \sqrt{WK_a/K_b} - \sqrt{WK_b/K_a}, \tag{66}$$

if $K_a > K_b$; if $K_a < K_b$ and the pure ampholyte solution is alkaline, the concentration of strong base required is

$$(b_s)_{ie} = \sqrt{WK_b/K_a} - \sqrt{WK_a/K_b}. \tag{67}$$

This fact, that the concentration of strong acid (or strong base) required to bring a pure ampholyte solution to its iso-electric point is independent of the concentration of the ampholyte, might seem to be a rather special consequence of the peculiar nature of an ampholyte. But it is here seen to be a general fact for all solutes, salts as well as ampholytes, which have a compound charge coefficient capable of having the value of zero. It will be seen, for example, that to bring a pure solution of the 1 : 1 salt MX to its iso-electric point requires a concentration of strong acid or strong base given by

$$g_{ie} = \sqrt{WA/B} - \sqrt{WB/A}; \tag{68}$$

and for the alkali acid salt of a dibasic acid, NaHX,

$$g_{ie} = \sqrt{K_1K_2} - W/\sqrt{K_1K_2}. \tag{69}$$

These values are characteristic of the salts or ampholytes and independent of their concentrations.

2. Buffer Capacity With Respect to Compound Solute, at Iso-electric Point. An interesting corollary of this fact is that the buffer capacity with respect to a compound solute must be infinite if the solution is at the iso-electric point ($H = H_{ie}$) with respect to that compound solute. Given a pure aqueous solution of strong acid (or strong base) of such concentration that $H = H_{ie}$ for a particular compound solute (salt or ampholyte), then we may increase the concentration of this solute indefinitely (assuming that a_s or b_s is not changed by dilution) without change of pH. Or more generally, given any solution (provided always that it is "non-complex" in the sense discussed in Chapter I), with $H = H_{ie}$ for a particular compound solute, then the buffer capacity of this solution with respect to that solute is infinite. It acts as a perfect buffer for that solute.

If the solution contains many solutes (Σc) and a particular compound solute (c_1), so that $D = \Sigma c\beta + c_1\beta_1$, then

$$\pi_{c_1} = dc_1/d\text{p}H = (1/\beta_1)d(D - \Sigma c\beta)/d\text{p}H - (d\beta_1/d\text{p}H)(D - \Sigma c\beta)/\beta_1^2. \tag{70}$$

But if $H = H_{ie}$, $\beta_1 = 0$, and hence $dc_1/d\text{p}H = \pm\infty$.

3. SIGNIFICANCE. A solute can change the "reaction" (the difference $H - OH$) of a solution only if the value of H of the solution is not that of the iso-electric point of the solute. The direction, moreover, in which a solute changes the reaction of a solution is always toward its iso-electric point. The sign of the effect of a compound solute (salt or ampholyte) on the reaction of a solution therefore depends on whether the value of H of the solution is greater or smaller than H_{ie} for the solute. A salt or an ampholyte is "neutral" to pure water if H_{ie} for the solute equals $\sqrt{W}$; it is otherwise always "acid" or always "basic," at any concentration, in pure water. But if the solution contains other substances the sign of the effect of the salt or ampholyte will depend on the value of H of the solution in which it is being dissolved. While the relative effects (in respect to sign) of two acids, of two bases, or of a base and an acid, are independent of the value of H of the solution in which they are compared, this will not be the case for the comparison of a simple solute with a compound solute, or of two compound solutes. A salt or ampholyte which is "acid to water" will appear "acid" relative to a "base" as long as they are compared in a solution with $H < H_{ie}$; but compared in a solution with $H > H_{ie}$, the two substances have effects of the same sign, both decreasing H.

This point raises the question of how two substances are to be compared with respect to their "acid-strengths" or their individual "acidities," if there is any meaning in such terms, or if there is any such property of a substance. We shall therefore briefly consider some of the possible criteria for such a comparison:

(a) The effect of substances (at equal concentrations) on the reaction (such as $H - OH$) of a medium is not a significant comparison. The relative effects as to sign and as to magnitude will vary with the following conditions:

(1) The concentration at which they are compared. Thus a salt, MX, in which $A = 10^{-6}$ and $B = 10^{-12}$ may be as "acid" or more "acid" than a monobasic acid with $A_1 = 10^{-7}$, at low concentration, but less "acid" at high concentration.(2)

(2) The nature of the medium in which they are tested. This is seen in that even within a single solvent system the relative effect will change even as to sign, depending on whether the comparison is made with initial H higher or lower than H_{ie} of one of the substances.

(b) The degree of ionization as a criterion is not definable in such a way as to be numerically comparable between a monobasic and a polybasic acid, or between a simple acid and a salt or an ampholyte.

(c) The ionization constant as criterion presents several difficulties:

(2) With $D = c[A/(A + H) - B/(B + OH)] = aA_1/(A_1 + H)$, the two solutions have the same value of H at $c = a$ when $H = -A_1 + [A_1^2 + (AW - ABA_1 - A_1W)/B]^{\frac{1}{2}}$ and $c = a = D(A_1 + H)/A_1$. In the numerical case mentioned $H = 9.48 \times 10^{-5}$ when $c = a = 0.0900$.

(1) The ratio of the constants of two substances may change by change of solvent, even to the point of reversal.

(2) We cannot define an ionization constant of a salt or ampholyte to be comparable with that of a simple acid or base.

(3) It is not clear how a monobasic acid is to be compared with an acid with several constants.

(4) It is not clear whether "acidity" is a property continuous with "basicity," and if not, what the inverse relation is between "acidity" and "basicity," i.e. between "acid-strength" and "base-strength." Do small basic ionization constants approach small acid ionization constants? This does not seem reasonable. Can we compare an acid (**I**) having a certain value of A, with a base (**II**) having a certain value of B, in respect either to "acid-strength" or to "base-strength"? Is the first infinitely stronger as an acid than the second regardless of the value of A? To take W/B and speak of the "acid-strength" of the "conjugate acid of the base" does not advance the problem, for we are interested in the comparison of the substance **I** with the substance **II**, not with the "conjugate of **II**," whatever that is as a substance if it is a substance.

(5) Finally, as will be discussed in Chapter VI, the ionization constants here involved are "apparent ionization constants," and their interpretation as measures of some fundamental acidic or basic property or strength of the substances requires such assumptions as to make the comparison no longer one depending solely upon the "ionization constants." The so-called "ionization constant" is found to be an over-all constant including processes of rearrangement, reactions with the solvent, solvation of molecules and ions, tautomerism even of the ions, etc.

Hence we may be led to abandon the idea of "acid-strength" as a property of a *substance* ("substances" including not merely acids and bases but also salts, ampholytes, and nonelectrolytes), because we may feel limited to the above possible methods of comparison and to think rather of the "acid-strength" of an individual group, atom, point, etc., in a molecule, to which a specific "ionization constant" may be assigned. Unfortunately, it is in general impossible to analyze the actual apparent ionization constants into individual constants characteristic of points in a molecule (Chapter VI).

The definitions of the words used in the present treatment are therefore to be understood as purely mathematical conveniences and requirements, each word being given a restricted definition for the purpose of mathematical clarity and consistency. An aqueous solution is neutral if $H = OH$; it is acid if $H > OH$ and basic if $H < OH$. A process involving production of negative solute ions and increase of H has been called an acid ionization process, described by an acid ionization constant; similarly a basic process, described by a base ionization constant, involves production of positive solute ions and increase of OH. But we have not had to speak of anything such as an inherent acidity or acid-strength of substances. Mathematically every substance which is not a salt is potentially an ampholyte, capable of reacting in both acidic and basic processes in water. If both K_a and K_b are zero it is a "nonelectrolyte" since it introduces no foreign ions. If all the K_b's are zero it is an acid, and if all the K_a's are zero it is a base. But we do not have to state what is meant by a substance which is inherently an "acid" or a "base." *Strength* refers to the magnitude of some ionization constant. But these constants describe complex phenomena involving more than any such single property that may be called the "acid-strength" of the solute. Finally, since the field of ionization equilibria does not directly deal with such an inherent property of *substances*, but only with *processes* described by these over-all "ionization constants," which can not be defined with reference to a salt as solute, we disregard a salt as a species in solution and deal with it as a solution of acid and base in stoichiometric proportions.

V

Application of the Exact Equations for the Determination of the Constants

THE general equation, $D = H - OH = \Sigma c\beta$, is now to be considered from three points of view for explicit use. The first is as an expression for the determination of the constants, W, A, B, etc., each constant being expressible as a function of H, concentrations, and the other constants; the second is as a relation between concentrations, as functions of H and the constants; and the third is as an analytical formula for the calculation of H as function of concentrations and constants. Since the second and third applications of the equation are possible only after the constants are determined and units are assigned to them, the present chapter will deal with the expressions to be used in the evaluation of the constants. The evaluation here discussed depends on the measurement of H as function of analytical concentrations in unsaturated solutions. The ionization constants, however, are also involved in the dependence of the solubility of various solutes, both simple and compound, upon H and upon the concentrations of other solutes. The calculation of ionization constants through solubility determinations will be considered in the final chapters, which deal with the relations in saturated solutions.

A. DETERMINATION OF **W** (ION-PRODUCT CONSTANT OF WATER)

(1) In a solution containing a simple acid and base, in which

$$D = H - OH = a\alpha - b\alpha' = aA/(A + H) - bB/(B + OH), \qquad (1)$$

we may write

$$W = H^2 - H(a\alpha - b\alpha'), \qquad (2)$$

$$= -\frac{H}{2}\left(\frac{aA}{A+H} + B - H\right) + \sqrt{(\quad)^2 + H^2B\left(b + H - \frac{aA}{A+H}\right)}. \qquad (3)$$

[Note: as already stated, the quantity ( ) under the root sign represents the whole of the term preceding the root, in such a quadratic solution; in this case, therefore, $(\quad)^2$ represents the square of $-H[aA/(A + H) + B - H]/2$.] For the solution of the salt, in which $a = b = c$,

$$W = -\frac{H}{2}\left(\frac{cA}{A+H} + B - H\right) + \sqrt{(\quad)^2 + \frac{H^3B(c + A + H)}{A + H}}. \qquad (4)$$

For pure acid alone,

$$W = H^2 - HaA/(A + H), \tag{5}$$

and for pure base alone,

$$W = -H(B - H)/2 + \sqrt{(\quad)^2 + H^2B(b + H)}. \tag{6}$$

If any of these functions of observed hydrogen ion activity or concentration and analytical concentrations of solutes is extrapolated or otherwise corrected by means of known or calculated activity coefficients to infinite dilution, it then gives the thermodynamic ion-product constant of water. But since these functions involve the α's and hence the constants A and B, it is clear that none of these constants can be determined independently. At least A (or B), through a solution of pure acid (or pure base), must be determined or evaluated simultaneously with W. Without reference to the behavior of solutes in water, there is no way to evaluate the activities or the concentrations of the ions of water and hence to evaluate W. No determination or combination of determinations of properties of pure water alone can be used to evaluate the ion-product constant W, which therefore has meaning only as an interaction constant involving the effects of solutes upon the ions of water.

(2) It may appear, of course, from familiar lists of "methods of determining W" as though such simultaneous evaluation of ionization constants of solutes were not necessary. It seems sufficient, for example, to use a "strong base" with or without a "strong acid" to make possible the "determination of W." In this case Eq. (2) becomes simply $W = H(b_s - a_s + H)$. But this is still a simultaneous and interdependent evaluation since we assume a value of A, namely ∞, in setting α for the acid, for example, equal to 1. Extrapolation to infinite dilution does not otherwise justify the assumption that $\alpha = 1$ even at infinite dilution. For a "strong" acid, $\alpha = 1$ at any concentration, but otherwise $\alpha_\infty = A/(A + \sqrt{W})$. It is furthermore usually also necessary to assume that certain salts, especially NaCl and KCl, are completely dissociated in these familiar evaluations of W; but this is justified in extrapolating to infinite dilution, whether or not the assumption holds at higher concentration, for the dissociation of the salt of strong acid and strong base in water solution does not involve, in any equilibrium constant, the ions of water itself. Even the evaluation of W from the conductivity of water, it must be noted, involves the mobilities of the water ions, and these can not be determined without the assumptions just enumerated concerning the strength of HCl and NaOH as reference acid and base and concerning the complete dissociation of the salt NaCl. The first point to note, then, is that the value of W is not independent of the values of ionization constants of solutes, and vice versa.

It appears, also, as though the strengths of acids and bases may be determined independently of W for subsequent use in the determination of W.

But this is in the nature of a process of successive approximations. We assume temporarily, even in deciding that HCl is "strong," that W is extremely small. As the acid becomes weaker, the precise value of W becomes increasingly important in the determination of the ionization constant of the acid. The "determination" of W, then, from measurements on the solution of a salt of strong base and weak acid, with A "known," is merely a second approximation or a refinement of the estimate of W assumed in the determination of A. Then for such a salt Eq. (4) becomes

$$W = H^2(c + A + H)/(A + H), \tag{7}$$

while for the salt of strong acid and weak base, with B "known,"

$$W = - H(c + B - H)/2 + \sqrt{(\quad)^2 + H^3B}. \tag{8}$$

(3) The next question in evaluating W, after it has been assumed that the degree of ionization of some reference solute such as HCl or NaOH is 1, is that of the *units* to be assigned to H whatever the form of measurement used. The activity of hydrogen ion must be evaluated and represented in multiples or subdivisions of an activity assigned the value of 1. This requires the choice of some reference solution in which the activity of hydrogen ion is taken as 1, and which forms part of what is called the "standard hydrogen electrode." With the convention that the ionic activity coefficient is 1 at "infinite dilution," the practical reference turns out to be a solution of HCl of such concentration c, that $[H^+]\gamma_c$ ($= a_{H^+}$, and in which γ_c is the activity coefficient of H^+ at that concentration) is 1. The potential of the half-cell H_2/H^+ then, used as a reference potential for the determination of hydrogen ion activity, is defined (practically) as 0 in a pure solution of HCl approximately 1.075M in concentration.

The resulting numerical value of W then obtained by extrapolation of the above functions to infinite dilution has a value fixed with reference to such a solution. All subsequent determinations of A's and B's then are also based on this reference, either through their dependence on W so evaluated or through the very units assigned to H in its experimental determination. These considerations hold whether we are dealing with potentials or with conductivities. In either case the values assigned to H are subdivisions of the unit given to H in the experimental reference solution of HCl mentioned above.[1] In E.M.F. methods,

$$-\ln a_{H^+} = \mathbf{FE}/\mathrm{R}T, \tag{9}$$

if **E** is the hydrogen potential relative to the "standard hydrogen electrode."

(1) The question of the various ways in which the experimental data are to be interpreted for the best definition of a "pH scale" is discussed by R. G. Bates, *Chem. Rev.*, **42,** 1 (1948). The difficulty involved is that of the thermodynamic definition of the quantity "a_{H^+}" and the impossibility of arriving at the activity coefficient of an individual ion.

In conductivity methods, H in a solution of a weak acid is taken, for example, ideally as

$$H \cong a\alpha = a\left(\frac{\Lambda_{HA}}{\Lambda_{HCl} + \Lambda_{NaA} - \Lambda_{NaCl}}\right)(\alpha_{HCl}), \tag{10}$$

in which it is assumed first that the two sodium salts are completely dissociated and then that α_{HCl} and $\alpha_{NaOH} = 1$ at all concentrations. Both of these α's (HCl and NaOH) are involved in the interpretations of the conductivities of the salts. The unit for H is again that chosen for the reference HCl solution. It may be pointed out here that while Eq. (9) is exact by definition, Eq. (10) is always approximate even when extrapolated to infinite dilution, for the definition of α is not H/a but $[X^-]/a$, and H is never equal to $[X^-]$.

(4) With a standard hydrogen electrode defined, it is presumably possible then to measure the activity of hydrogen ion in any solution by an E.M.F. comparison with the standard solution, or with some reference solution or half-cell already compared with the standard, such as an HCl solution at a concentration other than "standard," or a silver-silver chloride electrode, etc. But we cannot similarly measure directly the hydroxyl ion activity of a solution unless the ion-product is already known. The nearest we can come, therefore, to a direct measurement of W is the determination of a_{H^+} in a solution of a strong base in which the concentration of OH^- is assumed equal to b_s, the product then being extrapolated to infinite dilution. This is what is done through the cell (G. N. Lewis)(2) H_2, KOH (b_s) || HCl (a_s), H_2. If the two half-cells (distinguished as primed for the alkaline solution and as unprimed for the acid or reference solution) are compared either directly or indirectly through a common reference electrode such as calomel, a'_{H^+} is determined if a_{H^+} is known, since

$$\ln a'_{H^+} = \ln a_{H^+} - \mathbf{FE}/\mathrm{R}T. \tag{11}$$

Then for the thermodynamic constant

$$\mathbf{W} = a'_{H^+} b_s \gamma'_{OH^-}, \tag{12}$$

or

$$\ln \mathbf{W} = \ln a_{H^+} - \mathbf{FE}/\mathrm{R}T + \ln b_s + \ln \gamma'_{OH^-}. \tag{13}$$

Now extrapolation against $\sqrt{\mu}$ to $\mu = 0$ gives $\mathbf{W}$. If a_{H^+} is not known, but only the concentration of the HCl, or a_s, then

$$\ln \mathbf{W} = \ln (a_s b_s) - \mathbf{FE}/\mathrm{R}T + \ln \gamma_{H^+} + \ln \gamma'_{OH^-}. \tag{14}$$

In the cell (E. J. Roberts)(3) H_2, NaOH (b_s), NaCl (c), AgCl, Ag, the silver electrode is substituted for a hydrogen electrode as experimental reference, and if the relation between these two electrodes is known, the E.M.F. of

(2) G. N. Lewis, T. B. Brighton and R. L. Sebastian, *J. Am. Chem. Soc.*, **39**, 2245 (1917).
(3) E. J. Roberts, *ibid.*, **52**, 3877 (1930).

the cell again measures a'_{H^+} in the alkaline solution, and we may again use Eq. (12) to find **W**.

If the true constant is defined as $a_{H^+}a_{OH^-}/(a_{H_2O})^h$, the numerical value of **W** is independent of h if the activity of pure water is taken as unity, but if a_{H_2O} is taken as 55.35 (moles per liter), the value of **W** depends on h (cf. Chapter I, Section D).

(5) Finally, the numerical value of W is not to be compared in any way with the ionization constant of a solute in water either as to meaning or as to magnitude. An acid with $A < W$ is not to be considered a "weaker acid" than water, for the value of A merely represents the effect of the acid upon water relative to the effect of HCl at one end and of a nonelectrolyte at the other end. Such an acid with $A < W$ still gives an acid solution ($H > OH$) at any concentration in pure water. At the same time it does not follow that such a substance is "more acidic" or has a greater "acid-strength" (whatever these words mean) than water itself in any absolute sense, for if the substance and water are compared with respect to their ionization constants in some third substance as the solvent or medium, their order may for various reasons (steric effects, entropy, etc.) be reversed; such order may not be predicted from the values of A and W. In the water system the value of an "acid constant" A has meaning, numerically, with reference to a "strong acid" at one end and a nonacid ($A = 0$)—not water or a "strong base" etc.—at the other end; and the value of B, with reference to a "strong base" at one end and a nonbase ($B = 0$) at the other. A constant A is numerically comparable therefore neither with W nor with B.

B. Determination of Ionization Constant of Monobasic Acid

1. General Formulas. With strong base present,

$$D = H - OH = aA/(A + H) - b_s. \qquad (15)$$

The constant A may now be determined by measuring **H** or H (through E.M.F., conductivity, etc.) for known analytical concentrations of acid and base (a and b_s), provided W is known (and provided the "strong base," if present, is known or assumed to be completely ionized).

$$A = \frac{H(b_s + H - OH)}{a - b_s - H + OH}. \qquad (16)$$

The determination may be done on various solutions: (1) that of the pure acid ($b_s = 0$); (2) that of the pure salt ($b_s = a = c$); and (3) buffered solutions ($0 < b_s < a$).

For the pure acid,

$$A = H(H - OH)/(a - H + OH). \qquad (17)$$

If $OH \ll H$,

$$A \cong H^2/(a - H), \text{ or } \cong a\alpha^2/(1 - \alpha), \tag{18}$$

since, from Eq. III(3), $H = A\rho/\alpha$. Equation (18) in either form is the Ostwald dilution law. Whether or not OH is negligible in this sense is merely a numerical question, and the exact or full equation (17) is as easily used as any approximation, given the experimental data and the value of W to calculate OH from observed H.

For the salt, the usual point of view in the determination of A is that we determine the so-called "hydrolysis constant" of the salt, which is merely W/A; and since W is known, A is calculated "from" the "hydrolysis constant." But from Eq. (16), with $b_s = a = c$,

$$W/A = OH(OH - H)/(c - OH + H). \tag{19}$$

Now if $H \ll OH$ (and noting again that $H = A\rho/\alpha$), we have

$$W/A \cong (OH)^2/(c - OH) \cong c\rho^2/(1 - \rho), \tag{20}$$

an expression to be compared with Eq. (18) as a "dilution law" for the salt. In Eq. (20), ρ is the unionized fraction ($= 1 - \alpha$) of the weak acid of the salt. But again, whether or not H is negligible compared to OH in the exact expression of Eq. (19) is merely a matter of numerical values requiring no chemical explanation. Special approximation formulas such as Eq. (20) are obviously unnecessary since the data will clearly tell whether a particular term is negligible or not.

But the very idea of W/A as a "constant" separate from A itself is unnecessary since Eq. (19) may be written as

$$A = \frac{W(c - OH + H)}{(OH)^2 - W}. \tag{21}$$

If the solution is sufficiently alkaline, this becomes

$$A \cong W(c - OH)/(OH)^2, \tag{22}$$

or Eq. (20) again. In terms of H,

$$A = H^2(c - W/H + H)/(W - H^2) \cong H^2(c - OH)/W. \tag{23}$$

In the use of equivalent conductivities for these determinations, H is assumed to equal $a\alpha$ and it is evaluated as shown in Eq. (10), for the pure acid. For the salt, the assumption used is that $OH \cong c\rho$, and that

$$\rho_c \cong \left(\frac{\Lambda_{\text{observed}} - \Lambda_{\text{salt}}}{\Lambda_{\text{base}} - \Lambda_{\text{salt}}}\right)_c, \tag{24}$$

in which Λ_{salt} represents $(\Lambda_{M^+} + \Lambda_{X^-})$ and $\Lambda_{\text{base}} = (\Lambda_{M^+} + \Lambda_{OH^-})$.

For buffered solutions, the usual basis of accurate E.M.F. measurements, Eq. (16) is all that is needed if the measurements give H for known values of the analytical concentrations, a and b_s. The equation is exact, at least in terms of concentrations, and "approximations" are only numerical questions once W is known. These "approximations" merely depend on whether one of the quantities, H and OH, is negligible relative to the other.[4]

If the alkali salt of the weak acid, at concentration c, is titrated with strong acid, the expression becomes[4a]

$$A = H(c - a_s + H - OH)/(a_s - H + OH). \tag{24a}$$

More generally, however, if the solution of the weak acid a, containing strong base at the known concentration b_s, whether or not $b_s = a$ as in the pure salt solution, is titrated with a_s, then with $g = a_s - b_s$,

$$A = H(-g + H - OH)/(a + g - H + OH). \tag{24b}$$

(4) Nevertheless, the extent to which a simple exact equation like Eq. (16) is neglected may be seen in the following example illustrating the unnecessary complication and circumlocution introduced by starting with approximations which appear to have a primary derivation simply because an acid is thought of as "dissociating" rather than merely as ionizing in the sense defined in Chapter I of this treatment. Translating symbols into those of this discussion, we read in H. S. Harned and B. B. Owen, *The Physical Chemistry of Electrolytic Solutions*, Reinhold, N.Y., 1943, pp. 498–99, that whereas we may use the approximate formula

$$A = H(b_s + H)/(a - b_s - H), \tag{a}$$

—which is Eq. (16) with OH neglected—for the determination of the "dissociation" constant of an acid which is not too weak, we must somehow modify or refine the expression, if A is very small, as follows: "If, as in the case of boric acid, $A < 10^{-9}$, the hydrolysis of the acid anion, X^-, must be considered at high dilutions. Since H is quite negligible in solutions of such weak acids, we may write

$$A = H(b_s - OH)/(a - b_s + OH), \tag{b}$$

—which is again Eq. (16) with H now neglected—and estimate OH from the hydrolysis constant, 'K_h,' of the acid anion. A sufficiently accurate value of OH is given by $OH \simeq K_h b_s/(a - b_s)$, and a suitable value of K_h is obtained from the ratio of W and a rough estimate of A."

But the "rough estimate of A" of course turns out to be $A = Hb_s/(a - b_s)$, from the neglect of H as a term in Eq. (a); hence

$$OH = \frac{W(a - b_s)}{Hb_s}\left(\frac{b_s}{a - b_s}\right) = \frac{W}{H}. \tag{c}$$

If H is measured then, and if W is known, OH is estimated for use in Eq. (16) simply as W/H, with or without first computing a "rough estimate of A"; and whether or not either this quantity (OH), or H, is negligible in Eq. (16), to give either Eq. (a) or Eq. (b), is a matter of numerical values only. This is the obvious thing to do in the first place, without the necessity of mentioning or even thinking of the "hydrolysis" of anything at all.

From this "hydrolysis" point of view, the solution of an acid with or without strong base present would seem to involve two separate equilibrium constants other than W itself: the ionization constant of the acid and the "hydrolysis constant" of its anion. Not only do we find the concept of hydrolysis and hydrolysis constant unnecessary in all problems involving the hydrogen ion concentration of the solution, but if it has any effect, it leads to unnecessary confusion and apparent difficulty.

(4a) As used by E. Back and B. Steenberg, *Acta Chemica Scand.*, **4**, 810 (1950).

2. Introduction of Activity Coefficient. Finally, we shall illustrate the introduction of activity coefficients by means of Eq. (16) for the practical calculation of the thermodynamic constant **A**, assuming the experimental data to include the analytical concentrations, a and b_s, and the activity of hydrogen ion, **H**, determined potentiometrically; if the thermodynamic ion-product constant of water, or **W**, is known, the experimental determination of **H** implies direct determination also of **OH** and of **D** (= **H** − **OH**). Eq. (16), being derived from the equation of electroneutrality, is in terms of mass constants and concentrations. It therefore becomes

$$\mathbf{A} = A\,\frac{\gamma_{\mathrm{H}^+}\gamma_{\mathrm{X}^-}}{\gamma_{\mathrm{X}^\circ}} = \frac{\mathbf{H}\gamma_{\mathrm{X}^-}}{\gamma_{\mathrm{X}^\circ}}\left(\frac{b_s + (\mathbf{H}/\gamma_{\mathrm{H}^+}) - (\mathbf{OH}/\gamma_{\mathrm{OH}^-})}{a - b_s - (\mathbf{H}/\gamma_{\mathrm{H}^+}) + (\mathbf{OH}/\gamma_{\mathrm{OH}^-})}\right), \quad (25)$$

in which **A** is the thermodynamic ionization constant of the acid and the γ's are the various pertinent activity coefficients. This is complete and exact and may be simplified by the neglect of small terms if the data warrant it. If the solution is sufficiently dilute, $\gamma_{\mathrm{X}^\circ}$ may be taken as 1, and γ_{H^+}, γ_{OH^-}, and γ_{X^-} as having one common value, namely γ. Then

$$\mathbf{A} = \mathbf{H}\gamma\left(\frac{b_s + (\mathbf{D}/\gamma)}{a - b_s - (\mathbf{D}/\gamma)}\right), \quad (26)$$

with the understanding that $\gamma D = \mathbf{D}$.

For the estimation of γ on the basis of the Debye-Hückel limiting law, or

$$-\log\gamma = 0.505\sqrt{\mu}, \quad (27)$$

in which μ is the ionic strength of the solution, we note that since $2\mu = H + b_s + OH + a\alpha$, and $H - OH = a\alpha - b_s$, then μ may be taken either as

$$\mu = H + b_s = \mathbf{H}/\gamma + b_s, \quad (28)$$

or as

$$\mu = OH + a\alpha = OH + aA/(A + H), \quad (29a)$$

$$= \mathbf{OH}/\gamma + a\alpha = \mathbf{OH}/\gamma + a\mathbf{A}/(\mathbf{A} + \mathbf{H}\gamma). \quad (29b)$$

For given experimental values of **H** and b_s, γ may be calculated directly by successive approximations from

$$-\log\gamma = 0.505\sqrt{\mathbf{H}/\gamma + b_s}, \quad (30)$$

and then used in Eq. (26).

It is otherwise also possible, of course, to obtain the thermodynamic constant by extrapolation of the function of Eq. (26) to zero ionic strength.

For this purpose, with $\mathbf{pA} = -\log \mathbf{A}$, the expression becomes

$$\log\left(\frac{a - b_s - (\mathbf{D}/\gamma)}{b_s + (\mathbf{D}/\gamma)}\right) + \mathbf{pH} = \mathbf{pA} + \log \gamma, \tag{31a}$$

$$= \mathbf{pA} - 0.505\sqrt{\mu}, \tag{31b}$$

$$= \mathbf{pA} - \text{A}\sqrt{\mu}/(1 + \kappa\text{B}\sqrt{\mu}). \tag{31c}$$

Here A and B are constants of the Debye-Hückel theory, appropriate for water and for the temperature involved, and κ is a presumably constant parameter to be evaluated usually by trial and error, having the units of distance such as that of the closest approach of the ions involved. If a linear extrapolation to $\mu = 0$ according to Eq. (31b) is not justified because of high experimental values of μ, then the data have to be treated in the form of simultaneous equations to obtain constant values of both **A** and κ. In plotting the left hand member of Eq. (31) against $\sqrt{\mu}$, the value of γ in the term $\mathbf{D}/\gamma$ may usually be taken safely as unity since this quantity is usually small compared to either b_s or $a - b_s$.

Summarizing, therefore, we have from Eq. (26) or (31a):

For the pure acid, in which $H \gg OH$,

$$\mathbf{pA} = -\log \mathbf{A} \cong 2\mathbf{pH} + \log(a - \mathbf{H}/\gamma). \tag{32}$$

If $\text{H} \ll a$, $\mathbf{H}/\gamma$ may be taken as **H** in the last term.

For $b_s < a$,

$$\mathbf{pA} \cong \mathbf{pH} - \log\left[\frac{b_s + (\mathbf{D}/\gamma)}{a - b_s - (\mathbf{D}/\gamma)}\right] - \log \gamma. \tag{33}$$

If $\mathbf{D}/\gamma$ is small compared to b_s and to $a - b_s$, it may be taken as **D**.

For the salt, $b_s = a = c$, when $\text{H} \ll OH$,

$$\mathbf{pA} \cong 2\mathbf{pH} - \mathbf{pW} - \log(c - \mathbf{OH}/\gamma) - 2\log\gamma, \tag{34}$$

in which $\mathbf{OH}/\gamma$ may be taken as **OH** if small compared to c. But this is also the formula for the determination of **W** from the **pH** of the solution of the alkali salt, M_sX, of a weak monobasic acid of known ionization constant, the so-called "hydrolysis method" of determining **W**, or

$$\mathbf{pW} = -\log \mathbf{W} = 2\mathbf{pH} - \mathbf{pA} - \log(c - \mathbf{OH}/\gamma) - 2\log\gamma. \tag{35}$$

Extrapolation to $\mu = 0$, in other words, gives **A** if **W** is known or **W** if **A** is known.

C. Dibasic Acid (and Strong Base)

1. Fundamental Relations.

$$D = a\beta - b_s = a(1 + 2K_2/H)/(1 + H/K_1 + K_2/H) - b_s; \quad (36)$$

$$K_1 = H(b_s + D)/[(a - b_s - D) + (2a - b_s - D)K_2/H]; \quad (37)$$

$$K_2 = [H(b_s - a + D) + (b_s + D)H^2/K_1]/(2a - b_s - D). \quad (38)$$

With values of b_s such that $H \gg K_2$ (in this connection, for example, we note that $H \cong K_1$ when $b_s \cong a/2$), an approximate value of K_1 may be obtained from Eq. (37) simplified to

$$K_1 \cong H(b_s + D)/(a - b_s - D). \quad (39)$$

Then, with measurements at another range ($b_s \cong 3a/2$), Eq. (38) may be used for an accurate determination of K_2, with which the value of K_1 may then be refined through the full Eq. (37), etc. Similarly an approximate value of K_2 is obtainable from a simplified form of Eq. (38), when $H \ll K_1$ (we note here that $H \cong K_2$ when $b_s \cong 3a/2$), or

$$K_2 \cong H(b_s - a + D)/(2a - b_s - D); \quad (40)$$

and such a value of K_2 may then be used in Eq. (37) for K_1 and refined further through the full expression of Eq. (38).

Again all approximations are numerical, involving no chemical explanation and no speculation.

With $b_s = 0$, Eqs. (37) and (38) describe the effect of dilution on the pure acid solution. It is mathematically possible, therefore, although more difficult and less accurate, to determine the constants K_1 and K_2 without buffering and without base at all. Since, in general, $H > K_1$ when $a > 2K_1$, it is possible to start with Eq. (39) to estimate K_1 and then to use Eqs. (37) and (38) alternately and at different concentrations to refine the values of both constants. On the rough relation between a and K_1, Eq. (36), with $b_s = 0$, gives

$$a_{H=K_1} = (K_1 - W/K_1)(2K_1 + K_2)/(K_1 + 2K_2) \cong 2(K_1 - W/K_1), \quad (41)$$

provided $K_1 \gg K_2$; hence $H > K_1$ if $a > 2K_1$.

For a *dibasic acid "strong in K_1,"* Eq. (40) is the exact equation for the only constant, K_2. For the pure acid, ($b_s = 0$),

$$K_2 = H(D - a)/(2a - D), \quad (42)$$

which is therefore the "dilution law" for such an acid, to be compared with Eqs. (17) and (18).

2. Simplification With Regard to Activity Coefficients. In introducing activity coefficients into these equations for the purpose of expressing

the thermodynamic constants, we shall assume here, and on all subsequent occasions, as a *convention for simplification* that in dilute solution the activity coefficient for uncharged species is 1, the activity coefficient for all univalent ions, or $\gamma_{+ \text{ or } -}$, has one common value, γ, that for any divalent ion is γ^4, that for a trivalent ion is γ^9, etc., as in the limiting law. Hence

$$\mathbf{W} = W\gamma^2; \text{ and } \mathbf{K}_1 = K_1\gamma^2, \mathbf{K}_2 = K_2\gamma^4, \mathbf{K}_3 = K_3\gamma^6. \tag{43}$$

With this conventional understanding, Eq. (37) becomes

$$\mathrm{p}\mathbf{K}_1 = \mathrm{p}\mathbf{H} - \log\left(\frac{b_s + (\mathbf{D}/\gamma)}{a - b_s - (\mathbf{D}/\gamma)}\right) + \log\left[1 + \frac{\mathbf{K}_2}{\mathbf{H}\gamma^3}\left(\frac{2a - b_s - (\mathbf{D}/\gamma)}{a - b_s - (\mathbf{D}/\gamma)}\right)\right] - \log\gamma. \tag{44}$$

This is an exact expression, except for the convention about γ. For extrapolation to $\mu = 0$, we note that in the region $b_s \cong a/2$ the third term (which may be called the "correction term") is very small in general, since $H \cong K_1$, and that therefore the value of γ in this term may be taken as 1. Similarly, as long as $\mathbf{D}/\gamma$ is small compared to the analytical concentrations, we may also take γ as 1 in the second term.

From Eq. (38),

$$\mathrm{p}\mathbf{K}_2 = \mathrm{p}\mathbf{H} - \log\left(\frac{b_s - a + (\mathbf{D}/\gamma)}{2a - b_s - (\mathbf{D}/\gamma)}\right) - \log\left[1 + \frac{\mathbf{H}\gamma}{\mathbf{K}_1}\left(\frac{b_s + (\mathbf{D}/\gamma)}{b_s - a + (\mathbf{D}/\gamma)}\right)\right] - 3\log\gamma. \tag{45}$$

Again γ may be taken as 1 for the second and third terms, the third term being very small if the measurements are made at $b_s \cong 3a/2$, when $H \cong K_2$.

If the activity coefficient is to be calculated from μ, we shall in general assume the limiting law, Eq. (27), in which μ is either a known quantity for a problem or a quantity to be calculated from the concentrations of acid and base present. For such calculation the following considerations may be used. By definition, in a solution containing a polybasic acid and simple strong base,

$$2\mu = H + OH + a(\alpha_1 + 4\alpha_2 + 9\alpha_3 + 16\alpha_4 + \cdots) + b_s. \tag{46}$$

At the same time,

$$H - OH = a(\alpha_1 + 2\alpha_2 + 3\alpha_3 + 4\alpha_4 + \cdots) - b_s. \tag{47}$$

Hence we may write either

$$\mu = OH + a(\alpha_1 + 3\alpha_2 + 6\alpha_3 + 10\alpha_4 + \cdots), \tag{48}$$

or

$$\mu = H + b_s + a(\alpha_2 + 3\alpha_3 + 6\alpha_4 + \cdots). \tag{49}$$

When b_s has the values $a/2$, a, $3a/2$, etc., these functions are generally simple, provided that $K_1 \gg K_2 \gg K_3$, etc.; the *approximate* values are listed in Table 5–1.

Table 5–1. Ionic strength in titration of polybasic acid.

b_s	H	α_1	α_2	α_3	α_4	$(\mu - H)$ or $(\mu - OH)$
$a/2$	K_1	1/2	0	0	0	$a/2$
a	$\sqrt{K_1K_2}$	1	0	0	0	a
$3a/2$	K_2	1/2	1/2	0	0	$2a$
$2a$	$\sqrt{K_2K_3}$	0	1	0	0	$3a$
$5a/2$	K_3	0	1/2	1/2	0	$9a/2$

For pure alkali salt solutions, at concentration c, Eq. (48) becomes

$$\mu = OH + c(\alpha_1 + 3\alpha_2 + 6\alpha_3 + \cdots); \tag{50}$$

but in the form of Eq. (49) the expression varies with the "degree" of the salt, so that

$$\mu \text{ (primary)} = H + c(1 + \alpha_2 + 3\alpha_3 + \cdots), \tag{51a}$$

$$\mu \text{ (secondary)} = H + c(2 + \alpha_2 + 3\alpha_3 + \cdots), \tag{51b}$$

$$\mu \text{ (tertiary)} = H + c(3 + \alpha_2 + 3\alpha_3 + \cdots), \text{ etc.} \tag{51c}$$

In a determination of equilibrium constants, the uncorrected "mass constants" may be used to calculate a first approximation value of γ from Eqs. (48) or (49), which require the constants, the observed value of H and the concentrations a and b_s. With such a value of γ, the **K**'s are corrected, as in Eqs. (44, 45). Then a second approximation may be calculated for γ through the first corrections of the **K**'s, etc., until the desired precision in these constants is obtained.

3. Ratio (y) and Product (**p**) of the Two Constants. If we define y as the ratio of the two constants, K_2/K_1, and **p** as their product, K_1K_2, we may express D as function of any two of the independent constant parameters, K_1, K_2, y, **p**. Eq. (36) is therefore only one of the six possibilities. For certain purposes the other functions may be more convenient, but it must be realized that they are identical through the definitions of y and **p**. In terms of K_1 and y, so that

$$\beta = (1 + 2yK_1/H)/(1 + H/K_1 + yK_1/H), \tag{52}$$

the analogs of Eqs. (36–38) become

$$D = a(1 + 2yK_1/H)/(1 + H/K_1 + yK_1/H) - b_s, \tag{53}$$

$$K_1 = -\frac{H}{2y}\left(\frac{a - b_s - D}{2a - b_s - D}\right) + \sqrt{(\quad)^2 + \frac{H^2}{y}\left(\frac{b_s + D}{2a - b_s - D}\right)}, \tag{54}$$

$$y = \frac{H}{K_1}\left(\frac{b_s - a + D}{2a - b_s - D}\right)\left[1 + \frac{H}{K_1}\left(\frac{b_s + D}{b_s - a + D}\right)\right]. \tag{55}$$

For the use of Eq. (54) an approximate value of y may be at hand from the ratio of H at the two successive mid-points of titration of the dibasic acid, since $H \cong K_1$ at $b_s = a/2$ and $H \cong K_2$ at $b_s = 3a/2$. The relation between this ratio of H values and the constant y will be discussed further in Chapter XIV. If then an approximate value of y is available, Eq. (54) may be used for the accurate determination of K_1 through measurements at $b_s \cong a/2$. Then if H is also measured at $b_s \cong 3a/2$, the value of K_1 may be used in Eq. (55) to obtain y accurately and thence K_2. Whatever function is used to express D, the useful values of b_s for experimentation are always: $b_s \cong a/2$ for K_1, when $H \cong K_1$; $b_s \cong 3a/2$ for K_2, when $H \cong K_2$; $b_s = a$ for $\mathbf{p}$, when $H \cong \sqrt{\mathbf{p}}$ or $\sqrt{K_1K_2}$; $b_s \cong a/2$ for y if an approximate value of K_2 is known, and $b_s \cong 3a/2$ for y from an approximate value of K_1. With this in mind, we shall therefore simply express the remaining functions without discussion.

In terms of K_2 and y,

$$K_2 = \frac{H}{2}\left(\frac{b_s - a + D}{2a - b_s - D}\right) + \sqrt{(\quad)^2 + \frac{yH^2(b_s + D)}{2a - b_s - D}}; \tag{56}$$

$$y = \frac{K_2}{H}\left(\frac{a - b_s - D}{b_s + D}\right)\left[1 + \frac{K_2}{H}\left(\frac{2a - b_s - D}{a - b_s - D}\right)\right]. \tag{57}$$

In terms of K_1 and $\mathbf{p}$,

$$K_1 = H\left(\frac{b_s + D}{a - b_s - D}\right)\left[1 - \frac{\mathbf{p}}{H^2}\left(\frac{2a - b_s - D}{b_s + D}\right)\right]; \tag{58}$$

$$\mathbf{p} = H^2\left(\frac{b_s + D}{2a - b_s - D}\right)\left[1 + \frac{K_1}{H}\left(\frac{b_s - a + D}{b_s + D}\right)\right]. \tag{59}$$

When $b_s = a = c$, c being the concentration of the acid salt M_sHX, then

$$\mathbf{p} = H^2\left(\frac{c + D}{c - D}\right)\left[1 + \frac{K_1}{H}\left(\frac{D}{c + D}\right)\right]. \tag{60}$$

In terms of K_2 and $\mathbf{p}$,

$$\frac{1}{K_2} = \frac{2a - b_s - D}{H(b_s - a + D)}\left[1 - \frac{H^2}{\mathbf{p}}\left(\frac{b_s + D}{2a - b_s - D}\right)\right]; \tag{61}$$

$$\frac{1}{\mathbf{p}} = \frac{2a - b_s - D}{H^2(b_s + D)}\left[1 + \frac{H}{K_2}\left(\frac{a - b_s - D}{2a - b_s - D}\right)\right]. \tag{62}$$

When $b_s = a = c$, for the acid salt M_sHX, then

$$\frac{1}{\mathbf{p}} = \frac{1}{H^2}\left(\frac{c-D}{c+D}\right)\left[1 - \frac{H}{K_2}\left(\frac{D}{c-D}\right)\right]. \tag{63}$$

Eqs. (60) and (63) are useful for the accurate determination of $\mathbf{p}$ from the pH of the solution of the pure acid salt if an approximate value of either constant is available. Finally, in terms of $\mathbf{p}$ and y,

$$\sqrt{\mathbf{p}} = -\frac{H}{2\sqrt{y}}\left(\frac{a-b_s-D}{2a-b_s-D}\right) + \sqrt{(\quad)^2 + H^2\left(\frac{b_s+D}{2a-b_s-D}\right)}; \tag{64}$$

$$\frac{1}{\sqrt{y}} = \frac{H}{\sqrt{\mathbf{p}}}\left(\frac{b_s+D}{a-b_s-D}\right)\left[1 - \frac{\mathbf{p}}{H^2}\left(\frac{2a-b_s-D}{b_s+D}\right)\right]. \tag{65}$$

Now if an approximate value of y is known, the solution of the pure acid salt at concentration c may be used to find $\sqrt{\mathbf{p}}$ accurately from

$$\sqrt{\mathbf{p}} = \frac{H}{2\sqrt{y}}\left(\frac{D}{c-D}\right) + \sqrt{(\quad)^2 + H^2\left(\frac{c+D}{c-D}\right)}. \tag{66}$$

Since these formulas are all completely general and exact, no rules of approximation are necessary; the data themselves (analytical solute concentrations and H) will in every case indicate explicitly whether any term is negligible.

For the activity constants, the activity coefficients may easily be introduced as follows, for the case of $\mathbf{p}$ as an example, with the usual convention regarding γ and with D in place of $\mathbf{D}/\gamma$: From Eq. (60),

$$-\log \mathbf{K}_1\mathbf{K}_2 = 2\text{p}\mathbf{H} - \log\left(\frac{c+D}{c-D}\right) - \log\left[1 + \frac{\mathbf{K}_1\mathbf{D}}{\gamma^2\mathbf{H}(c+D)}\right] - 4\log\gamma. \tag{67}$$

From Eq. (63),

$$-\log \mathbf{K}_1\mathbf{K}_2 = 2\text{p}\mathbf{H} - \log\left(\frac{c+D}{c-D}\right) + \log\left[1 - \frac{\gamma^2\mathbf{HD}}{\mathbf{K}_2(c-D)}\right] - 4\log\gamma. \tag{68}$$

Here c is the concentration of the alkali acid salt M_sHX, and as usual γ in D and in the third term may be taken as 1 for extrapolation purposes. If the solution is pure M_sHX, a preliminary value of γ may be calculated with $\mu \cong c$, according to Table 5–1, and then refined through Eq. (51a) with the preliminary values of the constants so obtained.

4. Logarithmic Form of Solution of Quadratic Equation. In Eqs. (67) and (68) the logarithmic form is used for convenience in introducing γ as $\log \gamma$. Since logarithmic expressions become even more important later in the discussion of pH problems, we shall point out here a variation for the expression of the solution of a quadratic equation which is useful in connection

with the evaluation of logarithmic quantities generally. The quadratic equation,

$$\chi^2 + v\chi + w = 0, \tag{69}$$

is assumed to have coefficients satisfying the requirements for real values of χ, or $(v/2)^2 - w > 0$. Then

(a) If w is negative,

$$\chi = -v/2 \pm \sqrt{(\quad)^2 + |w|}, \tag{70}$$

which may be rearranged as follows for the absolute magnitude of χ:

$$|\chi| = \sqrt{|w|}(\pm x + \sqrt{x^2 + 1}), \tag{71}$$

in which

$$x = -v/2\sqrt{|w|}. \tag{72}$$

Also,

$$\ln |\chi| = \ln \sqrt{|w|} + \ln(\pm x + \sqrt{x^2 + 1}). \tag{73}$$

But since

$$\ln(x + \sqrt{x^2 + 1}) = \sinh^{-1} x, \tag{74}$$

with $\sinh^{-1} x = -\sinh^{-1}(-x)$, then

$$\log |\chi| = \log \sqrt{|w|} \pm (\sinh^{-1} x)/2.3. \tag{75}$$

Eq. (75) gives the two numerical values of χ but not their signs. But only one of the roots will be positive in this case, namely, that involving the positive sign in Eq. (70). It must be recalled in this connection that the maximum number of real positive roots of a polynomial is the number of changes of sign in the progression of its terms; and the signs of the terms of the quadratic corresponding to Eq. (75) are either $+ - -$ or $+ + -$.

According to Eq. (75), then, the solution of a quadratic equation in H may be expressed directly in terms of pH. In the usual real case such an equation for H will have only one positive root and will therefore pertain to the present class, or

$$H^2 \pm \mathbf{v}H - \mathbf{w} = 0, \tag{76}$$

in which the quantities $\mathbf{v}$ and $\mathbf{w}$ are positive. Then, with sign as in Eq. (76),

$$\text{p}H = -\log \sqrt{\mathbf{w}} \pm (1/2.3) \sinh^{-1} (\mathbf{v}/2\sqrt{\mathbf{w}}). \tag{77}$$

(b) If w is positive,

$$\chi = -v/2 \pm \sqrt{(\quad)^2 - w}, \tag{78}$$

$$\chi = \sqrt{w}(x \pm \sqrt{x^2 - 1}), \tag{79}$$

with x as in Eq. (72). Now since

$$\ln (x \pm \sqrt{x^2 - 1}) = \cosh^{-1} x, \tag{80}$$

then

$$\log |\chi| = \log \sqrt{w} \pm |(\cosh^{-1} x)|/2.3. \tag{81}$$

We note that $\cosh^{-1} x$ has two values differing only in sign; also, in $\cosh^{-1} x$, x must be greater than 1, which is equivalent to the requirement of $(v/2)^2 - w > 0$ for a real value of χ. In this case the roots of the quadratic are either both positive or both imaginary, since the signs of its terms are either $+ - +$ or $+ + +$.

We may also note for possible use the relation

$$\ln [(1 + x)/(1 - x)] = \tanh^{-1} x, \tag{82}$$

with $\tanh^{-1} x = \tanh^{-1} (- x)$.

In the form of Eq. (71), Eq. (66) becomes

$$\sqrt{\mathbf{p}} = H\sqrt{(c + D)/(c - D)}(\mathbf{Y} + \sqrt{\mathbf{Y}^2 + 1}), \tag{83}$$

with $\mathbf{Y} = D/2\sqrt{y(c^2 - D^2)}$. Hence, from Eqs. (73), (74), and (82),

$$- \log K_1K_2 = 2pH - (1/2.3) \tanh^{-1} (D/c) - (2/2.3) \sinh^{-1} \mathbf{Y}. \tag{84}$$

With activity constants,

$$- \log \mathbf{K}_1\mathbf{K}_2 = 2p\mathbf{H} - (1/2.3) \tanh^{-1} (\mathbf{D}/\gamma c) - (2/2.3) \sinh^{-1} \mathbf{Y}' - 4 \log \gamma, \tag{85}$$

with $\mathbf{Y}' = \mathbf{D}/2\sqrt{(c^2 - \mathbf{D}^2/\gamma^2)\mathbf{K}_2/\mathbf{K}_1}$. The hyperbolic functions are tabulated in ordinary mathematical tables.

D. Tribasic Acid (And Strong Base)

$$D = a(1 + 2K_2/H + 3K_{23}/H^2)/(1 + H/K_1 + K_2/H + K_{23}/H^2) - b_s. \tag{86}$$

1. Expressions for Individual Constants.

$$K_1 = \frac{H(b_s + D)}{(a - b_s - D) + (K_2/H)(2a - b_s - D) + (K_{23}/H^2)(3a - b_s - D)}; \tag{87}$$

$$K_2 = \frac{H(b_s - a + D) + (H^2/K_1)(b_s + D)}{(2a - b_s - D) + (K_3/H)(3a - b_s - D)}; \tag{88}$$

$$K_3 = \frac{H(b_s - 2a + D) + (H^2/K_2)(b_s - a + D) + (H^3/K_{12})(b_s + D)}{(3a - b_s - D)}. \tag{89}$$

These equations reduce to those of a dibasic acid and monobasic acid if K_3 or if both K_3 and K_2 equal zero.(5)

Near the mid-point of the addition of each of the three successive equivalents of base to the pure acid, or near $b_s \cong a/2$, $3a/2$, and $5a/2$, H is approximately equal, respectively, to K_1, K_2, and K_3. If $K_1 \gg K_2 \gg K_3$

(5) For application of approximate formulas in the case of citric acid, see R. G. Bates and G. D. Pinching, *J. Am. Chem. Soc.*, **71**, 1274 (1949).

then, these three exact equations may be simplified for those regions not near equivalent points to

$$K_1 \cong H(b_s + D)/(a - b_s - D), \text{ when } b_s \cong a/2; \quad (90)$$

$$K_2 \cong H(b_s - a + D)/(2a - b_s - D), \text{ when } b_s \cong 3a/2; \quad (91)$$

$$K_3 \cong H(b_s - 2a + D)/(3a - b_s - D), \text{ when } b_s \cong 5a/2. \quad (92)$$

When each K is taken equal to H at a mid-point of titration, we are neglecting not only the quantities involving the other ionization constants but also the quantity D, or $H - OH$, compared to the concentrations of the solutes. The usual approach to the problem of the calculation of all such ionization constants from the simple monobasic acid case to the most complicated is in fact virtually to *start* from this *final* simplest approximation,

$$K_m \cong H[b_s - (m - 1)a]/(ma - b_s), \quad (93)$$

and then to consider how to take account of the values of H and OH relative to the concentrations, and how to *introduce* the effect of the other ionization constants as special corrections, either when we are too close to the equivalent points or when the K's do not differ sufficiently in magnitude. But since it is difficult to know beforehand when these "corrections" are necessary, the result is that they are not expressed or applied uniformly by all investigators, and rarely, if ever, altogether exactly.

The point of view should be the very opposite. These are the exact equations (Eqs. [87–89]). They can be handled by the merest student very simply and directly. The question should be not how to introduce corrections into Eq. (93) but whether or not certain quantities are negligible in the exact equations, which should always be available. Clearly this is a simple question of numerical values. In this way there need never be any uncertainty about overlooking any significant effect in the final calculation, and the kind of circumlocution discussed above (Note 4) would never occur.

The complete titration curve is not necessary in determining any one of the constants very precisely, provided even very approximate values of the other constants of the problem are available. In the appropriate region for the determination of each constant ($a < b_s < 2a$, for example, for the determination of K_2), the effects of K_1 and K_3 will always be relatively small, according to Eq. (88) provided $K_1 > K_2 > K_3$, since here H is of the order of K_2 and therefore less than K_1 and more than K_3. But the effects of these two constants are given completely and exactly—and very simply—by the exact Eq. (88)—and it is merely a numerical question whether they are to be neglected in calculating K_2. Algebraically, with the appropriate exact equation applied at the appropriate values of a and b_s, a particular constant may be calculated from a single point (a determination of H at known values of a and b_s) as well as from the mid-point of a whole curve, provided the other two constants are known at least approximately.

Even without previous knowledge of approximate values of all but one constant, it is always (at least theoretically) possible to solve for all the constants involved by using the exact equation for any one of the constants (and involving them all) in connection with a sufficient number of simultaneous sets of data. The problem is nothing more than the ordinary algebraic solution of simultaneous equations in two or three unknowns.

With the usual convention for the introduction of activity coefficients, and with D in place of $\mathbf{D}/\gamma$, Eqs. (87–89) become, respectively:

$$\mathrm{p}\mathbf{K}_1 = \mathrm{p}\mathbf{H} - \log\left(\frac{b_s + D}{a - b_s - D}\right) + \log\left\{1 + \frac{\mathbf{K}_2}{\mathbf{H}\gamma^3}\left(\frac{2a - b_s - D}{a - b_s - D}\right)\left[1 + \frac{\mathbf{K}_3}{\mathbf{H}\gamma^5}\left(\frac{3a - b_s - D}{2a - b_s - D}\right)\right]\right\} - \log\gamma; \quad (94)$$

$$\mathrm{p}\mathbf{K}_2 = \mathrm{p}\mathbf{H} - \log\left(\frac{b_s - a + D}{2a - b_s - D}\right) - \log\left[\frac{1 + \dfrac{\mathbf{H}\gamma}{\mathbf{K}_1}\left(\dfrac{b_s + D}{b_s - a + D}\right)}{1 + \dfrac{\mathbf{K}_3}{\mathbf{H}\gamma^5}\left(\dfrac{3a - b_s - D}{2a - b_s - D}\right)}\right] - 3\log\gamma; \quad (95)$$

$$\mathrm{p}\mathbf{K}_3 = \mathrm{p}\mathbf{H} - \log\left(\frac{b_s - 2a + D}{3a - b_s - D}\right) - \log\left\{1 + \frac{\mathbf{H}\gamma^3}{\mathbf{K}_2}\left(\frac{b_s - a + D}{b_s - 2a + D}\right)\left[1 + \frac{\mathbf{H}\gamma}{\mathbf{K}_1}\left(\frac{b_s + D}{b_s - a + D}\right)\right]\right\} - 5\log\gamma. \quad (96)$$

The relation between these expressions and Eqs. (44) and (45) should be noted. As in those cases, each constant may be evaluated accurately if the other two are known approximately; and in extrapolation to zero μ, again, γ may be taken as 1 in D and in the third terms, which are generally very small at the values $b_s \simeq a/2$, $3a/2$, $5a/2$, respectively.

If γ is to be calculated, we may use the formulas of Eqs. (48) and (49) for μ.

2. Ratios and Products of Constants. While in Eq. (86) D is expressed in terms of K_1, K_2, K_3 as the three independent constant parameters, an equivalent expression may be written for D as function of any three independent constant parameters chosen from the following: K_1, K_2, K_3, y $(= K_2/K_1)$, y_1 $(= K_3/K_2)$, $\mathbf{p}$ $(= K_1K_2)$, $\mathbf{p}_1 = (K_2K_3)$.[6] In terms of $\mathbf{p}$, y, y_1, for example, Eq. (86) becomes

$$D = a\,\frac{1 + (2\sqrt{\mathbf{p}y}/H) + 3\mathbf{p}yy_1/H^2}{1 + H\sqrt{y/\mathbf{p}} + (\sqrt{\mathbf{p}y}/H) + \mathbf{p}yy_1/H^2} - b_s. \quad (97)$$

(6) For approximate formulas used in determination of the products K_1K_2 and K_2K_3, see R. G. Bates, *J. Am. Chem. Soc.*, **70**, 1579 (1948).

Solved for the constants, this equation is of first degree in y_1, second in $\sqrt{y}$, and third in $\sqrt{\mathbf{p}}$, so that, with $x = b_s + D$,

$$y_1 = \frac{H^2}{\mathbf{p}y}\left(\frac{x-a}{3a-x}\right)\left[1 - \frac{\sqrt{\mathbf{p}y}}{H}\left(\frac{2a-x}{x-a}\right) + H\sqrt{\frac{y}{\mathbf{p}}}\left(\frac{x}{x-a}\right)\right]; \qquad (98)$$

$$\sqrt{y} = -\frac{H\mathbf{p}(2a-x) - xH^3}{2\mathbf{p}(\sqrt{\mathbf{p}})y_1(3a-x)} + \sqrt{(\quad)^2 + \frac{H^2}{\mathbf{p}y_1}\left(\frac{x-a}{3a-x}\right)}; \qquad (99)$$

$$(\sqrt{\mathbf{p}})^3 yy_1(3a-x) + (\sqrt{\mathbf{p}})^2 H\sqrt{y}(2a-x) - \sqrt{\mathbf{p}}H^2(x-a) - xH^3\sqrt{y} = 0. \qquad (100)$$

Approximate values of y and y_1 may be obtained from the ratios of H at successive mid-points of titration; and the approximate value of $\sqrt{\mathbf{p}}$ comes from H in the pure solution of the primary acid salt, M_sH_2X, when $H \cong \sqrt{K_1K_2}$. With approximate values of two of the parameters, the other may be evaluated with considerable accuracy from one of these equations. When the concentration of the salt M_sH_2X is c, for example, Eq. (100) becomes

$$(\sqrt{\mathbf{p}})^3 yy_1(2c-D) + (\sqrt{\mathbf{p}})^2 H\sqrt{y}(c-D) - \sqrt{\mathbf{p}}H^2D - H^3\sqrt{y}(c+D) = 0, \qquad (101)$$

which may be solved by approximation methods for the value of $\sqrt{\mathbf{p}}$ from measured values of H and c and approximate values of y and y_1. Since $H \cong \sqrt{\mathbf{p}}$, the first term in this equation is generally small compared to the last three.

If D is expressed as function of two K's and one $\mathbf{p}$ the equation is of first degree for each of the three parameters; as f(two K's and one y), it is quadratic for the K involved in the y and of first degree for the other two parameters; as f(one K, one $\mathbf{p}$, and one y), it is quadratic only for the K, if K is involved in y, and quadratic for both $\mathbf{p}$ and K if K is not involved in the y; as f(two $\mathbf{p}$'s and one y), it is of first degree for one $\mathbf{p}$, quadratic for the square root of the other and of first degree for the square root of the y; as f(two y's and one $\mathbf{p}$), it is always cubic for the square root of the $\mathbf{p}$; as f(two y's and one K), it is cubic for the K; as f(two $\mathbf{p}$'s and one K), it is of first degree for all three parameters.

This means that an accurate value of a product, either K_1K_2 or K_2K_3, may be obtained from the pH of the solution of the appropriate acid salt if two other suitable parameters are known approximately. Given, for example, rough values of y_1 ($= K_3/K_2$) and of K_2, then we may obtain $\mathbf{p}$ ($= K_1K_2$)

accurately from the solution of the primary salt M_sH_2X. The expression for D in terms of y_1, K_2, and $\mathbf{p}$, solved for $\mathbf{p}$, gives (again with $x = b_s + D$),

$$\frac{1}{\mathbf{p}} = \frac{1}{H^2}\left(\frac{2a - x}{x}\right)\left[1 + \frac{H}{K_2}\left(\frac{a - x}{2a - x}\right) + \frac{y_1K_2}{H}\left(\frac{3a - x}{2a - x}\right)\right]. \qquad (102)$$

At $b_s = a = c$, or for the salt M_sH_2X, then,

$$-\log \mathbf{K}_1\mathbf{K}_2 = 2\mathbf{pH} - \log\left(\frac{c + D}{c - D}\right) + \log\left[1 - \frac{\gamma^2\mathbf{HD}}{\mathbf{K}_2(c - D)} + \frac{y_1\mathbf{K}_2}{\gamma^5\mathbf{H}}\left(\frac{2c - D}{c - D}\right)\right] - 4\log\gamma. \qquad (103)$$

If the known approximate parameters are $\mathbf{K}_2$ and $\mathbf{K}_3$, then we simply write $\mathbf{K}_3$ in place of $y_1\mathbf{K}_2$ in the third term.

Similarly, to determine $\mathbf{p}_1$ or K_2K_3 from approximate values of K_1 and K_2, we start with D as $f(\mathbf{p}_1, K_1, K_2)$ and obtain

$$\mathbf{p}_1 = H^2\left(\frac{x - a}{3a - x}\right)\left[1 + \frac{H}{K_1}\left(\frac{x}{x - a}\right) + \frac{K_2}{H}\left(\frac{x - 2a}{x - a}\right)\right]. \qquad (104)$$

At $b_s = 2a = 2c$, for the salt $(M_s)_2HX$,

$$-\log \mathbf{K}_2\mathbf{K}_3 = 2\mathbf{pH} - \log\left(\frac{c + D}{c - D}\right) - \log\left[1 + \frac{\mathbf{H}\gamma}{\mathbf{K}_1}\left(\frac{2c + D}{c + D}\right) + \frac{\mathbf{K}_2\mathbf{D}}{\gamma^4\mathbf{H}(c + D)}\right] - 8\log\gamma. \qquad (105)$$

If, however, we have approximate values of y and y_1, then for $\mathbf{p}$ we have Eqs. (97–101), above, while with D as $f(\mathbf{p}_1, y, y_1)$ on the other hand, we have as the analog of Eq. (100):

$$(\sqrt{\mathbf{p}_1})^3\sqrt{y_1}(3a - x) + (\sqrt{\mathbf{p}_1})^2H(2a - x) - \sqrt{\mathbf{p}_1}H^2\sqrt{y_1}(x - a) - xH^3yy_1 = 0. \qquad (106)$$

Then for the salt $(M_s)_2HX$, now with $x = 2a + D = 2c + D$,

$$(\sqrt{\mathbf{p}_1})^3\sqrt{y_1}(c - D) - (\sqrt{\mathbf{p}_1})^2HD - \sqrt{\mathbf{p}_1}H^2\sqrt{y_1}(c + D) - \mathbf{H}^3yy_1(2c + D) = 0; \qquad (107)$$

with $H = W/OH$, this is seen to be the "OH" analog of Eq. (101).

Finally, if the cubic equations (101) and (107) have to be extrapolated to $\mu = 0$ for the thermodynamic products $\mathbf{K}_1\mathbf{K}_2$ and $\mathbf{K}_2\mathbf{K}_3$ respectively, it is necessary to use slight approximations. Eq. (101) may be rearranged to give

$$\mathbf{p} = H^2\left(\frac{c + D}{c - D}\right)\left[1 + \frac{\sqrt{\mathbf{p}}D}{H\sqrt{y}(c + D)} - \frac{(\sqrt{\mathbf{p}})^3y_1\sqrt{y}(2c - D)}{H^3(c + D)}\right]. \qquad (108)$$

The second and third terms in the bracket are correction terms in which, therefore, we shall set $\mathbf{p} \cong H^2(c + D)/(c - D)$; we may thereupon extrapolate the following function to $\mu = 0$:

$$- \log \mathbf{K}_1\mathbf{K}_2 \cong 2\mathbf{pH} - \log \left(\frac{c + D}{c - D}\right) - \log \left[1 + \frac{\mathbf{D}}{\sqrt{y}\sqrt{c^2 - D^2}} - \frac{y_1\sqrt{y}}{\gamma^3}\left(\frac{2c - D}{c - D}\right)\sqrt{\frac{c + D}{c - D}}\right] - 4 \log \gamma. \quad (109)$$

Similarly for Eq. (107), we have

$$\mathbf{p}_1 = H^2 \left(\frac{c + D}{c - D}\right) \left[1 - \frac{Hy\sqrt{y_1}(2c + D)}{\sqrt{\mathbf{p}_1}(c + D)} + \frac{\sqrt{\mathbf{p}_1}D}{H\sqrt{y_1}(c + D)}\right]. \quad (110)$$

Again if we assume that $\mathbf{p}_1 \cong H^2(c + D)/(c - D)$ in the correction terms, we may extrapolate the following expression to $\mu = 0$:

$$- \log \mathbf{K}_2\mathbf{K}_3 \cong 2\mathbf{pH} - \log \left(\frac{c + D}{c - D}\right) - \log \left[1 + \frac{\mathbf{D}}{\sqrt{y_1}\sqrt{c^2 - D^2}} - \frac{y\sqrt{y_1}}{\gamma^3}\left(\frac{2c + D}{c + D}\right)\sqrt{\frac{c - D}{c + D}}\right] - 8 \log \gamma. \quad (111)$$

In Eqs. (103), (109), and (111) y and y_1 represent ratios of activity constants.

E. Ampholytes

1. Simple Ampholyte.

a. Expressions for individual constants. For a simple ampholyte, with strong acid and strong base present at concentrations such that $a_s - b_s = g$, which may be positive or negative,

$$D = c(K_aOH - K_bH)/(K_aOH + K_bH + W) + g. \quad (112)$$

Then

$$W/K_b = H(c - g + D)/[(g - D) + (c + g - D)K_a/H]; \quad (113)$$

$$K_a = [H(- g + D) + (c - g + D)H^2K_b/W]/(c + g - D). \quad (114)$$

These formulas have been arranged so as to be comparable with Eqs. (37) and (38) respectively; when the hydrochloride of such an ampholyte at concentration $c = a_s$ is titrated with strong base, then $g = c - b_s$ and the formulas become identical, respectively, with Eqs. (37) and (38), K_a being identified with K_2^* and W/K_b with K_1^* of the hydrochloride considered as a dibasic acid. But we may equally well consider the sodium salt of the ampholyte as a diacid base with base constants $K_1'^* = W/K_a$ and $K_2'^* = K_b$.

Then for titration of the sodium salt (at $c = b_s$) with strong acid, we may write

$$K_1'^* = W/K_a = OH(c - g' + D')/[(g' - D') + (c + g' - D')K_2'^*/OH]; \tag{115}$$

$$K_2'^* = K_b = [OH(-g' + D') + (c - g' + D')(OH)^2/K_1'^*]/(c + g' - D'). \tag{116}$$

These are the "base analogs" of Eqs. (37) and (38) respectively, with $D' = -D = OH - H$, and $g' = -g = b_s - a_s$. When the sodium salt ($b_s = c$) is titrated with strong acid, $g' = c - a_s$ so that $c - g' = a_s$, etc.

Hence the determination of the constants of a 1 : 1 ampholyte is identical with the determination of the constants of a dibasic acid (in which the ampholyte corresponds to the alkali acid salt of the acid and in which K_a corresponds to K_2^* and W/K_b to K_1^* of the hydrochloride of the ampholyte as the dibasic acid)—*or* with the determination of the constants of a diacid base (in which the ampholyte corresponds to the basic salt of the base with strong acid and in which $K_b = K_2'^*$ and $W/K_a = K_1'^*$ of the sodium salt of the ampholyte as the diacid base).

For the introduction of activity coefficients into Eqs. (113) and (114), we must consider the question of the possible existence of both molecular and dipolar-ion forms of the ampholyte, besides the simple ions, U^+ and U^-. If $[U^\pm] = 0$, and if $\gamma_{U^0} = 1$, as usual, then $\mathbf{K}_2^* = \mathbf{K}_a = K_a\gamma^2$ and $\mathbf{K}_1^* = \mathbf{W}/\mathbf{K}_b = W/K_b$, whereas for the constants of a true dibasic acid we have $\mathbf{K}_2 = K_2\gamma^4$ and $\mathbf{K}_1 = K_1\gamma^2$. If, on the other hand, $[U^0] = 0$, then $\mathbf{K}_a = K_a\gamma^2/\gamma_{U\pm}$ and $\mathbf{W}/\mathbf{K}_b = (W/K_b)\gamma_{U\pm}$; and to take $\gamma_{U\pm}$ as unity, as for the uncharged molecule, is certainly questionable. If both forms have to be considered present, the consequent difficulty is discussed in Chapter VI.

If, then, we assume the species $U^\pm$ to be absent, we have from Eq. (113) for the titration of the ampholyte with strong acid,

$$\mathrm{p}\mathbf{W} - \mathrm{p}\mathbf{K}_b = \mathrm{p}\mathbf{H} - \log\left(\frac{c - a_s + D}{a_s - D}\right) + \log\left[1 + \frac{\mathbf{K}_a\,(c + a_s - D)}{\gamma\mathbf{H}(a_s - D)}\right] + \log\gamma, \tag{117}$$

while for titration with strong base, Eq. (114) becomes

$$\mathrm{p}\mathbf{K}_a = \mathrm{p}\mathbf{H} - \log\left(\frac{b_s + D}{c - b_s - D}\right) - \log\left[1 + \frac{\mathbf{H}\mathbf{K}_b\,(c + b_s + D)}{\gamma\mathbf{W}(b_s + D)}\right] - \log\gamma. \tag{118}$$

Except for the activity coefficients, Eq. (117) may be identified with Eq. (44) for $\mathbf{K}_1$ of a dibasic acid, with a_s of Eq. (117) in place $a - b_s$ of Eq. (44). Similarly Eq. (118), giving $\mathbf{K}_2^*$, is identical with Eq. (45), with b_s in place of $b_s - a$.

If we assume the species U^0 to be absent, the last two terms become, for Eq. (117)

$$\cdots + \log\left[1 + \frac{\gamma_{U\pm}\mathbf{K}_a\,(c + a_s - D)}{\gamma\mathbf{H}(a_s - D)}\right] + \log\gamma - \log\gamma_{U\pm}, \qquad (119)$$

and for Eq. (118),

$$\cdots - \log\left[1 + \frac{\gamma_{U\pm}\mathbf{H}\mathbf{K}_b\,(c + b_s - D)}{\gamma\mathbf{W}(b_s + D)}\right] - \log\gamma + \log\gamma_{U\pm}. \qquad (120)$$

For the use of these expressions, we note that if the pure ampholyte solution at concentration c is titrated with strong acid, $H \cong W/K_b$ at $a_s\ (= g) = c/2$; when it is titrated with strong base, $H \cong K_a$ at $b_s\ (= -g) = c/2$. (This involves the important point that $K_aK_b < W$ for all known ampholytes, a relation which will be discussed in Chapter XIV.) At these ranges of g, then, Eqs. (113) and (114), or (117) and (118), or (119) and (120), are useful as simultaneous equations for the determination of the constants.

b. Ratio and product of constants. The product K_aK_b corresponds to WK_2^*/K_1^* of the dibasic acid, or to Wy^*. Hence the ampholyte relations may be expressed in terms of the product K_aK_b and one of the individual constants as the two parameters. The resulting expressions may be written directly from those of a dibasic acid or Eqs. (54–57) by appropriate substitution. For example, in terms of K_a and $y'\ (= K_aK_b = Wy^*)$,

$$K_a = \frac{H(b_s + D)}{2(c - b_s - D)} + \sqrt{(\ \ )^2 + \frac{y'H^2}{W}\left(\frac{c + b_s + D}{c - b_s - D}\right)}, \qquad (121)$$

$$y' = W\frac{K_a}{H}\left(\frac{a_s - D}{c - a_s + D}\right)\left[1 + \frac{K_a}{H}\left(\frac{c + a_s - D}{a_s - D}\right)\right]. \qquad (122)$$

Eq. (121) is used at $b_s \cong c/2$, when $H \cong K_a$, and Eq. (122) at $a_s \cong c/2$, when $H \cong W/K_b$ and $y' \cong WK_a/H$.

Finally, the ratio K_aK_b corresponds to $K_1^*K_2^*/W$ or $\mathbf{p}^*/W$. Now the relations may be considered in terms of the ratio K_a/K_b and one of the individual constants as the two parameters. In terms of K_a and $\mathbf{p}'\ (= K_a/K_b = \mathbf{p}^*/W)$, for example, we may write from Eqs. (61) and (62),

$$\frac{1}{K_a} = \frac{(c - b_s - D)}{H(b_s + D)}\left[1 - \frac{H^2}{\mathbf{p}'W}\left(\frac{c + b_s + D}{c - b_s - D}\right)\right], \qquad (123)$$

$$\frac{1}{\mathbf{p}'} = \frac{W}{H^2}\left(\frac{c - b_s - D}{c + b_s + D}\right)\left[1 - \frac{\mathbf{H}}{K_a}\left(\frac{b_s + D}{c - b_s - D}\right)\right]. \qquad (124)$$

Again Eq. (123) is used near $b_s = c/2$, but Eq. (124), for $\mathbf{p}'$, is used in connection with the solution of the pure ampholyte, with $b_s = 0$, when

$H \cong \sqrt{WK_a/K_b}$, and hence $\mathbf{p}' \cong H^2/W$. With $b_s = 0$, Eq. (124) becomes, as in Eq. (63),

$$\frac{1}{\mathbf{p}'} = \frac{W}{H^2}\left(\frac{c-D}{c+D}\right)\left[1 - \frac{H}{K_a}\left(\frac{D}{c-D}\right)\right]. \tag{125}$$

Hence with an approximate value of K_a, Eq. (125) may be used for the accurate determination of $\mathbf{p}'$, in expressions similar to Eq. (68). If $[U^{\pm}] = 0$, then

$$-\log \mathbf{K}_a/\mathbf{K}_b = 2\mathbf{pH} - \mathbf{pW} - \log [(c+D)/(c-D)] + \log [1 - \mathbf{HD}/\mathbf{K}_a\,(c-D)]; \tag{126}$$

if $[U^0] = 0$, the expression remains unchanged except for the substitution of $\gamma_{U\pm}\mathbf{K}_a$ in place of simply $\mathbf{K}_a$ in the correction term.

In Eqs. (113–126), c is the concentration of the ampholyte being titrated either with strong base, b_s, or with strong acid, a_s. If the hydrochloride, at concentration c, is being titrated with strong base, then, as already mentioned, $g = c - b_s$ in Eqs. (113) and (114), Eqs. (118), (120), (121), and (123) are used with $b_s - c$ in place of b_s, and Eqs. (117), (119), (122), and (124) with $c - b_s$ in place of a_s. The pure ampholyte now represents the first equivalent point in such titration.

If the activity coefficient to be used in Eqs. (117), (118), and (121–126) is to be calculated from the concentrations in the usual way, the species $U^{\pm}$ must be assumed either to be absent or not to contribute to the ionic strength. Then

$$\mu = H + b_s + c\alpha_+ = OH + a_s + c\alpha_-. \tag{127}$$

These expressions then hold whether or not the solution contains *both* a_s and b_s, in equal or unequal concentrations.

2. Higher Order Ampholyte. The constants of a higher order ampholyte are similarly determined, usually as the successive constants of the hydrochloride of the ampholyte considered as a polybasic acid. Histidine, for example, forms a di-hydrochloride which, when titrated with strong base, shows evidence of three successive equivalent points. The di-hydrochloride is therefore mathematically a tribasic acid and the histidine itself is equivalent to its secondary alkali salt. The histidine may therefore be said to have two base constants, $K_{b_1} > K_{b_2}$, and one acid constant, K_a. If these are transformed to successive acid constants of a tribasic acid, with $K_1^* = W/K_{b_2}$, $K_2^* = W/K_{b_1}$ and $K_3^* = K_a$, then Eq. (86) describes the solution of the di-hydrochloride at concentration a, together with strong base, b_s; and Eqs. (87–89) would be used for the determination of the constants in the titration of the di-hydrochloride with strong base. In general, if the highest hydrochloride of such an ampholyte, U, has the formula U·m HCl, then there are m "base constants" referred to the substance U itself; if

$m = 3$, the observed successive acid constants of the tri-hydrochloride would have the following interpretation as constants of the substance U as a "higher order ampholyte": $K_1^* = W/K_{b_3}$, $K_2^* = W/K_{b_2}$, $K_3^* = W/K_{b_1}$, $K_4^* = K_{a_1}$, $K_5^* = K_{a_2}$, etc.

F. Salt of Weak Acid and Weak Base

1. Titration with Strong Acid or Strong Base. The solution of the 1 : 1 salt MX, of weak acid and weak base, is mathematically identical with a solution of two weak acids or two weak bases at equal concentrations at the first equivalent point with strong base or strong acid, respectively Hence, titration with strong base and with strong acid should theoretically make possible the determination of both ionization constants. With $g = a_s - b_s$ as before,

$$D = H - OH = c[A/(A + H) - B/(B + OH)] + g, \tag{128}$$

so that

$$A_1^* = A = \frac{H(c - g + D) - (g - D)\,W/B}{(g - D) + (c + g - D)\,W/HB}, \tag{129}$$

$$A_2^* = \frac{W}{B} = \frac{(c - g + D)H^2/A - H(g - D)}{(c + g - D) + (g - D)H/A}. \tag{130}$$

The constants of the "two acids" would be $A_1^* = A$ and $A_2^* = W/B$ respectively; and if $AB > W$, then $A_1^* > A_2^*$. The question of the relation between the ratio AB/W and the possibility of determining the constants by titration will be discussed in Chapter XIII.

Each of these equations may be used to determine one of the constants, if the other is known, through measurements of $\mathbf{H}$, c, and g at suitable ranges of values of the experimental variables c and g. In titration with strong acid, for example, we have $H \cong A$ at $g = a_s \cong + c/2$, if $AB > W$; then if $\mathbf{B}$ is known even approximately, Eq. (129) may be used for the determination of $\mathbf{A}$ in the form

$$\mathrm{p}\mathbf{A} = \mathrm{p}\mathbf{H} - \log\left(\frac{c - g + D}{g - D}\right) - \log\left[\frac{1 - \dfrac{\gamma\mathbf{W}}{\mathbf{HB}}\left(\dfrac{g - D}{c - g + D}\right)}{1 + \dfrac{\gamma\mathbf{W}}{\mathbf{HB}}\left(\dfrac{c + g - D}{g - D}\right)}\right] - \log\gamma. \tag{131}$$

As usual, $D = \mathbf{D}/\gamma$. As long as $AB > W$, then with $g = c/2$ and $H \cong A$ the second and third terms will be small correction terms, and the extrapolation of the sum of the first three terms, which consist of experimental observations, to zero ionic strength should give p**A**. In titration with strong base $H \cong W/B$ at $g = -b_s \cong -c/2$, if $AB > W$; then an approximate value of **A** may be used for the determination of **B** through Eq. (130) in the form

$$\mathrm{p}\mathbf{B} = \mathrm{p}\mathbf{W} - \mathrm{p}\mathbf{H} + \log\left(\frac{-g + D}{c + g - D}\right) - \log\left[\frac{1 + \dfrac{\gamma\mathbf{H}}{\mathbf{A}}\left(\dfrac{g - D}{c + g - D}\right)}{1 - \dfrac{\gamma\mathbf{H}}{\mathbf{A}}\left(\dfrac{c - g + D}{g - D}\right)}\right] - \log \gamma. \quad (132)$$

Again as long as $AB > W$ the second and third terms will be small correction terms when $g = -c/2$ and $H \cong W/B$.

If the salt is that of a strong base and weak acid, M_sX, at concentration c, the constant A of the weak acid may be determined by titration with strong acid, a_s. The problem is now mathematically that of the titration of a weak base with ionization constant $K_b^* = W/A$, with strong acid. From the base analog of Eq. (16), or

$$K_b^* = OH(a_s + D')/(c - a_s - D'), \quad (133)$$

we then have

$$A = H(c - a_s + D)/(a_s - D). \quad (134)$$

2. The Pure Salt Solution. If $g = 0$, Eqs. (129) and (130) allow the calculation of one of the ionization constants, if the other is known, from the value of **H** in the pure salt solution. For known **A**, for example, Eq. (130) gives

$$\mathrm{p}\mathbf{B} = \mathrm{p}\mathbf{W} + \mathrm{p}\mathbf{A} - 2\mathrm{p}\mathbf{H} + \log\left(\frac{c + \mathbf{D}/\gamma}{c - \mathbf{D}/\gamma}\right) + \log\left[\frac{1 + \dfrac{\mathbf{AD}}{\gamma^2\mathbf{H}(c + \mathbf{D}/\gamma)}}{1 - \dfrac{\mathbf{HD}}{\mathbf{A}(c - \mathbf{D}/\gamma)}}\right]. \quad (135)$$

Since $H \cong \sqrt{WA/B}$ for appreciable values of c, the last two terms will in general be low.

The mixed salt of a dibasic acid strong in K_1, with an equivalent of strong base and an equivalent of weak base or the salt of type "$NaNH_4SO_4$" at the molar concentration c, is mathematically identical, except for the activity coefficients involved, with the simple 1 : 1 salt here considered. The term p**A** becomes $\mathrm{p}\mathbf{K}_2 + 2 \log \gamma$, and the quantity **A** in the last term of Eq. (135) becomes $\mathbf{K}_2/\gamma^2$. As usual, "γ" represents the activity coefficient of a univalent ion, to be evaluated according to the expressions in Eqs. (31b) and (31c).

For the salt of weak base and strong acid, with $A = \infty$, Eq. (130), reducing to the base analog of Eq. (21), leads to the base analog of Eq. (34), or

$$\mathbf{pB} \cong \mathbf{pW} - 2\mathbf{pH} - \log (c - \mathbf{H}/\gamma) - 2 \log \gamma. \qquad (136)$$

On the other hand, if the acid is extremely weak so that $A \stackrel{\rightarrow}{=} 0$, the solution is essentially a pure base at concentration $c = b$, and Eq. (130) reduces to the base analog of Eq. (16), leading to the base analog of Eq. (32), or

$$\mathbf{pB} = 2\mathbf{pW} - 2\mathbf{pH} + \log (b - \mathbf{OH}/\gamma). \qquad (137)$$

VI

Interpretation of Ionization Constants

An ionization constant, K_1, defined and determined as a function of H, W, and concentrations of solutes, is an over-all equilibrium constant, or an *apparent* ionization constant, covering all the individual specific effects comprising the total phenomenon of ionization in aqueous solution. These specific processes include the possible hydration of the solute, tautomerizations both of uncharged solute and of solute ions, and actual ionization processes of the individual species (charged and uncharged) in solution. For simplicity, in accordance with the limitation set in Chapter II, we shall not consider the association and polymerization of solute species in any way, giving rise to complex species, either homo- or hetero-polymeric in nature; that is, we shall restrict ourselves as usual to the relations in what we have defined as "noncomplex" solutions.

In such solutions it is found that whatever may be the multiplicity of the individual processes involved in ionization, the basicity of an acid is, as defined in Chapter I, always simply the valence of its most highly charged ion. Consequently the number of "apparent ionization constants" (constant parameters other than W) required for the mathematical description of the relations between H, W, and the concentration is (in these noncomplex cases) always simply the "basicity" of the acid thus defined.

We shall here consider the question of the possible analysis or interpretation of these apparent or over-all ionization constants into individual equilibrium constants of the several specific processes making up the total effect of ionization. For the purpose of such analysis of the constants it will in general be necessary to introduce measurements (and interpretations of these measurements) based on colorimetry, spectrophotometry, determination of dipole moments, etc. But it must be observed that this "resolution" of the constants is not concerned in any way with the question of the mechanism of the *specific process* of ionization itself on the part of any particular species. The thermodynamic definition of ionization constants (over-all or specific) and the thermodynamic methods for their determination can not decide among the various possible mechanisms or explanations for the electrolytic properties of these solutions.[1]

(1) cf. J. E. Ricci, *J. Am. Chem. Soc.*, **70,** 109 (1948).

A. Monobasic Acid

1. Two Simple Schemes. Only if the original solute species, X^0, undergoes no rearrangement and no hydration before becoming ionic by reaction with the water, and only if the anion does not itself suffer any rearrangement, does the ionization constant pertain to a single specific process. But if any such accessory reactions are involved, many combinations are possible, of which we shall consider two simple schemes:

Scheme (1):

$$X^0 \overset{k_T}{\rightleftarrows} X_T^0 \overset{k_i}{\rightleftarrows} X_T^- \overset{k_T^-}{\rightleftarrows} X_T'^- \qquad (1)$$

In this k_T is an equilibrium constant for either a rearrangement or a hydration of X^0 to X_T^0, k_T^- applies to a rearrangement of the anion, and k_i is a "true" ionization constant for the species X_T^0.

Scheme (2):

$$\begin{array}{ccc} X^0 & \overset{k_i}{\rightleftarrows} & X^- \\ k_T \downarrow\uparrow & & \\ X_T^0 & \overset{k_{i_T}}{\rightleftarrows} & X_T^- \end{array} \qquad (2)$$

2. Relation of Specific Constants to Apparent Ionization Constant of the Acid. In either scheme, D depends in the usual way simply on the concentration a of the acid and on a single over-all or apparent ionization constant, so that $D = H - OH = aK/(K + H)$ always, as for any monobasic acid. In Scheme (1), however, the meaning of K is easily shown to be

$$K = H([X_T^-] + [X_T'^-])/([X^0] + [X_T^0]) = k_i k_T(1 + k_T^-)/(1 + k_T), \qquad (3)$$

with symbols referring to Eq. (1). If k_T is very large (the solute completely hydrated, for example), $K \cong k_i(1 + k_T^-)$, and then $K \cong k_i$ if in addition $k_T^- \cong 0$. If k_T is very small, the acid "appears" to be weak, but $K \cong k_i k_T$ (if k_T^- is still assumed to be approximately 0). But even for constant k_T an acid may "appear" strong if k_T^- is large, even though k_i is small.

In Scheme (2), with symbols from Eq. (2),

$$K = H([X^-] + [X_T^-])/([X^0] + [X_T^0]) = (k_i + k_{i_T} k_T)/(1 + k_T). \qquad (4)$$

With $k_i = 0$, this reduces to

$$K = H[X_T^-]/([X^0] + [X_T^0]) = k_T k_{i_T}/(1 + k_T). \qquad (5)$$

Such a scheme would apply in the interpretation of the behavior of certain acids and bases as depending on an equilibrium between a "pseudo form" assumed to be nonionizing and a "true" or ionizing form as, possibly, in the case of CO_2, and for indicators according to Kolthoff's "theory of indicators."[2] Thus, if the color of a one-color indicator is that of the ionizing form, then

$$r_{\mathrm{col}} = ([\mathrm{X}_T^0] + [\mathrm{X}_T^-])/[\mathrm{X}^0] = k_T(1 + k_{i_T}/H). \quad (6)$$

The scheme would also apply if the formation of the ionizing form from the nonionizing form is not a rearrangement but a solvation, as has been suggested both for CO_2 and for NH_3. In any of these cases the apparent ionization constant may be small despite a high specific ionization constant k_{i_T} if the equilibrium constant for the rearrangement or for the solvation, k_T, is small.

3. Relation of Mass Constant to Thermodynamic Constant. Concerning Eq. (3) we note that an apparent or over-all ionization constant is essentially, by its very definition, a *mass constant.* It can not be given a primary definition in terms of activities because it involves the *sum* of concentrations of species. While concentrations may be added, there is strictly no meaning in the addition of activities. We are accustomed to express all ionization constants by the homogeneous relation

$$\log \mathbf{K} = \log K + \log f(\gamma), \quad (7)$$

in which K involves concentrations of individual species, but not the general ionic strength, the effect of which is supposed to appear only in the second term. But this is evidently true, strictly, only if K is a simple constant, like one of the k's of this chapter, referring to a single individual process. If we introduce activity constants ($\mathbf{k}$'s) and activity coefficients for the mass constants (k's) in Eq. (3), we may write

$$K = (\mathbf{k}_i \mathbf{k}_T \gamma_{\mathrm{X}^0}/\gamma_{\mathrm{H}^+}\gamma_{\mathrm{X}_T^-})(1 + \mathbf{k}_T^- \gamma_{\mathrm{X}_T^-}/\gamma_{\mathrm{X}'^-_T})/(1 + \mathbf{k}_T \gamma_{\mathrm{X}^0}/\gamma_{\mathrm{X}_T^0}), \quad (8)$$

and this can not be put into the form of Eq. (7), or the familiar type of relation,

$$\log \mathbf{K} = \log K + \log (\gamma_{\mathrm{H}^+}\gamma_{\mathrm{X}^-}/\gamma_{\mathrm{X}^0}), \quad (9)$$

implied in the usual extrapolation of data to infinite dilution. If the numerator and denominator of Eq. (3) are each uniform in respect to the type of ion involved, the difficulty may be considered negligible on the basis of the γ convention assumed in Chapter v, in which one assumes for Eq. (8), for example, that the ratios $\gamma_{\mathrm{X}_T^-}/\gamma_{\mathrm{X}'^-_T}$ and $\gamma_{\mathrm{X}^0}/\gamma_{\mathrm{X}_T^0}$ both have the value 1. But in the case of an ampholyte, the denominator in the definition of K, as in Eqs. (3) and (4), may be the sum of molecular and dipolar-ion forms, as will be discussed below, and then the effect of ionic strength is not simple.

4. Problem of Determination of Specific Constants. The problem of evaluating the individual constants which together determine the apparent

(2) I. M. Kolthoff and C. Rosenblum, *Acid-Base Indicators*, Macmillan, N.Y., 1937; p. 231.

ionization constant is essentially that of determining, as functions of H, various ratios of the species involved other than that ratio involved in the definition of K, such as in Eqs. (3) and (4). The ratio involved, for example, in the m'th ionization constant of a polybasic acid is, in terms of the function r of Eq. III(41),

$$r_m = K_m/H = \Sigma[\mathrm{X}^{-m}]/\Sigma[\mathrm{X}^{-(m-1)}], \tag{10}$$

or an "apparent ionization ratio." Thus for a monobasic acid, as in Eqs. (3) and (4),

$$r_1 = K_1/H = \Sigma[\mathrm{X}^-]/\Sigma[\mathrm{X}^0], \tag{11}$$

whatever the actual nature and number of the processes involved. If it is then possible, by colorimetry, spectrophotometry, etc., to determine, in this case, any two other ratios, the three individual constants (of either of the above simple schemes) could be evaluated: thus, for Scheme (1), ratios such as $r_a = ([\mathrm{X}_T^\circ] + [\mathrm{X}_T^-] + [\mathrm{X'}_T^-])/[\mathrm{X}^0]$; $r_b = [\mathrm{X'}_T^-]/([\mathrm{X}^0] + [\mathrm{X}_T^-] + [\mathrm{X}_T^-])$; $r_c = [\mathrm{X}_T^\circ]/([\mathrm{X}^0] + [\mathrm{X}_T^-] + [\mathrm{X'}_T^-])$; etc., in which, for example, $r_a = k_T [1 + k_i(1 + k_T^-)/H]$ and $r_b = k_i k_T k_T^-/(k_i k_T + Hk_T + H)$. From three such ratios, for example r_1, r_a and r_b, the individual constants could be obtained. But if the colorimetric determination gives the same ratio as that involved in defining K, Eqs. (10) and (11), it adds nothing to the data.

B. Dibasic Acid

For a dibasic acid the possibilities are much more numerous, but it is important to consider at least a few because of the interest in the possibility of "assigning" the two constants of a dibasic acid to various "groups" or points in the molecule.

1. Simplest Case. In the simplest case, in which it is assumed that the unionized species is either not hydrated at all or completely hydrated and that there is no rearrangement of any kind for either molecule or ions, there are nevertheless always two primary processes to be considered since there are in general two different possible univalent anions, although only one divalent anion, to be formed:

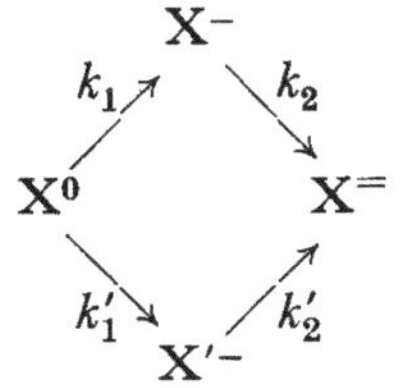

$$
\begin{array}{ccc}
 & \mathrm{X}^- & \\
k_1 \nearrow & & \searrow k_2 \\
\mathrm{X}^0 & & \mathrm{X}^= \\
k_1' \searrow & & \nearrow k_2' \\
 & \mathrm{X'}^- &
\end{array}
\tag{12}
$$

If there are two ionizable points or groups in the molecule, X^- and $\mathrm{X'}^-$ are the two different univalent ions formed from the uncharged molecule by processes described by the specific ionization constants k_1 and k_1' respectively.

The process described by k_1 can then occur upon the anion X'^-, forming the divalent anion $X^=$; the equilibrium involved is described by the constant k_2'. It is clear that $k_2' < k_1$ since k_2' refers to the ionization of the same point or group as involved in k_1, but upon an already charged species. Similarly when the process described by k_1' occurs upon the anion X^- forming the divalent ion $X^=$, the constant applying is k_2 and k_2 must be $< k_1'$. The individual constants therefore have the following meanings:

$$k_1 = H[X^-]/[X^0]; \qquad k_2' = H[X^=]/[X'^-];$$
$$k_1 = H[X'^-]/[X^0]; \qquad k_2 = H[X^=]/[X^-]. \tag{13}$$

Hence,

$$k_1'/k_1 = k_2/k_2' = x, \tag{14}$$

and

$$k_2'/k_1 = k_2/k_1' = x_1. \tag{15}$$

While x must simply be greater than 0, x_1 must be less than 1 but still positive. The ratio x is that of the "strengths" of the two "acid-groups" or acidic processes of the dibasic acid, and it follows from the definitions of the k's that this ratio is the same whether or not the ionizing species already has a charge upon it; x equals 1 if the two groups are identical, as presumably in sulfuric acid, hydrogen sulfide, and oxalic acid. The ratio x_1 represents the effect of charge (presumably at one point) on the ionization at another point, and the effect, according to the definitions of the k's, is the same on either primary process; x_1 equals 1 if there is no such effect or if the ionization is the same both upon charged and upon uncharged species.

With

$$a = [X^0] + [X^-] + [X'^-] + [X^=], \tag{16}$$

and

$$D = H - OH = [X^-] + [X'^-] + 2[X^=], \tag{17}$$

$$D = a(1 + x + 2xx_1k_1/H)/(1 + x + H/k_1 + xx_1k_1/H). \tag{18}$$

If we now define "apparent ionization constants" as follows,

$$K_1 = H([X^-] + [X'^-])/[X^0] = k_1(1 + x), \tag{19}$$

$$K_2 = H[X^=]/([X^-] + [X'^-]) = xx_1k_1/(1 + x), \tag{20}$$

Eq. (18) becomes the usual expression for a dibasic acid, involving only two "ionization constants" regardless of the complexity of the actual process:

$$D = a(1 + 2K_2/H)/(1 + H/K_1 + K_2/H). \tag{21}$$

It follows, then, that the apparent constants do not pertain to actual or individual ionization processes and that their ratio is not necessarily the ratio of the constants for two individual processes. The ratio of apparent constants is

$$K_1/K_2 = (1 + x)^2/xx_1. \tag{22}$$

The ratios x and x_1 can not be obtained from K_1 and K_2 alone, then, except by special assumptions or in limiting cases. From Eq. (22),

$$d(K_1/K_2)/dx = (x^2 - 1)/x^2x_1, \tag{23}$$

and K_1/K_2 is a minimum when $x = 1$, i.e. when the two "groups" of the dibasic acid are identical, whatever the values of x_1. Since x_1 must be less than 1, then the minimum observable ratio of K_1/K_2, in this simplest possibility for a dibasic acid, is seen from Eq. (22) to be 4. If (K_1/K_2) is greater than 4, therefore, the simple Scheme, Eq. (12), may or may not apply; but if (K_1/K_2) is less than 4, the total ionization scheme must be more complicated than what has here been assumed.

In the limiting case, in which both x and x_1 equal 1 (and hence $K_1/K_2 = 4$) or when the two processes are both equivalent and independent, the result should be mathematically the case of two independent monobasic acids of equal ionization constant, at equal concentrations, or (its equivalent) the case of a single monobasic acid at double the concentration of the dibasic acid and with ionization constant equal to $\sqrt{K_1K_2}$ and hence to $K_1/2$. Eq. (21) does in fact reduce to

$$D = (2a)(K_1/2)/(K_1/2 + H), \tag{24}$$

when $K_1/K_2 = 4$. The significance of this is that the titration curve of such a dibasic acid (with $K_1/K_2 = 4$) is identical, in every respect, with that of a monobasic acid of half its molecular weight, and with ionization constant $K_1/2$. If $x = 1$ and $0 < x_1 < 1$, the curve is still distinctive of a dibasic acid; now $x_1 = 4K_2/K_1$, $k_1 = k_1' = K_1/2$ and $k_2 = k_2' = 2K_2$. If $x_1 = 1$ and $0 < x < 1$, the acid is mathematically a mixture, at equal concentrations, of two independent and different acids, with ionization constants k_1 and xk_1. It may easily be shown that Eq. (18), with $x_1 = 1$, is equivalent to $D = a[k_1/(k_1 + H) + xk_1/(xk_1 + H)]$, which is the equation of two independent acids each at the concentration a.

With $K_1 \gg K_2$, the pH titration curve of a dibasic acid being titrated with strong base shows an inflection near the equivalent point $b_s = a$ and near $H = \sqrt{K_1K_2}$. It will be shown later that the appearance of this inflection requires $K_1 > 16K_2$, and we have here seen that when $K_1 = 4K_2$

the curve is indistinguishable from that of a monobasic acid at the concentration $2a$. The shape of the titration curve as a function of the ratio K_2/K_1 is discussed in Chapter XIV.[3]

2. Other Simple Possibilities. Variations[4] arise if there is tautomerism of the acid in any of its forms (uncharged, univalent, or divalent ion). For example:

Scheme a.

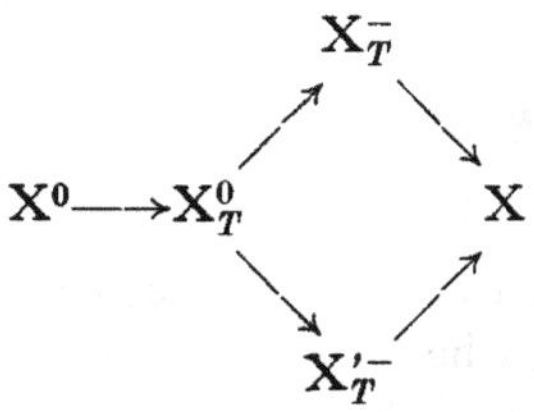

(25)

Scheme b.

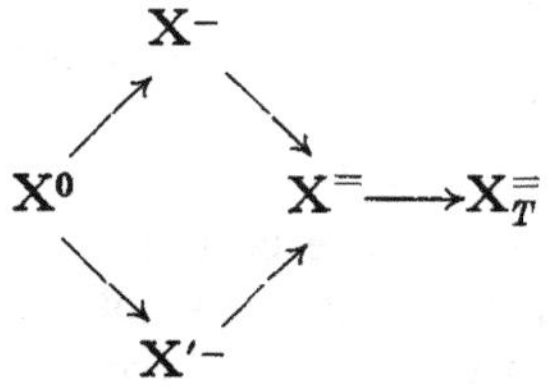

(26)

Scheme c.

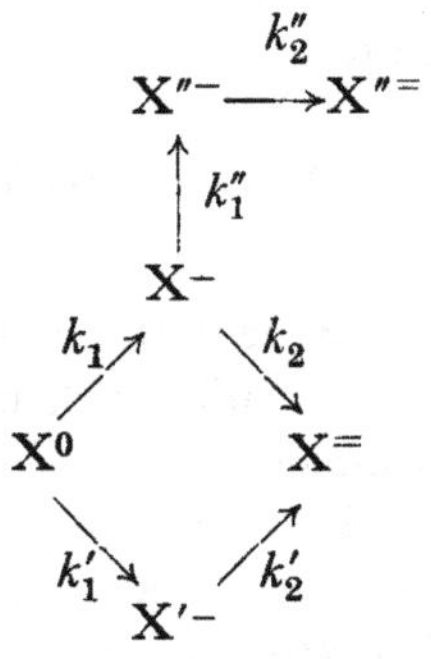

(27)

(3) For similar considerations on both dibasic and tribasic acids, see N. Bjerrum, *Z. physik. Chem.*, **106,** 219 (1923); also, J. Greenspan, *Chem. Rev.*, **12,** 339 (1933).

(4) For the scheme

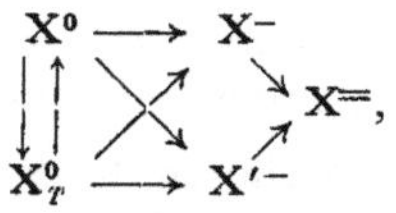

see G. Schwarzenbach and A. Willi, *Helv. Chim. Acta,* **30,** 1303 (1947).

Scheme d.

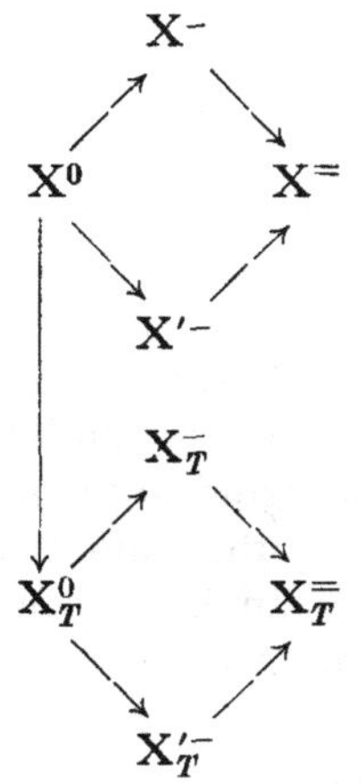

(28)

Scheme e. All these in combination:

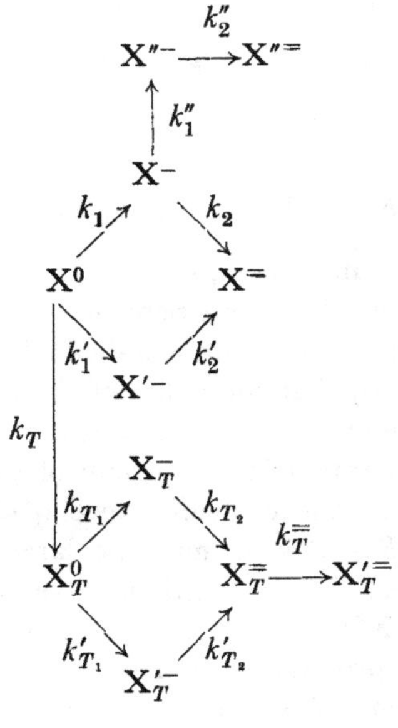

(29)

Here

$$k_T = [X_T^0]/[X^0];$$

k_1, k_1', k_2, k_2' as in Eq. (13);

$$k_1'' = [X''^-]/[X^-];$$

$$k_2'' = H[X''^=]/[X''^-];$$

$$k_{T_1} = H[X_T^-]/[X_T^0]; \quad k'_{T_1} = H[X_T'^-]/[X_T^0];$$

$$k_{T_2} = H[X_T^=]/[X_T^-]; \quad k'_{T_2} = H[X_T^=]/[X_T'^-];$$

$$k_T^= = [X_T'^=]/[X_T^=]. \qquad (30)$$

Now if

$$K_1 = H\Sigma[X^-]/\Sigma[X^0] = [k_1 + k_1k_1'' + k_1' + k_T(k_{T_1} + k_{T_1}')]/(1 + k_T), \quad (31)$$

and

$$K_2 = H\Sigma[X^=]/\Sigma[X^-] = [k_1k_2 + k_1k_1''k_2'' + k_Tk_{T_1}k_{T_2}(1 + k_T^=)]/K_1(1 + k_T), \quad (32)$$

then the dibasic acid is again described by Eq. (21), as always.

3. Scheme (c), Tautomerization of a Univalent Ion. Scheme (c), or Eq. (27), is sometimes applied in the interpretation of the constants of certain dibasic acids, and the significance and limitations of this application will here be discussed. When the ratio of apparent ionization constants, K_1/K_2, is observed to be less than 4, it is sometimes assumed in explanation that one of the univalent anions (here X^-) rearranges to produce another univalent anion, X''^-, which is "stronger" as an acid than the original acid, X^0, itself. In Scheme (c) the apparent constants defined as in Eqs. (31) and (32) simplify to

$$K_1 = k_1(1 + x + k_1''), \quad (33)$$

$$K_2 = (k_2 + k_1''k_2'')/(1 + x + k_1''), \quad (34)$$

in which $x = k_1'/k_1 = k_2/k_2'$, as in Eq. (14). Now with x_1 defined as in Eq. (15),

$$K_1/K_2 = (1 + x + k_1'')^2/(xx_1 + k_1''k_2''/k_1). \quad (35)$$

If $k_1'' = 0$, there is no tautomerism, and the case simplifies to that considered under Scheme (12); but the simple Scheme (12) is impossible if $K_1/K_2 < 4$, when, therefore, some complication is indicated.

It is clear, of course, that the value of the ratio K_1/K_2 expressed as in Eq. (35) can not be used without special assumptions, or except in limiting cases, to determine the ratio of the constants of any two individual processes of the supposed scheme. Nevertheless, since $x_1 < 1$, Eq. (35) may be used to show that if $(K_1/K_2) < 4$, k_2'' must be larger than both k_1 and k_1'.[5] Hence if K_1/K_2 is observed to be less than 4, we may infer that if Scheme (c) applies, the new ion, X''^-, is "stronger as an acid" than the original X^0 in respect to either of its primary ionization constants, k_1 and k_1'.[6]

Such rearrangement of an ion, however, is not suspected unless the observed ratio, K_1/K_2, is found to be less than 4. Nevertheless the possibility of tautomerism of an ion is not ruled out even if this ratio is greater than 4. If, for example, $x = 1$, $x_1 = 10^{-2}$, $k_2''/k_1 = 10$ and if k_1'', the tautomerization constant is 10^3, the observed value of the ratio K_1/K_2 will be 100.4, from Eq. (35), with very little dependence upon x and x_1.

(5) For example, with $K_1/K_2 < 4$ and $x_1 < 1$, Eq. (35) requires $4k_1''k''/k_1 + 4x > (1 + x + k_1'')^2$. Hence $4k_1''k_2''/k_1 > (1 - x - k_1'')^2 + 4k_1''$, and $k_2''/k_1 > 1 + (1 - x - k_1'')^2/4k_1''$.

(6) See, for example, G. Schwarzenbach, E. Kampitsch, and R. Steiner, *Helv. Chim. Acta*, **28,** 828 (1945).

C. Tribasic Acid

For a tribasic acid, we shall consider only one very simple possibility: that there is no tautomerism of any species at all; that there are three different univalent anions formed by primary processes with the specific ionization constants k, xk, and $x'k$; that there are three different divalent anions, the three specific ionization constants being weakened by the uniform factor x_1, for the formation of the divalent ions; and that for the formation of the single trivalent anion, the primary ionization constants are weakened by the uniform factor x_2. In other words, the specific ionization constant k for the formation of the first univalent anion becomes x_1k as the result of the effect of a single charge on the molecule, and x_2k as the result of the effect of two charges; $0 < x_2 < x_1 < 1$; x and x' are greater than 0. Then, as for any tribasic acid,

$$D = a(1 + 2K_2/H + 3K_{23}/H^2)/(1 + H/K_1 + K_2/H + K_{23}/H^2), \quad (36)$$

with

$$K_1 = H\Sigma[X^-]/[X^0] = k(1 + x + x'), \quad (37)$$

$$K_2 = H\Sigma[X^=]/\Sigma[X^-] = x_1k(x + x' + xx')/(1 + x + x'), \quad (38)$$

$$K_3 = H[X^{\equiv}]/\Sigma[X^=] = xx'x_2k/(x + x' + xx'). \quad (39)$$

If $x' = 0$, $K_3 = 0$, and these expressions reduce to those of a dibasic acid, or Eqs. (19) and (20), applicable for similar assumptions. If $x = x' = 1$ (as in H_3PO_4), the minimum value of $K_1/K_2 = 3/x_1$ and the minimum value of $K_2/K_3 = 3x_1/x_2$. [Eq. (36) results, in form, even without the restrictions assumed.]

D. Simple Ampholyte

The case of a simple ampholyte, with two apparent ionization constants, is of course mathematically essentially the same problem as that in the case of a dibasic acid. We may consider the original, or primary, solute species to be either (1) a molecule of zero charge or (2) a dipolar ion of no net charge; or:

(1)

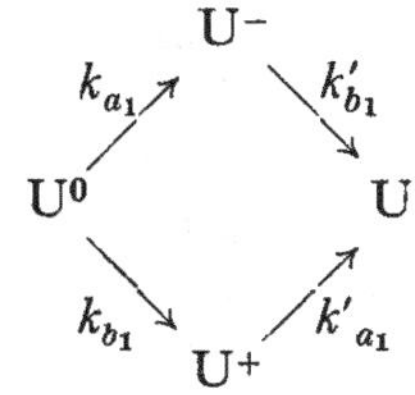

(40)

(2)

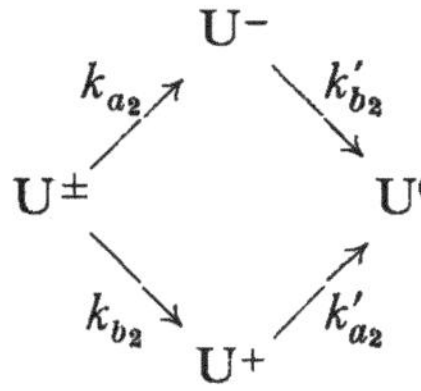

(41)

Analytically and mathematically the two schemes are indistinguishable. In either case the ratio of $[U^0]$ to $[U^{\pm}]$ is the same, a constant, independent of the concentration and of pH. In Scheme (1),

$$k_{a_1} = H[U^-]/[U^0], \qquad k'_{b_1} = OH[U^{\pm}]/[U^-],$$
$$k_{b_1} = OH[U^+]/[U^0], \qquad k'_{a_1} = H[U^{\pm}]/[U^+], \tag{42}$$

and

$$[U^0]/[U^{\pm}] = W/k_{a_1}k'_{b_1}. \tag{43}$$

In Scheme (2),

$$k_{a_2} = H[U^-]/[U^{\pm}], \qquad k'_{b_2} = OH[U^0]/[U^-],$$
$$k_{b_2} = OH[U^+]/[U^{\pm}], \qquad k'_{a_2} = H[U^0]/[U^+], \tag{44}$$

and

$$[U^0]/[U^{\pm}] = k_{a_2}k'_{b_2}/W, \tag{45}$$

which is the same as Eq. (43), since $k_{a_2} = W/k'_{b_1}$ and $k'_{b_2} = W/k_{a_1}$.

In terms of apparent ionization constants, we have for Scheme (1),

$$D = c(K_{a_1}OH - K_{b_1}H)/(K_{a_1}OH + K_{b_1}H + W), \tag{46}$$

with

$$K_{a_1} = k_{a_1}/(1 + k_{a_1}k'_{b_1}/W), \tag{47}$$

$$K_{b_1} = k_{b_1}/(1 + k_{a_1}k'_{b_1}/W). \tag{48}$$

Hence $K_{b_1}/K_{a_1} = k_{b_1}/k_{a_1}$, so that the ratio of the apparent ionization constants is here the same as the ratio of the specific primary ionization constants; but $K_{a_1} \neq k_{a_1}$ and $K_{b_1} \neq k_{b_1}$ unless $k'_{b_1} = 0$, or unless the only species present are U^0, U^-, and U^+. In Scheme (2),

$$D = c(K_{a_2}OH - K_{b_2}H)/(K_{a_2}OH + K_{b_2}H + W), \tag{49}$$

with

$$K_{a_2} = k_{a_2}/(1 + k_{a_2}k'_{b_2}/W), \tag{50}$$

$$K_{b_2} = k_{b_2}/(1 + k_{a_2}k'_{b_2}/W). \tag{51}$$

But again $K_{b_2} = k_{b_2}$ and $K_{a_2} = k_{a_2}$ only if $k'_{b_2} = 0$ or only if $[U^0] = 0$. Through the definitions of the k's, Eqs. (50) and (51) are identical with Eqs. (47) and (48). The relation still remains identical if the k's of Scheme (2), Eq. (41), are defined after the fashion of "hydrolysis constants," with $k_{a_2} = [U^-]/OH[U^{\pm}]$, $k'_{b_2} = [U^0]/H[U^-]$, etc.

With either of the ionization schemes considered, moreover, we would expect, according to Eq. (23), that the ratio W/K_aK_b can not be less than 4.[7]

If we introduce activity coefficients into either Eq. (47) or Eq. (50), we have of course an identical dependence upon ionic strength:

$$K_a = (\mathbf{k}_{a_1}\gamma_{U^0}/\gamma_{H^+}\gamma_{U^-})/(1 + \mathbf{k}_{a_1}\mathbf{k}'_{b_1}\gamma_{U^0}/\mathbf{W}\gamma_{U^{\pm}}); \tag{52}$$

(7) This limiting value for a simple ampholyte was pointed out by Bjerrum; N. Bjerrum, *Z. physik. Chem.*, **104**, 147 (1923).

and from Eq. (48) or (51),

$$K_b = (\mathbf{k}_{b_1}\gamma_{U^0}/\gamma_{OH^-}\gamma_{U^+})/(1 + \mathbf{k}_{a_1}\mathbf{k}'_{b_1}\gamma_{U^0}/\mathbf{W}\gamma_{U\pm}). \tag{53}$$

It is clear that neither of these expressions can be put into the form of Eq. (7) for extrapolation. Such simple extrapolation is possible only if we assume either U^0 or $U^{\pm}$ to be absent, or if we assume the ratio $\gamma_{U^0}/\gamma_{U\pm}$ to be constant.

Furthermore, it follows that the study of the mathematical relations involving the hydrogen ion concentration of ampholyte solutions, including dependence on ionic strength, can not distinguish between the two Schemes (1) and (2). The distinction may therefore be said to be meaningless. The ratio $[U^0]/[U^{\pm}]$ may be of interest and experimentally significant in other types of physicochemical experimentation, but the question whether the ampholyte is a "dipolar substance" to begin with, as in Scheme (2), or whether the dipolar species results only after ionization processes in solution, as in Scheme (1), is of no significance here because the distinction has no mathematical consequences.[8] Whether a substance is ionic or not before dissolving, in other words, is a question outside the scope of all the present considerations.

(8) A substance like glycine is usually considered to exist as an "inner salt," with "$U^{\pm}$" as the form with no net charge, but S. Kilpi [*Suomen Kemistilehti*, **19B,** 51 (1946), cited in *C.A.*, **41**, 5362 (1947)] concludes, on the basis of electrolyte effects, that it may be simply a "neutral molecule," in the form here called "U^0." In either case, however, the experiment may determine the ratio $[U^{\pm}]/[U^0]$ but can not distinguish between the two ionization schemes here discussed.

VII

Some Direct Applications of the General Equations as Relations of the Concentrations

A. Linearity in Respect to Concentrations

The exact equations defining H in solutions of various compositions illustrated in Chapter I, Section H, are in general of high degree in respect to H as the dependent variable, or as a function of analytical concentrations and ionization constants. Even in the case of simple strong acid or base, the equation is quadratic in H. If, however, the general equation, $H - OH = \Sigma c\beta$, is so considered that the concentrations are expressed as functions of H taken as the independent variable, the relation is found to be always simple and linear in respect to the concentrations, provided only that the solutions are "noncomplex" in the sense defined in Chapter I, Section F. Consequently, although only approximate values may be calculated for the hydrogen ion concentration fixed by the known composition (analytical concentrations and ionization constants) of a solution, we can find directly and exactly the values and relations of the concentrations corresponding to any specified value of H. If we consider, for example, the case of a solution containing a weak acid and a weak base, generally regarded as a rather difficult problem in "pH" and in titration curves, the equation is $H - OH = aA/(A + H) - bB/(B + OH)$ or

$$H^4B + H^3(bB + AB + W) + H^2[W(A - B) - AB(a - b)] - HW(aA + AB + W) - AW^2 = 0. \quad (1)$$

This expression will require considerable attention later for the purpose of calculating H as $f(a,b,A,B,W)$. But with b as $f(a,H,A,B,W)$, for example, the equation is linear with respect to the relations of a and b in its full and exact form,

$$b = aA(B + OH)/B(A + H) - (H - OH)(B + OH)/B. \quad (2)$$

It is consequently a simple matter to calculate corresponding values of b and a for any specified value of H, given the constants involved.

Eq. (2) is written in terms of concentrations and "mass constants." For its use in terms of activity constants we shall assume as usual that $\gamma_{X^0} = \gamma_{M^0} = 1$ and that $\gamma_+ = \gamma_- = \gamma$, and obtain

$$b = a\mathbf{A}(\mathbf{B} + \mathbf{W}/H)/\mathbf{B}(\mathbf{A} + \gamma^2H) - (H - \mathbf{W}/\gamma^2H)(\mathbf{B} + \mathbf{W}/H)/\mathbf{B}, \quad (3)$$

in which H stands for $[H^+]$, or

$$b = a\mathbf{A}(\mathbf{B} + \gamma\mathbf{W}/\mathbf{H})/\mathbf{B}(\mathbf{A} + \gamma\mathbf{H}) - \mathbf{D}(\mathbf{B} + \gamma\mathbf{W}/\mathbf{H})/\gamma\mathbf{B}, \qquad (4)$$

for a specified hydrogen ion activity **H**. For the solution of Eq. (3) for specified values of a and H, and on the assumption of the limiting law, γ may be calculated directly, by successive approximations, from

$$-\log\gamma = 0.505[\mathbf{W}/H\gamma^2 + a\mathbf{A}/(\mathbf{A} + H\gamma^2)]^{1/2}. \qquad (5)$$

For specified **H**, we have

$$-\log\gamma = 0.505[\mathbf{W}/\mathbf{H}\gamma + a\mathbf{A}/(\mathbf{A} + \mathbf{H}\gamma)]^{1/2}. \qquad (6)$$

B. The General Titration Curve of Acid Against Base

1. Explicit Calculation. It follows at once that we can, in general and exactly, subject only to the uncertainties in activity coefficients, calculate and plot any titration curve for any combination of acids and bases, not by considering the pH as a function of the percentage of titration taken as the independent variable, but by calculating the ratio b/a $(= n)$ and hence the percentage of titration as a function of H or of pH as the independent variable. If a solution containing j acids and j' bases is being titrated with a base, the concentration of which is represented by $b_{j'}$, Eq. 1(28) gives

$$b_{j'} = \frac{1}{\beta'_{j'}}[\Sigma_j(a\beta) - \Sigma_{j'-1}(b\beta') - (H - OH)]. \qquad (7)$$

Since the charge coefficients are functions of constants and H, the value of $b_{j'}$ for any specified value of H is readily calculated, thus yielding the complete curve. The slope of the curve at any point is the reciprocal of π_b, the buffer capacity, which again is readily estimated for specified H but not for specified concentrations (in the exact sense).

A direct and exact calculation, then, through equations such as (2) and (7) is always possible and ultimately available as a check on other more familiar methods of calculating titration curves. But it seems more desirable, for practical purposes, to try to obtain usable formulas giving H or pH, and hence the feasibility of titrations, as functions of the concentrations and constants. This explicit formulation, in terms of first and second approximations, will be left to later chapters.

2. Consideration of Volumes Involved. Eq. (7) allows the explicit and direct calculation of the *concentration* of a particular base required to give a specified value of H in a solution containing any number of other acids and bases. It would be equally easy to calculate explicitly the required *volume*, V_B, of a solution of the particular base of specified concentration, c_B, if the

volume, V_{orig}, of the original solution is known together with the concentrations, $c_1, \cdots, c_j$, of its solutes, which may be a mixture of acids, bases, and ampholytes, provided that all the ionization constants are known. Since

$$D = [V_{\text{orig}}\Sigma_j(c\beta) - V_B c_B \beta'_B]/(V_{\text{orig}} + V_B), \tag{8}$$

we have

$$V_B = V_{\text{orig}}[\Sigma_j(c\beta) - D]/(c_B\beta'_B + D). \tag{9}$$

Thus the volume, V_B, of a solution of monoacid base of concentration c_B required to give a specified value of H when added to the volume V_A of a solution of monobasic acid of concentration c_A, is

$$V_B = V_A \frac{c_A A/(A + H) - H + W/H}{c_B B/(B + W/H) + H - W/H}. \tag{10}$$

If both the acid and the base are strong, this becomes simply

$$V_B = V_A(c_A - H + W/H)/(c_B + H - W/H). \tag{11}$$

If the activity coefficient is to be considered on the assumption of the Debye-Hückel limiting law, then for a specified value of H Eq. (10) becomes

$$V_B = V_A \frac{c_A \mathbf{A}/(\mathbf{A} + H\gamma^2) - H + \mathbf{W}/H\gamma^2}{c_B \mathbf{B}/(\mathbf{B} + \mathbf{W}/H) + H - \mathbf{W}/H\gamma^2}, \tag{12}$$

while for a specified value of $\mathbf{H}$,

$$V_B = V_A \frac{c_A \mathbf{A}/(\mathbf{A} + \mathbf{H}\gamma) - \mathbf{H}/\gamma + \mathbf{W}/\mathbf{H}\gamma}{c_B \mathbf{B}/(\mathbf{B} + \gamma\mathbf{W}/\mathbf{H}) + \mathbf{H}/\gamma - \mathbf{W}/\mathbf{H}\gamma}. \tag{13}$$

A first approximation of V_B is obtained from either of these expressions with $\gamma = 1$, and then μ is estimated either through Eq. v(29b) with $a = c_A V_A/(V_A + V_B)$ or as

$$\mu = H + b\mathbf{B}/(\mathbf{B} + \mathbf{W}/H) = \mathbf{H}/\gamma + b\mathbf{B}/(\mathbf{B} + \gamma\mathbf{W}/\mathbf{H}), \tag{14}$$

with $b = c_B V_B/(V_A + V_B)$, whereupon the value of V_B may be refined, etc.

C. The Special Significance of Neutralization

The concentration of any given base required to *neutralize* any solution (*literally* to neutralize it, or to make $H = OH$) is, according to Eq. (7),

$$(b_{j'})_{\mathbf{n}} = [\Sigma_j(a\beta_{\mathbf{n}}) - \Sigma_1^{j'-1}(b\beta'_{\mathbf{n}})]/(\beta'_{j'})_{\mathbf{n}}. \tag{15}$$

It follows then that for any acidic material (simple or multi-component containing any number of acids and bases in any fixed proportions, at a total concentration t, in any units, even empirical), the neutralization ratio

$(b/t)_n$, where b is the concentration of any neutralizing material (basic, but also of any arbitrary composition), is a constant independent of the concentration of the solution. This is so since the numerator in Eq. (15) is linear in the concentrations, and since the β_n's are constants for neutrality. Hence a change in t involves the same proportionate change in each individual term and hence also in $(b_{j'})_n$.

If titrations, then, could be carried out to the true neutral point, they would be free of dilution errors, although the calculations would depend on empirical titers rather than on the stoichiometry of chemical equivalents. When the titration is carried to some particular value of H other than $H = \sqrt{W}$, in the attempt to stop at an equivalent point, there is a continuous effect of dilution on the ratio $(b/t)_{H \neq \sqrt{W}}$, which must always be taken into account, as may be seen in Eq. (7). But of course the actual "feasibility" of titration depends on the slope of the curve of pH against n (or b/t) at the ratio chosen as the end-point, and this is in general too low for sharp determination anywhere but at the equivalent point.

It is nevertheless important to note the fundamental and exact constancy of the specific ratio $(b/t)_n$, for neutralization independent of concentration. The addition or subtraction of water to or from a solution which is not neutral causes a continuous change in the value of D $(= H - OH)$, which approaches zero as dilution increases. But a neutral solution of any number of solutes remains neutral through all changes of concentration because of the constancy of $(b/t)_n$. This then is actually the proof of a special case of the theorem of isohydric solutions, already considered in general form; water will mix with a solution of any composition but isohydric with water, without change of H. In the process, all the concentrations of solute species, charged and uncharged, undergo change, but the concentrations H and OH remain constant. At the same time all the α's and β's remain constant, depending only on the ionization constants and H itself, so that the distribution of the forms of a solute (X^0, X^-, $X^=$, $\cdot\cdot\cdot$, X^{-z}, for a polybasic acid) remains constant although the total concentration, a, changes. The relative proportions of $[PO_4^{\equiv}]$, $[HPO_4^=]$, $[H_2PO_4^-]$, and $[H_3PO_4]$, for example, provided that no "complex" or polymeric species are involved, must remain fixed during change of concentration of a neutral solution containing phosphoric acid, and the relation holds whatever the base or bases, weak or strong, used for the neutralization. The distribution of the ions of a polybasic acid is constant for neutrality whatever the other components of the solution and whatever the concentration. All this holds provided always that the solution is "noncomplex."

D. Relation Between a and b for Certain Specified Values of H

We shall illustrate in the remaining sections of this chapter the practical importance of the simple linear and exact relation between a and b for the calculation of the values and ratios of a and b for any specified particular value of H; essentially, this means the application of Eqs. (7) and (15).

The most important general case is that of a solution containing a simple acid and a simple base (both monobasic) described by Eq. (2). If the base is strong, this becomes

$$b_s = aA/(A + H) - D; \tag{16}$$

if the acid is strong,

$$b = a_s(B + OH)/B - D(B + OH)/B. \tag{17}$$

The conditions for $H = A$ (or $OH = B$) are as follows:

For the pure acid,

$$a_A = 2(A - W/A) \cong 2A, \tag{18}$$

provided $A \nless \sqrt{W}$; otherwise impossible. It is interesting to note that Eq. (18) is general, applying even to very high values of A.

In titration of a weak acid with b_s, $H = A$ when

$$b_s = a/2 + W/A - A, \tag{19}$$

which may be significantly different from the usual approximate value of $a/2$, when A is either very small or very large. Similarly, for the titration of weak base with a_s, $OH = B$ when

$$a_s = b/2 + W/B - B. \tag{20}$$

The ratio $n(= b/a)$ for neutrality, or $n_{\mathbf{n}}$, is

$$n_{\mathbf{n}} = (1 + \sqrt{W}/B)/(1 + \sqrt{W}/A), \tag{21}$$

so that the pure salt solution ($n = 1$) is neutral if and only if $A = B$. For strong acid, $n_{\mathbf{n}} = 1 + \sqrt{W}/B$; for strong base, $n_{\mathbf{n}} = 1/(1 + \sqrt{W}/A)$. If both are strong, $n_{\mathbf{n}} = 1$; if both are extremely weak (A and $B \lll \sqrt{W}$), $n_{\mathbf{n}} \cong A/B$.

The ratio, n, for any combination of A and B is independent of the concentration itself for neutrality and only for neutrality. This means that the titration curves of a given pair of acid and base at various concentrations all pass through the same point (at $H = \sqrt{W}$, where $b/a = n_{\mathbf{n}}$) when pH is plotted against n (or against percentage of titration, 100 n). This point is the only intersection of the family of curves for the pair.

This independence of the concentration on the part of $n_{\mathbf{n}}$ holds strictly only if the activity coefficients are unity. Otherwise, with the usual convention regarding γ, Eq. (21) becomes

$$n_{\mathbf{n}} = (1 + \gamma\sqrt{\mathbf{W}}/\mathbf{B})/(1 + \gamma\sqrt{\mathbf{W}}/\mathbf{A}), \tag{22}$$

so that $(dn_{\mathbf{n}}/d\gamma) = (\mathbf{A} - \mathbf{B})\sqrt{\mathbf{W}}/\mathbf{AB}(1 + \gamma\sqrt{\mathbf{W}}/\mathbf{A})^2$. Since γ ordinarily increases with dilution, then if $\mathbf{A} > \mathbf{B}$, $n_{\mathbf{n}}$ increases with dilution and a neutral solution becomes acid on dilution, while if $\mathbf{A} < \mathbf{B}$, n_n decreases with dilution and a neutral solution becomes alkaline on dilution. But if **A** and **B** are both greater or both smaller than $\sqrt{\mathbf{W}}$, there is, as seen from Eq. (22), very little effect of dilution upon a neutral solution.

E. The Titration Error

1. Types of End-Point. The titration of an acid with strong base always involves a certain error, the magnitude of which depends on the type of observation used as the end-point of the titration. We shall consider four types of end-points and hence four types of titration error:

(a) Titration to a predetermined value of H ($= H_{\text{end}}$) without the use of an indicator;

(b) Titration to a predetermined value of H recognized by the color of an indicator;

(c) Titration to a predetermined value of H recognized by the point of maximum relative color change of an indicator;

(d) Titration to an inflection point in the pH titration curve. [We shall distinguish H_e as the value of H at the equivalent point, H_{end} as that at the actual end-point of titration, and H_i as that at the inflection point of the pH curve.]

The first three types of error are ultimately arbitrary since they depend on the value of H predetermined as the end-point. The second and third of these types will also depend for their magnitude on the nature and concentration of the indicator used. The first type of end-point is the only one which, theoretically at least, can be altogether free of error, that is, when the titration is carried to the value of H corresponding exactly to the equivalent point of the titration, or H_e. The fourth method of titration involves an error which is not arbitrary but inherent, and characteristic of the acid being titrated, its magnitude depending on the ionization constant of the acid and on its concentration, a_i, at the end-point. This inherent error, which will later be called simply *the* titration error, will be discussed in later chapters in connection with the consideration of the shape of titration curves. The inflection point in the curve of pH against b_s does not in general occur at the equivalent point except in the case of the titration of pure strong acid with pure strong base, for which $D_i = D_e = 0$, and for which the "titration error" so defined is therefore zero.

a. Type 1. In this case a solution of an acid of known ionization constant is titrated by measurement of the concentration of strong base, $(b_s)_{\text{end}}$, required to give a particular value of H, H_{end}. The concentration of the acid present at that point may be calculated from Eq. (16) as follows:

$$a_{\text{end}} = [(b_s)_{\text{end}} + D_{\text{end}}]/\alpha_{\text{end}}. \tag{23}$$

Then if $100E$ is the relative percentage error of the titration, we have $E = [(b_s)_{\text{end}} - a_{\text{end}}]/a_{\text{end}}$, or

$$E = -[\rho_{\text{end}} + D_{\text{end}}/(b_s)_{\text{end}}]/[1 + D_{\text{end}}/(b_s)_{\text{end}}]. \tag{24}$$

The error for this type of titration may therefore be calculated exactly, subject only to activity coefficient corrections, from the value of the ionization constant, the specified value of H_{end} used for the end-point, and the observed value of b_s required to reach the end-point. It is to be noted that it is not necessary to know the value of H to which the titration *should* be carried (or H_e) in order to calculate the error. Also, if the acid being titrated is strong, then $\rho = 0$ and the error is merely

$$E = -D_{\text{end}}/(b_s)_{\text{end}}[1 + D_{\text{end}}/(b_s)_{\text{end}}]. \tag{25}$$

In the expressions of Eqs (24) and (25), which are exact, $(b_s)_{\text{end}}$ represents the concentration of the strong base in the solution at the end-point, calculated from the concentration c_B and the volume V_B of the titrating solution used, and the total volume V at the end-point. Thus $(b_s)_{\text{end}} = c_B V_B/V$, an experimental quantity. The quantity $c_B V_B$ is presumably measured with analytical accuracy, but the final volume V and hence $(b_s)_{\text{end}}$ need be measured only roughly, since the error to be calculated through Eq. (24) is small in any significant case. Eq. (23), in other words, is used experimentally to calculate the number of millimoles of the acid as follows:

millimoles of acid

$$= (c_B V_B)_{\text{end}} + [(c_B V_B)_{\text{end}} H_{\text{end}}/A + D_{\text{end}}(1 + H_{\text{end}}/A)V], \tag{26}$$

the error of the titration being the quantity in the bracket; V is in milliliters. As pointed out under Eq. (15), the relative error is seen to depend on the volume V only if $D_{\text{end}} \neq 0$; with $D_{\text{end}} = 0$, $\rho_{\text{end}} = \rho_n =$ a constant.

If we wish to predict the relative error to be expected in the titration of an acid solution of concentration c_A, we may take $(b_s)_{\text{end}}$ as approximately $c_A V_A/V$, provided that the error is small. Then if the total volume is approximately $V_A + V_B$, this becomes $(b_s)_{\text{end}} \cong c_A c_B/(c_A + c_B)$; strictly, this is $(b_s)_e$, the value at the equivalent point. In this way we may calculate the error for the titration of an acid solution of concentration c_A with a strong base solution of concentration c_B, as

$$E = -(\rho_{\text{end}} + D_{\text{end}}/\mathbf{c}_e)/(1 + D_{\text{end}}/\mathbf{c}_e), \tag{27}$$

in which

$$\mathbf{c}_e = c_A c_B/(c_A + c_B). \tag{28}$$

For an approximation we may write for Eq. (27), with $D_{\text{end}} \cong - OH_{\text{end}}$,

$$E \cong - (\rho_{\text{end}} - OH_{\text{end}}/\mathbf{c}_e)/(1 - OH_{\text{end}}/\mathbf{c}_e), \tag{29}$$

$$\cong - (\rho_{\text{end}} - OH_{\text{end}}/\mathbf{c}_e)(1 + OH_{\text{end}}/\mathbf{c}_e),^{(1)} \tag{30}$$

$$\cong - \rho_{\text{end}} + \alpha_{\text{end}} W/\mathbf{c}_e H_{\text{end}}, \tag{31}$$

$$\cong - H_{\text{end}}/(A + H_{\text{end}}) + AW/\mathbf{c}_e H_{\text{end}}(A + H_{\text{end}}). \tag{32}$$

With activity constants, Eq. (27) becomes, in terms of a specified activity of hydrogen ion as the end-point,

$$E = - [\mathbf{H}_{\text{end}}/(\mathbf{A}/\gamma + \mathbf{H}_{\text{end}}) + \mathbf{D}_{\text{end}}/\mathbf{c}_e\gamma]/(1 + \mathbf{D}_{\text{end}}/\mathbf{c}_e\gamma). \tag{33}$$

If the solution contains only the acid and base, we may take $-\log \gamma = 0.505\sqrt{\mathbf{c}_e + \mathbf{H}_{\text{end}}}$.

[For titration of polybasic acid with strong base, Eq. (23) becomes $a_{\text{end}} = [(b_s)_{\text{end}} + D_{\text{end}}]/\beta_{\text{end}}$, in which a is the molarity of the acid. In place of Eq. (24), we then have

$$E = - [(1 - \beta_{\text{end}}) + D_{\text{end}}/(b_s)_{\text{end}}]/[1 + D_{\text{end}}/(b_s)_{\text{end}}]. \tag{34}$$

For the monobasic case, $1 - \beta = \rho$, but for the general case, $1 - \beta = \rho - \alpha_2 - 2\alpha_3 - 3\alpha_4 - \cdots$.]

b. Type 2. With a given fixed concentration, c, at the end-point, of an indicator acid with ionization constant K, the solution is titrated to a particular predetermined intensity of color or to a predetermined ratio of two colors if it is a two-color indicator. The end-point is therefore determined by a specified value of the degree of ionization of the indicator. If the ratio ρ/α for the indicator is represented as i (reciprocal of r of Eq. III[44]), this means that the titration is carried to $H_{\text{end}} = iK$. Then with

$$D = a\alpha + (c\alpha)_{\text{ind}} - b_s, \tag{35}$$

Eq. (23) becomes

$$a_{\text{end}} = [(b_s)_{\text{end}} - c/(1 + i) + D_{\text{end}}](1 + iK/A), \tag{36}$$

in which $D_{\text{end}} = iK - W/iK$. The percentage error is then $[100(b_s - a)/a]_{\text{end}}$, or

$$E \cong [c/(1 + i) - \mathbf{c}_e iK/A + W/iK]/\mathbf{c}_e, \tag{37}$$

in which it is assumed that $D_{\text{end}} \cong - OH_{\text{end}}$ and that $c \ll \mathbf{c}_e$. Since the equivalent point is at $H_e \cong \sqrt{AW/\mathbf{c}_e}$, a correctly chosen indicator has this value of K, whereupon, if this is the case,

$$E \cong [c/(1 + i) - \sqrt{\mathbf{c}_e W/A}(i - 1/i)]/\mathbf{c}_e; \tag{38}$$

then if $i = 1$, $E \cong c/2\mathbf{c}_e$.

(1) Eq. (30) follows from Eq. (29) through the approximation $1/(1 \pm \varepsilon) \cong 1 \mp \varepsilon$, for $\varepsilon \ll 1$.

If the acid being titrated is strong, $iK/A = 0$ in Eq. (36), so that in place of Eq. (37) we have

$$E = [c/(1 + i) - iK + W/iK]/\mathbf{c}_e. \tag{39}$$

If an indicator with p$K < 7$ is used, then $E \cong [c/(1 + i) - iK]/\mathbf{c}_e$, while if p$K > 7$, $E \cong [c/(1 + i) + W/iK]/\mathbf{c}_e$. If the "correct" indicator is used, with $K = \sqrt{W}$, $E = [c/(1 + i) - \sqrt{W}(i - 1/i)]/\mathbf{c}_e$, and now if $i = 1$, $E = c/2\mathbf{c}_e$.

c. Type 3. We now assume that the titration is carried to the point of maximum rate of change of the degree of ionization of the indicator with respect to the concentration of the titrating strong base. Whereas $d\alpha/dpH$ for a weak acid is a maximum at $H = K$, the maximum of $d\alpha/db_s$ occurs at some lower value of H, even in the absence of other acids, weak or strong. This value of H_{max} for an indicator acid, or the value of H for the maximum of $d\alpha/db_s$ for the indicator, is independent of the concentration c of the indicator itself; it is also independent of any strong acids or strong bases present, but it does vary with the concentrations and strengths of weak electrolytes present.

With the derivative $(dH/d\alpha)_{ind} = 1/(d\alpha/dH)_{ind} = -(K + H)^2/K$, we have, from Eq. (35),

$$\left(\frac{db_s}{d\alpha}\right)_{ind} = \frac{aA(K + H)^2}{K(A + H)^2} + c + \frac{(K + H)^2}{K}\left(1 + \frac{W}{H^2}\right), \tag{40}$$

$$\left(\frac{d^2b_s}{d\alpha^2}\right)_{ind} = \left(\frac{dH}{d\alpha}\right)_{ind}\left(\frac{K + H}{K}\right)\left[\frac{2aA(A - K)}{(A + H)^3} + \left(2 + \frac{W}{H^2} - \frac{KW}{H^3}\right)\right]. \tag{41}$$

*c*1. *Strong Acid.* If the acid being titrated is strong or if its concentration is zero, the first term in the bracket is zero and the value of H_{max} is given by the equation

$$2H^3 + HW - KW = 0. \tag{42}$$

This gives

$$H^3_{max} \cong KW/2 - (W/2)\sqrt[3]{KW/2} \cong KW/2, \tag{43}$$

if $K > \sqrt{W}$; and if $K < \sqrt{W}$,

$$H_{max} \cong K(1 - 2K^2/W) \cong K. \tag{44}$$

With this value of H_{max} as H_{end}, the error of the titration of the strong acid becomes

$$E \cong [(c\alpha)_{ind} - D_{end}]/\mathbf{c}_e. \tag{45}$$

Some illustrative calculations are listed in Table 7–1. With $K > \sqrt{W}$, $E \cong (c - \sqrt[3]{KW/2})/\mathbf{c}_e$; with $K < \sqrt{W}$, $E \cong (c/2 + W/K)/\mathbf{c}_e$.

Table 7–1. Error in Titration of strong acid; Type 3.

K	$H_{\max}$ (= H_{end})	E (× 100$\mathbf{c}_e$)
10^{-5}	3.64×10^{-7}	$96c - 3.4 \times 10^{-5}$
10^{-6}	1.61×10^{-7}	$86c - 1.0 \times 10^{-5}$
10^{-7}	5.9×10^{-8}	$63c + 1.1 \times 10^{-5}$
10^{-8}	9.8×10^{-9}	$50c + 10^{-4}$
10^{-9}	1.0×10^{-9}	$50c + 10^{-3}$

*c*2. *Weak acid.* For titration of weak acid at concentration a, the maximum of $(d\alpha/db_s)_{\text{ind}}$ occurs when, from Eq. (41),

$$2H^6 + 6H^5A + H^4(6A^2 + W) + H^3[2aA(A - K) + 2A^3 + 3AW - KW] + 3H^2AW(A - K) + HA^2W(A - 3K) - A^3KW = 0. \quad (46)$$

With $a \gg A$, $A > K$, $K < \sqrt{W}$, $H < A$ and $H < \sqrt{W}$, the largest terms are those in H^3, H, and H^0; approximately, then,

$$2H^3a + HAW - AKW \cong 0. \quad (47)$$

With reference to the equivalent point at which $H_e \cong \sqrt{AW/a}$: if $K > H_e$, then

$$H^3_{\max} \cong AKW/2a - (AW/2a)\sqrt[3]{AKW/2a}, \quad (48)$$

while if $K < H_e$,

$$H_{\max} \cong K(1 - 2aK^2/AW). \quad (49)$$

Table 7–2. Error in Titration of weak acid with $A = 10^{-7}$ at $\mathbf{c}_e = 0.1$; Type 3.

K	$H_{\max}$ (= H_{end})	E (× 100)
10^{-8}	3.64×10^{-10}	$960c - 0.337$
10^{-9}	1.61×10^{-10}	$860c - 0.099$
10^{-10}	5.9×10^{-11}	$630c + 0.110$
10^{-11}	9.8×10^{-12}	$500c + 1.01$
10^{-12}	1.0×10^{-12}	$500c + 10.0$

Some calculations for the case of $A = 10^{-7}$ and $a(= \mathbf{c}_e$ of Eq. [28]) $= 0.1$ (with $H_e \cong 10^{-10}$) are given in Table 7–2, based on the error according to the following expression:

$$E = [(c\alpha)_{\text{ind}} - \mathbf{c}_e H_{\text{end}}/A + OH_{\text{end}}]/\mathbf{c}_e. \quad (50)$$

F. The "Reaction" ($H \gtrless OH$) of Salt Solutions (Equivalent Points)

1. The Ratio $(b/a)_n$ or n_n, for Neutrality. Whether any given solution containing j acids and j' bases is "acid" or "basic" can be determined by comparing the concentration of the (j')'th base required for neutrality, or $(b_{j'})_n$ of Eq. (15), with the actual value of $b_{j'}$ in the solution. With actual $b_{j'}$ more than (or less than) $(b_{j'})_n$, the solution is alkaline (or acid). There seems to be no easy way (if any) to answer such a *general* question by the usual methods of treating H as the dependent variable.

For the combination of a single acid and a single base (both of any "basicity"), the ratio n_n for neutrality is, as discussed above, independent of the concentration since then Eq. (15) becomes simply

$$n_n = (\beta_n/\beta'_n), \tag{51}$$

and the β_n's are constants. The point made above, then, for the family of titration curves of a simple acid at various concentrations, applies perfectly generally for any pair of acid and base. For any value of H other than $H = \sqrt{W}$, $D \neq 0$, and

$$n = \beta/\beta' - D/a\beta', \tag{52}$$

thus varying with the concentration.

2. Application to Equivalent Points. For any particular equivalent point or pure salt solution when n is a stoichiometric integer or integral ratio, the "reaction" depends on whether this value of n is more than or less than n_n, defined as in Eq. (51). Since n is defined as b/a, the solution is alkaline if $n > n_n$.

(a) For example, while the normal alkali salt, Na_2X, of a general dibasic acid is always basic, the corresponding acid salt, NaHX, may be acid or alkaline, depending on the ionization constants.

Here

$$n_n = (1 + 2K_2/\sqrt{W})/(1 + \sqrt{W}/K_1 + K_2/\sqrt{W}). \tag{53}$$

For $n_n = 1$, K_1K_2 must equal W. The salt solution is thus acid if $K_1K_2 > W$, alkaline if $K_1K_2 < W$, at any concentration. One obtains the same result from the familiar approximation that $H \cong \sqrt{K_1K_2}$ for such a salt solution, but the answer here given is exact and general and does not depend on whether or not the approximation in question actually applies; the limitations of this approximation, especially as $f(c)$, will be discussed later (Chapter xiv). We may see at once, in fact, that in order that H may equal $\sqrt{K_1K_2}$ in a solution of dibasic acid and strong base, the condition is, according to Eq. v(36), that

$$b_s = a - (\sqrt{K_1K_2} - W/\sqrt{K_1K_2}). \tag{54}$$

Hence $b_s \cong a - \sqrt{K_1K_2}$ if $K_1K_2 > W$, and $\cong a + W/\sqrt{K_1K_2}$ if $K_1K_2 < W$;

$b_s = a$, or $n_{\sqrt{K_1K_2}} = 1$, only if $K_1K_2 = W$. In other words, $H = \sqrt{K_1K_2}$ exactly, in the solution of NaHX, only if $K_1K_2 = W$, whereupon $H = \sqrt{W}$ at any concentration.

Introducing activity coefficients into Eq. (53), and using our γ convention, we obtain

$$n_{\mathbf{n}} = (1 + 2\mathbf{K}_2/\gamma^3\sqrt{\mathbf{W}})/(1 + \gamma\sqrt{\mathbf{W}}/\mathbf{K}_1 + \mathbf{K}_2/\gamma^3\sqrt{\mathbf{W}}), \tag{55}$$

which gives $dn_{\mathbf{n}}/d\gamma < 0$. Hence $n_{\mathbf{n}}$ decreases with dilution and a neutral solution of a dibasic acid and strong base becomes alkaline upon dilution.

If the dibasic acid is strong in respect to K_1, then

$$n_{\mathbf{n}} = (2 + \sqrt{W}/K_2)/(1 + \sqrt{W}/K_2); \tag{56}$$

since this expression is always more than 1 and less than 2, the acid salt, when $n = 1$, is always acid and the normal salt ($n = 2$) always alkaline.

(b) When a mixture of two weak monobasic acids is titrated with b_s, the solution is neutral when

$$(b_s/a_1)_{\mathbf{n}} = 1/(1 + \sqrt{W}/A_1) + a_2/a_1(1 + \sqrt{W}/A_2). \tag{57}$$

At the first equivalent point, then, when $b_s/a_1 = 1$, the solution is acid if

$$a_2/a_1 > (1 + \sqrt{W}/A_2)/(1 + A_1/\sqrt{W}); \tag{58}$$

the second equivalent point (when $b_s = a_1 + a_2$) is of course always alkaline.

(c) The "reaction" of a general salt solution may sometimes be determined more advantageously without considering the "neutralization" ratio. From Eq. I(65), or $D = c(\nu_a\beta - \nu_b\beta')$, the salt solution is neutral when

$$\beta_{\mathbf{n}} = (\nu_b/\nu_a)\beta'_{\mathbf{n}}, \tag{59}$$

acid if greater and basic if less. Eq. (59) gives the relationship of all the ionization constants involved in the salt solution required for neutrality. In this way, with no need of approximation or speculation (except to assume at least temporarily that $\gamma = 1$), we may determine the sign of the reaction of a salt solution of any composition, including mixed salts. If the salt, for example, is that of a polybasic acid and three different bases,

$$D = c(\nu_a\beta - \nu_{b_1}\beta'_1 - \nu_{b_2}\beta'_2 - \nu_{b_3}\beta'_3); \tag{60}$$

and again it is acid, neutral, or alkaline, according to whether $\nu_a\beta_{\mathbf{n}}$ is greater than, equal to, or less than $\Sigma\nu_b\beta'_{\mathbf{n}}$.

3. Mixtures of Two Simple Salts in Any Proportion. The "reaction" of mixtures of two simple salts (1 : 1 type) in any proportion, or the answer to the question whether a solution containing the salt MX (of the base MOH with ionization constant B_1 and the acid HX with ionization constant A_1) at the concentration c_1, and the salt NY (NOH with B_2 and HY with A_2) at concentration c_2, is acid or alkaline, may also easily be determined as a

special case of Eq. (15), in which $a_1 = b_1 = c_1$, and $a_2 = b_2 = c_2$. The general equation becomes

$$D = c_1(\alpha_X - \alpha'_M) + c_2(\alpha_Y - \alpha'_N). \quad (61)$$

If both acids and both bases are strong (all α's $= 1$), the solution is of course neutral for any ratio of c_1 to c_2; similarly, the mixture is acid if only the acids are strong and alkaline if only the two bases are strong. Otherwise, neutrality requires

$$\mathbf{R}_n = \left(\frac{c_1}{c_2}\right)_n = \left(\frac{\alpha'_N - \alpha_Y}{\alpha_X - \alpha'_M}\right)_n, \quad (62)$$

$$= \left(\frac{B_2 - A_2}{A_1 - B_1}\right) \frac{(A_1B_1 + W) + \sqrt{W}(A_1 + B_1)}{(A_2B_2 + W) + \sqrt{W}(A_2 + B_2)}. \quad (63)$$

If only HX and MOH are strong, the result depends simply on $A_2 \gtrless B_2$; if only HY and NOH are strong, simply on $A_1 \gtrless B_1$. If HX alone is strong,

$$\mathbf{R}_n = \left(\frac{\alpha'_N - \alpha_Y}{1 - \alpha'_M}\right)_n = \frac{(B_2 - A_2)(B_1 + \sqrt{W})}{(A_2B_2 + W) + \sqrt{W}(A_2 + B_2)}; \quad (64)$$

if only HX and NOH are strong,

$$\mathbf{R}_n = \left(\frac{1 - \alpha_Y}{1 - \alpha'_M}\right)_n = \frac{B_1 + \sqrt{W}}{A_2 + \sqrt{W}}; \text{ etc.} \quad (65)$$

If the given value of $\mathbf{R}$ ($= c_1/c_2$) is in any case greater (or less) than the required value of $\mathbf{R}_n$ for neutrality, the solution is acid (or basic)—for $A_1 > B_1$ and $A_2 < B_2$; vice versa if $A_1 < B_1$ and $A_2 > B_2$. With the ordinary methods of estimating H in such mixtures, the problem would appear too formidable—in the general case of Eq. (63), for example, involving four unrelated ionization constants—for anything but guesses and approximations.

G. "Isohydric" Salt Solutions and Indicator Solutions

Since it is possible to calculate directly and exactly by Eq. (52) the ratio n, or b/a, to give any desired value of H, it follows that it is possible to calculate very simply the ratio of concentrations of successive "salts" of a polybasic acid required to give solutions of any specified value of H, or "isohydric" with any solution of known H, which they are intended to buffer, for example. The most important special value of H is $\sqrt{W}$, for solutions "isohydric" with water itself, and in this case the problem is particularly simple since n_n is independent of the concentrations themselves. In this section we restrict our considerations to strong base plus dibasic acid.

1. Ratio of Acid to Acid Salt. The ratio of the acid itself to the acid salt for neutrality, $(c/c_1)_n$, or $([H_2X]/[NaHX])_n$, is found as follows: since $a = c + c_1$ and $b_s = c_1$,

$$n = b_s/a = c_1/(c + c_1). \tag{66}$$

Hence

$$\mathbf{R} = c/c_1 = (1 - n)/n. \tag{67}$$

Using the value of n_n in Eq. (53), we have

$$\mathbf{R}_n = (W - K_1K_2)/K_1(2K_2 + \sqrt{W}), \tag{68}$$

which is possible only if $W > K_1K_2$, and is of course impossible if the acid is "strong" in respect to K_1, since Eq. (56) would then give $\mathbf{R}_n < 0$.

2. Ratio of Acid Salt to Normal Salt. For the ratio of acid salt to normal salt for neutrality, or $([NaHX]/[Na_2X])_n$ at concentrations c_1 and c_2, we have $a = c_1 + c_2$, $b_s = c_1 + 2c_2$. Then

$$n = (c_1 + 2c_2)/(c_1 + c_2), \tag{69}$$

$$\mathbf{R} = c_1/c_2 = (2 - n)/(n - 1), \tag{70}$$

$$\mathbf{R}_n = (K_1\sqrt{W} + 2W)/(K_1K_2 - W), \tag{71}$$

possible only if $K_1K_2 > W$.

In this case, if the acid is strong in respect to K_1, Eq. (71) becomes

$$\mathbf{R}_n = \sqrt{W}/K_2. \tag{72}$$

The practical utility of these formulas will be evident by comparison with the laborious procedures necessary by ordinary methods in which such problems are attacked through equations dealing with H as the dependent variable. As an example we may mention the problem of calculating the ratio of the two forms of an "indicator salt," the "acid-form" ($[HI^-]$ or the salt NaHI), and the "alkaline form" ($[I^=]$ or the salt Na_2I), required for the preparation of an indicator solution isohydric with water ($H = \sqrt{W}$). For bromthymol blue, a dibasic acid which may be considered "strong" in respect to K_1 and with $K_2 = 5.5 \times 10^{-8}$, the required ratio is immediately and exactly, by Eq. (72), 10/5.5, and is independent of concentration.[2]

(2) The very same problem is treated by Kolthoff and Rosenblum, (*op.cit.*, Ref. VI–2; pp. 325–28). The procedure involves the calculation of pH as function of concentrations and equilibrium constants through approximate forms of the exact equation for H to find the ratio of $[HI^-]$ to $[I^=]$, at low concentrations, giving a pH of 7. Although giving the same numerical answer (10/5.5), the method requires much calculation, gives the answer as of some degree of accuracy (but not "exact" since it involves the approximate solution of a cubic equation), and leaves open the question of how the ratio (for pH = 7) depends on concentration. (Application of the theorem of isohydric solutions, of course, would tell us that the ratio for neutrality is independent of concentration.) The calculation would obviously have been much more difficult still if the indicator had been weak in respect to K_1 too, but Eq. (71) would give an immediate answer even in this case.

The form of Eqs. (68), (71), and (72), furthermore, permits the easy introduction of activity coefficients. For the simple case of Eq. (72),

$$\mathbf{R}_n = \gamma^3\sqrt{\mathbf{W}}/\mathbf{K}_2, \tag{73}$$

whence it follows that the ratio for neutrality increases as the solution is diluted. For Eq. (71),

$$\mathbf{R}_n = (\mathbf{K}_1\sqrt{\mathbf{W}}/\gamma + 2\mathbf{W})/(\mathbf{K}_1\mathbf{K}_2/\gamma^4 - \mathbf{W}). \tag{74}$$

In either case the ratio required at any given ionic strength is found at once by taking $-\log \gamma \cong 0.505\sqrt{\mu}$. If the ratio is desired at a given concentration, c_2, of the salt Na_2X, a preliminary uncorrected value of $\mathbf{R}_n$ is used to calculate γ as in Eq. v(48) as follows:

$$-\log \gamma = 0.505[\sqrt{\mathbf{W}} + c_2(1 + \mathbf{R}_n)(\alpha_1 + 3\alpha_2)_n]^{\frac{1}{2}}, \tag{75}$$

in which $\sqrt{\mathbf{W}}$ will certainly be negligible; α_1 and α_2 are the first and second ionization fractions of the dibasic acid and are constants, of course, at $H = \sqrt{W}$. For a given concentration, c_1, of the salt NaHX, Eq. (75) becomes

$$-\log \gamma = 0.505[\sqrt{\mathbf{W}} + (\alpha_1 + 3\alpha_2)_n c_1(1 + \mathbf{R}_n)/\mathbf{R}_n]^{\frac{1}{2}}. \tag{76}$$

H. Special Buffers and Isohydric Solutions

1. Ratios of Successive Alkali Salts of Polybasic Acid. The ratios of the successive alkali salts of a polybasic acid for neutrality may similarly be expressed. But since it may also be valuable to express the ratios of such salts for any specified value of H, including $H = \sqrt{W}$ as a special case, we may proceed more generally as follows. For a polybasic acid, "H_zX," at concentration a, plus strong base, $D = a\beta - b_s$, whence

$$n = b_s/a = \beta - D/a. \tag{77}$$

Now designating the concentrations of the successive salts of the acid as c for H_zX, c_1 for $NaH_{z-1}X$, c_2 for $Na_2H_{z-2}X$, c_3 for $Na_3H_{z-3}X$, etc., we have, as for Eqs. (66), (67), and (69–70),

(1)
$$\mathbf{R}_{0,1} = c/c_1 = (1 - n)/n; \tag{78}$$

$$a = c + c_1 = c/(1 - n) = c_1/n. \tag{79}$$

Hence, through Eq. (77), n may be expressed in terms of c, of c_1, or of $(c + c_1)$; similarly, therefore, the ratio $\mathbf{R}_{0,1}$ may be expressed as a function of c, of c_1, or of $(c + c_1)$. The result is

$$\mathbf{R}_{0,1} = (1 - \beta)/(\beta - D/c), \tag{80a}$$

$$= (1 - \beta + D/c_1)/\beta, \tag{80b}$$

$$= [1 - \beta + D/(c + c_1)]/[\beta - D/(c + c_1)]. \tag{80c}$$

Here β is a function of the specified H and the ionization constants of the acid. If D/c, etc., is small,

$$\mathbf{R}_{0,1} \cong (1-\beta)/\beta. \tag{81}$$

Finally, if $H = \sqrt{W}$, $D = 0$, and $\mathbf{R}$, now $\mathbf{R}_n$, is independent of the concentration.

(2)
$$\mathbf{R}_{1,2} = c_1/c_2 = (2-n)/(n-1). \tag{82}$$

$$a = c_1 + c_2 = c_1/(2-n) = c_2/(n-1). \tag{83}$$

Then

$$\mathbf{R}_{1,2} = (2-\beta)/(\beta - 1 - D/c_1), \tag{84a}$$

$$= (2-\beta+D/c_2)/(\beta-1), \tag{84b}$$

$$\doteqdot [2-\beta+D/(c_1+c_2)]/[\beta-1-D/(c_1+c_2)]; \tag{84c}$$

$$\cong (2-\beta)/(\beta-1). \tag{85}$$

[For the ratio, c/c_2, of the acid and its secondary salt, such as dibasic acid and normal salt, Eqs. (80a,b,c) and (81) hold with 2 in place of 1 in the numerator.]

(3) Obviously, then,

$$\mathbf{R} = \frac{c_{m-1}}{c_m} = \frac{m-\beta+D/(c_{m-1}+c_m)}{\beta-(m-1)-D/(c_{m-1}+c_m)}, \text{ etc.}; \tag{86}$$

$$\cong (m-\beta)/[\beta-(m-1)]. \tag{87}$$

2. Special Cases. Some special cases will be considered for illustration.

a. Weak monobasic acid and its alkali salt, M_sX. The ratio for $H = A$, where $\beta = 1/2$ and $D = (A - W/A)$, is

$$(c/c_1)_A = 1/(1-2D/c), \tag{88a}$$

$$= 1 + 2D/c_1, \tag{88b}$$

$$= [1+2D/(c+c_1)]/[1-2D/(c+c_1)]. \tag{88c}$$

b. Dibasic acid and strong base. Eqs. (68), (71), and (72) apply for $H = \sqrt{W}$.

(1) For $H = K_1$, $\beta = (1+2y)/(2+y)$, with $y = K_2/K_1$. Then for the ratio of acid to acid salt,

$$(c/c_1)_{K_1} = (1-y)/[1+2y-f(c)], \tag{89a}$$

$$= [1-y+f(c_1)]/(1+2y), \tag{89b}$$

$$= [1-y+f(c+c_1)]/[1+2y-f(c+c_1)], \tag{89c}$$

in which

$$f(c) = (2+y)D/c, \text{ etc.} \tag{90}$$

If D is very small, then $(c/c_1)_{K_1} \cong (1-y)/(1+2y)$.

(2) For $H = K_2$, $\beta = 3/(2+y)$. Then for the ratio of acid salt to normal salt,

$$(c_1/c_2)_{K_2} = (1+2y)/[1-y-f(c_1)], \tag{91a}$$

$$= [1+2y+f(c_2)]/(1-y), \tag{91b}$$

$$= [1+2y+f(c_1+c_2)]/[1-y-f(c_1+c_2)], \tag{91c}$$

with $f(c)$ defined again as in Eq. (90).

(3) For dibasic acid "strong in K_1," β at $H = K_2$ is 3/2; hence

$$(c_1/c_2)_{K_2} = 1/(1-2D/c_1), \tag{92a}$$

$$= 1 + 2D/c_2, \tag{92b}$$

$$= [1+2D/(c_1+c_2)]/[1-2D/(c_1+c_2)]. \tag{92c}$$

c. Tribasic acid and strong base.

(1) For neutrality,

$$(\mathbf{R}_{0,1})_\mathbf{n} = \left(\frac{c}{c_1}\right)_\mathbf{n} = \frac{W - K_1K_2(1+2K_3/\sqrt{W})}{K_1[\sqrt{W}+K_2(2+3K_3/\sqrt{W})]}, \tag{93}$$

$$(\mathbf{R}_{1,2})_\mathbf{n} = \left(\frac{c_1}{c_2}\right)_\mathbf{n} = \frac{W(K_1+2\sqrt{W}) - K_1K_2K_3}{K_1K_2(\sqrt{W}+2K_3) - W\sqrt{W}}, \tag{94}$$

$$(\mathbf{R}_{2,3})_\mathbf{n} = \left(\frac{c_2}{c_3}\right)_\mathbf{n} = \frac{K_1(2\sqrt{W}+K_2) + 3W}{K_1K_2K_3/\sqrt{W} - K_1\sqrt{W} - 2W}. \tag{95}$$

These expressions, like Eqs. (68), (71), and (72), are useful not only in the preparation of neutral buffers but in the determination of the "reaction" ($H \gtrless OH$) of mixtures of the salts, or of acid and salt, in each case, in any proportion. If the actual ratio is greater than $\mathbf{R}_\mathbf{n}$, regardless of the concentration itself, the solution is acid; if less than $\mathbf{R}_\mathbf{n}$, alkaline.

(2) For other special values of H, we note that while the ratio for neutrality is independent of the concentration (at least ideally, with $\gamma = 1$), the ratio of the forms of the acid for any other specified value of H depends on the concentrations and may again be expressed in terms of the concentration of either of the two forms involved, or of their total. Illustrating only with $H = K_2$, we have

$$\mathbf{R}_{K_2} = (c_1/c_2)_{K_2}$$

$$= (1+2y-y_1)/[1-y+2y_1-f(c_1)], \tag{96a}$$

$$= [1+2y-y_1+f(c_2)]/(1-y+2y_1), \tag{96b}$$

$$= [1+2y-y_1+f(c_1+c_2)]/[1-y+2y_1-f(c_1+c_2)], \tag{96c}$$

with

$$f(c) = (2 + y + y_1)D/c, \tag{97}$$

and $y = K_2/K_1$, $y_1 = K_3/K_2$.

In these expressions, derived directly from the equation of electroneutrality, the constants are strictly mass constants, and they may be taken as the activity constants only for unity activity constants. With this limitation, or in the "ideal" sense, the formulas in this chapter are general and exact. Although apparently cumbersome, they are all actually simple and of course explicit and *direct* for the calculation of the relations of the concentrations corresponding to any specified H, provided the ionization constants are known. Approximations and simpler expressions are easily formulated in any actual case if the numerical data justify the neglect of any of the terms involved. But consideration of the data in connection with the full exact equation will always tell definitely whether or not any approximation is valid. Incidentally, we have not had to use the word "hydrolysis" at all in this chapter dealing principally with salts.

Again, the introduction of activity coefficients will be illustrated in connection with Eq. (96b) to find the ratio $[NaH_2X]/[Na_2HX]$ to give $H = K_2$ at a fixed concentration, c_2, of the salt Na_2HX and at a given total ionic strength, μ. With the usual convention about activity coefficients,

$$\mathbf{R}_{H=K_2} = [1 + (2y' - y_1')/\gamma^2 + f(c_2)]/[1 - (y' - 2y_1')/\gamma^2], \tag{98}$$

in which the primed y's now represent ratios of activity constants, and in which

$$f(c_2) = [2 + (y' + y_1')/\gamma^2](\mathbf{K}_2/\gamma^4 - \mathbf{W}\gamma^2/\mathbf{K}_2)/c_2. \tag{99}$$

Eq. (98) gives the condition, therefore, in terms of activity constants for a value of the hydrogen ion concentration, H, equal to the mass constant K_2 at specified value of γ. If μ is specified, then $-\log \gamma$ may be taken as $0.505\sqrt{\mu}$. But if μ is not specified and the solution originally contains only the salt Na_2HX at the concentration c_2, then a preliminary value of $\mathbf{R}_{K_2}$ calculated from Eq. (96b) is used to evaluate γ for use in Eq. (98). This may be done through Eq. v(48), which gives

$$-\log \gamma = 0.505[\mathbf{W}/\mathbf{K}_2 + c_2(1 + \mathbf{R}_{K_2})(\alpha_1 + 3\alpha_2 + 6\alpha_3)_{K_2}]^{\frac{1}{2}}, \tag{100}$$

in which the α's are taken as constants at $H = K_2$. Less accurately, from Table 5–1, and with OH neglected,

$$-\log \gamma \cong 0.505\sqrt{2c_2(1 + \mathbf{R}_{K_2})}. \tag{101}$$

For the condition in which the hydrogen ion concentration, H, equals the activity constant $\mathbf{K}_2$, Eq. (84b) gives

$$\mathbf{R}_{H=\mathbf{K}_2} = [1 + 2y'\gamma^2 - y_1'/\gamma^{10} + f(c_2)]/(1/\gamma^4 - y'\gamma^2 + 2y_1'/\gamma^{10}), \quad (102)$$

with

$$f(c_2) = (1 + 1/\gamma^4 + y'\gamma^2 + y_1'/\gamma^{10})(\mathbf{K}_2 - \mathbf{W}/\mathbf{K}_2\gamma^2)/c_2. \quad (103)$$

Finally, for the condition in which the hydrogen ion activity, $\mathbf{H}$, equals the activity constant $\mathbf{K}_2$, Eq. (84b) gives

$$\mathbf{R}_{\mathbf{H}=\mathbf{K}_2} = [1 + 2y'\gamma - y_1'/\gamma^8 + f(c_2)]/(1/\gamma^3 - y'\gamma + 2y_1'/\gamma^8), \quad (104)$$

with

$$f(c_2) = (1 - 1/\gamma^3 + y'\gamma + y_1'/\gamma^8)(\mathbf{K}_2 - \mathbf{W}/\mathbf{K}_2)/\gamma c_2. \quad (105)$$

VIII

Calculation of the Numerical Value of H from the General Equations

A. General Procedures

When solved for H, the general equations of Chapter 1, Section H, are of degree equal to one plus the number of ionization constants (including W) involved. Most of them, therefore, are of high degree. For the solution of a simple ampholyte, for example, Eq. 1(67) gives

$$H^4K_b + H^3(cK_b + W) + H^2W(K_a - K_b) - HW(cK_a + W) - K_aW^2 = 0. \quad (1)$$

This may further be written in various equivalent forms such as

$$H^2K_b + H(cK_b + W) + W(K_a - K_b) - OH(cK_a + W) - (OH)^2K_a = 0, \quad (2)$$

showing the symmetry of the relations. Or, using the transformations of Eqs. 1(68) and 1(71), we have, in terms of acid constants,

$$H^4 + H^3(c + K_1^*) + H^2(K_1^*K_2^* - W) - HK_1^*K_2^*(c + W/K_2^*) - K_1^*K_2^*W = 0, \quad (3)$$

and in terms of base constants,

$$(OH)^4 + (OH)^3(c + K_1'^*) + (OH)^2(K_1'^*K_2'^* - W) - OHK_1'^*K_2'^*(c + W/K_2'^*) - K_1'^*K_2'^*W = 0. \quad (4)$$

Such general equations, whether in the form $D = f(c,\alpha)$ or $D = f(c,H,K\text{'s})$ as in the two parts of Eq. 1(67), or in the above expanded forms, are merely the most elementary combination of the simple definitions of the quantities H, OH, a, b, c, K's, W. In a sense, nothing is accomplished in writing these equations unless they are accompanied by some simple guiding principle for their numerical solution. If they are left as such in their full form for "brute solution" by numerical approximation methods, there is generally little point in writing them. One method of handling such equations is to express them as series in powers of certain functions of the concentrations and constants,[1] a particular series being applicable for

(1) As in the papers by P. S. Roller, *J. Am. Chem. Soc.*, **50,** 1 (1928); **54,** 3485 (1932); **57,** 98 (1935).

specified conditions. While capable of high accuracy when applicable, this procedure does not readily lead to what we may call "familiar" or easily remembered and recognized expressions. Another method is to arrange the general expression into a quadratic with H as function of itself, again to be estimated by means of successive approximations. Thus for a cubic,

$$H^3 + uH^2 + vH + w = 0, \tag{5}$$

we may write either

$$\text{(a)} \qquad uH^2 + vH + (w + H^3) = 0, \tag{6}$$

solving first with neglect of H^3 and then substituting the approximate value of H in the last term for a second approximation, etc.,[2] or

$$\text{(b)} \qquad H^2 + uH + (v + w/H) = 0, \tag{7}$$

now neglecting w/H in the first approximation, etc. We still require guiding principles, however, as to whether (a) or (b) will be the better procedure; and furthermore the method becomes very cumbersome for equations of still higher degree. The problem is discussed rather thoroughly by T. B. Smith,[3] who gives a more or less systematic method for the approximate numerical solution of such equations; the limits of applicability of the approximations obtained, however, are in terms of the unknown itself (H or OH). It seems desirable to find, if possible, approximate solutions with conditions of applicability expressed in terms of the known quantities of the problem, so that the choice of approximate formulas may be made without trial and error.

Consideration of a great number and variety of equations like Eq. (1) leads to the observation that a working approximation suitable to any actual set of values of the parameters will almost always be available as a quadratic solution of three consecutive terms of the equation. The nature and relation of the physical quantities involved makes for the predominance in importance or order of magnitude of certain sequences of terms, quite regularly. With this as an expectation, we can deal with the complete general equations and choose an approximation suitable to any given real case which does not overlook any important effect of any of the variables involved. Such coverage is impossible unless the approximation is always based on the full and general equation, taking all variables into account.

It will be rare, for any real data, that a fair approximation is not possible by the quadratic solution of three such consecutive terms correctly chosen. The guiding principles for the choice will become evident and will be stated explicitly whenever possible as we proceed with specific cases so that, for example, a set of terms may be used with confidence if the concentration is

(2) A similar procedure was used in dealing with Eqs. v(101) and v(107) in the form of v(108) and v(110) respectively. The method will be used again later for a number of cases.

(3) T. B. Smith, *Analytical Processes*, Arnold, London, 1940.

greater (or smaller) than a certain limiting value which is a function of the constants. But even without such "conditions of applicability" of approximations stated in terms of the known parameters, it is possible rather easily to make a check on the correctness of the choice of the three consecutive terms to be used in any actual numerical problem. This is done by substituting the calculated (approximate) value in the full equation and estimating merely the orders of magnitude of all the terms. If the choice was correct, the highest orders of magnitude must appear in the three terms used in the approximation. This gives a direct numerical verification of the fact that we have taken the best approximation, and it does not involve judgment or chemical reasoning—the "neglect" of this or that *process*.

Furthermore, it will almost always be found, in any particular case, that certain sequences of terms will, because of signs, give no real quadratic solution, so that they are eliminated as possibilities at once; this follows since H must be greater than 0.

Even some idea of the accuracy of the approximation finally chosen will be obtainable from considerations of the differences in order of magnitude between the terms used and those neglected. One may thereby roughly estimate the number of dependable or significant figures in the calculated value even without knowledge of the exact value.

While such more or less "mechanical" procedure is then always available for the comparatively easy numerical solution of any specific real problem, it will nevertheless be important to find for the various possible approximations for any given equation (three quadratic solutions, for example, for a fourth degree equation such as Eq. [1]) conditions of applicability stated explicitly in terms of the parameters (concentrations and ionization constants) so that a useful approximate formula may be readily available, when required, without the necessity of testing its correctness each time.

B. Approximate Solutions of the Equation for a General (1 : 1) Salt, MX

We shall illustrate the general method, which will then be followed throughout, with the equation for the aqueous solution of the salt, MX, of simple (monobasic) acid and base, which is typical of the form and content of the equations to be considered in the sequel. From Eq. I(53),

$$H^4B + H^3(cB + AB + W) + H^2W(A - B) - HW(cA + AB + W) - AW^2 = 0. \quad (8)$$

Like the equation for an ampholyte, this also is symmetrical, as may be seen by dividing by H^2:

$$H^2B + H(cB + AB + W) + W(A - B) - OH(cA + AB + W) - (OH)^2A = 0. \quad (9)$$

The familiar "solution" of this equation is the formula $H \cong \sqrt{WA/B}$, which ignores the concentration altogether and gives obviously absurd results for H if $c < \sqrt{WA/B}$. The full equation has been used by Griffith, (Ref. IV–1), and it is also discussed by T. B. Smith (Ref. VIII–3, pp. 188 *et seq.*). Griffith solved it directly without simplifications in the form of Eq. I(53) by numerical approximation methods. Smith simplifies the equation only to the usual $H \cong \sqrt{WA/B}$ and to the more accurate formula $H \cong \sqrt{WA(c + B)/B(c + A)}$. There is also the formula given by Clark (Ref. III–2, p. 336), $H = \sqrt{WA(B + [M^+])/B(A + [X^-])}$, which, however, is not explicit, and when solved for H is actually the above full quartic equation.

In referring to the approximate solutions of an equation such as Eq. (8), the symbol H^n will mean the sequence of three terms starting with that in H^n, the rest of the terms of the general polynomial being neglected; the approximate answer is the quadratic solution of the equation

$$H^n + vH^{n-1} + wH^{n-2} = 0, \tag{10}$$

or

$$H = -v/2 \pm \sqrt{(v/2)^2 - w} = Q^n_J. \tag{11}$$

For brevity, the full quadratic solution of a particular sequence, H^n, may not always be written out, but may be represented by the symbol Q^n_J, which means, then, the function of Eq. (11) for the nth sequence of equation number J, as illustrated below in Eqs. (14–16). Incidentally, the formula for OH may be obtained either by first substituting W/OH for H in Eq. (10) or by transforming a formula such as Eq. (11) directly to OH as follows:

$$OH = -Wv/2w \pm \sqrt{(\quad)^2 - W^2/w}, \tag{12}$$

because

$$1/(-x \pm \sqrt{x^2 + x_1}) = (x \pm \sqrt{x^2 + x_1})/x_1. \tag{13}$$

For Eq. (8), the H^4 sequence gives

$$H \cong -(cB + AB + W)/2B + \sqrt{(\quad)^2 + W(B - A)/B} = Q^4_8; \tag{14}$$

the H^3 sequence

$$H \cong -\frac{W(A - B)}{2(cB + AB + W)} + \sqrt{(\quad)^2 + W\left(\frac{cA + AB + W}{cB + AB + W}\right)} = Q^3_8; \tag{15}$$

the H^2 sequence

$$H \cong (cA + AB + W)/2(A - B) + \sqrt{(\quad)^2 + AW/(A - B)} = Q^2_8. \tag{16}$$

If $A > B$, Eq. (14), or Q^4_8, is eliminated since it can give no real answer. According to Eq. (8), then, the choice in this case between Eq. (15) and

Eq. (16) depends on whether $H^3(cB + AB + W) \gtrless AW^2$. Since we are dealing with approximations, introducing here the approximate value of H as $\sim \sqrt{WA/B}$, we see that this means $c + A + W/B \gtrless \sqrt{WB/A}$. Hence Eq. (15) is better than Eq. (16) as long as $c \gg \sqrt{WB/A}$, and with $A > B$ it follows that Eq. (16) can not be of any importance.

If $A < B$, Eq. (16) is impossible, and the choice between Eq. (14) and Eq. (15) depends on $H^4B \gtrless HW(cA + AB + W)$, or, if $H \cong \sqrt{WA/B}$, on whether $c + B + W/A \gtrless \sqrt{WA/B}$. Again, Eq. (15) is better than Eq. (14) if $c \gg \sqrt{WA/B}$, and with $A < B$, Eq. (14) can hardly ever be applicable.

Hence Eq. (15) must be of wide applicability.

Furthermore, however, if the H^2 term is very small compared to those in H^3 and H, it too may be neglected, and we have from the terms in H^3 and H the convenient and useful formula

$$H \cong \sqrt{W(cA + AB + W)/(cB + AB + W)}. \tag{17}$$

This holds well when both $H^3(cB + AB + W)$ and $HW(cA + AB + W) \gg | H^2W(A - B) |$. On the basis of the approximate value of $H \sim \sqrt{WA/B}$, the condition for the first of these requirements is

$$c + A + W/B \gg c_{crit}, \text{ or } |\sqrt{WA/B} - \sqrt{WB/A}\,|, \tag{18}$$

and that for the second requirement is $c + B + W/A \gg c_{crit}$. Hence with $A > B$, Eq. (15) may be simplified to Eq. (17) as long as $c \gg \sqrt{WA/B}$, and with $A < B$ as long as $c \gg \sqrt{WB/A}$. It will be necessary to use Eq. (15) rather than Eq. (17) only if c is not larger than the critical value c_{crit} defined in Eq. (18).

Finally, if Eq. (17) is applicable, it may be further simplified in actual use, depending on the relative magnitudes of cA (or cB), AB, and W. If $AB \ll W$,

$$H \cong \sqrt{W(cA + W)/(cB + W)}; \tag{19}$$

if $AB \gg W$,

$$H \cong \sqrt{WA(c + B)/B(c + A)}; \tag{20}$$

and if c is high enough so that $AB + W$ is negligible,

$$H \cong \sqrt{WA/B}, \tag{21}$$

in which the effect of c disappears altogether.

Now the limitations for the applicability of Eq. (17) and hence of its simplifications, Eqs. (19–21), become apparent in that, if $c < \sqrt{WA/B}$, H is calculated to be greater than c, which is absurd unless c is also less than $\sqrt{W}$. The value of H in formula (21), recognized as H_* or H_{1e} of Chapter IV, is the limiting value approached as c increases indefinitely and is never attained in any finite concentration.

Not only does Eq. (17) reduce to $H = \sqrt{W}$ if $A = B$ or if $c = 0$, but it appears to do so also if both A and B are very large compared to c. This result for very small c, however, is not significant since the very form of this equation is applicable only if $c \gg c_{\text{crit}}$; while the limit, in other words, seems correct, the equation begins to give wrong results as c falls to $\sim c_{\text{crit}}$ or less. Correct values for such low concentrations may be obtained only through Eq. (15). But if c_{crit} is extremely small, e.g. when $A \cong B$, then Eq. (17) will apply very well for the whole practical range of c; hence its wide usefulness.

In summary then, Eq. (15), except for vanishingly small values of c, is always dependable and quite precise for the calculation of H as $f(c,A,B)$ in the solution of a 1 : 1 salt of weak acid and weak base. Eq. (17) is almost as good if c is large compared to both $\sqrt{WA/B}$ and $\sqrt{WB/A}$. In either case, simplifications in actual use are possible, depending on the data ($AB \gtrless W$, etc.). These rules of applicability have been tested and verified by comparing values calculated through these approximations with the theoretical value obtained by trial and error numerical solution of Eq. (8). Such "exact" calculation may be performed by starting with the best approximation and then dealing with the equation in the form Σ (positive terms) $= \Sigma$ (negative terms), continuing the approximation until the equality is established to any desired degree of precision. Some values of H so calculated for salts of various combinations of acids and bases (both weak) are shown in Table 8–1

Table 8–1. Effect of concentration on $[H^+]$ in solutions of salts of weak acids and weak bases.

Constants		Concentration ($= c$)			$[H^+]$ from Eq. (21)
A	B	1	10^{-2}	10^{-4}	
1	10^{-4}	7.07×10^{-6}	1.00×10^{-6}	1.41×10^{-7}	10^{-5}
1	10^{-10}	7.04×10^{-3}	9.47×10^{-4}	6.18×10^{-5}	10^{-2}
1	10^{-14}	0.415	8.15×10^{-3}	1.00×10^{-4}	1
10^{-2}	10^{-7}	3.15×10^{-5}	2.23×10^{-5}	3.10×10^{-6}	3.16×10^{-5}
10^{-4}	10^{-8}	1.00×10^{-5}	9.95×10^{-6}	6.70×10^{-6}	10^{-5}
10^{-4}	10^{-10}	1.00×10^{-4}	9.80×10^{-5}	4.15×10^{-5}	10^{-4}
10^{-4}	10^{-14}	7.04×10^{-3}	9.47×10^{-4}	6.18×10^{-5}	10^{-2}
10^{-6}	10^{-12}	9.95×10^{-5}	7.03×10^{-5}	9.47×10^{-6}	10^{-4}
10^{-10}	10^{-14}	7.07×10^{-6}	1.00×10^{-6}	1.41×10^{-7}	10^{-5}

for various concentrations. Values according to the familiar formula, Eq. (21), are given for comparison to show the sign and magnitude of the error possible with such an approximation; the error in H (since it is calculated as H_*) is always positive if $A > B$ and negative if $A < B$. These and all subsequent numerical values are based on $W = 10^{-14}$.

Finally, introducing activity coefficients in the usual way, we have from Eq. (15)

$$H \cong -\frac{\mathbf{W}}{2\gamma^2}\left(\frac{\mathbf{A}-\mathbf{B}}{c\mathbf{B}+\mathbf{AB}/\gamma^2+\mathbf{W}}\right) + \sqrt{(\quad)^2 + \frac{\mathbf{W}}{\gamma^2}\left(\frac{c\mathbf{A}+\mathbf{AB}/\gamma^2+\mathbf{W}}{c\mathbf{B}+\mathbf{AB}/\gamma^2+\mathbf{W}}\right)}, \tag{22}$$

and from Eq. (17),

$$H \cong (1/\gamma)\sqrt{\mathbf{W}(c\mathbf{A}+\mathbf{AB}/\gamma^2+\mathbf{W})/(c\mathbf{B}+\mathbf{AB}/\gamma^2+\mathbf{W})}. \tag{23}$$

When multiplied through by γ these expressions, of course, give the hydrogen ion activity **H**.

For these equations to be used in the calculation of H or of **H** for a salt solution for given values of c, **A**, **B**, a preliminary uncorrected value of H (H_1) is used to find the activity coefficient as $-\log \gamma = 0.505\sqrt{H_1 + c\alpha'}$, in which $\alpha' = \mathbf{B}/(\mathbf{B} + \mathbf{W}/H_1)$. Then if necessary a second value of γ may be calculated from a second value of H, etc., until the quantities are constant.

C. On the Idea of the "Hydrolysis" of Salts

Eq. (8) is based on the fundamental definitions of Chapter I for the quantities H, W, A, B and involves, in addition, only the assumption that a salt is completely dissociated in solution. It is, in other words, only a special case of the general equation

$$D = aA/(A+H) - bB/(B+OH). \tag{24}$$

The solution of the 1 : 1 salt is one in which $a = b = c$. Clearly, if $A = B$ the solution is not neutral unless $a = b$. Just as clearly, if $a = b$, the solution is not neutral unless $A = B$. There is no more need of the idea of the "hydrolysis of the salt" to explain the second of these statements than to explain the first. Nor is it necessary to consider a "process of hydrolysis," a "degree of hydrolysis," or a "hydrolysis constant" in calculating the disturbance of neutrality in a salt solution. We have just seen in Section B how various formulas for the hydrogen ion concentration in such a salt solution are established without the idea of "hydrolysis" at all. It is clear that nothing has been omitted in our treatment, and that nothing more can be achieved by introducing the idea.

Almost without exception, however, the formula $H \cong \sqrt{WA/B}$, or Eq. (21), is "derived" on the basis of the "degree of hydrolysis of the salt."

Unless reached as the utmost simplification of the exact Eq. (8), however, its "derivation," although frequently presented as quite involved, is extremely simple. We have merely to combine the three definitions $W = H(OH)$, $A = H[X^-]/[X^0]$ and $B = OH[M^+]/[M^0]$, *assuming* that $[X^-]/[X^0] = [M^+]/[M^0]$ to obtain the result $H = \sqrt{WA/B}$. The effect of the concentration, c, of the salt then can not possibly be estimated, although we know that H must equal $\sqrt{W}$ when c is 0 for any value of A/B. Essentially this is the same as assuming $\beta = 0$ to find H_* or H_{ie} except that in so doing we realize that the resulting simple formula can not apply to any finite concentration of the salt and that its applicability increases with the concentration.

The formula of Eq. (21) is generally also called the "hydrolysis constant" of the salt, or the "degree of hydrolysis" of the salt, even though obviously enough it would give the same "degree" for any fixed ratio of A/B, i.e. for a strong-strong combination as well as for a weak-weak combination. It does not seem to be generally realized that it is mathematically impossible to define a significant "degree of hydrolysis" for a general salt.

The apparent need for the idea of the "hydrolysis" of salts arises, it seems, from the erroneous implication of the *trilogy* of "acids, bases, and salts," in which there is a suggestion that a salt solution per se is or should be neutral. "Hydrolysis" is then introduced to explain the "exceptional" cases of salt solutions which are not neutral. Actually the only salt solutions which are neutral are possibly those of the alkali salts of the strong monobasic acids, perhaps some twenty-five salts in all. It is clear that salt solutions *in general* are not neutral for obvious reasons, and it is the exceptional case that is neutral. An elementary error involved here is the confusion of the term "normal salt" with "neutral salt solution."

The use of the idea of the "hydrolysis of salts" leads to unnecessary circumlocution. The emphasis on "hydrolysis" leads, in fact, to the explanation that a solution of ammonium chloride is acid not because ammonia is a week base and HCl a strong acid but because "ammonium ion" is a strong acid and "chloride ion" a weak base. Although it may be maintained that these two statements are equivalent, the fact is that the second is useless except in terms of the experimental data of the first statement. No matter how we speak of the "acid-strengths" of ions, the mathematics of the problem must reduce to Eq. (24), in which everything is expressed in terms of ammonia and HCl as *solutes*. The emphasis on "hydrolysis" in fact leads to the curious point of view (mentioned in Chapter v) that a solution of acetic acid, for example, involves two constants (other than W): the ionization constant of the acid and the hydrolysis constant of the anion.

D. "Mechanical" Method of Solving Eq. (8)

A satisfactory numerical solution of Eq. (8) is possible, for almost any given problem, without knowledge of the rules of applicability of the approximations given in Section B, through the simple procedure explained in Section A.

Consider a salt with $A = 10^{-14}$, $B = 10^{-4}$, for which Eq. (21) gives $H = 10^{-12}$ at any concentration:

(1) At $c = 1$. The H^4 sequence gives $H = 10^{-14}$, and the orders of magnitude of the terms of Eq. (8) with this value of H (expressed as the negative of the exponent of the base 10) are 60, 46, 46, 42, 42. The choice was therefore wrong. The H^3 sequence gives $H = 1.419 \times 10^{-12}$, with magnitudes 52, 40, 42, 40, 42. The magnitude must be right and the accuracy good. Eq. (17), easier to handle than Eq. (15) but less accurate, must be good too since $c \gg \sqrt{WB/A}$; it gives $H = 1.414 \times 10^{-12}$, with the same magnitudes of the terms as for the H^3 sequence. Since the terms used are the two largest terms, this value must also be fairly accurate. The H^2 sequence is impossible since $A < B$. The "exact" value is 1.422×10^{-12}.

(2) At $c = 10^{-5}$. The H^4 sequence gives $H = 1.00 \times 10^{-9}$, with magnitudes 40, 36, 36, 37, 42. It includes the two largest terms but not the sequence of the three largest terms; hence the order of magnitude may be right but it can not be accurate. The H^3 séquence gives $H = 1.092 \times 10^{-9}$, with the same magnitudes (evidently correct and very accurate). Eq. (17) gives $H = 3.16 \times 10^{-10}$, but the magnitudes are 42, 38, 37, 38, 42, so that this approximation does not use the two largest terms, as it should, to be applicable. (But one would not use this formula in this case since c is now less than $\sqrt{WB/A}$.) The H^2 sequence is impossible. The "exact" value is 1.092×10^{-9}.

E. Failure of Method in "Abnormal" Cases

The method has been tested in a great number of different real problems, as will be evident in subsequent chapters, and there seems to be always some sequence of three terms giving at least a fair approximation in the form of a quadratic solution, and applicable over a useful range of concentration for given values of the constants. But when "abnormal" values are assigned to the constants, somehow the method does not work. For example, the equation for a dibasic acid at concentration a is:

$$H^4 + H^3K_1 + H^2(K_1K_2 - W - aK_1) - HK_1(W + 2aK_2) - K_1K_2W = 0. \tag{25}$$

For the usual or "normal" relation of the constants, $K_1 > K_2$, the method applies as expected. Of the three sequences the H^2 is possible only if $a < K_2$ and hence is not important. That in H^3 will apply for small H and hence, in general, at low concentrations. The H^4 sequence will predominate at high H and hence for high a; in fact it is impossible if the H^2 term is positive. In general the formula from the H^4 sequence applies as long as $a \gg \sqrt{K_1K_2}$, and it is necessary to use the H^3 formula only when this is not the case; see Chapter XIV, Section A2. This condition of applicability of the H^4 sequence has been tested at three concentrations ($a = 1, 10^{-2}, 10^{-4}$) for five combinations of constants: $K_1, K_2 = 10^{-2}, 10^{-4}$; $10^{-2}, 10^{-8}$; $10^{-2}, 10^{-12}$; $10^{-6}, 10^{-8}$; $10^{-6}, 10^{-12}$. The H^4 formula is satisfactory in every case except for the first of these acids at $a = 10^{-4}$, or when a is not $\gg \sqrt{K_1K_2}$; then the H^3 sequence gives a good answer, $H = 1.414 \times 10^{-4}$ as compared to the "exact" value of 1.405×10^{-4}.

But when K_1 is assumed to be less than K_2, which may be called an "abnormal" dibasic acid, no sequence of three terms of Eq. (25) gives a dependable formula applicable for any appreciable range of concentration. The results of the calculations for two such hypothetical acids are shown in Table 8–2; the H^2 sequence is not used at all since it does not give real answers for the data assumed. It is seen that neither the H^4 nor the H^3 series "behaves" at all; the test by orders of magnitude shows them to be of no value.

Nevertheless, the numerical mechanical method already explained may be modified, apparently, to apply even more generally. A good approximate solution of such an equation, if not available in the quadratic solution of a sequence of three terms, may turn up in some different combination of terms. In the present "abnormal" problem, if all terms but those in H^4 and H are dropped, we have

$$H \simeq \sqrt[3]{K_1(2aK_2 + W)} \simeq \sqrt[3]{2aK_1K_2}. \tag{26}$$

The orders of magnitude for the six sets of values listed in the table show this approximation to have some dependability, and this is borne out by comparison of the last two columns of Table 8–2.

Table 8–2. Values of [H+] for hypothetical dibasic acid with $K_1 < K_2$.

K_1	K_2	a	H^4 seq.	H^3 seq.	Exact.	Eq. (26)
10^{-6}	10^{-2}	1	9.9×10^{-4}	1	2.83×10^{-3}	2.71×10^{-3}
		10^{-2}	9.9×10^{-9}	1.4×10^{-2}	5.85×10^{-4}	5.85×10^{-4}
		10^{-4}	impossible	2.0×10^{-4}	1.00×10^{-4}	1.26×10^{-4}
10^{-12}	10^{-6}	1	1.0×10^{-6}	1.01	1.52×10^{-6}	1.26×10^{-6}
		10^{-2}	1.4×10^{-7}	0.02	2.96×10^{-7}	2.71×10^{-7}
		10^{-4}	1.0×10^{-7}	0.0101	1.09×10^{-7}	5.85×10^{-8}

F. Titration Curves

Section B developed the approximate solutions of a high degree equation pertaining to an equivalent point in which, for a given set of constants, the only variable is the single concentration, c, that of the salt characterizing the equivalent point. The same situation holds for the solution of any single solute—polybasic acid, ampholyte, etc. But in a titration curve there are two variable concentrations, the principal one, e.g. a, that of a polybasic acid being titrated with a strong base, and b_s, the concentration of the strong base. The full equation in such a case may be of high degree, of course, depending on the number of ionization constants of the acid, but the choice of the best sequence for the description of the various parts of the titration curve turns out to be quite simple (always provided $K_1 > K_2 > K_3$, etc.).

1. Tribasic Acid and Strong Base. For illustration we shall consider the titration of a tribasic acid with strong base, for which the polynomial expression for H, from Eq. v(86), is

$$H^5 + H^4(b_s + K_1) - H^3K_1(a - b_s - K_2 + W/K_1) - H^2K_{12}$$
$$(2a - b_s - K_3 + W/K_2) - HK_{123}(3a - b_s + W/K_3) - K_{123}W = 0. \quad (27)$$

In this problem there are mathematically four points of stoichiometric significance, fixed by $b_s = (0, 1, 2, 3)a$. When $b_s = 0$, we have the equation for the pure acid; when $b_s = a$, the first equivalent point or the solution of the salt NaH_2X, etc. Unless a is extremely small, these four stoichiometric points are best defined successively by the four possible "sequences" of the equation in the order H^5, H^4, H^3, H^2 respectively. Thus the best approximation for H in the solution of the pure acid, at any appreciable concentration, is (from the H^5 sequence)

$$H \cong - K_1/2 + \sqrt{K_1^2/4 + K_1(a - K_2 + W/K_1)}; \quad (28)$$

for the salt NaH_2X ($b_s = a = c$), the H^4 sequence gives

$$H \cong - (K_{12} - W)/2(c + K_1) + \sqrt{(\quad)^2 + K_{12}(c - K_3 + W/K_2)/(c + K_1)}; \quad (29)$$

for the salt Na_2HX ($b_s = 2a = 2c$), the H^3 sequence gives

$$H \cong - (K_{23} - W)/2(c + K_2 - W/K_1)$$
$$+ \sqrt{(\quad)^2 + K_{23}(c + W/K_3)/(c + K_2 - W/K_1)}; \quad (30)$$

for the salt Na_3X ($b_s = 3a = 3c$), or the last equivalent point, the H^2 sequence gives

$$H \cong W/2(c + K_3 - W/K_2) + \sqrt{(\quad)^2 + K_3W/(c + K_3 - W/K_2)}. \quad (31)$$

The applicability of each of these formulas for the particular solution involved depends, as in the case of Eq. (15) for the salt MX, on the magnitude of c as a function of the constants. The limits in the present case will not be considered in detail, but the above formulas are all quite accurate provided c is not extremely small.

The very start of the titration, when b_s is still small compared to a, must therefore be described by the H^5 sequence since when $b_s = 0$ the titration curve must simplify to Eq. (28), the expression for the pure acid. The end of the titration of the first equivalent point, including the equivalent point and even the start of the addition of the second equivalent of base, must be described by the H^4 sequence since when $b_s = a$ it must reduce to Eq. (29) for the pure salt involved. The middle range of the titration (of this first equivalent) is best described approximately by the two terms in H^4 and H^3 alone, leading to the simple form of the Henderson equation,[(4)] or the "common ion effect," for the region far from stoichiometric points. Hence, at the start ($a \gg b_s$),

$$H \cong -(b_s + K_1)/2 + \sqrt{(\quad)^2 + K_1(a - b_s - K_2 + W/K_1)} = Q_{27}^5, \quad (32)$$

or $H \cong \sqrt{K_1(a - b_s)}$. In the middle range ($0 < b_s < a$),

$$H \cong K_1(a - b_s - K_2 + W/K_1)/(b_s + K_1), \quad (33)$$

or $H \cong K_1(a - b_s)/b_s$ as in the Henderson equation. Through the first equivalent point ($b_s \cong a$),

$$H \cong \frac{K_1(a - b_s - K_2 + W/K_1)}{2(b_s + K_1)} + \sqrt{(\quad)^2 + \frac{K_{12}(2a - b_s - K_3 + W/K_2)}{b_s + K_1}} = Q_{27}^4, \quad (34)$$

or $H \cong \sqrt{K_1 K_2}$ when $b_s = a$. In the next "middle range" ($a < b_s < 2a$),

$$H \cong K_2(2a - b_s - K_3 + W/K_2)/(b_s - a + K_2 - W/K_1), \quad (35)$$

or $H \cong K_2(2a - b_s)/(b_s - a)$. Through the second equivalent point ($b_s \cong 2a$),

$$H \cong \frac{K_2(2a - b_s - K_3 + W/K_2)}{2(b_s - a + K_2 - W/K_1)} + \sqrt{(\quad)^2 + \frac{K_{23}(3a - b_s + W/K_3)}{b_s - a + K_2 - W/K_1}} = Q_{27}^3, \quad (36)$$

or $H \cong \sqrt{K_2 K_3}$ when $b_s = 2a$. In the "middle range" ($2a < b_s < 3a$),

$$H \cong K_3(3a - b_s + W/K_3)/(b_s - 2a + K_3 - W/K_2), \quad (37)$$

(4) L. J. Henderson, *J. Am. Chem. Soc.*, **30**, 954 (1908).

or $H \cong K_3(3a - b_s)/(b_s - 2a)$. Through the third equivalent point ($b_s \cong 3a$),

$$H \cong \frac{K_3(3a - b_s + W/K_3)}{2(b_s - 2a + K_3 - W/K_2)} + \sqrt{(\quad)^2 + \frac{K_3 W}{b_s - 2a + K_3 - W/K_2}} = Q_{27}^2, \quad (38)$$

or $H \cong \sqrt{K_3 W/a}$ when $b_s = 3a$. Sufficiently past the last equivalent point, the last two terms of Eq. (27) are used, giving

$$H \cong W/(b_s - 3a - W/K_3) \cong W/(b_s - 3a). \quad (39)$$

These are of course all "ideal" equations, based on unity activity coefficients. The introduction of activity coefficients will be illustrated with Eq. (38) for the curve through the third equivalent point. As usual, if $\gamma_0 = 1$, $\gamma_{+\text{ or }-} = \gamma$, $\gamma_= = \gamma^4$ and $\gamma_\equiv = \gamma^9$,

$$H \cong \frac{\mathbf{K}_3(3a - b_s + \gamma^4\mathbf{W}/\mathbf{K}_3)}{2\gamma^6(b_s - 2a + \mathbf{K}_3/\gamma^6 - \gamma^2\mathbf{W}/\mathbf{K}_2)} + \sqrt{(\quad)^2 + \frac{\mathbf{K}_3\mathbf{W}/\gamma^8}{b_s - 2a + \mathbf{K}_3/\gamma^6 - \gamma^2\mathbf{W}/\mathbf{K}_2}}. \quad (40)$$

To calculate an accurate value of the hydrogen ion concentration, H, for a pure solution of the salt Na_3PO_4, then, given the concentration a and the activity constants, we proceed as follows. An approximate value of H is taken as $H = \sqrt{K_3 W/a}$, under Eq. (38). This is used to calculate the ionic strength from Eq. v(51c), in which the α's are functions of H and the three ionization constants and in which H is certainly negligible as a term. With γ calculated from μ, Eq. (40), with $b_s = 3a$ for the pure salt, is used for a corrected value of H and if necessary the activity coefficient is modified, etc., until the desired precision is reached.

2. Two Acids (One Strong, One Weak) and Strong Base. One more example will be given to bring out another principle. We shall consider the titration, with strong base, of a mixture of two acids, one strong and one weak:

$$D = a_s + aA/(A + H) - b_s; \quad (41)$$

$$H^3 + H^2[A - a_s + b_s] - H[A(a_s + a - b_s) + W] - AW = 0. \quad (42)$$

Whenever strong acid or strong base is involved, the approximate equation for *excess* of either one or the other is generally given by the first two or the last two terms of the polynomial in H respectively, as seen above in Eq. (39). Thus if in this case $b_s < a_s$, so that there is present excess of strong acid, the first two terms of the equation give

$$H \cong (a_s - b_s) - A \cong (a_s - b_s). \quad (43)$$

As b_s approaches a_s, or as we near the first equivalent point, the curve is then best described by the H^3 sequence of Eq. (42), and

$$H \cong (a_s - b_s - A)/2 + \sqrt{(\quad)^2 + A(a_s + a - b_s) + W} = Q_{42}^3. \quad (44)$$

When $b_s = a_s$ this reduces to the equation for a mixture of the salt M_sX_s (mathematically of no effect, since $g = a_s - b_s = 0$) plus weak acid, and hence for the weak acid alone, or

$$H \cong -A/2 + \sqrt{A^2/4 + aA + W} \cong \sqrt{aA}. \quad (45)$$

Eq. (44) of course holds through the first equivalent point and describes the beginning of titration of a weak acid.

The middle range of the titration of the weak acid is then given by the H^2 and H terms,

$$H \cong [A(a_s + a - b_s) + W]/(b_s - a_s + A). \quad (46)$$

If we write g' for $-g$, or for the excess of strong base over strong acid, this is simply $H \cong A(a - g')/g'$. The approach to the second equivalent point, where $g' = a$, is given by the H^2 sequence,

$$H \cong [A(a - g') + W]/2(g' + A) + \sqrt{(\quad)^2 + AW/(g' + A)} = Q_{42}^2. \quad (47)$$

At the equivalent point ($g' = a$), this is the solution of the salt M_sX (in presence of M_sX_s) at concentration $c = a$, or

$$H \cong W/2(a + A) + \sqrt{(\quad)^2 + AW/(a + A)}. \quad (48)$$

In excess of strong base ($g' > a$), the last two terms are used, and

$$H \cong W/(g' - a - W/A) \cong W/(g' - a). \quad (49)$$

Finally, Eq. (42), with $b_s = 0$, describes the mixture of two acids, one strong and one weak,

$$H^3 + H^2(A - a_s) - H[A(a_s + a) + W] - AW = 0. \quad (50)$$

Except for extremely small values of a_s (i.e. much less than A), the last term may be neglected and

$$H \cong (a_s - A)/2 + \sqrt{(\quad)^2 + A(a_s + a) + W} = Q_{50}^3, \quad (51)$$

which may be simplified further in actual use; roughly, $H \cong a_s$.

With two or more independent acids (especially if two or more of them are weak), the choice of the best approximation even at the stoichiometric points ($b_s = 0$, a_1, $a_1 + a_2$, etc.) is more difficult than in the case of a polybasic acid because of the added variation of the ratio of the independent concentrations of the acids and because of the lack of restriction on the relative values of the ionization constants of the acids. The possible useful approximate solutions for problems involving two or more weak acids will therefore be considered under the specific problems themselves.

3. Relation Between Approximate Solutions and Titration Curve. An explanation of the applicability of successive sequences of three terms for the approximate value of H in successive ranges of the titration is to be found in the familiar shape of the pH titration curve, as a plot of pH against b_s. If strong base is added to a solution containing, for example, strong acid (a_s) and weak tribasic acid (a), H decreases as b_s increases for fixed values of

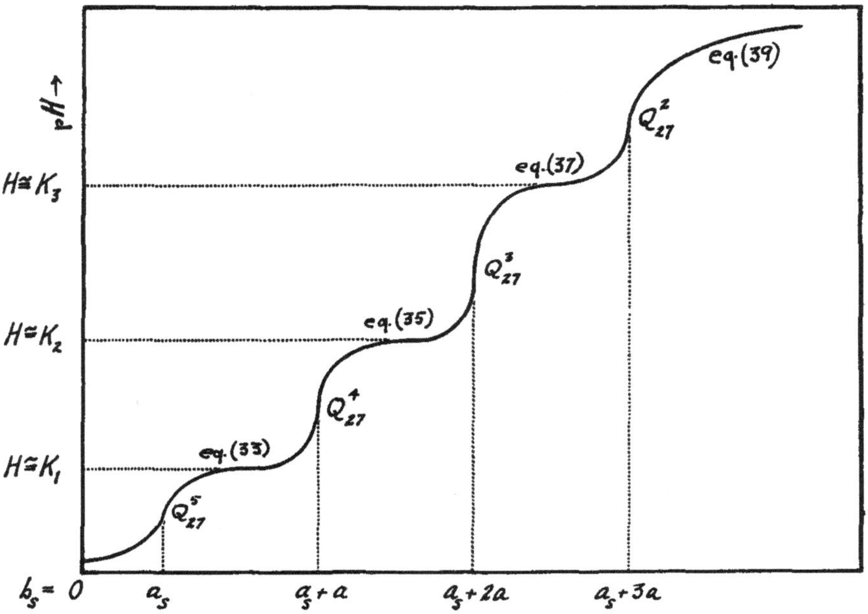

Fig. 8–1. Applicability of Eqs. (32–39) in titration with strong base of strong acid and tribasic acid.

the concentrations a_s and a and of the constants involved, so that it is to be expected that the important terms of the general equation move progressively from H^5 and H^4 to H^1 and H^0. At the same time the pH titration curve runs through a step-wise course as b_s passes through the four successive equivalent points, given by $b_s = a_s, a_s + a, a_s + 2a, a_s + 3a$, as shown in Fig. 8–1, for $a_s < a$. The portion of the curve through each equivalent point is roughly of the shape of a hyperbolic sine curve, with $\Delta b_s = \Delta \sinh \mathrm{p}H$. According to Eq. v(77), then, this means that as b_s varies through an equivalent point, the quantity $\mathbf{w}$ of Eq. v(76) must remain approximately constant, so that pH varies as the inverse hyperbolic sine of the function $\mathbf{v}/2\sqrt{\mathbf{w}}$ of the coefficients of a quadratic equation. We shall return to this point more fully in Chapter IX, in which the relation is applied to the problem of the feasibility of titration. For the present we note simply that the curve is expected to be described approximately by successive quadratic equations as it passes through successive equivalent points; these successive sequences of three terms are indicated directly on Fig. 8–1. Near the mid-points between equivalent points, the curve is described approximately by two consecutive terms of overlapping sequences, which give an equation of the Henderson type. With a_s present, b_s in these equations (27 and 32–39) is to be replaced by $g'(= b_s - a_s)$.

IX

Feasibility (Sharpness) of Titration; Exact Case of Strong Acid and Strong Base (Pure)

A. The Exact Case of (Pure) Strong Acid and Strong Base

If a pure solution of a strong acid is titrated with pure strong base, the whole process is described by a quadratic equation and the problem is capable of a complete exact solution. This case will therefore be studied as a preliminary to the general problem of the feasibility or sharpness of titrations.

1. The Pure Acid. The discussion of the pure base would be symmetrically related to that of the pure acid and hence will be omitted.

$$H - OH = a_s, \tag{1}$$

$$H^2 - Ha_s - W = 0, \tag{2}$$

$$H = a_s/2 + \sqrt{a_s^2/4 + W}. \tag{3}$$

In extrapolation to "infinite dilution," then, "[H+]" does not remain "equal" to "[HCl]," as we ordinarily assume, but diverges from it at very low concentration. With H assumed equal to a_s for a strong acid, the limiting value of the derivative, dH/da_s, would be 1, but it is actually 1/2.

If a_s is not very small, the familiar approximation is

$$H \cong a_s. \tag{4}$$

It is interesting to ask what is being "neglected" in this formula. The usual statement that we are neglecting "the H^+ coming from the water" has meaning only from the point of view of a special theory of ionization, that of dissociation, in which we have to assume that the solute in aq. SO_3, for example, is not SO_3 but "H_2SO_4," etc. Mathematically and more generally, it is clearer to say that we neglect OH (not a part of the quantity H) as compared to a_s or to H itself. The quantity neglected is $(\sqrt{a_s^2/4 + W} - a_s/2)$, or $\sim W/a_s$ for large a_s and $\sim \sqrt{W}$ for vanishingly small a_s.

2. Titration Curve (Acid Titrated With Base).

$$H - OH = a_s - b_s = g. \tag{5}$$

$$H = (a_s - b_s)/2 + \sqrt{(\quad)^2 + W}. \tag{6}$$

These are all still identical with Eqs. (1–3), since mathematically $b_s = -a_s$. Similarly then, as in Eq. (4), $H \cong a_s - b_s$.

3. Buffer Capacity. For the relation between the buffer capacity, π, and the nature of the titration curve, we have, from Eq. (5)

$$\pi = db_s/dpH = 2.3(H + OH); \tag{7}$$

and at the equivalent point, when $a_s = b_s$ and $H = \sqrt{W}$,

$$\pi_e = 2.3(2\sqrt{W}). \tag{8}$$

Also, $d\pi/dpH = (2.3)^2(-H + OH) = 0$ when $a_s = b_s$; and $d^2\pi/dpH^2 = (2.3)^3(H + OH)$, always positive. Hence π is always a minimum at the

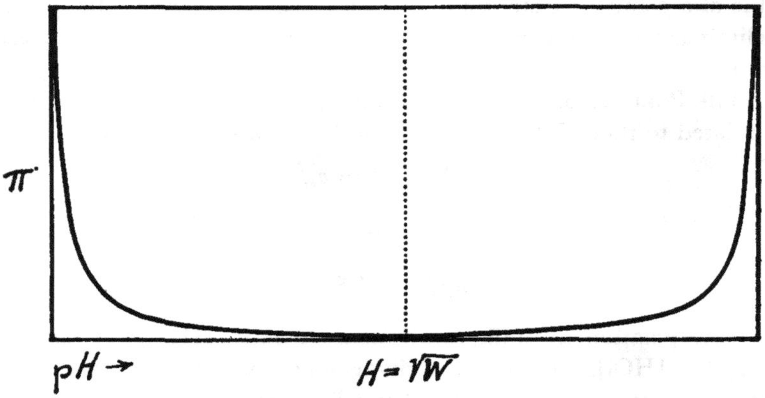

Fig. 9–1. Buffer capacity, strong acid and strong base.

equivalent point whatever the value of a_s. The relations are shown graphically in Fig. 9–1. Since the slope of the titration curve, pH against b_s, is simply $1/\pi$, this means that there will be an inflection of the titration curve at π_{min}, and hence at the equivalent point, as shown in Fig. 9–2. Both curves are symmetrical with respect to the equivalent point.

The slope of the titration curve (Fig. 9–2) at the equivalent point is $1/2.3(2\sqrt{W})$, independent of the concentration. But since this is a point of inflection, this slope does not describe the practical feasibility of the titration, which must depend also on the concentration, a_s, of the acid being titrated.

4. Feasibility of Titration. We shall now define the *feasibility of titration* as a quantity $(\Delta pH)_p$ or Δ_p, the change in pH caused by the addition of p per cent of titrating material (strictly, by an increase of p per cent of the stoichiometrically required *concentration* of titrating reagent) between $p/2$ per cent short of the equivalent point and $p/2$ per cent beyond the equivalent point. This definition, we shall find, leads to an unusually simple and

uniform analytical expression for the feasibility of most titrations. By definition then,

$$\Delta_p = (\text{p}H)_{+p/2} - (\text{p}H)_{-p/2} = \log (H_{-p/2}/H_{+p/2}). \tag{9}$$

The most direct approach to the formulation of an expression for Δ_p in the present exact case is to note that pH, as plotted in Fig. 9–2, varies as the inverse hyperbolic sine of a quantity, F, which will be called the *feasibility function* of the titration and which is here some simple function of g or $a_s - b_s$. From Eq. (6), with $F = g/2\sqrt{W}$, we have, through Eq. v(77),

$$\text{p}H = (\text{p}W)/2 - (\sinh^{-1} F)/2.3. \tag{10}$$

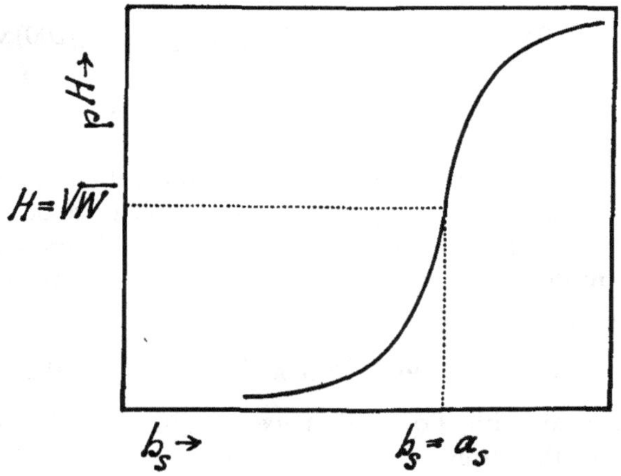

Fig. 9–2. Titration curve through the equivalent point, strong acid and strong base.

Eq. (10), identical of course with Eq. (6), is merely a form very useful in connection with titration curves in terms of pH; in its application it is to be recalled that $\sinh^{-1}(x) = -\sinh^{-1}(-x)$.

At the equivalent point, $g = 0$, $F_e = 0$ and p$H = (\text{p}W)/2$. At $p/2$ per cent short of the equivalent point, $g = pa_s/200$ and $F_- = pa_s/400\sqrt{W}$; at $p/2$ per cent past the equivalent point, $F_+ = -pa_s/400\sqrt{W}$. If a_s is assumed not to be changed by dilution during the addition of p per cent of the base through the equivalent point, then $|F_-| = |F_+| = F$. For mathematical purposes we shall assume therefore that the concentration of the titrating reagent is being changed without change of volume. In any given case, if p is very small, this assumption will be quite justifiable. It must be noted, however, that a_s represents the concentration at the equivalent point and not the initial concentration before start of titration; it must also be remembered always that although g changes, the quantity a_s is not changed, except through dilution, by the addition of b_s.

On this basis, with

$$F = pa_s/400\sqrt{W}, \tag{11}$$

$$\Delta_p = (2/2.3)\sinh^{-1}(pa_s/400\sqrt{W}). \tag{12}$$

Then if $F \gg 1$, $\Delta_p \cong \log(4F^2)$. Moreover, the feasibility increases with a_s since $d\Delta_p/da_s = (2/2.3)p/400\sqrt{W(F^2+1)} \cong 2/(2.3a_s)$ if $F^2 \gg 1$.

5. Effect of Volume Change. The effect of volume change is also easily introduced. If c_A and V_A are the concentration and volume of the original acid solution being titrated with a base solution of concentration c_B, then $(V_B)_{\mp}$, the volume of base added at $p/2$ per cent short or in excess of the equivalent point, is $c_AV_A(1 \mp p/200)/c_B$. Then if the total volume equals $V_A + B_B$, we have $g_{\mp} = a_s - b_s = \pm pc_AV_A/200(V_A + V_B)$. Hence

$$F_{\mp} = g_{\mp}/2\sqrt{W} = \pm pc_Ac_B/400(c_A + c_B \mp pc_A/200)\sqrt{W}, \tag{13}$$

to be used in

$$\Delta_p = (\sinh^{-1}F_- - \sinh^{-1}F_+)/2.3. \tag{14}$$

This equation gives the feasibility in terms of the actual concentrations of the solutions being mixed, with p equal to the percentage of the volume of titrating solution used. If p is small, which is the only interesting condition, then we may neglect the quantity $pc_A/200$ in the denominator of Eq. (13), whereupon Eq. (14) reduces to

$$\Delta p \cong (2/2.3)\sinh^{-1}[pc_Ac_B/400(c_A + c_B)\sqrt{W}], \tag{15}$$

which is identical with Eq. (12) since $c_Ac_B/(c_A + c_B) = \mathbf{c}_e = a_s$, the concentration of the strong acid at the equivalent point; cf. Eq. VII(28). Eq. (13) also shows that the feasibility increases with the ratio c_B/c_A for fixed c_A.

6. Effect of Ionic Strength. The foregoing formulas assume unity activity coefficients if W is a thermodynamic constant. Otherwise, with the usual simplification concerning activity coefficients, Eq. (11) becomes

$$F = pa_s\gamma/400\sqrt{\mathbf{W}}. \tag{16}$$

Hence the feasibility, Δ_p, decreases for a given value of a_s as the total ionic strength increases, other things being equal, as long as γ decreases with increasing ionic strength. For the titration of pure acid, on the other hand, Eq. (16), with $-\log\gamma \cong 0.505\sqrt{a_s}$ at the equivalent point, would predict a minimum in the feasibility at $a_s \cong 3.0$; but this result has no significance since the limiting law is no longer valid at such high ionic strength.

Finally, the feasibility in terms of hydrogen ion activity, or $(\Delta p\mathbf{H})_p$, is related to the feasibility Δ_p, which is in terms of hydrogen ion concentration, as follows: $(\Delta p\mathbf{H})_p = \Delta_p - \log(\gamma_{+p/2}/\gamma_{-p/2})$, and as long as p is small the two quantities are obviously practically identical.

B. "Titration" Distinguished from "Neutralization"

These terms are widely interchanged in the literature and in instruction with resulting confusion. But in a mathematical subject it is important to distinguish them if, as is the case, they can not be given the same mathematical meaning. "Neutralization" is the process of making the concentrations of H and OH of a solution equal since neutrality means $H = OH$. It is a process done upon a solution. The "neutralization" of a solute of any kind has no mathematical meaning whatever. We may undo the effect of one solute (upon the relations of H and OH of the solution) by means of another solute, and in this sense we may speak of "neutralizing" such an *effect* but never of neutralizing a solute. We may also change the degree (or degrees) of ionization of a solute but we are to remember that such degrees are defined as $[X^-]/a$, $[X^=]/a$, etc., and never as H/a. But if we are dealing with a "strong acid" as a solute in solution, there is simply nothing we can do to it in the way of changing its degree of ionization or of "neutralizing" it. We may, however, be able to *measure* it by titration with some base. "Titration" is a process of measurement done upon, or at least attempted upon, a solute, and it may or may not be accompanied by the neutralization of the solution. Even the titration of strong acid by strong base does not necessarily end in a neutral solution, for this depends on the other solutes present in the solution.

The neglect of this distinction leads to the familiar but mathematically inconsistent point of view that various bases and acids "react" (i.e. "neutralize each other") to *various degrees* in solution. What is called the "degree of hydrolysis" of the salt "formed" (although we insist at the same time that no salt is "formed" at all since a salt is completely dissociated in the solution) is then considered to be the "degree of incompleteness" or "degree of reversal" (given the symbol α_r, by T. B. Smith, Ref. VIII–3, p. 215) of the "reaction" and it is customary to "explain" the various shapes of titration curves in terms of such a "degree of reversal." While such a point of view seems to be satisfactory in the simple case of strong-weak monobasic titration, this "degree of reversal" for a weak-weak titration has to be defined (in order to avoid two separate and unequal "degrees" for the same salt) as $([M^0] + [OH^-])/c = ([H^+] + [X^0])/c$, which then reduces to H/c or OH/c for the case of strong-strong titration. On such a point of view, T. B. Smith has presented the "sharpness of titration" as being inversely proportional to this "degree of reversal" or α_r.

While there may be a mathematical significance in H/c for a solution (pure or otherwise) of sodium chloride, it is certainly and obviously neither a "degree of hydrolysis" of the salt nor a "degree of reversal" of the reaction of HCl and NaOH in water. It is in fact not a "degree" at all, for the limits of a "degree" are 0 to 1, whereas with $H = \sqrt{W}$ in pure aq. NaCl, the quantity H/c has a limit of infinity when $c = 0$ and a lower limit depending on c. But whatever its meaning, the quantity is still inadequate to account for actual titration curves of strong acid plus strong base. According to such a point of view the titration is always uniformly very sharp (for a given concentration) if H at the equivalent point equals $\sqrt{W}$, so that α_r or H/c is very small. This is not always true, however; as will be shown later, it is possible to have $H = OH = \sqrt{W}$ at an equivalent point of "titration" of strong acid with strong base and yet have such buffering action in the solution as to make the titration impossible. This point will be explained and illustrated in detail in Chapter XII in connection with the buffer capacity of salt solutions.

The emphasis therefore must be placed not on the substances being titrated against each other, nor upon any specific "chemical reaction" between them, but upon the buffer capacity of the solution at the equivalent point for the titration. In the titration of *pure* a_s with b_s, the feasibility is seen to be an exact and explicit function of F, with $F = pa_s/400\sqrt{W}$. From the point of view of the "degree of reversal" as controlling the feasibility, it would seem possible to identify F with the expression $p/400\alpha_r$, since here $\alpha_r = H_e/a_s = \sqrt{W}/a_s$. But the expression $p/400\alpha_r$ is not expected to have any *general* meaning in connection with the feasibility of titration. The nature of the process rather suggests that the feasibility function will be proportional, for a given value of p, to the concentration of the acid being titrated, at the equivalent point, and inversely proportional to the buffer capacity of the solution at that point since the slope of the titration curve is actually $1/\pi$. Hence, using Eq. (8), we shall identify F with the expression

$$F = (2.3pa_s/200)/\pi_e. \qquad (17)$$

We can then understand that it is possible for the equivalent point in titration of a_s with b_s to be neutral ($H_e = \sqrt{W}$) and for the feasibility still to be very poor if π_e, which is a function not only of H_e but of the total composition of the solution, is high.

C. General Relation Between π and the Titration Curve

If a solution containing various solutes is being titrated with a strong base so that $b_s = \Sigma(c\beta) - D = \mathbf{Y}$, then the buffer capacity of the solution with respect to the strong base is $\pi = d\mathbf{Y}/d\text{p}H$, which is always positive; also, $d\pi/d\text{p}H = d^2\mathbf{Y}/d\text{p}H^2$ and $d^2\pi/d\text{p}H^2 = d^3\mathbf{Y}/d\text{p}H^3$. The reciprocal of π is the slope of the "titration curve" (course of pH against b_s). When $d\pi/d\text{p}H = 0$ (maximum or minimum of π) there is an inflection in the titration curve: a horizontal inflection (when pH is plotted vertically against b_s) at π_{max} and a vertical inflection at π_{min}. The first type is a "titration mid-point" always occurring between two vertical inflections or "titration end-points." When $d^2\pi/d\text{p}H^2 = 0$ there is an inflection in the curve of π against pH. This is in general of little significance in the titration curve except when it occurs at a maximum or minimum of π itself, whereupon it indicates the merging of a maximum and a minimum of π and hence the merging of two adjacent inflections (one horizontal and one vertical) of the titration curve.

Thus the titration curve of Fig. 9–3b corresponds to the π curve of Fig. 9–3a. If the two π maxima at points (2) and (4) merge, obliterating the minimum at point (3), the titration curve becomes that of Fig. 9–3c. Since $\pi = db_s/dpH$, the quantity b_s represents, in Fig. 9–3a, the area under the

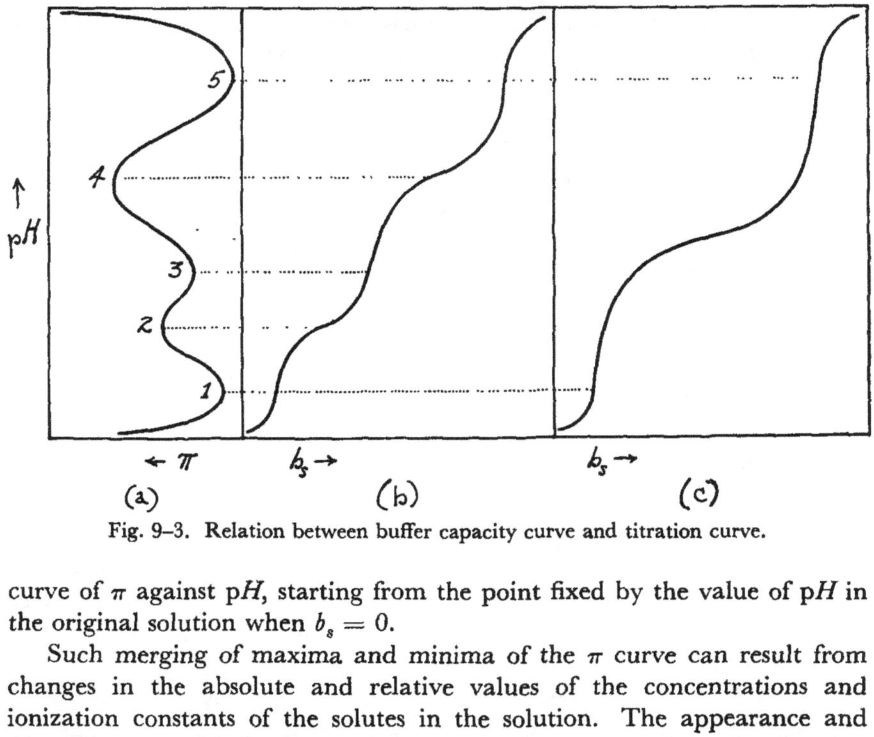

Fig. 9–3. Relation between buffer capacity curve and titration curve.

curve of π against pH, starting from the point fixed by the value of pH in the original solution when $b_s = 0$.

Such merging of maxima and minima of the π curve can result from changes in the absolute and relative values of the concentrations and ionization constants of the solutes in the solution. The appearance and disappearance of inflection points (horizontal or vertical) in the titration curve will therefore depend on the conditions, $f(c,K)$, which determine the existence and merging of maxima and minima of the π curve. The actual position (in respect to pH) of any inflection in the titration curve is of course obtained from the value of H for $d\pi/dpH = 0$, which will fix a maximum of π and hence a horizontal pH inflection (mid-point type) if $d^2\pi/dpH^2 < 0$, and a minimum of π and hence a vertical pH inflection (end-point type) if $d^2\pi/dpH^2 > 0$. When $d^2\pi/dpH^2 = 0$ together with $d\pi/dpH = 0$, two such inflections of pH merge and disappear.

The feasibility of titration has to do with the vertical (end-point type) inflection in the pH curve. If such an inflection exists in a particular titration there will in general be two important questions to consider: (1) the difference between the inflection point and the particular equivalent point involved, or the "error of the titration"; and (2) the "feasibility of the titration" as already defined. The "error of the titration" is of course zero for the pure strong-strong case already considered and only approximate answers are possible in all other cases. This question of the "titration error" has been considered by others, and quite accurate formulas and answers are available for several important cases, such as the titration of a weak acid, of a dibasic acid, etc. (Ref. VIII–1). A few special cases of general interest will be considered in subsequent chapters. The feasibility similarly can be expressed exactly only for the pure strong-strong case, and all other cases, which are the truly important ones, must be dealt with in approximations.

D. The Feasibility of a Titration in General

The feasibility, Δ_p, is defined in Eq. (9). But since H is in general a cubic or higher degree function of concentrations and ionization constants, only approximate forms of Eq. (9) are practical except for the simple case of strong-strong titration in absence of all kinds of "weak" or buffering solutes.

In order to arrive at the useful forms of the approximations required, we shall assume, as explained in Chapter VIII, that the titration curve through the equivalent point is best described by the H^n sequence of the full exact polynomial in H for the problem in question and that the H^n sequence, as usual, means $H^n + vH^{n-1} + wH^{n-2} = 0$. The feasibility is then given by the expression

$$\Delta_p = \log \frac{[-v/2 + \sqrt{(\quad)^2 - w}]_{-p/2}}{[-v/2 + \sqrt{(\quad)^2 - w}]_{+p/2}} = \log \frac{[Q^n]_{-p/2}}{[Q^n]_{+p/2}}. \tag{18}$$

The coefficients v and w are of course simple explicit functions of concentrations and constants, and they are given the values corresponding to $-p/2$ per cent and $+p/2$ per cent excess of titrating agent, with respect to the equivalent point, for numerator and denominator respectively. In the form of Eq. (18) the feasibility calculated will be as accurate as the dependability of the H^n sequence, which is usually high for appreciable concentrations.

Since w is negative for values of the coefficients applying near the equivalent point, as may be seen by examination of Eqs. VIII(34), (36), (38), (44), and (47), then less accurately

$$\Delta_p \cong \log \frac{[-v/2\sqrt{-w} + \sqrt{(\quad)^2 + 1}]_{-p/2}}{[-v/2\sqrt{-w} + \sqrt{(\quad)^2 + 1}]_{+p/2}}. \tag{19}$$

This is obtained from Eq. (18) when and if

$$(-w)_{-p/2} \cong (-w)_{+p/2}, \tag{20}$$

which is quite regularly the case if $pc/200$ (c being the concentration, at the equivalent point, of the substance being titrated) is small compared to the other concentrations of significance in the coefficient w. In fact, the final approximate formula for H_e (H at the equivalent point or when $p = 0$) is always, from the H^n sequence under consideration,

$$H_e \cong (\sqrt{-w})_{p=0}, \tag{21}$$

so that the quantities of the expression (20) are very nearly equal because each is very nearly equal to $(H_e)^2$.

If the *feasibility function* is now defined as follows,

$$F_{\mp} = (-v/2\sqrt{-w})_{\mp p/2}, \tag{22}$$

Eq. (19) becomes

$$\Delta_p = \log\,[(F_- + \sqrt{F_-^2 + 1})/(F_+ + \sqrt{F_+^2 + 1})], \tag{23}$$

or

$$\Delta_p = (\sinh^{-1} F_- - \sinh^{-1} F_+)/2.3. \tag{24}$$

As will be seen later, F_- and F_+ (whether or not equal in magnitude) are usually of opposite sign. Eqs. (19), (23), and (24) are identical.

The quantity Δ_p will of course be positive in titration with base and negative in titration with acid, but as in the case of the buffer capacity, only the absolute value is significant.

Furthermore, when Eq. (20) holds, it also results, as will be seen later, that except in some unusual situations in regard to independently variable concentrations, $|F_-| \cong |F_+| = F$, whereupon, but now still less accurately,

$$\Delta_p \cong (2 \sinh^{-1} F)/2.3. \tag{25}$$

Finally if, as is generally the case at least where the feasibility is good, $F \gg 1$, then Eq. (25) may be simplified to

$$\Delta_p \cong \log\,(4F^2). \tag{26}$$

In probably all cases of practical interest, formula (25), with its simplification in Eq. (26), is applicable and usually more than sufficiently accurate. In this case the estimation of the feasibility requires the writing of the simple expression, F, the feasibility function defined as in Eq. (22). In problems involving several ionization constants this expression may be quite involved. Since we are dealing all along, however, with approximations, the involved expression can be simplified step by step to leave ultimately, usually, some very simple expression applicable to some general type of problem; and the final simple expression is such that it may easily be remembered and applied to specific cases of that type of problem.

Eq. (25), in which one value, F, is taken for both $|F_-|$ and $|F_+|$, assumes not only that the shape of the titration curve through the equivalent point is that of a hyperbolic sine curve but in addition that it is symmetrical for the region $-p/2$ per cent to $+p/2$ per cent, whereas Eq. (24) takes into account the asymmetry of the curve on the two sides of the equivalent point.

The curve actually has this shape and this symmetry only in the case of pure strong-strong titration.

Finally we see, by referring to the exact case of strong-strong titration, Eq. (17), that the approximate hyperbolic sine relation involved in Eq. (25) must be one in which

$$F = (2.3pc/200)(dpH/dc')_e = (2.3pc/200)/(\pi_{c'})_e, \tag{27}$$

in which c is the concentration of the substance being titrated and c' the concentration of the titrating substance, both at the equivalent point. In titration with strong base, $\pi_{c'}$ is simply the quantity π, buffer capacity with respect to strong base.

Formula (27) is important in that it allows us to write the approximate feasibility function, F, directly without going through the procedure of simplification detailed in the beginning of this section. We require only an approximate value of the buffer capacity, π_e, at the equivalent point, in respect to the titrating substance; and this being a function of (c, H, K) requires merely an approximate value of H_e, or H at the equivalent point, which is usually known in the form of some familiar first approximation, or which may, at any rate, generally be taken according to Eq. (21).

The condition of applicability of the simple expression of Eq. (27) for the parameter to be used in the inverse hyperbolic sine relation of Eq. (25) is perhaps best stated as follows: it is applicable when the pH titration curve possesses a vertical type of inflection corresponding to the equivalent point in question. This follows since the curve $x = \sinh y$, corresponding to $F = \sinh \text{p}H$, has an inflection point at x (or F) $= 0$. The condition for the applicability of Eq. (27) is therefore the mathematical condition for an inflection point in the titration curve. ("Inflection point," unmodified, will hereafter always mean a vertical type of inflection.) The pure strong-strong titration curve, as already seen, always possesses such an inflection, and exactly at the equivalent point. The conditions involving the concentrations and the equilibrium constants in other important cases will be developed later.

E. Example; Two Weak Acids and Strong Base

We have, therefore, two ways of proceeding for the evaluation of the feasibility function, F. Procedure (i) is to find the sequence, H^n, describing the titration curve through the equivalent point and to use its coefficients then through the expressions (18–26), simplifying as desired and as permissible. Procedure (ii) is to evaluate the pertinent buffer capacity at the equivalent point through an approximate formula for H_e, and then to write F through Eq. (27).

Procedure (i) is more laborious but capable of higher accuracy in difficult cases as it does not assume that $|F_-| = |F_+|$, thus allowing for the asymmetry of the titration curve through the equivalent point. But Procedure

(II) has the advantage of leading more directly to simple formulas for the feasibility of typical titration cases, easily remembered and directly applied in any particular problem. The assumption that $|F_-| = |F_+|$, it may be pointed out again, is quite generally permissible as long as $pc/200$ is small compared to other significant concentrations in the solution.

Both procedures, however, will lead by appropriate simplifications of the function F to the same ultimate simple formula which will be of greatest practical use for any given type of problem. We shall now illustrate the use of both procedures with the calculation of the feasibility of the titration with strong base of a weak acid in the presence of a weaker acid, with concentrations a_1 and a_2, and constants A_1 and A_2 respectively; $A_1 > A_2$.

PROCEDURE I. For two acids and strong base,

$$D = (a\alpha)_1 + (a\alpha)_2 - b_s; \tag{28}$$

$$= a_1A_1/(A_1 + H) + a_2A_2/(A_2 + H) - b_s; \tag{29}$$

$$H^4 + H^3(A_1 + A_2 + b_s) + H^2[A_1A_2 - W + (A_1 + A_2)b_s - a_1A_1 - a_2A_2] - H[A_1A_2(a_1 + a_2 - b_s) + W(A_1 + A_2)] - A_1A_2W = 0. \tag{30}$$

The titration curve through the first equivalent point ($b_s = a_1$) is best described for appreciable and comparable concentrations by the H^3 sequence (according to the explanation in Chapter VIII). Hence what may be called the most accurate (though not "exact") estimate of the feasibility of the titration of one acid in presence of the other would be that calculated through Eq. (18) with the full coefficients of the H^3 sequence. With $x = pa_1/200$,

$$\Delta_p = \log \frac{\dfrac{A_1x + A_2(a_2 - a_1 + x) - A_{12} + W}{2(a_1 - x + A_1 + A_2)} + \sqrt{(\quad)^2 + \dfrac{A_{12}(a_2 + x) + (A_1 + A_2)W}{a_1 - x + A_1 + A_2}}}{\dfrac{-A_1x + A_2(a_2 - a_1 - x) - A_{12} + W}{2(a_1 + x + A_1 + A_2)} + \sqrt{(\quad)^2 + \dfrac{A_{12}(a_2 - x) + (A_1 + A_2)W}{a_1 + x + A_1 + A_2}}}. \tag{31}$$

The coefficients may be simplified as the numerical data warrant. In this way the feasibility calculated will be as dependable as the H^3 sequence of Eq. (30), which may mean actually an accuracy much higher than would ordinarily be required.

If we neglect A_2 in $(A_1 + A_2)$,

$$\Delta_p = \log \frac{\dfrac{A_1x + A_2(a_2 - a_1) - A_{12} + W}{2(a_1 - x + A_1)} + \sqrt{(\quad)^2 + \dfrac{A_{12}(a_2 + x) + A_1W}{a_1 - x + A_1}}}{\dfrac{-A_1x + A_2(a_2 - a_1) - A_{12} + W}{2(a_1 + x + A_1)} + \sqrt{(\quad)^2 + \dfrac{A_{12}(a_2 - x) + A_1W}{a_1 + x + A_1}}}, \tag{32}$$

an equation for future reference (Chapter XIV).

As long as $pa_1/200$ is small compared to both a_1 and a_2, the two values of $-w$ in Eq. (18) may be assumed to be equal, whereupon we may express the feasibility functions as follows, further neglecting A_{12} and W in v,

$$F_{\mp} = -v/2\sqrt{-w} = [\pm A_1x + A_2(a_2 - a_1)]/2\sqrt{A_1(a_1 + A_1)(a_2A_2 + W)}, \tag{33}$$

to be used in the asymmetric formula, Eq. (24). If $pa_1/200 \gg (a_2 - a_1)A_2/A_1$, then

$$F_- = -F_+ = F = (p/400)\sqrt{a_1A_1/a_2A_2(1 + A_1/a_1)(1 + W/a_2A_2)}, \tag{34}$$

to be used in Eq. (25), or in its simplification Eq. (26) if $F \gg 1$. Provided, finally, that $a_1 \gg A_1$ and that $a_2A_2 \gg W$, this becomes

$$F = (p/400)\sqrt{a_1A_1/a_2A_2}. \tag{35}$$

The simple expression for F in Eq. (35), to be used in Eq. (25), is still very accurate for a considerable variation of the relative magnitudes of both constants and concentrations. We have, in other words, in the expression

$$\Delta_p = (2/2.3)\sinh^{-1}[(p/400)\sqrt{a_1A_1/a_2A_2}], \tag{36}$$

a direct, widely applicable, and fairly accurate formula for the feasibility of the titration with strong base of a weak acid in presence of a weaker acid. The formula, however, shows no dependence upon the absolute values of either the concentrations or the constants. Such dependence is brought out in the more elaborate but nevertheless direct and completely explicit asymmetric formula, Eq. (24), involving the separate feasibility functions, F_- and F_+, defined by Eq. (33), and to some extent by the symmetrical formula of Eq. (34).

PROCEDURE II. The buffer capacity with respect to strong base is, from Eq. (28),

$$\pi = 2.3[(a\alpha\rho)_1 + (a\alpha\rho)_2 + H + OH], \tag{37}$$

$$= 2.3[a_1A_1H/(A_1 + H)^2 + a_2A_2H(A_2 + H)^2 + H + OH]. \tag{38}$$

For H_e, at the equivalent point, the H^3 sequence of Eq. (30) may be simplified to give

$$H_e \cong -\frac{A_2(a_1 - a_2)}{2(a_1 + A_1)} + \sqrt{(\quad)^2 + \frac{A_1A_2a_2}{a_1}\left(\frac{1 + W/a_2A_2}{1 + A_1/a_1}\right)}, \tag{39}$$

$$\cong \sqrt{(A_1A_2a_2/a_1)(1 + W/a_2A_2)/(1 + A_1/a_1)}, \tag{40}$$

$$\cong \sqrt{A_1A_2a_2/a_1}. \tag{41}$$

With H_e from Eq. (40) introduced into Eq. (38), in which we take $(A_1 + H_e) \cong A_1$ and $(A_2 + H_e) \cong H_e$ since $A_1 > H_e > A_2$, the result is

$$\pi_e \cong 2(2.3)\sqrt{(a_1a_2A_2/A_1)(1 + A_1/a_1)(1 + W/a_2A_2)}, \tag{42}$$

or, corresponding to Eq. (41),

$$\pi_e \cong 2(2.3)\sqrt{a_1a_2A_2/A_1}. \tag{43}$$

Hence, according to Eq. (27) we have again Eqs. (34) and (35) for the feasibility function, F.

The final simple approximation of Eq. (35) shows a feasibility not only independent of the actual concentrations involved but also of the activity coefficients and hence of the general ionic strength, always on the assumption that $\gamma_0 = 1$ and $\gamma_+ = \gamma_-$. The conditions for the applicability of the formula are that the concentrations a_1 and a_2 are both appreciable, that both acids are weak, with $A_1 > A_2$, and that the first is not too strong and the second not too weak, or $a_1 \gg A_1$ and $a_2A_2 \gg W$.

Nevertheless this simple type of formula is practical and important and since it is obtained so readily through Procedure II, by substitution of the approximations $H_e \cong \sqrt{A_1A_2a_2/a_1}$ and $\pi_e = 2(2.3)\sqrt{a_1a_2A_2/A_1}$, into Eq. (27), this method will be used later to arrive at simple formulas for the feasibility function, F, for the titration with strong or weak base of any acid in the presence of any number of other acids and bases. It is to be noted that the function F, so obtained, is then to be used to calculate the feasibility, Δ_p, through formula (25) by means of a table of hyperbolic sines or through its simplification, Eq. (26), if this is warranted.

The final simple expression for the function F will always be an approximate one applicable only for certain values and relations of the parameters involved. If in the present case the first acid is fairly strong so that we may not neglect A_1 relative to a_1, then

$$H_e \cong \sqrt{A_1A_2a_2/(a_1 + A_1)}, \tag{44}$$

$$\stackrel{\rightarrow}{=} \sqrt{a_2A_2}; \tag{45}$$

and

$$F \cong (p/400)\sqrt{a_1^2A_1/a_2A_2(a_1 + A_1)}, \tag{46}$$

$$\stackrel{\rightarrow}{=} (p/400)\sqrt{a_1^2/a_2A_2}. \tag{47}$$

The limits reached with increasing A_1 then, or Eqs. (45) and (47), must be the approximations applying for the titration of strong acid (now a_1 in) presence of a weak acid (now a_2). If the second acid is extremely weak so that we may not neglect W relative to a_2A_2, then

$$H_e \cong \sqrt{A_1(a_2A_2 + W)/a_1}, \tag{48}$$

$$\stackrel{\rightarrow}{=} \sqrt{A_1W/a_1}; \tag{49}$$

and

$$F \cong (p/400)\sqrt{(a_1A_1)/(a_2A_2 + W)}, \tag{50}$$

$$\stackrel{\rightarrow}{=} (p/400)\sqrt{a_1A_1/W}. \tag{51}$$

In this case the limits reached with decreasing value of a_2A_2, or Eqs. (49) and (51), must be the approximations for the titration of a weak acid (a_1) alone since the second weak acid (a_2) may now be ignored.

X

Problems Involving One Ionization Constant

A. Weak Acid and Strong Base

This case will be considered first, in detail, as the typical problem of this chapter.

1. The Pure Acid. At concentration a,

$$H - OH = a\alpha = aA/(A + H); \tag{1}$$

$$H^3 + H^2A - H(aA + W) - AW = 0. \tag{2}$$

(The corresponding equation for a pure base is, by symmetry,

$$(OH)^3 + (OH)^2B - (OH)(bB + W) - BW = 0.) \tag{3}$$

The approximate solutions of Eq. (2) depend primarily on whether $A \lesseqgtr \sqrt{W}$.

(a) With $A < \sqrt{W}$, the second and fourth terms of Eq. (2) are small compared to the other two since for pure acid, $H > \sqrt{W}$; then

$$H \cong \sqrt{aA + W}. \tag{4}$$

(b) With $A \nless \sqrt{W}$, the choice between the H^3 and H^2 *sequences* depends on the concentration.

(1) If $a \gg \sqrt[3]{AW}$, Eq. (1) may be used to show that the H^3 term of Eq. (2) is $> AW$; hence, from the H^3 sequence,

$$H \cong -A/2 + \sqrt{A^2/4 + aA + W}; \tag{5}$$

and since under these conditions aA is always greater than W, this becomes

$$H \cong -A/2 + \sqrt{A^2/4 + aA}. \tag{6}$$

Now if, furthermore, $a \gg A/4$, then as for H_* in Chapter III,

$$H \cong \sqrt{aA}. \tag{7}$$

an expression increasing in accuracy as the concentration increases. Consideration of Eq. (4) shows that as a becomes large enough we approach Formula (7) whatever the value of A.

(2) If $a < \sqrt[3]{AW}$, the H^2 sequence is better and

$$H \cong (a + W/A)/2 + \sqrt{(\quad)^2 + W}. \tag{8}$$

Now if $a > W/A$,

$$H \cong a/2 + \sqrt{a^2/4 + W}, \tag{9}$$

which is like the equation for a strong acid at very low concentration, giving $H \rightleftharpoons \sqrt{W}$ when $a \rightleftharpoons 0$.

(3) If $a \cong \sqrt[3]{AW}$, then the middle terms of Eq. (2) give

$$H \cong a + W/A. \tag{10}$$

2. The Salt. For the salt M_sX of strong base and weak acid, at concentration c,

$$OH - H = cp = cH/(A + H). \tag{11}$$

As pointed out in Chapter 1, this is then mathematically equivalent to a base with ionization constant W/A. Hence

$$OH - H = c(W/A)/(W/A + OH), \tag{12}$$

and

$$(OH)^3 + (OH)^2W/A - (OH)(cW/A + W) - (W/A)W = 0, \tag{13}$$

which is to be compared with Eq. (3) for the real base. The approximate solutions may obviously be written down by analogy with Eqs. (4–10):

(a) If $A > \sqrt{W}$, then

$$OH \cong \sqrt{W(c + A)/A}, \text{ or } H \cong \sqrt{AW/(c + A)}, \tag{14}$$

which corresponds to Eq. (4).

(b) If $A \not> \sqrt{W}$, and

(1) $c > \sqrt[3]{W^2/A}$,

$$OH \cong - W/2A + \sqrt{(\quad)^2 + W(c + A)/A}, \tag{15}$$

$$\cong - W/2A + \sqrt{(\quad)^2 + cW/A}; \tag{16}$$

and if $c \gg W/4A$,

$$OH \cong \sqrt{cW/A}, \text{ or } H \cong \sqrt{AW/c}, \tag{17}$$

which, like Eq. (7) for the acid, is also $(OH)_*$ or the formula approached as $c \rightleftharpoons \infty$.

(2) $c < \sqrt[3]{W^2/A}$,

$$OH \cong (c + A)/2 + \sqrt{(\quad)^2 + W}, \tag{18}$$

$$\cong c/2 + \sqrt{c^2/4 + W}, \tag{19}$$

the analog of Eq. (9).[(1)]

Finally, for the salt MX_s of weak base and strong acid, the relations are mathematically those for a pure weak acid with ionization constant W/B. Hence, from Eq. (2),

$$H^3 + H^2 W/B - H(cW/B + W) - (W/B)W = 0, \tag{20}$$

with approximate solutions exactly as in Eqs. (4–10), with W/B in place of A.

The final rough formulas, therefore, are $H = \sqrt{aA}$ and $OH = \sqrt{bB}$ for the acid or base, and $OH = \sqrt{cW/A}$, $H = \sqrt{cW/B}$ for the corresponding salts, provided that the quantity "cK" $\gg W$. For very weak acid or base, or more generally when "cK" is of the order of W, the rough formulas become $H = \sqrt{aA + W}$, $OH = \sqrt{bB + W}$ for the acid or base and $OH = \sqrt{W(c + A)/A}$, $H = \sqrt{W(c + B)/B}$ for the corresponding salts.

Table 10–1 gives values of $[H^+]$ for solutions of salts of strong acid and weak base at various concentrations; these values also pertain to solutions of pure weak acids with $A = W/B$.

3. Titration Curve; Weak Acid Titrated With Strong Base.

$$H - OH = a\alpha - b_s = aA/(A + H) - b_s; \tag{21}$$

$$H^3 + H^2(b_s + A) - H[A(a - b_s) + W] - AW = 0. \tag{22}$$

a. Approximations.

(1) At the start (high H), AW may be neglected as the smallest term; then

$$H \cong -(b_s + A)/2 + \sqrt{(\quad)^2 + A(a - b_s) + W} = Q^3_{22}, \tag{23}$$

(1) Only Eqs. (14), (16), and (17) seem to be found in the literature; and because of the usual method of explaining or deriving these equations (through the unnecessary idea of the "hydrolysis" of the salt, as though the problem were quite different from that of a pure base or pure acid, mathematically), the interrelation and the conditions for the applicability of these equations are neither clearly nor consistently presented. Eq. (16), usually called the "exact" or the "more exact" equation, is said for example to "reduce" to Eq. (17) if $A > 10^{-10}$, if $A > W$, if $W/A \ll c$, if W/A is very small, if "the degree of hydrolysis" is small, or if $c \gg OH$. But these conditions are obviously neither all the same nor even related, and from what has already been said, it is clear that not one of them is generally correct; the nearest is the third, that $W/A \ll c$. Eq. (14), on the other hand, is found recommended for "cases of very slight hydrolysis" (but with no statement of critical values of the explicit parameters c and A leading to "very slight hydrolysis"); or it is called "necessary" for small values of c "of the magnitude of A or less."

which is further simplifiable in actual use by the neglect of either A or b_s in $(b_s + A)$, depending on their relative magnitudes, and by neglect of W.

Table 10-1. Effect of concentration on $[H^+]$ in solution of salt of strong acid and weak base, or in solution of weak acid, with $A = W/B$.

B	Concentration			
	10	0.1	10^{-3}	10^{-5}
10	1.41×10^{-7}	1.005×10^{-7}	1.00×10^{-7}	1.00×10^{-7}
10^{-2}	3.16×10^{-6}	3.32×10^{-7}	1.05×10^{-7}	1.00×10^{-7}
10^{-5}	1.00×10^{-4}	1.00×10^{-5}	1.005×10^{-6}	1.41×10^{-7}
10^{-8}	3.16×10^{-3}	3.16×10^{-4}	3.11×10^{-5}	2.70×10^{-6}
10^{-11}	9.95×10^{-2}	9.51×10^{-3}	6.18×10^{-4}	9.90×10^{-6}
10^{-14}	2.70	9.16×10^{-2}	9.99×10^{-4}	1.00×10^{-5}
10^{-15}	6.18	9.90×10^{-2}	1.00×10^{-3}	1.00×10^{-5}

(2) Somewhere in mid-titration, H^3 and AW are both negligible compared with the concentration terms, and then

$$H \cong [A(a - b_s) + W]/(b_s + A) \cong A(a - b_s)/(b_s + A). \qquad (24)$$

If $A \ll b_s$,

$$H \cong A(a - b_s)/b_s, \text{ or } \cong A(1 - n)/n, \qquad (25)$$

with $n = b_s/a$ or the fraction of analytical titration. According to Eq. (24), H at the mid-point, when $b_s = a/2$, $\cong Aa/(a + 2A)$, $\cong A$ if $a \gg 2A$. Eq. (25), the simplest and most useful of the approximations in titration curves, may also be obtained directly from Eq. (21) by setting $H - OH \cong 0$.

(3) Near the end, when the solution is alkaline, H^3 may be neglected in Eq. (22), leaving

$$H \cong [A(a - b_s) + W]/2(b_s + A) + \sqrt{(\quad)^2 + AW/(b_s + A)} = Q_{22}^2, \quad (26)$$

which is further simplifiable by neglect of A and W as terms, if justified. When $b_s = a = c$, Eq. (26) becomes Eq. (15) for the pure salt solution (M_sX), the identity following through the relations of Eqs. VIII(12) and (13). Since Eq. (26) holds through the equivalent point, or for $a \gtrless b_s$, it also represents the effect of the weak acid or of strong base on the salt of the weak acid. In terms of OH this would be

$$OH \cong -(a - b_s + W/A)/2 + \sqrt{(\quad)^2 + W(b_s + A)/A}. \qquad (27)$$

When the excess of strong base $(b_s - a)$ is large enough, Eq. (27) becomes $OH \cong b_s - a$. Roughly, by consideration of Eq. (22), this may be shown to be dependable when $(b_s - a) \gg \sqrt{aW/A}$, or much greater than the value of OH at the equivalent point.

In general then, Eq. (25) should be a fair approximation in the middle range of titration, Eq. (23) better at the start, and Eq. (26) or (27) better at the end. The relative advantages of these equations will depend not on the concentration or on the strength of the acid alone, nor on the pH alone, but on the interrelation of the quantities a, b_s, and A (or a, n, and A). Strictly, unless the solution is very nearly neutral, one of the "second approximations," Eqs. (23) and (26), will be better than the simple formula (25), and the choice between Eq. (23) and Eq. (26) depends on whether $H^3 \gtrless AW$. The critical value of n $(= b_s/a)$, or n_{crit}, is then, from Eq. (21), $n_{\text{crit}} = 1/(1 + \sqrt[3]{W/A^2}) - (\sqrt[3]{AW} - \sqrt[3]{W^2/A})/a$. When $n < n_{\text{crit}}$, Eq. (23) is better than both Eqs. (25) and (26); when $n > n_{\text{crit}}$, Eq. (26) is best. But this is so cumbersome a function that it is better in the present case to proceed differently and to determine the merits and limitations of the very simple Eq. (25) itself.

b. Applicability of the Henderson equation, Eq. (25). The simple, convenient, and widely used equation, Eq. (25), known as Henderson's equation (Ref. VIII-4), or the "common ion effect," is strictly applicable only when the solution is neutral, since from Eq. (21)

$$H = A(a - b_s)/b_s - D(A + H)/b_s, \tag{28}$$

so that Eq. (25) follows only if $D = 0$. Otherwise Eq. (23) and its simplifications will usually be better than Eq. (25) if the solution is acid, and Eq. (26) and its simplifications will be better than Eq. (25) if the solution is alkaline.

But since Eq. (25) is so convenient, it may be useful to be able to judge beforehand the error of a calculation based upon it so as to avoid the use of the more cumbersome second approximations, Eqs. (23) and (26), unless they are necessary.

The criterion for the applicability of Eq. (25) can be stated in useful form through the following considerations. According to Eq. (28) the Henderson equation is applicable with some degree of approximation as long as $A(a - b_s) \gg |(A + H)D|$, and hence, by substitution of the approximate value of H as $A(1 - n)/n$, as long as $an(1 - n) \gg |D|$. While the solution is acid $(D \cong H)$, or if, from Eq. VII(21), $n < n_n$ or $1/(1 + \sqrt{W/A})$, then the requirement is

$$n \gg \sqrt{A/a}. \tag{29}$$

When the solution is alkaline ($D \cong -$ OH and $n > n_n$), the requirement is

$$n \ll 1 - \sqrt{W/aA}. \tag{30}$$

If these conditions are satisfied, or if n lies between the limits of Eqs. (29) and (30), Eq. (25) is dependable and fairly accurate. Furthermore, if these conditions are almost satisfied, we may calculate the error in the calculation and still use the simple equation with some ultimate accuracy. If e is the percentage error in H to be allowed in calculation through Eq. (25) (the percentage being referred to the calculated, rather than to the true, value), the final conditions for the applicability of the simple equation become: $n = \sqrt{100A/ea}$ to n_n while the solution is still acid, and n_n to $n = 1 - \sqrt{100W/eaA}$ while the solution is alkaline. These are quite wide limits of applicability for ordinary values of a and A as long as e is not to be very large —when Eq. (25) might fail even to give the order of magnitude correctly. When these conditions are not satisfied, Eqs. (23) and (26) are to be used in acid and in alkaline solution respectively. It must be emphasized that the applicability of Eq. (25) is not a function of a, A, H, or n alone, but of a, b_s (or n), and A together. For small errors then,

$$e \cong 100A/an^2 \text{ in acid solution;} \tag{31}$$

$$e \cong -100W/(1-n)^2aA \text{ in alkaline solution.} \tag{32}$$

Example. A solution is $0.275M$ in formic acid and $0.025M$ in sodium formate ($A = 2 \times 10^{-4}$, $a = 0.300$, $b_s = 0.025$). Here $n = 0.083$; hence the solution is acid, but n is not much greater than $\sqrt{A/a}$, or 0.026, as demanded by Eq. (29) for Eq. (25) to apply accurately. According to Eq. (31) the error e will be +9.7 per cent. By Eq. (25), $H = 2.2 \times 10^{-3}$, and corrected for the error, $H = 2.2 \times 10^{-3}$ (0.903) $= 1.99 \times 10^{-3}$. The correct "second approximation" for this case, Eq. (23), gives 2.02×10^{-3}, which is also the exact value calculated from Eq. (21).

4. Buffer Capacity and the Titration Curve.

a. General equations and relations. For the buffer capacity with respect to strong base,

$$\pi = 2.3(a\alpha\rho + H + OH) = 2.3[aAH/(A+H)^2 + H + OH], \tag{33}$$

$$d\pi/dpH = (2.3)^2[a\alpha\rho(\rho - \alpha) - H + OH], \tag{34}$$

$$d^2\pi/dpH^2 = (2.3)^3[a\alpha\rho(\alpha^2 + \rho^2 - 4\alpha\rho) + H + OH]. \tag{35}$$

Graphically, the relations are shown in Fig. 10–1. In this figure the solid curve represents $\pi/2.3$ for unit concentration of the acid ($a = 1$). This "π curve" has in general a maximum and two minima, one in acid solution (minimum a) and one on the alkaline side (minimum b).

Mathematically the "π curve," meaning π as $f(\text{p}H)$, for titration of weak acid solution with strong base is the sum of two curves, one of which is the function $2.3a\alpha\rho$ while the other is the fixed curve of Fig. 9–1 or the function $2.3(H + OH)$. As in Eqs. I(112) and I(113) we call the first π_j, the contribution of the weak acid. That of Fig. 9–1, to which the π_j curve is added, we shall call π_w because it involves only the water ions and the equilibrium constant W; this curve is the same whether or not strong acid is also present. The π curve for titration of strong and weak acids together is therefore $\pi = \pi_w + \Sigma_j \pi_j$.

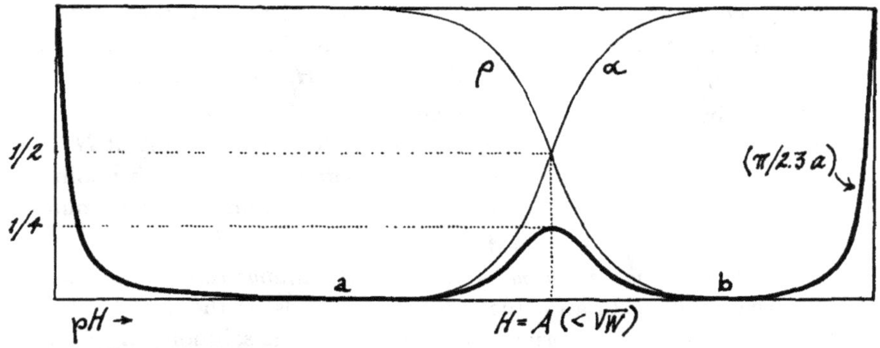

Fig. 10–1. Buffer capacity of solution containing weak acid (schematic).

The π_w curve passes through a minimum at $H = \sqrt{W}$, with $(\pi_w)_{\min} = 2.3(2\sqrt{W})$. The π_j curve of a monobasic weak acid has no minima but a maximum at $H = A$ and $(\pi_j)_{\max} = 2.3\ a/4$. It is a symmetrical bell-shaped curve approaching the value zero at either extremity, as shown in Fig. 10–2. Since both α and ρ are less than unity, the value of $\pi_j/2.3a$ is always smaller

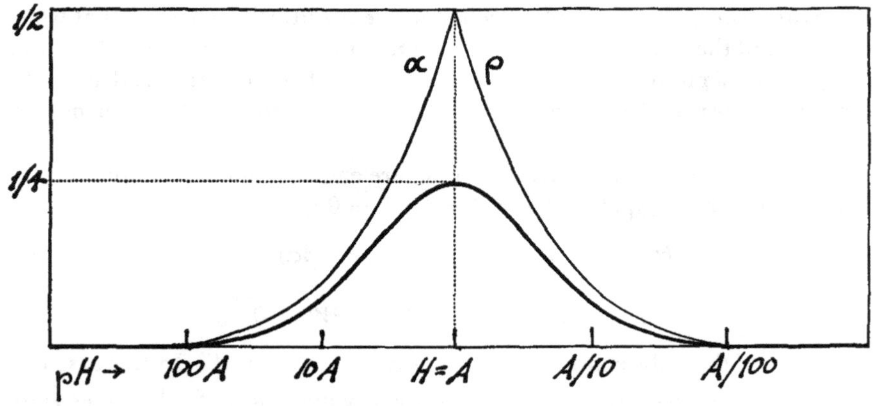

Fig. 10–2. The function $\pi_j/2.3a = \alpha\rho$.

than either α or ρ. When the curves are added, the result, Fig. 10–1, is a curve with a maximum and two minima unless A is either too large or too small. The curve degenerates to π_w if $a = 0$, $A = \infty$ or $A = 0$. If $A = \sqrt{W}$, the π curve is symmetrical; the maximum is at $H = A = \sqrt{W}$, with $\pi_{\max} = 2.3(a/4 + 2\sqrt{W})$ and the two lateral minima are equidistant in pH from $H = \sqrt{W}$; the value of π at either minimum, whether or not the curve is symmetrical, is always greater than $2.3(2\sqrt{W})$. The value of π at $H = A$ is always $2.3(a/4 + A + W/A)$.

As A decreases below $A = \sqrt{W}$, π_{max} increases and moves to the right, occurring at $H < A$. At the same time minimum b is raised and moved to lower H, and when A falls below a certain critical relation the maximum and this minimum of the curve merge and vanish. In this process minimum a is flattened, lowered, and moved to lower H, approaching the limit $H = \sqrt{W}$ when $A = 0$. If A increases above $A = \sqrt{W}$, these changes occur in reverse; π_{max} increases and moves to the left, occurring at $H > A$. Above a certain critical relation for A the maximum merges with minimum a and both vanish, leaving a curve with a single minimum, now minimum b, which approaches $H = \sqrt{W}$ as $A \rightleftharpoons \infty$.

b. Conditions for an "end-point" inflection in the titration curve. The start of the π curve applying to an actual titration depends on the initial value of H, H_{orig}, and hence on the original composition of the solution. In titration of pure weak acid, $H_{orig} \cong \sqrt{aA}$, which is on the left of π_{max} if there is a maximum but on the right of minimum a.

In titration of pure weak acid then, π passes through both a maximum and a minimum (minimum b) if A is neither too large nor too small. If A is too large there will be no maximum but only minimum b near $H = \sqrt{W}$. If A is too small there will be neither a maximum nor a subsequent minimum; the π curve rises continuously from the start. The minimum b in the π curve causes an "end-point" type of inflection in the pH titration curve so that what we shall call the "possibility of the titration" of the weak acid depends on the occurrence of minimum b. If A is very large, therefore, the acid is easily titrated, almost as if it were a strong acid; if A is too small it can not be titrated.

The conditions for the lateral minima are easily estimated. Combination of the equations $d\pi/d\text{p}H = 0$ and $d^2\pi/d\text{p}H^2 = 0$ gives

$$H(1 - \alpha)(1 - 3\alpha) + OH\alpha(2 - 3\alpha) = 0, \tag{36}$$

or

$$H^4 - 2H^3A + 2HAW - A^2W = 0. \tag{37}$$

If $A = \sqrt{W}$ the only solution of Eq. (37) is $H = A = \sqrt{W}$. In this case the maximum of π, occurring at $H = \sqrt{W}$, can vanish only if a becomes zero when the two minima merge simultaneously with the maximum at $H = A = \sqrt{W}$, leaving the curve π_w.

If $A \neq \sqrt{W}$, however, we may have either the vanishing of minimum b by merging with the maximum in basic solution for $A < \sqrt{W}$ or the vanishing of minimum a by merging with the maximum in acid solution for $A > \sqrt{W}$. In either case the critical value of H is given by Eq. (37) and is seen to be independent of a. For $A < \sqrt{W}$ we may neglect H relative to

OH in Eq. (36), which is the same as neglecting the first two terms of Eq. (37), obtaining the condition $\alpha \cong 2/3$ and $H \cong A/2$. Hence, from $d\pi/dpH = 0$, with H neglected relative to OH, the required condition for this merging of the maximum with minimum b is seen to be

$$aA \cong 27W. \tag{38}$$

Since this is the condition for minimum b there will be no end-point type of inflection in the pH curve for the titration of weak acid with strong base unless $aA > 27W$. (On the basis of the γ convention stated in Chapter v, Eq. (38) remains unchanged in terms of activity constants so that this condition for the occurrence of the inflection is substantially independent of the ionic strength.) At this critical point, moreover, the value of $\pi/2.3$, with $H = A/2$ and $A = 27W/a$, is given by Eq. (33) as $8a/27$. Hence both π_{max} and $\pi_{min\,b}$ have increased, with decreasing A, to this common value when the end-point inflection vanishes. [The factor 27 in the condition for the inflection point has already been derived by others through different approaches to the problem.]

It should be noted that the very possibility of titration of an acid, weak or strong, depends on the equilibrium relations of the water ions. The function $a\alpha\rho$, the contribution to the π curve characteristic of the weak acid itself, has no minimum to cause an end-point inflection in the titration curve.

By considerations of symmetry, we see that the merging of minimum a with π_{max} as A increases above $A = \sqrt{W}$ occurs when $\alpha \cong 1/3$, $H \cong 2A$, and

$$a \cong 27A. \tag{39}$$

There will be no inflection in the corresponding pH titration curve unless $a > 27A$. In terms of $\mathbf{A}$, the activity constant, this condition becomes $a > 27\mathbf{A}/\gamma^2$.

If $A > \sqrt{W}$ therefore, the inflection corresponding to minimum b will always occur while minimum a occurs only if $a > 27A$; and if $A < \sqrt{W}$ minimum a always occurs while minimum b occurs only if $aA > 27W$.

The schematic relation between the π curve and the corresponding pH titration curve, in titration of pure weak acid with strong base, is therefore that shown in Fig. 10–3, for a fixed value of a. With high A, Fig. 10–3(a), the π curve resembles that in Fig. 9–1 for strong acid; the titration curve has only one inflection, the vertical type corresponding to the equivalent point, since there is a minimum but no maximum in the π curve. With intermediate values of A the π curve has a maximum preceding the minimum and the pH curve becomes S-shaped with both a horizontal and a vertical inflection, the first near the mid-point of the titration and the second near the equivalent point. With very low A the maximum and the minimum have merged and the pH curve has neither type of inflection.

For a fixed value of a, finally, the π curves for acids of different strengths, with or without maxima, cross each other at a certain value of H. Equating the expressions π_{A_1} and π_{A_2}, according to Eq. (33), for the titration with strong base of each of two different acids, with $A_1 > A_2$, we find that the buffer capacities are equal at $H = \sqrt{A_1 A_2}$. At higher H (lower pH), $\pi_{A_1} > \pi_{A_2}$ and at lower H, $\pi_{A_1} < \pi_{A_2}$.

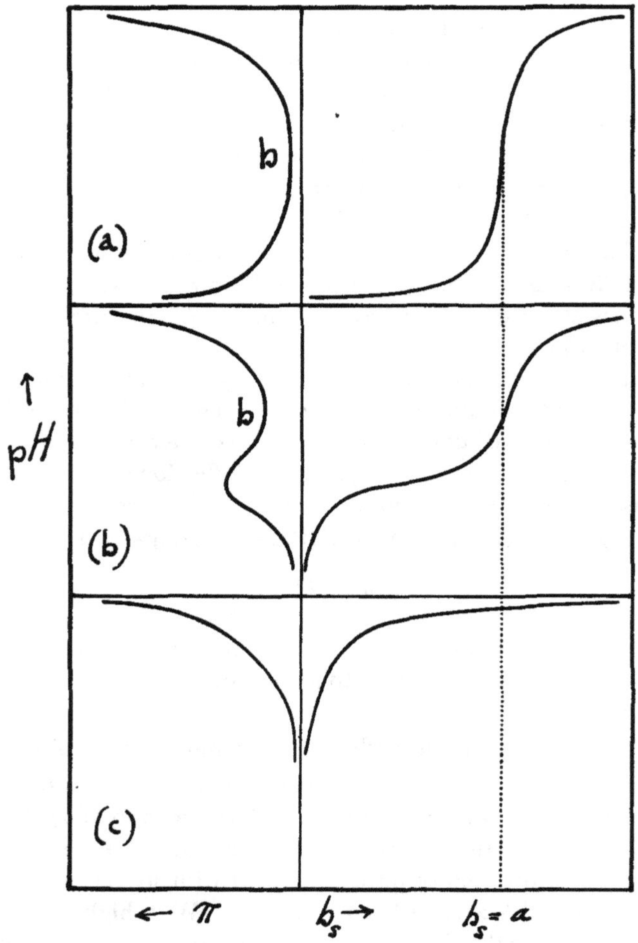

Fig. 10–3. Relation between π curve and titration curve for pure weak acid: (a) $A > a/27$; (b) $a/27 > A > 27W/a$; (c) $27W/a > A$.

c. Approximate values of π *during titration.* From Eq. (25),

$$\pi \cong 2.3b_s(a - b_s)/a, \text{ or } 2.3an(1 - n), \tag{40}$$

so that for a given value of n, π is proportional to a; it is a maximum, with $\pi \cong 2.3a/4$, when $n \cong 1/2$ and $H \cong A$. At the start, for the pure acid solution when $H_{\text{orig}} \cong \sqrt{aA}$, which is usually greater than A and always greater than OH, Eq. (33) gives $\pi_{\text{orig}} \cong (2.3)2\sqrt{aA}$. At the equivalent point or for the salt solution, M_sX, when $H_e \cong \sqrt{AW/a}$, which is smaller than both A and OH,

$$\pi_e \cong (2.3)2\sqrt{aW/A}; \tag{41}$$

in connection with Eq. (41), see the special note on p. 177.

d. Feasibility of titration. The titration curve through the equivalent point is given approximately by Eq. (26). Provided that $paA/200 > W$ we may then use Procedure II of Chapter IX. This means that with π_e given by Eq. (41), Eq. IX(27) leads to the feasibility function of Eq. IX(51):

$$F = (p/400)\sqrt{aA/W}. \tag{42}$$

Hence

$$\Delta_p \cong (2/2.3)\sinh^{-1} F \cong (2/2.3)\sinh^{-1}[(p/400)\sqrt{aA/W}], \tag{43}$$

and if $F \gg 1$, $\Delta_p \cong \log(p^2aA/4 \times 10^4 W)$. These formulas are quite accurate provided that aA is not too small.

From Eq. (43) finally, $d\Delta_p/da = (p/400)\sqrt{A/aW}/2.3\sqrt{F^2+1} \cong 1/2.3a$, and $d\Delta_p/dA = (p/400)\sqrt{a/AW}/2.3\sqrt{F^2+1} \cong 1/2.3A$; in both cases the final approximations apply for $F \gg 1$. With effect of ionic strength considered, on the basis of the limiting law, Eq. (42) becomes

$$F = (p/400)\sqrt{a\mathbf{A}/\mathbf{W}}, \tag{44}$$

substantially independent of ionic strength.

If A is very small and $paA/200 \not> W$, the accurate estimation of the feasibility requires the application of Procedure I of Chapter IX through Eq. (26). The result is

$$F_{\mp} \cong (W \pm paA/200)/2\sqrt{aAW}, \tag{45}$$

for the feasibility functions to be used in Eq. IX(24) for the calculation of Δ_p. In all subsequent considerations the feasibility function will be expressed by means of the simpler Procedure II on the assumption that $paA/200$ is large compared to other terms occurring with it.

e. Positions of maximum and minima of π.

e1. The maximum. For the condition $d\pi = 0$, Eq. (34) becomes

$$H^5 + 3H^4A - H^3[A(a-3A) + W] + H^2A[A(a+A) - 3W] - 3HA^2W - A^3W = 0. \tag{46}$$

If $A = \sqrt{W}$, one solution of this equation is easily seen to be $H = \sqrt{W}$ and it is clear that this represents the value of H for the maximum of π. The maximum therefore occurs exactly at $H = A = \sqrt{W}$ if $A = \sqrt{W}$, and in general very close to $H = A$ if $A \cong \sqrt{W}$. For the value of H when the maximum occurs at $H \cong A$, Eq. (46) is rearranged as follows:

$$H = A\left(1 + \frac{A}{a} - \frac{3W}{aA}\right) - \frac{AW}{Ha}\left(3 + \frac{A}{H}\right) + \frac{H}{a}\left(3A - \frac{W}{A}\right) + \frac{H^2}{a}\left(3 + \frac{H}{A}\right). \tag{47}$$

A maximum occurs only if A is neither too large nor too small since the conditions for the lateral minima of the π curve are at the same time the conditions for the occurrence of the maximum. With this in mind we may substitute the approximate value, A, for H in the right hand member of Eq. (47), obtaining

$$H_{\max} \cong A(1 + 8A/a - 8W/aA). \tag{48}$$

The π maximum therefore occurs at $H \cong A(1 + 8A/a)$ if $A > \sqrt{W}$, and at $H \cong A(1 - 8W/aA)$ if $A < \sqrt{W}$.

e2. The minimum b. For the minimum b, near the equivalent point, we note first that if the acid being titrated is strong, the minimum occurs exactly at the equivalent point, or at $H = \sqrt{W}$, as we saw in Chapter IX; there is then no titration error. This statement refers to a "titration error" defined mathematically as Type 4 in Chapter VII. It is the difference between $(b_s)_i$, the value of b_s required to reach the inflection point, or the actual titer, and the value of b_s corresponding to the stoichiometric or equivalent point. For the practical success of an actual titration one must consider not only the "error" but also the "feasibility" of the titration. For pure strong-strong titration the "error," so defined, is zero at any concentration, but the feasibility depends on the concentration, Eq. IX(11).

For titration of weak acid with strong base the titration error so defined is always negative and it increases in relative magnitude as the quantity aA decreases. The minimum b, in other words, and with it the inflection point of the pH titration curve, occurs early although at a value of $H < \sqrt{W}$. When, according to Eq. (38), aA is no longer greater than $27W$, the π minimum disappears, leaving a pH curve without inflection. From the point of view of the use of the inflection point as the end-point of the titration, we shall say that when the pH titration curve has no inflection to mark the end-point the titration is "impossible." The titration may still be experimentally possible, of course, if the end-point is determined by reference not to an inflection point but to a specified value of H at or near the equivalent point (Types 1, 2, and 3 of Chapter VII). Moreover, whether or not the titration curve possesses an inflection point for the end-point, one may still calculate a "feasibility" for the titration if this is defined as the change in pH for p per cent of titration through the equivalent point. But if the inflection is absent, the feasibility is very low and the approximate feasibility function defined in Eq. IX(27) does not apply. As pointed out in Chapter IX, such an approximate but useful feasibility function applies only when the conditions are such that there is a vertical or end-point type of inflection in the pH titration curve.

Consideration of Eqs. (14–17) shows that if either a or A is large, the value of H at the equivalent point may be written as $H_e \cong \sqrt{AW/(a+A)}$. For this approximate value of H, near which the π minimum b will occur, Eq. (46) is rearranged as follows:

$$H^2 = \frac{AW}{a+A}\left[1 + \frac{3H}{A} + \frac{3H^2}{A^2} + \frac{H^3 a}{A^2 W}\left(1 - \frac{3A}{a} + \frac{W}{aA}\right) - \frac{3H^4}{A^2 W} - \frac{H^5}{A^3 W}\right]. \quad (49)$$

If we now substitute the approximate value of $\sqrt{AW/(a+A)}$ for H in the right hand member, we have

$$H^2 \cong \frac{AW}{a+A}\left\{1 + \frac{4a}{a+A}\sqrt{\frac{W}{A(a+A)}} + \frac{aW}{A(a+A)^2}\left[3 + \sqrt{\frac{W}{A(a+A)}}\right]\right\}. \quad (50)$$

But since aA must be greater than $27W$ for this minimum to occur, the third term in the brace may be neglected. Finally then, since for $\varepsilon \ll 1$, $\sqrt{1+\varepsilon} \cong 1 + \varepsilon/2$,

$$H_{\min} \cong \sqrt{AW/(a+A)} + 2aW/(a+A)^2, \quad (51)$$

which is greater than H_e. With $A > a$, as for strong acid, this means $H_{\min} \cong \sqrt{W}$; but for weak acid we have

$$H_{\min} \cong \sqrt{AW/a} + 2W/a. \quad (52)$$

Through the relation $1/(1+\varepsilon) \cong 1 - \varepsilon$, Eq. (51) gives

$$OH_{\min} \cong \sqrt{(a+A)W/A} - 2aW/A(a+A), \quad (53)$$

whereupon Eq. (52) corresponds to

$$OH_{\min} \cong \sqrt{aW/A} - 2W/A. \quad (54)$$

e3. The minimum a. The minimum at point a corresponds to the end-point inflection for the titration of weak base with strong acid or of the alkali salt of a weak acid with strong acid. By symmetry then, we may write from Eqs. (52) and (54),

$$OH_{\min} \cong \sqrt{BW/b} + 2W/b, \quad (55)$$

$$H_{\min} \cong \sqrt{bW/B} - 2W/B, \quad (56)$$

for titration of weak base. Similarly, for titration of the salt at a concentration here called a, with $B^* = W/A$,

$$OH_{\min} \cong W/\sqrt{aA} + 2W/a, \tag{57}$$

$$H_{\min} \cong \sqrt{aA} - 2A. \tag{58}$$

5. THE TITRATION ERROR. If the titration of weak acid with strong base is stopped at the inflection point rather than at the value of H for the equivalent point, there will be a negative error since from Eq. (51) $H_i > H_e$. If the titration is carried to the inflection point and if both $(b_s)_i$ and H_i are measured, a_i may be calculated exactly through Eqs. VII(23) and VII(26) or the error through Eqs. VII(24) and VII(27). If as with an uncalibrated potentiometer only $(b_s)_i$ and not H_i is determined, then those same equations may be used with the theoretical value of H_i or OH_i from Eqs. (51) and (53).

From Eq. (21),

$$a_i = (b_s)_i + [(b_s)_i H/A - OH - W/A + H + H^2/A]. \tag{59}$$

The bracketed quantity constitutes the error. If this is small and if the final volume is approximately $V_A + V_B$, we may replace $(b_s)_i$ in the bracket by $\mathbf{c}_e$ or the quantity $c_A c_B/(c_A + c_B)$, as explained under Eq. VII(28). We may similarly then replace a_i by $\mathbf{c}_e$ in Eqs. (51) and (53). Then when these are substituted for H and OH in Eq. (59) we obtain, neglecting only the second and third terms of H^2/A, the following simple expression:

$$a_i \cong (b_s)_i + 3\mathbf{c}_e W/(\mathbf{c}_e + A)A. \tag{60}$$

Hence when A is large compared to $\mathbf{c}_e$ and $H_i \cong \sqrt{W}$, the error is negligible with $a_i \cong (b_s)_i + 3\mathbf{c}_e W/A^2$. Otherwise, for the important case of weak acid, with $\mathbf{c}_e > A$, we have

$$a_i \cong (b_s)_i + 3W/A. \tag{61}$$

Hence, the relative percentage error for titration of a weak acid ($\mathbf{c}_e > A$) is $100E$, with

$$E \cong -[3W/(b_s)_i A]/[1 + 3W/(b_s)_i A] \cong -3W/\mathbf{c}_e A. \tag{62}$$

Finally, by symmetry, we may write the error in the titration of weak base with strong acid as

$$E \cong -[3W/(a_s)_i B]/[1 + 3W/(a_s)_i B] \cong -3W/\mathbf{c}_e B. \tag{63}$$

Roller has given a different treatment for the problem of the titration error. It is based on the evaluation of H_i from a high degree equation such as Eq. (46) through expansion in series of the quantity $1/(A + H_i)$, for example, in powers of H_i/A. Eq. (62) however, although reached rather simply, gives results for the present case as accurate as those from this somewhat more involved method. As an example: for a titration in which

$c_eA = 10^{-11}$ Roller's method gives the titration error as $-$ 0.30 per cent,(2) the same as that calculated directly and simply through Eq. (62). This approximation, moreover, is seen to be independent of the ionic strength.

B. Other Simple "Strong-Weak" Titrations

By analogy now with the case of titration of weak acid with strong base and remembering that the salt MX_s (of weak base and strong acid) is mathematically a weak acid with ionization constant W/B, we may tabulate without further examination the following simple cases of related titrations; the "equivalent point" is abbreviated as "Eq.Pt."

1. Titration with Strong Base (inflection point at minimum b of Fig. 10-1).

a. Of weak acid:

1. Start: $H^3 + H^2A \quad - H(aA + W) \quad - AW = 0,$
2. Curve: $H^3 + H^2(b_s + A) - H[A(a - b_s) + W] - AW = 0,$
3. Eq.Pt: $H^3 + H^2(a + A) \quad - HW \quad - AW = 0.$

b. Of salt MX_s*:*

1. Start: $H^3 + H^2W/B \quad - H(cW/B + W) \quad - (W/B)W = 0,$
2. Curve: $H^3 + H^2(b_s + W/B) - H[(W/B)(c - b_s) + W] - (W/B)W = 0,$
3. Eq.Pt: $H^3 + H^2(c + W/B) \quad - HW \quad - (W/B)W = 0.$

The early part of the titration curve is given by the H^3 sequence; the middle range by the two middle terms or $H \cong A(a - b_s)/b_s$ and $H \cong (W/B)(c - b_s)/b_s$ respectively; the end and through the equivalent point, by the H^2 sequence. With appreciable excess of b_s, $OH \cong b_s - a$ or $b_s - c$. The quantity Δ_p is positive and the feasibility functions are $F = (p/400)\sqrt{aA/W}$ and $(p/400)\sqrt{c/B}$ respectively. In the titration of the salt there is no inflection unless $c > 27B$.

2. Titration with Strong Acid (inflection point at minimum a of Fig. 10–1, as pH decreases).

a. Of weak base:

1. Start: $H^3 + H^2(b + W/B) \quad - HW \quad - (W/B)W = 0,$
2. Curve: $H^3 + H^2(b - a_s + W/B) - H(a_sW/B + W) - (W/B)W = 0,$
3. Eq.Pt: $H^3 + H^2W/B \quad - H(bW/B + W) \quad - (W/B)W = 0.$

b. Of salt M_sX*:*

1. Start: $H^3 + H^2(c + A) \quad - HW \quad - AW = 0,$
2. Curve: $H^3 + H^2[(c - a_s) + A] - H(a_sA + W) - AW = 0,$
3. Eq.Pt: $H^3 + H^2A \quad - H(cA + W) \quad - AW = 0.$

(2) P. S. Roller, *J. Am. Chem. Soc.*, **50,** 1 (1928).

The early part of the titration curve is given by the H^2 sequence; the middle range by $H \cong (W/B)a_s/(b - a_s)$ and $H \cong Aa_s/(c - a_s)$ respectively; the end and through the equivalent point, by the H^3 sequence. With appreciable excess of a_s, $H \cong a_s - b$ or $a_s - c$. Δ_p is now negative and the feasibility functions are $(p/400)\sqrt{bB/W}$ and $(p/400)\sqrt{c/A}$ respectively. There is no inflection point unless $bB > 27W$ for the weak base or $c > 27A$ for the salt.

The equations for titration with strong acid have been put in terms of H to bring out their relation with those for titration with strong base and to express them all in terms of effective "acid constants," A or W/B. But those for titration with strong acid may be more familiar and useful in terms of OH as follows:

Weak base (plus strong acid):

1. Start: $(OH)^3 + (OH)^2 B \qquad\quad - (OH)(bB + W) \qquad\quad - BW = 0$,
2. Curve: $(OH)^3 + (OH)^2(a_s + B) \; - (OH)[B(b - a_s) + W] - BW = 0$,
3. Eq.Pt: $(OH)^3 + (OH)^2(b + B) \quad - (OH)W \qquad\qquad\quad - BW = 0$.
   Salt M_sX (plus strong acid):

1. Start:
$(OH)^3 + (OH)^2 W/A \qquad\quad - (OH)(cW/A + W) \qquad\quad - (W/A)W = 0$,
2. Curve:
$(OH)^3 + (OH)^2(a_s + W/A) - (OH)[(W/A)(c - a_s) + W] - (W/A)W = 0$,
3. Eq.Pt:
$(OH)^3 + (OH)^2(c + W/A) \; - (OH)W \qquad\qquad\qquad - (W/A)W = 0$.

Now, in terms of these equations, the early part of the titration curve with strong acid is given by the $(OH)^3$ sequence; the middle range by the two middle terms, or $OH \cong B(b - a_s)/a_s$ and $OH \cong (W/A)(c - a_s)/a_s$ respectively; the end and through the equivalent point, by the $(OH)^2$ sequence. The feasibility functions of course remain as already stated, $|\Delta pH| = |\Delta pOH|$.

C. Titration of Strong Acid in Presence of Weak Acid

The titration with strong base of a mixture of two acids, one strong (a_s) and one weak (a), has already been considered in Chapter VIII based on the general equation, Eq. VIII(42). In this titration the first equivalent point $(b_s = a_s)$ is mathematically the solution of a pure weak acid since the salt M_sX_s is of no effect in such equations; consequently the second part of the titration (that of the weak acid) is the case already discussed in Section A of the present chapter.

1. Buffer Capacity, Titration Curve, and Feasibility. The schematic relations between the π curve and the pH titration curve are shown in Fig. 10–4 for fixed values of a_s and a. If $a/27 < A$, there is only one minimum in π, minimum b in (very slightly) alkaline solution corresponding

to the equivalent point $b_s = a_s + a$. In this case only the sum of the two acids may be determined by the titration with an inflection at minimum b. The feasibility function for such titration, with p representing percentage of

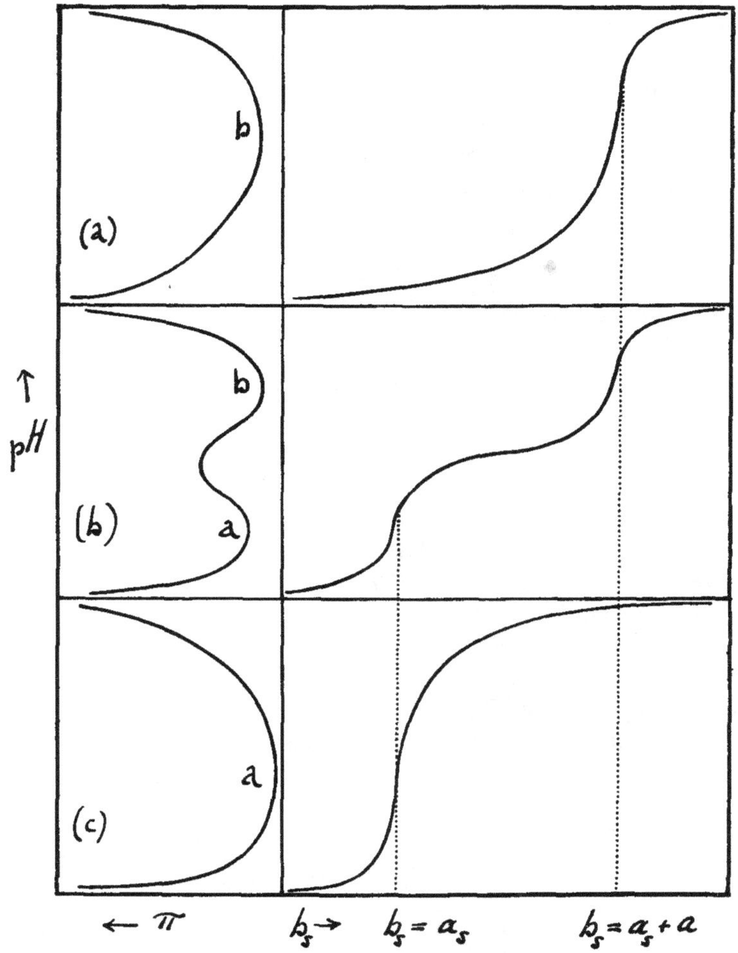

Fig. 10–4. Relation between π curve and titration curve for mixture of strong and weak acids: (a) $A > a/27$; (b) $a/27 > A > 27W/a$; (c) $27W/a > A$.

strong base relative to the sum $a_s + a$, is $F = 2.3p(a_s + a)/200\pi_e$. Since with high A the equivalent point is practically neutral, $\pi_e \cong (\pi_w)_{\min} = 2.3(2\sqrt{W})$; then

$$F = p(a_s + a)/400\sqrt{W}. \tag{64}$$

If, on the other hand, $A < 27W/a$, there is again only one minimum in π, minimum a in (very slightly) acid solution, again at $H \cong \sqrt{W}$ for the titration of the strong acid alone, with $F = pa_s/400\sqrt{W}$ as in Chapter IX. In this case no inflection occurs for the titration of the second acid or for the sum of the two acids.

For intermediate values of A there are two successive vertical inflections in the titration curve; as a result of the maximum which appears in the π curve, minimum a, for the titration of the strong acid alone, moves to higher H, and minimum b for total titration of both acids moves to lower H. As noted in Chapter VIII, the titration curve through the first equivalent point for the determination of strong acid in presence of weak acid is given by the H^3 sequence of Eq. VIII(42) simplified to

$$H^2 - H(a_s - b_s) - A(a_s + a - b_s) \cong 0. \tag{65}$$

A single feasibility function for this titration may be assumed only if the quantities $(w)_{-p/2}$ and $(w)_{+p/2}$ of this equation are approximately equal, or only if

$$pa_s/200 + a \cong - pa_s/200 + a. \tag{66}$$

These may not be equated for the purpose unless $a \gg pa_s/200$, when we may proceed to take F according to Eq. IX(22) as

$$F = (p/400)\sqrt{a_s^2/aA}. \tag{67}$$

Reference to Eq. VIII(42) shows that this also requires $pa_s/200 > A$, a condition corresponding to $paA/200 > W$ for Eq. (42). Otherwise, as in Eq. (45), two separate feasibility functions must be used, or

$$F_{\mp} = (A \pm pa_s/200)/2\sqrt{aA}. \tag{68}$$

The simple expression of Eq. (67), which was also reached as Eq. IX(47), may also be obtained through Eq. IX(27) since, for a solution of pure weak acid, $\pi_e \cong (2.3)2\sqrt{aA}$, from Eq. (41). But if Eq. (66) does not hold, then from Eq. (65)

$$\Delta_p = \log \frac{1 + [1 + A(400/pa_s)^2(a + pa_s/200)]^{\frac{1}{2}}}{-1 + [1 + A(400/pa_s)^2(a - pa_s/200)]^{\frac{1}{2}}}. \tag{69}$$

A single simple feasibility function, F, then, is not always to be assumed in problems involving independently variable concentrations unless $pc/200$ in Eq. IX(27) is small compared to the concentrations of significant buffering solutes; this condition must be remembered in applying the simple Procedure II of Chapter IX and Eq. IX(27) in writing the feasibility function.

2. TITRATION ERROR. The two consecutive inflection points I and II occur when $(b_s)_I \cong (a_s)_I$ with $(H_i)_I$ given by Eq. (58) and when $(b_s)_{II} \cong (a_s + a)_{II}$ with $(H_i)_{II}$ given by Eq. (52).

a. From values of both $(b_s)_i$ and H_i. If both H_i and $(b_s)_i$ are determined at each inflection point, then from $D = a_s + a\alpha - b_s$ we have

$$(a_s)_I = (b_s)_I - (a\alpha)_I + D_I. \tag{70}$$

If both the concentration, $(a)_I$, and the ionization constant, A, of the weak acid are known, it follows that $(a_s)_I$ may be calculated from the data. But if only A is known, then since $a\alpha$ is very small in acid solution, we may take $(a)_I$ approximately as

$$(a)_I \cong c_B(V_{B_{II}} - V_{B_I})/V_I, \tag{71}$$

in which V_{B_I} and $V_{B_{II}}$ are the volumes of the base solution used to reach the first and second inflection points respectively and V_I is the total volume of the solution at the first point; or

$$(a_s)_I \cong (b_s)_I - (\alpha)_I c_B(V_{B_{II}} - V_{B_I})/V_I + D_I. \tag{72}$$

For the weak acid,

$$(a)_{II} = [(b_s)_{II} - (a_s)_{II} + D_{II}]/(\alpha)_{II}, \tag{73}$$

in which all the concentrations refer to the second inflection point; $(a_s)_{II}$, if not already known, may be taken as $(a_s)_I V_I/V_{II}$ with $(a_s)_I$ from Eq. (72) and V_{II} the volume of the solution at the second inflection point.

b. Theoretical, from $(b_s)_i$ *alone.* If only the values of b_s at the two consecutive inflection points are noted without the actual values of H_i, then the theoretical values of the latter are introduced from Eqs. (58) and (52). With H_i from Eq. (58), $a\alpha = aA/(A + H) \cong aA/(\sqrt{aA} - A) \cong \sqrt{aA} + A$. Then neglecting OH in Eq. (70) and again using Eq. (58) for H_i, we have

$$(a_s)_I \cong (b_s)_I - 3A, \tag{74}$$

the relative error being

$$E \cong + 3A/(b_s)_I \tag{75}$$

with $(b_s)_I = c_B V_B/V_I$.

At the second inflection point the error for the titration of the weak acid is given by Eq. (62) if the concentration of the strong acid at that point is known so that $(b_s)_i$ is taken as $c_B V_{B_{II}}/V_{II} - (a_s)_{II}$. If the weak acid is determined from the quantity of strong base, $c_B(V_{B_{II}} - V_{B_I})$, used between the two inflection points, then since the first inflection occurs late and the second early, we have from Eqs. (61) and (74)

$$(a)_{II} \cong c_B(V_{B_{II}} - V_{B_I})/V_{II} + 3A + 3W/A; \tag{76}$$

and the relative error is

$$E \cong -(3W/A + 3A)/(a)_{II}, \tag{77}$$

in which $(a)_{II}$ has the experimental value $c_B(V_{B_{II}} - V_{B_I})/V_{II}$.

Finally, the titration with strong acid of strong base in presence of weak base is of course a problem identical with this and the necessary formulas, in OH, may be written by symmetry. The simple feasibility function corresponding to Eq. (67) is $F = (p/400)\sqrt{b_s^2/bB}$, and the analogs of Eqs. (75) and (77) are respectively $E \cong 3B/(a_s)_I$ and $E \cong -(3W/B + 3B)/(b)_{II}$.

D. Dibasic Acid "Strong in K_1" and Strong Base

Since only one ionization constant is involved, this problem is a special case of the general problem of this chapter. Here

$$D = a(1 + \alpha_2) - b_s = a(H + 2K_2)/(H + K_2) - b_s. \tag{78}$$

Graphically, the relations are those of Fig. 10–5. From Eq. (78),

$$\pi = 2.3(a\alpha_1\alpha_2 + H + OH); \tag{79}$$

$$d\pi/dpH = (2.3)^2[a\alpha_1\alpha_2(\alpha_1 - \alpha_2) - H + OH]. \tag{80}$$

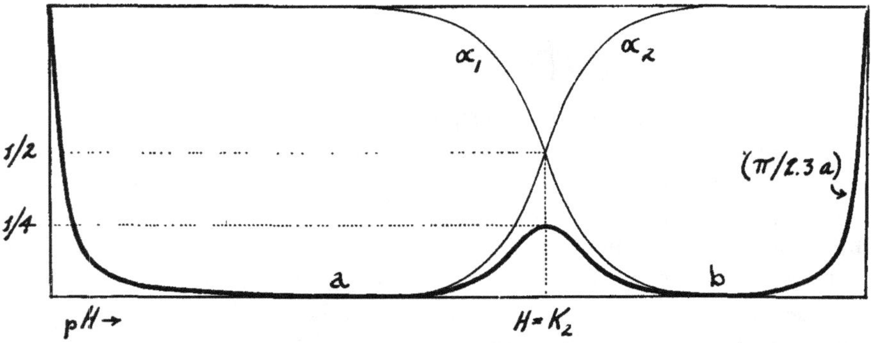

Fig. 10–5. Buffer capacity of solution containing dibasic acid strong in respect to K_1.

Since $\alpha_1 = 1 - \alpha_2 = H/(K_2 + H)$, then with H and OH neglected in Eqs. (79) and (80) π_{max} is found to occur at $\alpha_1 \cong \alpha_2 \cong 1/2$ with $H \cong K_2$, the value of π_{max} being approximately $2.3a/4$.

From Eq. (78),

$$H^3 + H^2[K_2 - a + b_s] - H[K_2(2a - b_s) + W] - K_2W = 0. \tag{81}$$

Hence this is mathematically the titration with strong base of a mixture of strong acid and weak acid with $a_s = a$. The pure dibasic acid ($b_s = 0$) is given by the H^3 sequence of Eq. (81),

$$H \cong (a - K_2)/2 + \sqrt{(\quad)^2 + 2aK_2 + W}, \tag{82}$$

or $H \cong a$. For the titration curve through the first equivalent point, the same sequence gives

$$H \cong (a - b_s - K_2)/2 + \sqrt{(\quad)^2 + K_2(2a - b_s) + W}. \tag{83}$$

At $b_s = a$ (first equivalent point or solution of salt of type $NaHSO_4$ for example) the solution is mathematically that of a weak acid with ionization constant K_2, at concentration a (or c, the molar concentration of the salt itself). During the titration of the second equivalent,

$$H \cong K_2(2a - b_s)/(b_s - a). \tag{84}$$

At the second equivalent point ($b_s = 2a$, salt type Na_2SO_4) the solution is equivalent to that of a salt M_sX (strong base and weak monobasic acid) at concentration equal to the molarity of Na_2SO_4 with $A = K_2$.

Since the first equivalent point is mathematically that for the titration of strong acid in presence of weak acid with the restriction of equal concentrations, the approximate feasibility function is

$$F = (p/400)\sqrt{a/K_2}. \tag{85}$$

At the second equivalent point,

$$F = (p/400)\sqrt{aK_2/W}. \tag{86}$$

In both cases p is here taken relative to the molarity, a. If the acid is titrated to the second equivalent point and p refers to percentage of strong base required for the total titration (or $b_s = 2a$), then

$$F = (p/200)\sqrt{aK_2/W}. \tag{87}$$

This would be the feasibility function for the titration of an acid such as sulfuric acid (assumed "strong in K_1") to the point $b_s = 2a$. It holds whether or not the titration curve shows an inflection at $b_s = a$, which requires $a > 27K_2$. From Eq. (62), the relative error for the total titration would be

$$E \cong -(3W/2aK_2). \tag{88}$$

E. Titration With Strong Base of Strong Acid in Presence of Weak Base

We have already considered the titration, with strong base, of the salt MX_s (salt of strong acid and weak base) as mathematically the titration of a weak acid with ionization constant W/B. But in a more general sense, this is in fact only a special case (in which $a_s = b$) of the titration of strong acid in presence of weak base. Whereas, in other words, the titration of the salt MX_s is usually "explained" as the "displacement titration" of the salt with emphasis on the "displacement" of the *base* present, it is actually more correctly the titration of the strong acid with strong base, the weak base present merely affecting the buffer capacity of the solution and hence the position and nature of the "end-point" and the feasibility of the titration. The problem will therefore now be considered as an aspect of the problem of titrations involving a single constant, and with all possible variations of the ratio of the concentrations of strong acid and weak base ($a_s \gtrless b$).

1. Inflection Points and Feasibility. The first thing to be noted is the identity of the π curve in the presence of weak base instead of weak acid, provided $B = W/A$. For a solution containing weak acid, $\pi = 2.3(a\alpha\rho + H + OH)$, and for weak base, $\pi = 2.3(b\alpha'\rho' + H + OH)$; here $\pi = db_s/d\text{p}H$,

as usual. Hence with $a = b$ and $B = W/A$, the value of π is the same for the two cases at any given value of pH since then $\alpha = \rho'$ and $\rho = \alpha'$. These relations are shown graphically in Fig. 10–6.

The maximum and the minima of π are therefore the same as in the case of weak acid, with W/B in place of A. These minima will occur as long as $b > 27B$ for minimum b and $bB > 27W$ for minimum a. Then there will be two inflections in the titration, with strong base, of a solution containing

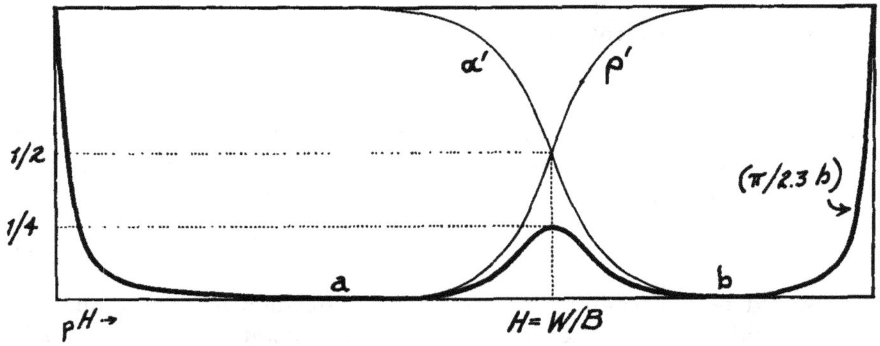

Fig. 10–6. Buffer capacity of solution containing weak base.

strong acid (a_s) in presence of weak base (b) provided $a_s > b$. The first equivalent point ($b_s = a_s - b$) is marked by the inflection corresponding to minimum a and the second ($b_s = a_s$) has its inflection at minimum b. The first titration is that of the "free" strong acid, or the excess of strong acid over that corresponding to the salt MX_s; hence, since the salt MX_s is mathematically a weak acid with ionization constant W/B, the titration is similar to that of strong acid in presence of a weak acid, with

$$F = (p/400)\sqrt{(a_s - b)^2 B/bW}, \tag{89}$$

in which p is taken relative to $(a_s - b)$. But if we ignore the inflection at minimum a, the inflection at minimum b becomes that for the titration of the total strong acid in presence of weak base, with

$$F = (p/400)\sqrt{a_s^2/bB}, \tag{90}$$

where p is now taken relative to the total a_s.

Hence if $B = \sqrt{W}$, it is as "feasible" to titrate the excess of strong acid (in presence of the salt MX_s) as it is to titrate the total strong acid in the presence of the base; but the first titration is more (or less) feasible than the second if B is greater than (or less than) $\sqrt{W}$.

While the first inflection point, at minimum a, occurs only if $a_s > b$, marking the titration of the excess, $(a_s - b)$, the second inflection, at minimum b, occurs regardless of the relative magnitudes of a_s and b. When a salt MX_s is titrated to the alkaline inflection point corresponding to minimum b, the total quantity of strong base required depends only on the quantity of strong acid in the "salt" and is the same whether or not the salt is exactly pure in the sense of the equality of a_s and b in its composition. With a_s greater than, equal to, or less than b, in other words, the total b_s required to reach the alkaline inflection point is independent of b, at least as a first approximation. Even the feasibility, from Eq. (90), is practically independent of b as long as b is of the same order of magnitude as a_s, and the same is true of the titration error, as will be shown in the next paragraph. When $b = a_s = c$, Eq. (90) of course gives F for the titration of the pure salt, MX_s, with strong base.

2. Titration Error. If the values of $(b_s)_I$ and $(b_s)_{II}$ or of $c_B V_{B_I}$ and $c_B V_{B_{II}}$ required to reach the two successive inflection points are determined together with the values of H_i, the calculation of both a_s and b is a simple matter of substitution in the equation $D = a_s - b\alpha' - b_s$. If H_i is not measured, then by analogy with Eqs. (74–77), and noting that W/B is now to be written for A, we have for the determination of the strong acid

$$(a_s)_{II} \cong (b_s)_{II} + 3B, \tag{91}$$

with a relative error $E \cong -3B/(b_s)_{II}$. For the determination of the weak base,

$$(b)_{II} \cong c_B(V_{B_{II}} - V_{B_I})/V_{II} + 3W/B + 3B, \tag{92}$$

the relative error being $E \cong -(3W/B + 3B)/(b)_{II}$.

3. Summary, for Feasibility of Titration of Strong Acid. In summary then, the feasibility function for the titration of the strong acid with strong base is given by the following (approximate) expressions:

(a) In unbuffered solution, $F = (p/400)\sqrt{a_s^2/W}$, Eq. IX(11).

(b) In presence of weak acid (provided $pa_s/200 \ll a$ and $\gg A$), $F = (p/400)\sqrt{a_s^2/aA}$, Eq. (67). Here the inflection is in acid solution, the equivalent point being mathematically a pure solution of the weak acid; a second inflection, in alkaline solution, is then possible for the titration of the weak acid.

(c) In presence of weak base (provided $pa_s/200 \ll b$ and $\gg B$), $F = (p/400)\sqrt{a_s^2/bB}$, Eq. (90). This refers to titration of total a_s, for $a_s \gtrless b$. The inflection occurs in alkaline solution, the equivalent point being mathematically the pure solution of the weak base. If $a_s > b$ there may be an earlier inflection, in acid solution, for the titration of the difference, $(a_s - b)$. When $b = a_s = c$, this is the titration of the salt MX_s.

If $a_s > b$ therefore, complete analysis is possible by titration with strong base since there may be two successive inflection points, one at $b_s = a_s - b$ and the second at $b_s = a_s$. When $b > a_s$ titration with strong base can measure only the total, a_s. Titration of a second sample of the original solution with strong acid however will measure $(b - a_s)$ with an inflection in acid solution or at minimum a, whereupon b may also be calculated.

Similarly for the analysis of mixtures of strong base and weak acid: if $b_s > a$ titration with strong acid can give two successive equivalent points; if $b_s < a$, b_s is determined by titration with strong acid and $(a - b_s)$ is determined by titration with strong base.

With $a_s > b$ the titration of $(a_s - b)$ by means of strong base would be illustrated by the determination of "free HCl" at concentration $(a_s - b)$ in presence of "NH_4Cl" at concentration b; with $a_s < b$ the titration of $(b - a_s)$ by means of strong acid would be illustrated by the determination of "free NH_3" in presence of "NH_4Cl." For either case the feasibility function is given by Eq. (89), in which b represents the concentration of total ammonia as a solute, as explained in Chapter I. If the concentration of "free HCl" is 1 and that of "NH_4Cl" also 1 for example, then the feasibility function for titration of the "free HCl" is $F_{\text{HCl}} = (p/400)\sqrt{B/W}$; but if ["free NH_3"] = 1 and ["NH_4Cl"] = 1, $F_{NH_3} = (F_{\text{HCl}})/\sqrt{2}$.

With $b_s > a$ the titration of $(b_s - a)$ by means of strong acid would be illustrated by the determination of "free NaOH" at concentration $(b_s - a)$ in presence of "sodium acetate" at concentration a; with $b_s < a$ the titration of $(a - b_s)$ by means of strong base would be illustrated by the determination of "free acetic acid" in presence of "sodium acetate." In a solution of pure borax, $Na_2B_4O_7$, at concentration c, $b_s = 2c$ while $a = 4c$. Now the base may be titrated with strong acid such as HCl; the equivalent point is a solution of boric acid at the concentration $4c$ in the presence of NaCl with $H_e \cong \sqrt{4cA}$; and the feasibility function of the titration is, from Eq. (90), $F = (p/400)\sqrt{c/A}$ with p relative to $a_s = 2c$ for 100 per cent. The excess of weak acid however, or $(a - b_s)$, can be titrated only with difficulty since the acid is extremely weak. The equivalent point, corresponding to the salt NaH_2BO_3 at concentration $4c$, is at $OH_e \cong \sqrt{4cW/A}$ and the feasibility function, from Eq. (89), is $F = (p/400)\sqrt{cA/W}$ with p relative to $b_s = 2c$ for 100 per cent. With $a = 0.1$ (or $c = 0.025$), $A = 6 \times 10^{-10}$ and $p = 2$, this gives $F = 0.19$ and, from Eq. IX (25), $\Delta_p = 0.17$ pH units for 2 per cent of titration through the equivalent point. This titration is improved by addition of mannitol, which forms complex borate ions, as discussed in Chapter XV; in that chapter furthermore, the question of polymerization to form tetraborate ions, here neglected, will also be discussed.

4. "Displacement Titrations." Whatever meaning seems to lie in the expression "displacement titration of the salt MX_s" vanishes when we consider that it is the strong acid that is being titrated in the presence of the weak base. Similarly for the titration of strong base (NaOH) with strong acid in presence of weak acid (such as HCN) with $b_s \gtrless a$. Given a solution of NaCN for example, with or without excess of HCN, its titration with strong acid is customarily called a "displacement titration." But mathematically one must ask what happens to the "displacement" as the acid involved becomes vanishingly weak. (As A decreases, minimum b finally

disappears by merging with the maximum of π and minimum a moves to lower and lower values of H with a limit of $H = \sqrt{W}$.) We still titrate the strong base, NaOH, but the solution is simply less and less buffered at the equivalent point until with A (the ionization constant of the weak acid of the salt) equal to zero the problem reduces to the titration of strong base with strong acid in unbuffered solution. The presence or absence of weak acid, or even its concentration relative to that of the strong base being titrated, has (practically) nothing to do with the occurrence and position of the inflection point in question (with its limit at $H = \sqrt{W}$). It is the titration of the difference between b_s and a, when $b_s \gtrless a$, which depends on the presence of the weak acid.

The titration of borax with strong acid is the titration of the two equivalents of strong base in the borax; two equivalents of HCl would be used per mole of "borax" for any general formula of "borax" such as $Na_2O \cdot xB_2O_3$. Only the pH of the equivalent point and the feasibility of the titration will depend, slightly, on the value of x in this formula. This is distinctly not the implication when the "titration reaction" is represented[(3)] as "$B_4O_7^{=} + 2H^+ + H_2O \rightleftarrows 4HBO_2$," with the emphasis on the ion of the weak acid rather than on the strong base itself as that which is being *titrated* (i.e. measured).

In connection with this carefully "balanced" chemical equation, we may point out that the whole of the present treatment does without such "equations," which often misrepresent reality and seldom convey information. Any number of such "equations" may be written for a given equilibrium constant for example, as will be pointed out in Chapter xv, but not one of them necessarily adds meaning to the constant. In the titration of aq. HCl with NaOH there is simply no "chemical equation" that represents the titration process. The "chemical reaction" for its titration is certainly not $HCl + NaOH \rightarrow NaCl + H_2O$ for no salt molecule is presumably formed and both HCl and NaOH are presumably completely ionized. Nor is it $H^+ + OH^- \rightarrow H_2O$ for several reasons. In the first place we are not adding OH^-, an ion which is already present. Secondly the concentration of H^+ to begin with is not equal to that of the HCl; nor is it diminished during the titration by an amount equal to the concentration of the HCl. Finally while the reaction $H^+ + OH^- \rightleftarrows H_2O$ is by definition incomplete, the titration, as a process carried to the inflection point of the pH titration curve, is potentially absolutely mathematically *exact*. As stated in Chapter IX, Section B, *nothing* happens to the dissolved HCl, a strong acid, during its titration with NaOH. We measure the HCl because of its effect on the equilibrium of the water ions and by means of the corresponding effect of the titrating base, as explained in connection with the buffer capacity curves in Section A. The stoichiometry involved is not representable by any "chemical equation."

(3) For example: W. C. Pierce and E. L. Haenisch, *Quantitative Analysis*, Wiley, N.Y., 1948; p. 104.

As to what happens to the weak acid during the titration of the strong base with strong acid, it merely becomes less ionized whatever its concentration relative to that of the base being titrated. It is "displaced" (if "displacement" means decrease of α or increase of ρ) quite completely in the following example for a considerable variation of the ratio of b_s to a. In the titration with strong acid of 1 MNaOH in presence of HCN (with $A = 4 \times 10^{-10}$ and with neglect of volume changes) we have the numerical relations shown in Table 10–2. The ratio ρ_e/ρ_{orig}, or the "relative displacement," is 1.11, 199, 3.6×10^4, while $\alpha_e/\alpha_{orig} = 6.32 \times 10^{-5}$, 2.01×10^{-5}, 6.32×10^{-5} respectively; the pure salt solution, NaCN, is of course the second case, with $b_s = a$. If

Table 10–2. Titration of NaOH in presence of HCN.

NaOH b_s	HCN a	a_s	H	α	ρ
1	10	0	3.6×10^{-9}	0.1	0.9
		1	6.32×10^{-5}	6.32×10^{-6}	0.999994
1	1	0	2.02×10^{-12}	0.995	0.00502
		1	2×10^{-5}	2×10^{-5}	0.99998
1	0.1	0	1.11×10^{-14}	0.99997	2.78×10^{-5}
		1	6.32×10^{-6}	6.32×10^{-5}	0.99994

instead of HCN we consider an acid with $A = 10^{-5}$, the corresponding ratios, ρ_e/ρ_{orig}, are 1.11, 3.15×10^4, and 8.92×10^8 while the ratios α_e/α_{orig} are 0.0100, 0.00315, 0.00995 respectively. (The calculations for H_{orig} and H_e are made in every case by means of Eq. VIII[42].)

In each titration in Table 10–2 the error in the measurement of the NaOH is altogether negligible, according to Eq. (91), which here becomes $(b_s)_i \cong (a_s)_i + 3A$. The feasibility moreover is relatively high in every case. From Eq. (90) transformed to $F = (p/400)\sqrt{b_s^2/aA}$, we calculate $\Delta p = 1.81$, 2.80, and 3.80 pH units for $p = 0.1$ respectively. Simple titration with HCl therefore can hardly distinguish the three solutions.

According to current Brønsted terminology, NaCN may be titrated with HCl because CN^- is a strong base (while Na^+ is a negligibly weak acid). The titration reaction is accordingly represented as $CN^- + H^+ \rightarrow HCN$. One would then expect the titer to depend in some fashion upon the analytical concentration of cyanide or at least on the quantity of cyanide ion changed to unionized HCN. Yet we see in Table 10–2 that NaOH at concentration

$1M$ may be titrated with almost equal feasibility in the presence of from 0.1 to $10M$ HCN and that the corresponding change in the concentration of the cyanide ion, CN^-, is 0.1, 0.995, and 1 respectively in the three cases considered. It seems again therefore more logical and instructive to discuss such a process in terms of properties of real solutes—the strength of the NaOH, which is the substance actually being measured or titrated, and the weakness of the HCN, the quantity of which affects not the titer but the feasibility of the titration.

To what extent the "equation" $B_4O_7^= + 2H^+ + H_2O \rightarrow 4HBO_2$ fails to represent reality in the titration of borax may similarly be appreciated by reference to the concentrations of the individual species in the solution before titration, as calculated in Chapter xv, Section D1, from the available values of the equilibrium constants involved. With borax at the molar concentration 0.015 for example, $[BO_2^-] \cong 0.028$, $[H_2B_4O_7] \cong 2.2 \times 10^{-9}$, $[HB_4O_7^-] \cong 3.7 \times 10^{-4}$, $[B_4O_7^=] \cong 6.1 \times 10^{-4}$, and $[HBO_2] \cong 0.028$.

Special note on Eq. (41). Eq. (41) gives at once a rough estimate of A from the rate of change of pH at the equivalent point; see also S. Kilpi, *Z. phys. Chem.* **A177**, 116, 427 (1936), and E. Grunwald, *Jour. Am. Chem. Soc.*, **73**, 4934 (1951). With $b_s = na$ and $\pi = a(dn/dpH)$, the exact relation, from Eq. (33), is

$$(dn/dpH)_e/2.3 = H_eA/(H_e + A)^2 + (H_e + OH_e)/a_e, \qquad (93)$$

a simple quadratic in A in which, however, $H_e = f(A, a_e)$. With $OH_e \cong \sqrt{a_e W/A}$ and $OH_e \gg H_e$, a first approximation for A is obtainable as

$$A \cong [2(2.3)(dpH/dn)_e - 1]^2 W/a_e, \qquad (94)$$

reducing to Eq. (41) if $a_eA \gg W$. Then OH_e may be refined through Eq. (16) and A itself through Eq. (93) until satisfactory constancy is reached. These relations are as suggested by Grunwald, who also considers, however, the volume change depending on c_B, the concentration of titrating strong base. The exact relation, then, with all quantities but c_B referring to the equivalent point, is

$$\frac{1}{2.3}\left(\frac{dn}{dpH}\right) = \left(\frac{c_B}{c_B + D}\right)\left[\frac{HA}{(H + A)^2} + \frac{H + OH}{a}\left(1 - \frac{a}{c_B}\right)\right] + \frac{H + OH}{(c_B + D^2)}\left[\frac{c_BA}{H + A} + D\left(1 - \frac{c_B}{a}\right)\right], \qquad (95)$$

still a quadratic in A reducing to Eq. (93) if c_B is large. Again with $OH_e \cong \sqrt{a_e W/A}$, the first approximation for A is Eq. (94) with $-a/2c_B$ as an added relatively insignificant term in the bracket.

XI

Two Weak Acids and Strong Base

A. Analytical Equations for H

In a solution containing two monobasic acids (with $A_1 > A_2$) and strong base, we have, repeating Eqs. IX(28–30),

$$H - OH = (a\alpha)_1 + (a\alpha)_2 - b_s, \tag{1}$$

$$H - OH = a_1A_1/(A_1 + H) + a_2A_2/(A_2 + H) - b_s; \tag{2}$$

$$H^4 + H^3(A_1 + A_2 + b_s) - H^2[a_1A_1 + a_2A_2 - b_s(A_1 + A_2) + W - A_1A_2] - H[A_1A_2(a_1 + a_2 - b_s) + W(A_1 + A_2)] - A_1A_2W = 0. \tag{3}$$

For some special values of H, we note from Eq. (2) with $b_s = 0$ that the value of a_1 required for $H = A_1$ in the pure solution of the acids is

$$(a_1)_{H=A_1} = 2(A_1 - W/A_1) - 2a_2/(1 + 1/y), \tag{4}$$

in which $y = A_2/A_1$. Hence $H > A_1$ if $a_1 > 2A_1$ provided that $A_1 > \sqrt{W}$. With strong base present,

$$(b_s)_{H=A_1} = a_1/2 + a_2/(1 + 1/y) - (A_1 - W/A_1) \cong a_1/2; \tag{5}$$

$$(b_s)_{H=\sqrt{A_1A_2}} = a_1/(1 + \sqrt{y}) + a_2/(1 + 1/\sqrt{y}) - (\sqrt{A_1A_2} - W/\sqrt{A_1A_2}) \cong a_1; \tag{6}$$

$$(b_s)_{H=A_2} = a_1/(1 + y) + a_2/2 - (A_2 - W/A_2) \cong a_1 + a_2/2. \tag{7}$$

1. The Pure Acids ($b_s = 0$).

$$H^4 + H^3(A_1 + A_2) - H^2(a_1A_1 + a_2A_2 + W - A_1A_2) - H[A_1A_2(a_1 + a_2) + W(A_1 + A_2)] - A_1A_2W = 0. \tag{8}$$

For approximate solutions of this equation we note first a certain analogy with Eq. x(2) for a single weak acid. Since H must be greater than $\sqrt{W}$, it is clear that if A_1 (which is greater than A_2) is less than $\sqrt{W}$, the largest terms are those in H^4 and H^2 so that

$$H \cong \sqrt{a_1A_1 + a_2A_2 + W}, \tag{9}$$

the analog of Eq. x(4).

With $A_1 > \sqrt{W}$ the H^2 sequence is seen to be of no importance since a real solution through this sequence requires very small concentrations; it could apply only when $A_1A_2 > (a_1A_1 + a_2A_2 + W)$, to make the H^2 term positive. Of the two other sequences, the H^4 applies in general for higher and the H^3 for lower values of a_1. Since the quantities a_1, a_2, A_1, and A_2 are independent, it would be difficult to formulate a general critical function of the parameters determining the choice between these sequences. A rough analogy however is available with the simpler problem of a dibasic acid, in which only one concentration is involved. For a dibasic acid with $K_1 > K_2$, the H^4 sequence is better than the H^3 as long as the concentration is large compared to $\sqrt{K_1K_2}$ (Chapter VIII, Section E, and Chapter XIV, Section A2). In the present case therefore, if $a_1 \gg \sqrt{A_1A_2}$ and if a_2 is not too much greater than a_1, we may use the H^4 sequence, which gives

$$H \cong -(A_1 + A_2)/2 + \sqrt{(\quad)^2 + a_1A_1 + a_2A_2 + W - A_1A_2} = Q_8^4. \quad (10)$$

If $a_1 < \sqrt{A_1A_2}$, the H^3 sequence is better, giving

$$H \cong \frac{a_1A_1 + a_2A_2 + W - A_1A_2}{2(A_1 + A_2)} + \sqrt{(\quad)^2 + \frac{A_1A_2(a_1 + a_2) + W(A_1 + A_2)}{A_1 + A_2}} = Q_8^3, \quad (11)$$

provided again that a_2 is not very large compared to a_1. If a_1 is roughly equal to $\sqrt{A_1A_2}$ or if $a_2 \gg a_1$, neither sequence is dependable and all four terms from H^4 to H must be used for solution by approximation methods. These approximate conditions of applicability have been verified by numerical examples covering a wide range of absolute and relative values of the four parameters involved.

With sufficiently high values of the concentrations, Eq. (10) becomes

$$H \cong \sqrt{a_1A_1 + a_2A_2}, \quad (12)$$

which is the familiar "first approximation" formula. This formula may be said to represent what has been defined as H_* in Chapter III, or the value of H approached as the concentrations increase indefinitely; it is obtainable directly from Eq. (2) by the neglect of OH and of the ionization constants as terms relative to H.

2. The First Equivalent Point ($b_s = a_1$). From Eq. (3),

$$H^4 + H^3(a_1 + A_1 + A_2) - H^2[A_2(a_2 - a_1) + W - A_1A_2] - H[a_2A_1A_2 + W(A_1 + A_2)] - A_1A_2W = 0. \quad (13)$$

The H^4 sequence is important only when $H^4 > Ha_2A_1A_2$ and hence $a_2 < \sqrt{A_1A_2}$ since $H \cong \sqrt{A_1A_2}$; and the H^2 sequence only when $a_1 \ll W/\sqrt{A_1A_2}$

and $(a_2 - a_1) < A_1$. The best general approximation is therefore the H^3 sequence and its simplifications:

$$H = \frac{A_2(a_2 - a_1) + W - A_1A_2}{2(a_1 + A_1 + A_2)} + \sqrt{(\quad)^2 + \frac{a_2A_1A_2 + W(A_1 + A_2)}{a_1 + A_1 + A_2}} = Q_{13}^3. \tag{14}$$

Various simplifications of this expression are possible, one of which is

$$H \cong A_2(a_2 - a_1)/2a_1 + \sqrt{(\quad)^2 + A_1A_2a_2/a_1}; \tag{15}$$

this is obtainable directly from Eq. (13) as the value of H_* approached as the concentrations become so high that all terms not involving a_1 or a_2 may be neglected. Also, especially if the two concentrations are high and nearly equal, Eq. (14) may be written as

$$H \cong \sqrt{A_1A_2(a_2 + W/A_2 + W/A_1)/(a_1 + A_1 + A_2)}. \tag{16}$$

A different simplification was discussed in Chapter IX as Eq. IX(39), which then led to Eq. IX(40) or

$$H \cong \sqrt{(A_1A_2a_2/a_1)(1 + W/a_2A_2)/(1 + A_1/a_1)}. \tag{17}$$

Now if $A_1A_2 > Wa_1/a_2$, then

$$H \cong \sqrt{A_1A_2a_2/(a_1 + A_1)}, \tag{18}$$

while if $A_1A_2 < Wa_1/a_2$, then

$$H \cong \sqrt{A_1A_2(a_2 + W/A_2)/a_1}. \tag{19}$$

Essentially this means that if the solution is acid, which may be determined from the data by application of Eq. VII(57), then Eq. (18) is better than Eq. (19) while if the solution is alkaline, Eq. (19) is better. Finally, in either case, with high enough values of the concentrations, we have as in Eq. IX(41)

$$H \cong \sqrt{A_1A_2a_2/a_1}. \tag{20}$$

With this approximate formula for H, a better idea of the conditions of applicability of Eq. (14) and hence of its final simplifications may now be obtained. If the H^2 sequence of Eq. (13) is ruled out, i.e. if $a_1 > W/\sqrt{A_1A_2}$ or $(a_2 - a_1) > A_1$, then the H^4 sequence is better than Eq. (14) only if roughly $H^3 > a_2A_1A_2$; from Eq. (20) this means only if $a_1 < \sqrt[3]{a_2A_1A_2}$. If the H^4 sequence is ruled out, with $(a_2 - a_1) < A_1$, then the H^2 sequence is better than Eq. (14) only if roughly $H^3(a_1 + A_1 + A_2) < A_1A_2W$ and, from Eq. (20), this means roughly $a_2 < \sqrt[3]{a_1W^2/A_1A_2}$. Unless at least one of the concentrations then is quite low, Eq. (14) and its simplifications must be widely applicable; it will describe the solution of the alkali salt of a weak acid mixed with a weaker acid.

The conditions of applicability of Eq. (14) have been verified by comparing H calculated through it for a variety of combinations of values of the constants and of the concentrations, with the exact value obtained from Eq. (2). Table 11–1 shows some examples together with calculations based on the simpler and less accurate approximations of Eqs. (20), (17), (16), and (15). Eq. (14) is in general quite satisfactory while its simplifications, Eqs. (17) and (16), are not. Eq. (15), for H_*, while still quite simple, is seen to be far superior to the familiar formula of Eq. (20), which is generally poor.

Since the alkali salt of a weak acid is mathematically equivalent to a weak base, the solution at this equivalent point corresponds to a mixture of weak base (concentration "b" $= a_1$, ionization constant "B" $= W/A_1$) and weak acid (concentration "a" $= a_2$, ionization constant "A" $= A_2$). Hence the considerations for the numerical solution of Eq. (13) are identical after appropriate substitutions with those for the equation of a solution containing weak base and weak acid, Eq. XIII(17), to be treated later. It follows then that for the case of two acids with constants A_1' and A_2' so related to those of Table 11–1 that $A_1' = W/A_2$ and $A_2' = W/A_1$, the values of H tabulated are the values of OH for reversed values of the concentrations, or for $a_1' = a_2$ and $a_2' = a_1$. This follows since "B" now has the value of A_2 and "A" the value W/A_1. Thus if $A_1 = 10^{-10}$ and $A_2 = 10^{-12}$, then at $a_1 = 10^{-4}$ and $a_2 = 10^{-1}$ the first line of the Table gives OH; the "exact" value, from the last column, is $OH = 8.38 \times 10^{-6}$.

3. Special Case of the First Equivalent Point When $a_1 = a_2$. With $a_1 = a_2 = c$, Eq. (13) becomes

$$H^4 + H^3(c + A_1 + A_2) + H^2(A_1A_2 - W) - H[cA_1A_2 + W(A_1 + A_2)] - A_1A_2W = 0. \quad (21)$$

The solution contains, in equal concentrations, a weak acid and the alkali salt of another weak acid. (It would be meaningless now to say that it contains the salt of the stronger rather than that of the weaker acid; the two points of view are mathematically equally useful and consistent.) Since the alkali salt is mathematically a base, as mentioned in the preceding section, then this special equivalent point is mathematically identical with a solution of the salt of weak acid and weak base. If we set either of the constants of Eq. (21) equal to W/"B" it becomes Eq. VIII(8). With $a_1 = a_2 = b_s = c$, then for example, with A_2 for "A" and with W/A_1 for "B", the values of H in such solutions are those listed in Table 8–1. The condition "A" = "B" for Eq. VIII(8) here corresponds to $A_1A_2 = W$.

The approximations derived in Chapter VIII for the case of such a salt solution and there treated from the point of view of the salt could therefore be obtained from the approximations here discussed for the more general

Table 11–1. Value of H at the first equivalent point for two weak acids and strong base.

A_1	A_2	a_1	a_2	Eq. (20)	Eq. (17)	Eq. (16)	Eq. (15); H_*	Eq. (14)	Eq. (13); exact
10^{-2}	10^{-4}	0.1	10^{-4}	3.16×10^{-5}	3.02×10^{-5}	3.01×10^{-5}	9.17×10^{-6}	8.38×10^{-6}	8.38×10^{-6}
		0.1	0.11	1.05×10^{-3}	1.00×10^{-3}	1.00×10^{-3}	1.05×10^{-3}	1.00×10^{-3}	9.95×10^{-4}
		10^{-3}	10^{-4}	3.16×10^{-4}	9.53×10^{-5}	9.49×10^{-5}	2.74×10^{-4}	5.78×10^{-5}	5.77×10^{-5}
		10^{-4}	0.1	3.16×10^{-2}	3.15×10^{-3}	3.13×10^{-3}	0.1091	3.60×10^{-3}	3.10×10^{-3}
10^{-6}	10^{-8}	1.0	10^{-3}	3.16×10^{-9}	3.16×10^{-9}	3.16×10^{-9}	9.17×10^{-10}	9.18×10^{-10}	9.27×10^{-10}
		1.1	1	9.53×10^{-8}	9.53×10^{-8}	9.53×10^{-8}	9.49×10^{-8}	9.49×10^{-8}	9.49×10^{-8}
		10^{-3}	10^{-2}	3.16×10^{-7}	3.16×10^{-7}	3.16×10^{-7}	3.64×10^{-7}	3.64×10^{-7}	3.64×10^{-7}
		10^{-3}	1	3.16×10^{-6}	3.16×10^{-6}	3.16×10^{-7}	1.091×10^{-5}	1.090×10^{-5}	1.089×10^{-5}

case of Eq. (13) by setting $a_1 = a_2 = c$ and A_1 (or A_2) $= W/B$. The conditions for the applicability of Eq. (14), namely $a_1 > \sqrt[3]{a_2A_1A_2}$ and $a_2 > \sqrt[3]{a_1W^2/A_1A_2}$, are then seen to be identical with those for the applicability of Eq. VIII(15) becoming respectively $c > \sqrt{WA/B}$ and $c > \sqrt{WB/A}$. With the condition $a_1 = a_2 = c$ and with the transformation $A_1 = W/B$ (and A_2 as "A"), Eqs. (14), (16), and (20) of the present chapter become identical with Eqs. VIII(15), (17), and (21).

Eq. (20), for equal concentrations, gives $H \cong \sqrt{A_1A_2}$, corresponding to $H \cong \sqrt{WA/B}$ for the salt. Also *exactly* the solution is neutral if $A_1A_2 = W$, as it is for the salt if $A = B$.

Finally the approximation of Eq. (20) is one that has been used in connection with the experimental determination of the ratio of the ionization constants of two acids, HX with A_1 and HY with A_2. At this "first equivalent point" with strong base for a solution containing two monobasic acids at the same concentration (or the solution of the alkali salt of one acid and a second "free" acid, at the same concentration) we have

$$\frac{[X^-]}{[Y^-]} = \frac{(\alpha)_1}{(\alpha)_2} = \frac{A_1(A_2 + H)}{A_2(A_1 + H)} \cong \frac{1 + \sqrt{A_1/A_2}}{1 + \sqrt{A_2/A_1}}, \tag{22}$$

if $H \cong \sqrt{A_1A_2}$. Now if $A_1 \gg A_2$ then

$$[X^-]/[Y^-] \cong \sqrt{A_1/A_2}. \tag{23}$$

Hence the measurement of the ratio $[X^-]/[Y^-]$, as by colorimetry, etc., at such an equivalent point, gives an approximate value of $\sqrt{A_1/A_2}$.

4. THE SECOND EQUIVALENT POINT ($b_s = a_1 + a_2$). At the second equivalent point the solution is mathematically one of two bases, the stronger with concentration $b_1 = a_2$ and ionization constant $B_1^* = W/A_2$, the second with concentration $b_2 = a_1$ and ionization constant $B_2^* = W/A_1$. Therefore the relations will be defined by the base analog of Eq. (8), or

$$(OH)^4 + (OH)^3(B_1^* + B_2^*) - (OH)^2(b_1B_1^* + b_2B_2^* + W - B_1^*B_2^*) \\ - (OH)[B_1^*B_2^*(b_1 + b_2) + W(B_1^* + B_2^*)] - B_1^*B_2^*W = 0. \tag{24}$$

The best general approximation is then the analog of Eq. (10), or

$$OH \cong -(B_1^* + B_2^*)/2 + \sqrt{(\quad)^2 + b_1B_1^* + b_2B_2^* + W - B_1^*B_2^*}, \tag{25}$$

or finally as in Eq. (12)

$$OH \cong \sqrt{b_1B_1^* + b_2B_2^*}. \tag{26}$$

In terms of the actual acids at this equivalent point this means

$$H \cong \sqrt{A_1A_2W/(a_2A_1 + a_1A_2)}. \tag{27}$$

5. THE TITRATION CURVE. Using the H^3 sequence of Eq. (3) as the most applicable approximation, and dropping A_1A_2, W, and $W(A_1 + A_2)$ in the H^2 and H terms, we have

$$H \cong \frac{A_1(a_1 - b_s) + A_2(a_2 - b_s)}{2(A_1 + A_2 + b_s)} + \sqrt{(\quad)^2 + \frac{A_1A_2(a_1 + a_2 - b_s)}{A_1 + A_2 + b_s}}, \quad (28)$$

or neglecting $(A_1 + A_2)$ relative to b_s,

$$H \cong [A_1(a_1 - b_s) + A_2(a_2 - b_s)]/2b_s + \sqrt{(\quad)^2 + A_1A_2(a_1 + a_2 - b_s)/b_s}, \quad (29)$$

which may be obtained directly by assuming $H - OH \cong 0$ in Eq. (2). Eq. (29) has been derived (differently) and used by Michaelis[1] for the calculation of the "error in pH due to indicator." Eq. (29) itself may be quite accurate and holds on both sides of the first equivalent point.

If A_1/A_2 is very large while A_1 is itself small, the Henderson equation, or Eq. x(25), applies roughly near the mid-point of each half of the titration (i.e. $b_s \cong a_1/2$ and $b_s \cong a_1 + a_2/2$). This is seen by simplifying the H^3 sequence of Eq. (3) to

$$H^3b_s - H^2[A_1(a_1 - b_s) + A_2(a_2 - b_s)] - HA_1A_2(a_1 + a_2 - b_s) \cong 0. \quad (30)$$

In the first half of the titration ($b_s < a_1$) $H \cong A_1$ so that the third of these terms is the smallest; then $H \cong A_1(a_1 - b_s)/b_s$. In the second half ($b_s > a_1$), $H \cong A_2$, and the first term is the smallest; then $H \cong A_2(a_1 + a_2 - b_s)/(a_1 - b_s)$.

Near and through the second equivalent point the titration curve will be given by the H^2 sequence of Eq. (3), or

$$H \cong \frac{[A_1A_2(a_1 + a_2 - b_s) + (A_1 + A_2)W]/2}{A_1(b_s - a_1) + A_2(b_s - a_2) + A_1A_2 - W} + \sqrt{(\quad)^2 + \frac{A_1A_2W}{\mathbf{Y}}} = Q_3^2, \quad (31)$$

in which $\mathbf{Y}$ represents the denominator of the first term.

Less accurately,

$$H \cong \frac{A_1A_2(a_1 + a_2 - b_s)/2}{A_1(b_s - a_1) + A_2(b_s - a_2)} + \sqrt{(\quad)^2 + \frac{A_1A_2W}{A_1(b_s - a_1) + A_2(b_s - a_2)}}. \quad (32)$$

With $b_s = a_1 + a_2$, these equations describe the equivalent point itself, reducing to Eq. (27) as a final approximation. Furthermore, if $A_1 \gg A_2$ and if the concentrations a_1 and a_2 are comparable, Eq. (32) becomes

$$H \cong A_2(a_1 + a_2 - b_s)/2(b_s - a_1) + \sqrt{(\quad)^2 + A_2W/(b_s - a_1)}, \quad (33)$$

which is essentially the curve for the titration of the weak acid A_2 alone.

6. FEASIBILITY OF TITRATION.

(1) *op. cit.*, Ref. II–3; pp. 46–48.

(a) For the first equivalent point the case has already been considered as the illustrative example in Chapter IX.

(b) For the second equivalent point we may use the H^2 sequence of Eq. (3) with coefficients simplified as in Eq. (32). Then with $A_1 > A_2$ and if p is small, it is seen that $(-w)_{-p/2} \cong (-w)_{+p/2}$ in this sequence, as required in Eq. IX(20). Hence the feasibility function may be written through either of the procedures of Chapter IX. Through Eq. IX(22),

$$F = (p/400)\sqrt{a_2^2 A_1 A_2/(a_2 A_1 + a_1 A_2)\,W}. \quad (34)$$

For use in Eq. IX(27) we take π_e from Eq. IX(38) with H_e from Eq. (27); neglecting H relative to OH, A_1 and A_2 in Eq. IX(38), we thereby obtain

$$\pi_e = 2(2.3)\sqrt{(a_2 A_1 + a_1 A_2)\,W/A_1 A_2}, \quad (35)$$

which, with Eq. IX(27), then gives the expression for F in Eq. (34). Finally, if the concentrations are comparable, with $A_1 \gg A_2$, so that $H_e \cong \sqrt{A_2 W/a_2}$, then $F \cong (p/400)\sqrt{a_2 A_2/W}$, as for a weak acid alone.

B. Relation Between Buffer Capacity and Titration Curve

From Eq. IX(37),

$$d\pi/dpH = (2.3)^2\{[a\alpha\rho(\rho - \alpha)]_1 + [a\alpha\rho(\rho - \alpha)]_2 - H + OH\}; \quad (36)$$

$$d^2\pi/dpH^2 = (2.3)^3\{[a\alpha\rho(\alpha^2 + \rho^2 - 4\alpha\rho)]_1 + [a\alpha\rho(\alpha^2 + \rho^2 - 4\alpha\rho)]_2 + H + OH\}. \quad (37)$$

Also, with π_j defined as $2.3(a\alpha\rho)_j$, then

$$\pi = 2.3\,[(a\alpha\rho)_1 + (a\alpha\rho)_2 + H + OH] = \pi_1 + \pi_2 + 2.3\,(H + OH). \quad (38)$$

If, therefore, we are dealing with that portion of the π curve in which the quantity $(H + OH)$ is negligible, the curve may be considered as the sum of two independent curves, π_1 and π_2, representing the "separate" contributions of the two acids, as explained under Eqs. I(112) and I(113). Graphically, the relations are shown in Fig. 11–1 for a case in which $a_2 > a_1$. [*Note:* Like subsequent similar diagrams, this figure is drawn with unit height in order to show the relations of the ionization fractions, which are helpful for orientation. The values of the fractions depend simply on pH. But the value of π, whether an individual contribution such as π_1, π_2, etc., or a total actual buffer capacity, depends on concentrations of solutes and may therefore be either greater or less than $1(\times 2.3)$.]

Since the π curve has three minima, there will accordingly be three vertical inflections ("end-points") in the titration curve of pH against b_s, as shown in Fig. 9–3b; and the quantity b_s at any value of pH is the area under the π curve and is equal to $a_s + (a\alpha)_1 + (a\alpha)_2 - D = \int \pi dpH$. This refers to Eq. (1) with an additional term in a_s.

The first inflection, at minimum a, is in acid solution and represents the end-point, for the titration with strong base, of strong acid in presence of two weak acids. The central minimum, minimum c, represents the inflection point for the titration of the first weak acid, and the third, minimum b, in alkaline solution, the final titration of the second weak acid. The conditions for the appearance and disappearance of the two terminal inflection points (minimum a and minimum b) are essentially the same as in the case of a single weak acid (Chapter x) since, for example, the second weak acid hardly affects the inflection at minimum a.

1. Condition for Inflection for Titration of First Acid. We shall here examine the requirements, as $f(a_1, a_2, A_1, A_2)$, for the central minimum

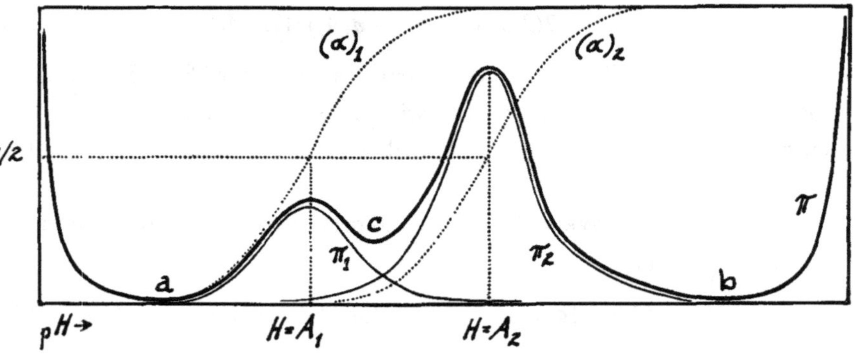

Fig. 11–1. Buffer capacity, two weak acids at unequal concentrations.

in the π curve or for the possibility of an inflection indicating the titration of one weak acid in presence of a second. It is clear from Fig. 11–1 that as the two ionization constants approach each other for fixed values of the concentrations, this minimum will disappear leaving a single maximum between minimum a and minimum b. Similarly, for fixed values of the constants, the central minimum will disappear if the ratio of the concentrations becomes either too large or too small. The condition for the disappearance of minimum c and its merging with one or both of the maxima is that both $d\pi/dpH$ and $d^2\pi/dpH^2$ equal 0.

The considerations will be approximate to the extent of neglecting H and OH in the expression for π in Eqs. (36–38). This is generally permissible if the concentrations are appreciable and if A_1A_2 is neither very large nor very small, so that the equivalent point may be taken as $H_e \cong \sqrt{A_1A_2a_2/a_1}$. If these conditions are satisfied, the relations derived are quite accurate.

a. The case of equal concentrations ($a_1 = a_2$). With H and OH neglected in π as expressed in Eq. (38), the curve in this case is symmetrical with a minimum or a maximum always at $H = \sqrt{A_1A_2}$, as shown in Fig. 11–2. While the maxima for the individual curves of π_1 and π_2 are at $H = A_1$ and $H = A_2$ respectively, the corresponding maxima of the total π curve occur at $H < A_1$ and $H > A_2$ in each case; and as the values of A_1 and A_2 approach each other, these maxima move toward each other at a faster rate, merging (disappearance of the central minimum) at a certain critical value of the ratio A_1/A_2 or $(A_1/A_2)_{\mathrm{crit}}$.

In this simple symmetrical case, we note that at $H = \sqrt{A_1A_2}$, $(\alpha)_1 = 1 - (\alpha)_2 = (\rho)_2$ and $(\rho)_1 = (\alpha)_2$. On introducing these relations into $d^2\pi/d\mathrm{p}H^2$, from Eq. (37), with $a_1 = a_2$ and with H and OH neglected, and noting that $(\rho)_1 = 1 - (\alpha)_1$, we obtain the condition $(\alpha)_1^2 - (\alpha)_1 + 1/6 = 0$; and since $(\alpha)_1 > 1/2$ at this point, then $(\alpha)_1 = (1 + \sqrt{3})/2\sqrt{3}$. Hence since, from Eq. III(3), $H = (A\rho/\alpha)_1 = [A(1 - \alpha)/\alpha]_1$ and since the value of H is here $\sqrt{A_1A_2}$, then $\sqrt{A_1A_2} = A_1(\sqrt{3} - 1)/(\sqrt{3} + 1)$. Finally then,

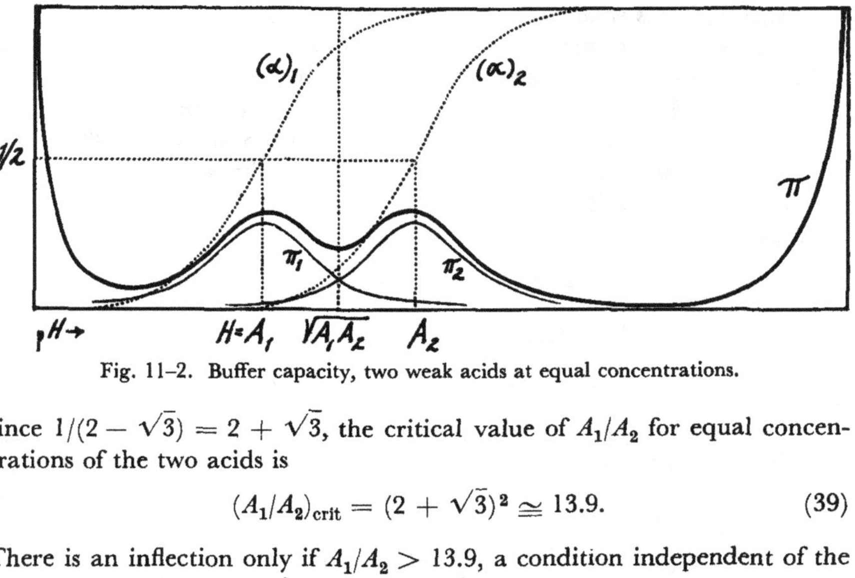

Fig. 11–2. Buffer capacity, two weak acids at equal concentrations.

since $1/(2 - \sqrt{3}) = 2 + \sqrt{3}$, the critical value of A_1/A_2 for equal concentrations of the two acids is

$$(A_1/A_2)_{\mathrm{crit}} = (2 + \sqrt{3})^2 \cong 13.9. \tag{39}$$

There is an inflection only if $A_1/A_2 > 13.9$, a condition independent of the ionic strength if $\mathbf{A}_1 = A_1\gamma^2$ and $\mathbf{A}_2 = A_2\gamma^2$.

b. Unequal concentrations. This case is shown in Fig. 11–1. Let $y = A_2/A_1$ (and hence $y < 1$) and $x = a_2/a_1$. Now, setting $d\pi/d\mathrm{p}H = 0$ in Eq. (36) we have, always with the neglect of $H + OH$,

$$x = [\alpha\rho(\rho - \alpha)]_1/[\alpha\rho(\alpha - \rho)]_2. \tag{40}$$

Setting $d^2\pi/d\mathrm{p}H^2 = 0$, Eq. (37), we have

$$x = [\alpha\rho(\alpha^2 - 4\alpha\rho + \rho^2)]_1/[\alpha\rho(4\alpha\rho - \alpha^2 - \rho^2)]_2. \tag{41}$$

Equating, and noting that $(\alpha)_1 = A_1/(A_1 + H)$ and $(\alpha)_2 = yA_1/(yA_1 + H)$ while $(\rho)_1 = 1 - (\alpha)_1$ and $(\rho)_2 = 1 - (\alpha)_2$, we find the value of H_{crit} at the disappearance of the central minimum to be

$$H_{\mathrm{crit}} = A_1[(1 + y)/4 \pm \sqrt{(\quad)^2 - y}]. \tag{42}$$

Given A_1 and y, H_{crit} so calculated (two values) may then be used in Eq. (40) to determine the critical values of x for the disappearance of minimum c: one on the left toward $H = A_1$ and one on the right toward $H = A_2$. The two values of x_{crit} will be the reciprocal of each other. This means, as expected, that the π minimum disappears at the same critical value of either a_2/a_1 or a_1/a_2 for given y.

Inspection of Eq. (42) shows that no real solution is possible unless $y < (2 - \sqrt{3})^2$, which is recognized as the limiting value of $(A_2/A_1)_{crit}$ for equal values of a_1 and a_2, according to Eq. (39). With $y > (2 - \sqrt{3})^2$, there is no minimum c whatever the ratio of the concentrations, and the more nearly equal are the concentrations, the longer does the minimum c persist as y approaches 1.

Roughly, if y is small enough, the two values of H_{crit} from Eq. (42) are $A_1/2$ and $2yA_1$. With these (approximate) values of H, Eq. (40) may be used to find the (approximate) critical values of x as $f(y)$. With $H_{crit} \cong A_1/2$, then $(\alpha)_1 \cong 2/3$, $(\rho)_1 \cong 1/3$, $(\alpha)_2 \cong 2y$, and $(\rho)_2 \cong 1$; hence $x_{crit} = 1/27y$. With $H_{crit} \cong 2yA_1$, $(\alpha)_1 \cong 1$, $(\rho)_1 \cong 2y$, $(\alpha)_2 \cong 1/3$ and $(\rho)_2 \cong 2/3$, so that $x_{crit} = 27y$. This means that there will be no minimum c unless $27y < x < 1/27y$; or there will be no inflection for the titration of a_1 in presence of a_2 unless the ratio a_2/a_1 is within the limits

$$27A_2/A_1 < a_2/a_1 < A_1/27A_2 \tag{43}$$

or unless the ratio A_2/A_1 satisfies the requirements

$$A_2/A_1 < a_2/27a_1 \text{ and } < a_1/27a_2. \tag{44}$$

The limiting value of A_2/A_1 for unit ratio of a_2/a_1 however is not 1/27 as in Eq. (44) but 1/13.9 as in Eq. (39). The relation in Eqs. (43) and (44) is an approximation holding only when A_2/A_1 is already small compared to the critical value of Eq. (39).

As we have just seen, Eqs. (40) and (42) may be used to find the critical ratio x (or a_2/a_1) for given values of A_1 and y, the relations being exact except for the neglect of H and OH in the equations for π and its derivatives. If the problem is simply that of finding x_{crit} for given y or, in reverse, y_{crit} for given x, Eqs. (40) and (41) may first be combined into an equation involving $(\alpha)_1$ and y. This is possible through the relations $(\rho)_2 = 1 - (\alpha)_2$ and $(\alpha)_2 = yA_1/(yA_1 + H) = y/\{y + [(1 - \alpha)/\alpha]_1\}$ since $H = (A\rho/\alpha)_1$. The resulting expression may be written in two forms, with α standing for $(\alpha)_1$:

$$\alpha = (5 + y)/6(1 + y) \pm \sqrt{(\quad)^2 - 2/3(1 + y)}, \tag{45a}$$

$$y = (1 - \alpha)(3\alpha - 2)/\alpha(3\alpha - 1). \tag{45b}$$

At the same time, Eq. (40) becomes

$$x = (2\alpha - 1)(1 - \alpha + \alpha y)^3/y(1 - \alpha - \alpha y), \tag{46}$$

and combination with Eq. (45b) gives

$$16\alpha^6 - 48\alpha^5 + 56\alpha^4 - 32\alpha^3 + 9\alpha^2(1 + x/27) - \alpha(1 + x/3) + 2x/27 = 0. \tag{47}$$

For given x, α_{crit} is calculated through Eq. (47) and then y_{crit} through Eq. (45b). For given y, α_{crit} is obtained from Eq. (45a) and then x_{crit} from Eq. (46).

In illustration, for $y = 0.01$, α_{crit} from Eq. (45a) is either 0.979 or 0.674, so that x_{crit} is either 0.25 or 4.0. For $x = a_2/a_1 = 1/2$, Eq. (47) gives $\alpha_{crit} = 0.955$ and Eq. (45b) gives $y_{crit} = 0.0219$ or 1/46; from Eq. (43), which is an approximation, we would have roughly $y = 1/54$ for $x = 1/2$ (or $a_2/a_1 = 1/2$).

2. VALUE OF H AT THE CENTRAL MINIMUM OF π; TITRATION ERROR. In considering the position of the central minimum of π or the value of H_i at the inflection point between two acids and hence the error of the titration, we shall continue to neglect H and OH in π. When x (or a_2/a_1) $= 1$, the inflection is of course always at $H = \sqrt{A_1A_2}$, corresponding exactly to the the equivalent point (with neglect of H and OH) so that the error would then be zero.

Otherwise, Eq. (40) gives for $d\pi/dpH = 0$

$$H^4(1 + xy) - H^3A_1(1 - 3xy - 3y + xy^2) - 3H^2A_1^2y(1 - x + xy - y) + HA_1^3y(x - 3xy + y^2 - 3y) - A_1^4y^2(y + x) = 0. \tag{48}$$

But since an inflection occurs only if $y \ll 1$ and, according to Eqs. (43) and (44), only if in general $y \ll x$, then Eq. (48) becomes

$$H^4 - H^3A_1 - 3H^2A_1^2y(1 - x) + HA_1^3yx - y^2xA_1^4 \cong 0. \tag{49}$$

Since at the equivalent point $A_1 > H > yA_1$ while $H \cong A_1\sqrt{xy}$, we proceed as in the treatment of Eqs. x(49–51) obtaining

$$H_i \cong A_1\sqrt{xy} + 2A_1y(x - 1). \tag{50}$$

Hence if x (or a_2/a_1) > 1, $H_i \cong A_1\sqrt{xy}(1 + 2\sqrt{xy})$, while if $x < 1$, $H_i = A_1\sqrt{xy}(1 - 2\sqrt{y/x})$. The inflection point therefore occurs too soon and the error is negative if $a_2 > a_1$; it occurs too late, causing a positive error, if $a_2 < a_1$.

The titration error may be expressed through Eq. (1), with neglect of H and OH, so that at the inflection point

$$a_1 \cong [(b_s)_i - (a\alpha)_2]/(\alpha)_1. \tag{51}$$

Combined with Eq. (50) this gives the relative error, or $E = [(b_s)_i - a_1]/a_1$, as

$$E \cong 3y(1 - x)\left[\frac{1 - (2/3)\sqrt{y/x}(1 - x)}{1 - \sqrt{y/x}(1 - 2x) - 3y(1 - x) + 2y\sqrt{y/x}(1 - x)^2}\right]. \tag{52}$$

(This assumes that we may take $a_2/(b_s)_i \cong x$ in expressing the error from Eq. [51].) As expected, the error, with neglect of H and OH, is zero if $x = 1$. It is positive if $x < 1$ and negative if $x > 1$. Less accurately, Eq. (52) becomes

$$E \cong 3y(1-x)[1 + (1/3)\sqrt{y/x}(1-4x)], \tag{53}$$

or roughly

$$E \cong 3y(1-x). \tag{54}$$

The necessary approximate values of x and y, for use in these expressions, are obtainable from the titration itself—from the values of b_s required for the two consecutive inflection points for x, and from the ratio of the values of H at the two consecutive mid-points for y.

Some illustrative values calculated through these equations to bring out their relative applicability are given in Table 11–2.

Table 11-2. Relative percentage error in titration of weak acid in presence of a weaker acid.

x	y	Eq. (54)	Eq. (52) or (53)
0.1	10^{-4}	+ 0.027	+ 0.027
0.25	10^{-3}	+ 0.225	+ 0.225
0.1	10^{-3}	+ 0.270	+ 0.276
10	10^{-4}	− 0.27	− 0.259
4	10^{-3}	− 0.9	− 0.829
10	10^{-3}	− 2.7	− 2.35

(If the titration is carried not to the inflection point but to an arbitrary known value of H as H_{end} near the inflection, then the actual value of a_1 may be calculated directly through Eq. (1) from H_{end} and the value of b_s used if a_2, A_1, and A_2 are known at least approximately.)

Summarizing, therefore, on the problem of the titration of a weak acid in presence of a still weaker one:

(a) The titration curve shows an inflection point for the measurement of the stronger acid only if $27y < x < 1/27y$, from Eq. (43), with x as either a_2/a_1 or a_1/a_2.

(b) The feasibility of the titration is $\Delta_p = 2/2.3 \sinh^{-1} (p/400\sqrt{xy})$, from Eq. IX(36).

(c) If two successive inflection points are noted for the complete titration so that x can be estimated and if y is known at least approximately, then the error of the titration is given by Eq. (53), or roughly as $E \cong 3y(1 - x)$.

(d) If the titration is done with even an uncalibrated potentiometer so that the ratio of the values of H at the two successive mid-points of the titration may be estimated, then the error may be calculated with these equations through the rough values of x and y given by the titration curve itself.

XII

Titration (With Strong Base or Strong Acid) in Mixtures Involving Any Number of Independent Ionization Constants

A. Titration With Strong Base of a Mixture of Any Number of Monobasic Acids

1. The General Equation. In a solution containing $\mathbf{z}$ weak acids, together with strong acid, a_s, and strong base, b_s, so that $g = a_s - b_s$,

$$H - OH = (a_s - b_s) + \Sigma_\mathbf{z}[aA/(A + H)]. \tag{1}$$

As a polynomial in H, the degree of this equation is $\mathbf{z} + 2$:

$$H^{\mathbf{z}+2} + (\Sigma_\mathbf{z} A - g)H^{\mathbf{z}+1} + \mathbf{C}_0 H^\mathbf{z} + \mathbf{C}_1 H^{\mathbf{z}-1} + \mathbf{C}_2 H^{\mathbf{z}-2} + \cdots + \mathbf{C}_j H^{\mathbf{z}-j} + \cdots + \mathbf{C}_{\mathbf{z}-1} H + \mathbf{C}_\mathbf{z} = 0. \tag{2}$$

The coefficients, $\mathbf{C}_0 \cdots \mathbf{C}_\mathbf{z}$, have the following meaning in terms of concentrations and constants. Each consists of three general summations, some of which may be zero.

$$\mathbf{C}_j = \Sigma_\mathbf{z}(\Pi_{j+2} A) - W\Sigma_\mathbf{z}(\Pi_j A) - \Sigma_\mathbf{z}[(g + \sigma a_{j+1})\Pi]. \tag{3}$$

Here the expression $(\Pi_x A)$ means the product of x different A's and $\Sigma_\mathbf{z}(\Pi_x A)$ means the sum of all such possible products, considering all the constants, from A_1 to $A_\mathbf{z}$. The value of $\Sigma(\Pi_x A)$ is ΣA for $x = 1$, 0 for $x > \mathbf{z}$, and as an arbitrary definition 1 for $x = 0$. In the third summation σa_{j+1} means the sum of $j + 1$ different a's, and the quantity $(g + \sigma a_{j+1})$ is then multiplied by Π, the product of the constants of the same $j + 1$ acids; here $\Sigma_\mathbf{z}$ means that all possible combinations of $j + 1$ different acids are to be used, from a_1 to $a_\mathbf{z}$, in this third summation of the coefficient $\mathbf{C}_j$. This summation equals zero when $(j + 1) > \mathbf{z}$. $\mathbf{C}_\mathbf{z}$ is therefore seen to be always simply $-WA_{1\ldots\mathbf{z}}$. The units of the terms of Eq. (2) are $c^{(\mathbf{z}+2)}$ since an ionization constant is equivalent to c and W to c^2, in units.

When $b_s = 0$ and $g = a_s$, Eq. (2) is the equation for a mixture of pure acids, weak acids alone if $a_s = 0$ or strong acid alone if $\mathbf{z} = 0$. For the equivalent points in titration with b_s, we set b_s equal successively to a_s, $a_s + a_1$, $a_s + a_1 + a_2$, etc., if $A_1 > A_2 > A_3 > \cdots > A_\mathbf{z}$. This means that the number of equivalents of base present at a particular equivalent point equals the sum of the equivalents of acids titrated or that the concentration of the base equals the sum of the concentrations of these acids *at the equivalent point.* The actual concentration of a particular acid, or the quantity represented as a_j, is a constant during the titration only if the volume is assumed to remain unchanged.

2. Approximate Formulas for H. We here consider approximate formulas for H for appreciable values of the concentrations or, better, for the case in which each quantity "aA" is large compared to W. When $b_s < a_s$ ("free strong acid" present), H is given approximately by the first two terms of Eq. (2), or $H \cong a_s - b_s - \Sigma_z A = a_s - b_s$. Similarly, when $b_s > (a_s + \Sigma_z a)$, or with "free strong base" present, H is given approximately by the last two terms so that $H \cong - \mathbf{C}_z/\mathbf{C}_{z-1}$. Now from only those terms of $\mathbf{C}_{z-1}$ involving the concentrations, $OH \cong b_s - (a_s + \Sigma_z a)$.

For the equivalent points, and for the parts of the titration curve through each equivalent point, the best sequences of three terms are as follows: that starting with H^{z+2} for $b_s = a_s$ or $g = 0$; the H^{z+1} sequence for $b_s = a_s + a_1$ or $g = - a$; that in H^{z-j+2} for $b_s = a_s + \Sigma_j a$ or $g = - \Sigma_j a$; that in H^{z-j} for $b_s = a_s + \Sigma_{j+2} a$ or $g = - \Sigma_{j+2} a$; and that in H^2 for $b_s = a_s + \Sigma_z a$ or $g = - \Sigma_z a$.

At each equivalent point an approximate formula for H_e, which may be called $(H_e)_*$ or the value of H_e approached as all the concentrations become very large, is obtainable from the first and third terms of the sequence chosen as just indicated; or

$$(H_e)_j \cong \sqrt{- \mathbf{C}_j/\mathbf{C}_{j-2}}, \tag{4}$$

in which $\mathbf{C}_{j-2}$ and $\mathbf{C}_j$ are respectively the coefficients of the first and third terms of the sequence and which are furthermore assumed to be *simplified to contain only the terms involving the concentrations* so that "$\mathbf{C}_j$" $= - \Sigma_z[(g + \sigma a_{j+1})\Pi]$.

This means therefore that at the equivalent point for the titration of the j'th weak acid,

$$(H_e)_j^2 \cong \frac{\Sigma_z[(- \Sigma_j a + \sigma a_{j+1})\Pi]}{\Sigma_z[(\Sigma_j a - \sigma a_{j-1})\Pi]}. \tag{5}$$

For further simplification, this is first rearranged as follows:

$$(H_e)_j^2 \cong \frac{A_{1\cdots j}\, \Sigma_{j+1}^{z}\,(aA) + \Sigma_{>j}^{z}[(- \Sigma_j a + \sigma a_{j+1})\Pi]}{A_{1\cdots(j-1)} a_j + \Sigma_{>(j-1)}^{z}[(\Sigma_j a - \sigma a_{j-1})\Pi]}. \tag{6}$$

While Σ_{j+1}^{z} in the first term of the numerator means, as usual, summation from $(j + 1)$ to z, $\Sigma_{>j}^{z}$ in the second term means that we are to consider all combinations in σa_{j+1} of $j + 1$ acids not containing *together* the acids 1 through j since these combinations have been segregated in the first term. As a result the second summation of the numerator will be neglected compared to the first since its A products do not contain the largest A's, $A_{1\ldots j}$, in combination, which have been collected in the first summation. The largest member of the second summation for comparable values of the concentrations will be of about the magnitude of the third member of the first summation. Dropping the second summation in the numerator then, and dividing both numerator and denominator by $A_{1\ldots j}$, we have

$$(H_e)_j^2 \cong [\Sigma_{j+1}^{\mathbf{z}}(aA)] \Big/ \left\{\frac{a_j}{A_j} + \frac{1}{A_{1\ldots j}} \Sigma_{>(j-1)}^{\mathbf{z}} [(\Sigma_j a - \sigma a_{j-1})\Pi]\right\}. \quad (7)$$

Now selecting the largest terms of the summation of the denominator, we have the following simple formula,

$$(H_e)_j^2 \cong \Sigma_{j+1}^{\mathbf{z}}(aA)/\Sigma_j(a/A), \quad (8)$$

or finally least accurately,

$$(H_e)_j \cong \sqrt{A_j A_{j+1} a_{j+1}/a_j}. \quad (9)$$

These are to be used for equivalent points corresponding to $0 < j < \mathbf{z}$.

The equivalent point for the titration of the strong acid ($b_s = a_s$, $g = 0$) in the presence of the $\mathbf{z}$ weak acids (corresponding to $j = 0$) and that for the titration of the last weak acid ($j = \mathbf{z}$) will be called *terminal equivalent points*. For $j = 0$, $\mathbf{C}_{j-2} = 1$ as seen from Eq. (2) and hence, from Eq. (4), $(H_e)_0^2 \cong \Sigma_{\mathbf{z}}(aA) \cong a_1 A_1$, for the equivalent point in titration of the strong acid. The first value, or $\Sigma_{\mathbf{z}}(aA)$, is H_* in a solution containing $\mathbf{z}$ weak acids. For $j = \mathbf{z}$, $\mathbf{C}_j = \mathbf{C}_{\mathbf{z}} = -WA_{1\ldots\mathbf{z}}$, and hence, from Eq. (4) and the denominator of Eq. (6),

$$(H_e)_{\mathbf{z}}^2 \cong WA_{1\ldots\mathbf{z}}/\{a_{\mathbf{z}}A_{1\ldots(\mathbf{z}-1)} + \Sigma_{>(\mathbf{z}-1)}^{\mathbf{z}}[(\Sigma_{\mathbf{z}}a - \sigma a_{\mathbf{z}-1})\Pi]\}; \quad (10)$$

less accurately then, $(H_e)_{\mathbf{z}}^2 \cong W/\Sigma_{\mathbf{z}}(a/A) \cong WA_{\mathbf{z}}/a_{\mathbf{z}}$, or $(OH_e)_{\mathbf{z}}^2 \cong W\Sigma_{\mathbf{z}}(a/A) \cong Wa_{\mathbf{z}}/A_{\mathbf{z}}$.

3. Feasibility of Titration. With these values of $(H_e)_j$ we may find $(\pi_e)_j$ or the (approximate) buffer capacity at the equivalent point and thence the feasibility function, F, for the titration of the acid j through Procedure II of Chapter IX, applicable when the concentrations are not too small and when $pa_j/200$ is small compared to other significant concentrations. (For more accurate evaluation of the feasibility, Procedure I of Chapter IX would be used to express $H_{-p/2}$ and $H_{+p/2}$ according to the appropriate sequence, here that in $H^{\mathbf{z}-j+2}$, for the titration curve through the equivalent point, whereupon the expression $\Delta_p = \log(H_{-p/2}/H_{+p/2})$ would then be simplified as far as the data permitted.)

We shall consider first the terminal equivalent points. The buffer capacity at any equivalent point is

$$\pi_e = 2.3\{\Sigma_{\mathbf{z}}[aAH_e/(A + H_e)^2] + H_e + OH_e\}. \quad (11)$$

For the titration of the strong acid, $(H_e)_0 > A_1$ and $> OH_e$; hence if we write H_e for $A + H_e$ and neglect OH_e, we obtain, using $(H_e)_0^2 \cong \Sigma_{\mathbf{z}}(aA)$, $\pi_e \cong 2(2.3)H_e \cong 2(2.3)\sqrt{\Sigma_{\mathbf{z}}(aA)}$, in which it is understood that $a_1 > 27A_1$, to give an inflection point.

Therefore, $F = 2.3pa_s/200\pi_e$ or

$$F = pa_s/400\sqrt{\Sigma_z(aA)} \cong pa_s/400\sqrt{a_1A_1}. \tag{12}$$

For the titration of the last weak acid $(H_e)_z < A_z$ (as long as $a_zA_z > 27W$) and less than OH_e; now writing A for $A + H_e$ and neglecting H_e compared to OH_e in Eq. (11) we have, with $OH_e \cong \sqrt{W\Sigma_z(a/A)}$, $\pi_e \cong 2(2.3)OH_e \cong 2(2.3)\sqrt{W\Sigma_z(a/A)}$. Then $F = 2.3pa_z/200\pi_e \cong pa_z/400(OH)_e$ or

$$F = pa_z/400\sqrt{W\Sigma_z(a/A)} \cong (p/400)\sqrt{a_zA_z/W}. \tag{13}$$

For the titration of the j'th weak acid $(0 < j < z)$, we neglect $H_e + OH_e$ in Eq. (11), and noting that $(H_e)_j < A_j$ and $> A_{j+1}$, we obtain, assuming always that conditions roughly corresponding to Eqs. XI(43) and XI(44) for inflection points are satisfied,

$$(\pi_e)_j \cong 2.3[\Sigma_j(aH_e/A) + \Sigma_{j+1}^{z}(aA/H_e)]. \tag{14}$$

With Eq. (8) this becomes

$$(\pi_e)_j \cong 2(2.3)H_e\Sigma_j(a/A) \cong 2(2.3)\sqrt{\Sigma_{j+1}^{z}(aA)\Sigma_j(a/A)}; \tag{15}$$

$$\cong 2(2.3)\sqrt{(aA)_{j+1}(a/A)_j}. \tag{16}$$

Finally,

$$F_j \cong pa_j/400H_e\Sigma_j(a/A) \cong pa_j/400\sqrt{\Sigma_{j+1}^{z}(aA)\Sigma_j(a/A)}, \tag{17}$$

$$\cong (p/400)\sqrt{(aA)_j/(aA)_{j+1}}. \tag{18}$$

4. Example: Strong Acid and Four Weak Acids. For the titration of the strong acid in presence of the four weak acids,

$$(H_e)_0^2 \cong \Sigma_4(aA) \cong a_1A_1;$$

$$F_0 = pa_s/400\sqrt{\Sigma_4(aA)} \cong pa_s/400\sqrt{a_1A_1}.$$

For the titration of a_1,

$$(H_e)_1^2 = (a_2A_2 + a_3A_3 + a_4A_4)/a_1A_1 \cong A_1A_2a_2/a_1;$$

$$F_1 = pa_1/400\sqrt{(a_2A_2 + a_3A_3 + a_4A_4)a_1/A_1} \cong (p/400)\sqrt{a_1A_1/a_2A_2}.$$

For a_2,

$$(H_e)_2^2 = (a_3A_3 + a_4A_4)/(a_1/A_1 + a_2/A_2) \cong A_2A_3a_3/a_2;$$

$$F_2 = pa_2/400\sqrt{(a_3A_3 + a_4A_4)(a_1/A_1 + a_2/A_2)} \cong (p/400)\sqrt{a_2A_2/a_3A_3}.$$

For a_3,

$$(H_e)_3^2 = a_4A_4/(a_1/A_1 + a_2/A_2 + a_3/A_3) \cong A_3A_4a_4/a_3;$$

$$F_3 = pa_3/400\sqrt{a_4A_4(a_1/A_1 + a_2/A_2 + a_3/A_3)} \cong (p/400)\sqrt{a_3A_3/a_4A_4}.$$

For a_4,

$$(H_e)_4^2 = W/\Sigma_4(a/A) \cong WA_4/a_4;$$

with $(OH)_e \cong \sqrt{Wa_4/A_4}$;

$$F_4 = pa_4/400\sqrt{W\Sigma_4(a/A)} \cong (p/400)\sqrt{a_4A_4/W}.$$

These simple formulas assume the conditions $a_1 \gg A_1 \gg A_2 \gg A_3 \gg A_4 \gg W/a_4$. If the first "weak" acid is fairly strong there may be no inflection for the equivalent point at $b_s = a_s$, while at $b_s = a_s + a_1$ we would have, as in Eqs. IX(44) and IX(46), $H_e \cong \sqrt{A_1A_2a_2/(a_1 + A_1)}$ and with respect to the total $a_s + a_1$, $F = (p/400)\sqrt{(a_s + a_1)^2A_1/a_2A_2(a_1 + A_1)}$. Similarly if the

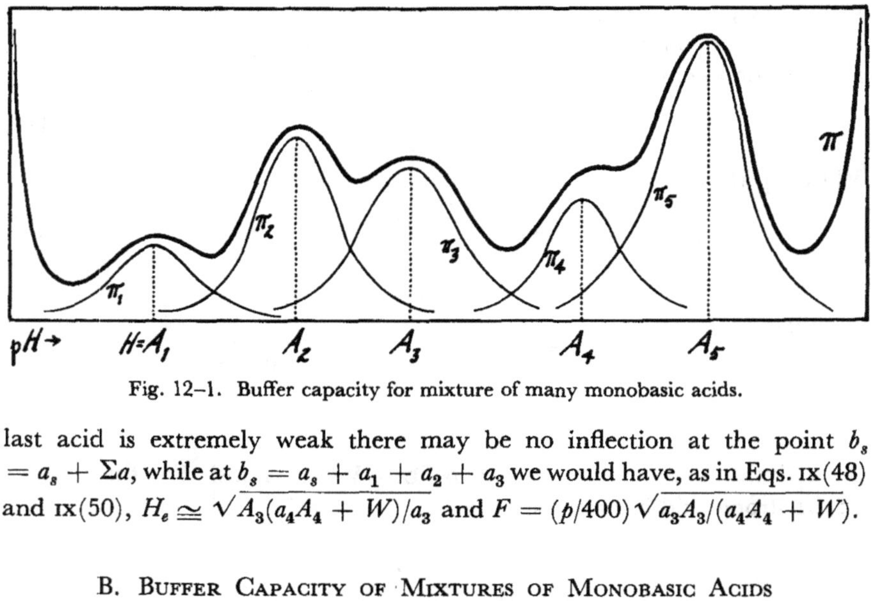

Fig. 12–1. Buffer capacity for mixture of many monobasic acids.

last acid is extremely weak there may be no inflection at the point $b_s = a_s + \Sigma a$, while at $b_s = a_s + a_1 + a_2 + a_3$ we would have, as in Eqs. IX(48) and IX(50), $H_e \cong \sqrt{A_3(a_4A_4 + W)/a_3}$ and $F = (p/400)\sqrt{a_3A_3/(a_4A_4 + W)}$.

B. Buffer Capacity of Mixtures of Monobasic Acids

1. General Curve. For a solution containing z monobasic acids together with only strong acid or strong base,

$$\pi = 2.3[\Sigma_z(a\alpha\rho)_z + H + OH]. \qquad (19)$$

The π curve is essentially (as treated in Chapter XI, Section B) the sum of the practically independent contributions of the individual acids; or, we shall write $\Sigma\pi = 2.3\,\Sigma_z(a\alpha\rho)_z = \Sigma_z\pi_j$, in which $\Sigma\pi$ stands for $\pi - 2.3\,(H + OH)$, or for the π curve with H and OH neglected. Each π_j has its maximum at $H = A_j$, where $(\alpha)_j = (\rho)_j = 1/2$. As long as the ionization constants are far enough apart and the concentrations appreciable, there will then be a maximum in the $\Sigma\pi$ curve corresponding to each of the individual maxima with intervening minima at $H \cong \sqrt{A_jA_{j+1}a_{j+1}/a_j}$, as shown graphically in Fig. 12–1. In such a plot the height of each maximum depends essentially on the concentration of some individual acid, while the pH of maximum or minimum depends essentially on the ionization constants.

2. Equal Concentrations, and Constant Ratio of the A's.[1] If all the concentrations are equal and if the ratio of the successive ionization constants ($y = A_{j+1}/A_j$, always less than 1) is constant, then $(\Sigma\pi)/a$ is a maximum at $H = A_j$ and a minimum at $H = A_j\sqrt{y}$, the values being given by the following formulas, in which π' is written for $(\Sigma\pi)/2.3a$.

$$\text{(a)}\quad \pi'_{\max} \text{ or } \pi'_{A_j} = \tfrac{1}{4} + 2y\left[\frac{1}{(1+y)^2} + \frac{y}{(1+y^2)^2} + \frac{y^2}{(1+y^3)^2} + \frac{y^3}{(1+y^4)^2} + \cdots\right]. \quad (20)$$

If the number of acids is finite, as many terms in the brackets are taken as there are acids on either side of A_j; thus for π' at $H = A_4$ in a series of six acids,

$$\pi'_{A_4}\,(\max) = \tfrac{1}{4} + y\left[\frac{2}{(1+y)^2} + \frac{2y}{(1+y^2)^2} + \frac{y^2}{(1+y^3)^2}\right]. \quad (21)$$

$$\text{(b)}\quad \pi'_{\min} \text{ or } \pi'_{A_j\sqrt{y}} = 2\sqrt{y}\left[\frac{1}{(1+\sqrt{y})^2} + \frac{y}{(1+y\sqrt{y})^2} + \frac{y^2}{(1+y^2\sqrt{y})^2} + \frac{y^3}{(1+y^3\sqrt{y})^2} + \cdots\right]. \quad (22)$$

For π' at $H = \sqrt{A_3A_4}$ in a series of five acids,

$$\pi'_{\sqrt{A_3A_4}}\,(\min) = \sqrt{y}\left[\frac{2}{(1+\sqrt{y})^2} + \frac{2y}{(1+y\sqrt{y})^2} + \frac{y^2}{(1+y^2\sqrt{y})^2}\right]. \quad (23)$$

(c) The ratio, $\pi'_{\max}/\pi'_{\min}$, for an infinite number of acids, at equal concentrations and with constant ratio (y) of the constants, is approximately, from only the first term of each of the series in Eqs. (20) and (22):

$$\frac{\pi'_{\max}}{\pi'_{\min}} \simeq \frac{(1+\sqrt{y})^2}{8\sqrt{y}}\left(1 + \frac{8y}{(1+y)^2}\right). \quad (24)$$

Some numerical values:

$y = 10^{-6}$	ratio $= 125.3$
10^{-4}	12.76
10^{-3}	4.24
10^{-2}	1.614
10^{-1}	1.013

The last two values in this table are calculated from the full formulas (20) and (22); the approximate values through Eq. (24) are 1.63 and 1.14 respectively.

(1) See also Kolthoff and Rosenblum, Ref. VI-2, pp. 29–30.

As the spacing between the acids decreases, or as y approaches 1, the quantity $\Sigma\pi$ increases; at the same time the curve of $\Sigma\pi$ becomes less wavy, the difference between maxima and minima decreasing until, for a series of an infinite number of acids, the $\Sigma\pi$ curve becomes a straight line, but with infinite value, when y becomes 1. For a finite number of acids, at equal concentrations, the distinction between maximum and minimum begins to disappear at the ends of the series as y increases, or as the ionization constants are squeezed together; as a limit the curve ends up with a single maximum at the middle of the series: at $H = A_{(z+1)/2}$ for an odd number of acids and at $H = A_{z/2}\sqrt{y}$ for an even number of acids. For an even number of acids the last minimum to disappear is the central one, at $H = A_{z/2}\sqrt{y}$.

3. Mixture of Acids and Bases. If, in the general case of Fig. 12–1, some of the acids are replaced by bases or, in other words, if we consider a solution of many weak acids and weak bases, the total effect remains the same. Every base is represented by $\pi_{j'}$, a curve with a maximum at $H = W/B_{j'}$ or at $H = A^*_{j'}$, in which $A^*_{j'}$ is the "conjugate" acid constant of the (j')'th base. The total $\Sigma\pi$ curve is still the sum of all the individual curves whether they are contributed by acids or by bases. The difference between a solution containing acids alone and one containing a mixture of acids and bases is one in respect to the pH of the solution; but the buffer capacity at any given pH may be the same if the constants are matched according to the requirement $W/B = A$. In Formulas (20–24), for example, y represents the (constant) ratio of successive acid constants regardless of whether they are real or "conjugate" acid constants; for a base, we take W/B, for the salt MX_s, W/B, and for the salt M_sX, A.

4. Special Cases.

a. Two acids. This case has already been treated (Chapter XI) and serves as a guide for the following considerations. The central minimum disappears if y, or A_2/A_1, is greater than $(2 - \sqrt{3})^2$, for equal concentrations of the two acids; for unequal concentrations, the central minimum (and hence the inflection for the titration of one acid in presence of the second) occurs only as long as (approximately) $27y < a_2/a_1 < 1/27y$.

b. Three acids.

b1. At equal concentrations. With constant y, the $\Sigma\pi$ curve is symmetrical with three maxima and two minima. The central maximum, at $H = A_2$, is higher than the lateral maxima; these are near $H = A_1$ and A_3, but always displaced toward the center, or toward $H = A_2$. As y increases and the constants approach each other, the central maximum rises and the side maxima merge with the adjacent minima. At a certain critical value of y then, there ceases to be any inflection in the titration of the three acids other than that for the total titration, when $b_s = 3a$. This critical value of y is difficult to estimate, but by comparison with the problem of two acids, it must be smaller than approximately 1/14, the value fixed by Eq. XI(39). The problem may presumably be solved by setting $d\pi'/d\text{p}H$ and $d^2\pi'/d\text{p}H^2$

both equal to 0, and solving the two equations simultaneously for the two unknowns, $(\alpha)_1$ and y: $\Sigma_1^3[\alpha\rho(\rho - \alpha)] = 0$ and $\Sigma_1^3[\alpha\rho(\alpha^2 + \rho^2 - 4\alpha\rho)] = 0$. Since $H = (A\rho/\alpha)_1$, these expressions may be written in terms of $(\alpha)_1$ (or α) and y: $(\rho)_1 = \rho = 1 - \alpha$; $(\alpha)_2 = y/(y + \rho/\alpha)$; $(\rho)_2 = 1 - (\alpha)_2$; $(\alpha)_3 = y^2/(y^2 + \rho/\alpha)$; $(\rho)_3 = 1 - (\alpha)_3$. The difficulty lies in finding some simplifying condition such as an expected value of H and hence of α to facilitate the solution. That the two minima disappear on the sides of the

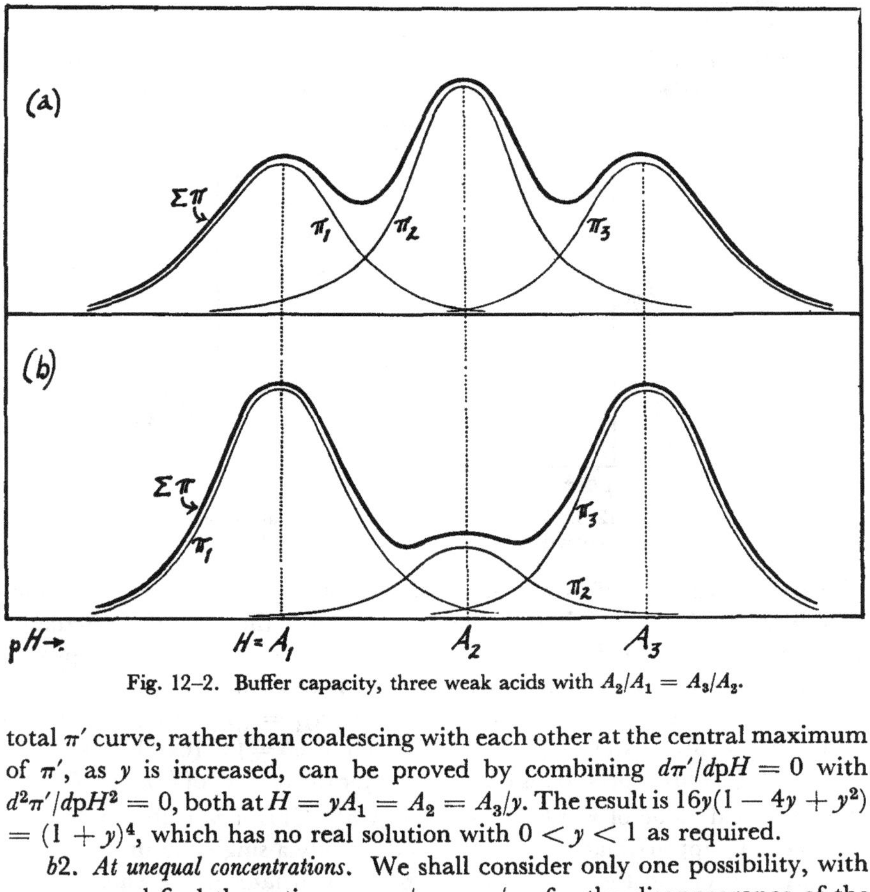

Fig. 12–2. Buffer capacity, three weak acids with $A_2/A_1 = A_3/A_2$.

total π' curve, rather than coalescing with each other at the central maximum of π', as y is increased, can be proved by combining $d\pi'/d\text{p}H = 0$ with $d^2\pi'/d\text{p}H^2 = 0$, both at $H = yA_1 = A_2 = A_3/y$. The result is $16y(1 - 4y + y^2) = (1 + y)^4$, which has no real solution with $0 < y < 1$ as required.

*b*2. *At unequal concentrations*. We shall consider only one possibility, with $a_1 = a_3$, and find the ratio x, or $a_1/a_2 = a_3/a_2$, for the disappearance of the inflections at a fixed value of y, with $y = A_2/A_1 = A_3/A_2$. With this restriction and neglecting H and OH in π, we have, with $\pi' = (\Sigma\pi)/2.3a_2$,

$$\pi' = (\pi_1 + \pi_2 + \pi_3)/2.3a_2 = x(\alpha\rho)_1 + (\alpha\rho)_2 + x(\alpha\rho)_3, \quad (25)$$

$$d\pi'/d\text{p}H = x[\alpha\rho(\rho - \alpha)]_1 + [\alpha\rho(\rho - \alpha)]_2 + x[\alpha\rho(\rho - \alpha)]_3, \quad (26)$$

$$d^2\pi'/d\text{p}H^2 = x[\alpha\rho(\alpha^2 + \rho^2 - 4\alpha\rho)]_1 + [\alpha\rho(\alpha^2 + \rho^2 - 4\alpha\rho)]_2 + x[\alpha\rho(\alpha^2 + \rho^2 - 4\alpha\rho)]_3. \quad (27)$$

If y is small enough so that at equal concentrations ($x = 1$) the two minima do exist, then as x decreases (less than 1), as in Fig. 12–2a, the side minima and hence the inflections will disappear at a certain critical value of x, for a given y. Below this ratio of a_1/a_2, the three acids behave in titration in this respect like a single acid, with no inflections between them. On the other hand, as x increases (greater than 1), as in Fig. 12–2b, the two minima will merge with the central maximum at a certain critical value of x. Above this ratio of a_1/a_2, the three acids behave somewhat like two acids, with an inflection at $b_s = a_1 + a_2/2$.

Case (a), decreasing x. In this case π_3 will have little effect on the minimum between π_1 and π_2. The conditions relating x and y will be roughly those

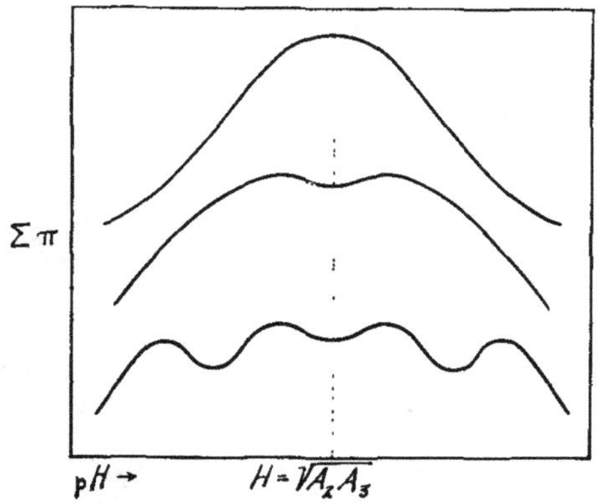

Fig. 12–3. Buffer capacity, four acids at equal concentrations with $A_2/A_1 = A_3/A_2 = A_4/A_3$.

discussed for the case of two acids in Chapter XI, Section B1b, the minimum vanishing at $H \cong A/2$ when x becomes smaller than $\sim 27y$; actually, however, the more acids are involved, the sooner the inflection disappears.

Case (b), increasing x. The condition here is $H = A_2$ when both $d\pi'$ and $d^2\pi' = 0$. But at $H = A_2$, $(\alpha)_2 = (\rho)_2 = 1/2$, $(\alpha)_1 = (\rho)_3 = 1/(1 + y)$, and and $(\alpha)_3 = (\rho)_1 = y/(1 + y)$. Hence when $d^2\pi' = 0$, $16xy(1 - 4y + y^2) = (1 + y)^4$. This expression is exact except for the neglect of H and OH in π; it gives a real value of x only if $y < (2 - \sqrt{3})$. This means that unless y is smaller than approximately 1/4, there will be only a single minimum of π in the titration of the three acids of this problem whatever the value of x if $x > 1$. But more generally, the expression gives explicitly, for any specific value of y, the value of x ($= a_1/a_2 = a_3/a_2$) above which the middle of the three acids cannot be detected by inflections in the titration curve. For example, if $y = 0.1$ (i.e. $pA_2 - pA_1 = 1$), $x = 1.50$; if $y = 0.01$ ($pA_2 - pA_1 = 2$), $x = 6.77$. With the concentration of the middle acid less than 1/6.77 of the concentrations of the two adjacent acids (these being equal) and with $A_1/A_2 = A_2/A_3 = 100$, there is then no inflection between the successive acids during titration with strong base, but only one at $b_s = a_1 + a_2/2$.

c. Four acids. We shall consider only the case of equal concentrations and fixed spacing of ionization constants ($y = A_2/A_1 = A_3/A_2 = A_4/A_3$). With small y there are four maxima in the π' curve, each displaced toward the central minimum, which is at $H = yA_1\sqrt{y}$. The value of π' at this central minimum is higher than π' at the two lateral minima, as in the lowest curve of Fig. 12–3. As the constants approach each other, the side minima merge with the extreme maxima, leaving simply two maxima and a central minimum, as in the middle curve. Then the four acids show only two inflection points in titration, the first when $b_s = a_1 + a_2$ and the second when $b_s = \Sigma a$. Finally, as y increases still further, the central minimum also disappears (uppermost curve) and there are no longer any inflections in the titration curve except that corresponding to $b_s = \Sigma a$.

To find the critical value of y for the disappearance of the central minimum, we set $d^2\pi'/dpH^2 = 0$ at $H = yA_1\sqrt{y}$, noting that at this central point, $\pi_1 = \pi_4$ and $\pi_2 = \pi_3$. Then,

$$[\alpha\rho(\alpha^2 + \rho^2 - 4\alpha\rho)]_1 + [\alpha\rho(\alpha^2 + \rho^2 - 4\alpha\rho)]_2 = 0. \qquad (28)$$

But with $H = yA_1\sqrt{y}$, we have $(\alpha)_1 = 1/(1 + y\sqrt{y})$, $(\rho)_1 = y\sqrt{y}/(1 + y\sqrt{y})$, $(\alpha)_2 = 1/(1 + \sqrt{y})$, $(\rho)_2 = \sqrt{y}/(1 + \sqrt{y})$. Using these values and expanding, we obtain a symmetrical polynomial equation in $\sqrt{y}$ of the fourteenth degree:

$$y^7 - 4y^6\sqrt{y} + 2y^6 + 8y^5\sqrt{y} - 10y^5 + 4y^4\sqrt{y} - 9y^4 - 48y^3\sqrt{y} - 9y^3$$
$$+ 4y^2\sqrt{y} - 10y^2 + 8y\sqrt{y} + 2y - 4\sqrt{y} + 1 = 0. \qquad (29)$$

Since y must be between 0 and 1, the solution required is $y \cong 0.125$, or $A_2/A_1 \cong 1/8$. Since the individual minima between A_1 and A_2 and between A_3 and A_4 may be expected to vanish at $y < 1/14$, which is the requirement for two acids alone, we may take this result as proof that these lateral minima disappear before the central minimum of the π' changes to a central maximum. Practically however, the lone central minimum would hardly be detectable in titration, for with $y \cong 0.1$ Eq. (22) gives $\pi_e \cong 2.3(0.4)$ for unit concentration, and with such high buffer capacity the pH titration curve would be practically horizontal through this "vertical type" of inflection.

C. Titration With Strong Base of Weak Acid in Presence of Weak Base

1. General Relations. We shall consider a solution containing a weak acid at concentration a and with ionization constant A together with a base, at concentration b and with ionization constant B, to be titrated with strong base. The graphical relations for the buffer capacity of the solution with

respect to strong base are shown in the accompanying figures: Fig. 12–4 for $AB \gg W$, Fig. 12–5 for $AB = W$, and Fig. 12–6 for $AB \ll W$. If the "weak base" were actually "strong," the point W/B of Fig. 12–4 would be moved

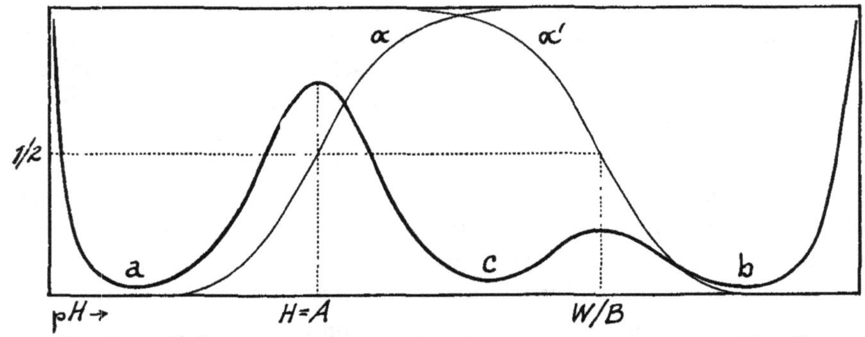

Fig. 12–4. Buffer capacity, mixture of weak acid and weak base with $AB \gg W$.

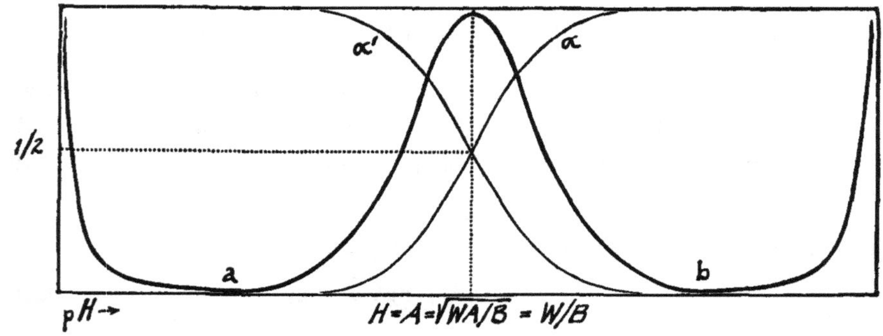

Fig. 12–5. Buffer capacity, mixture of weak acid and weak base with $AB = W$.

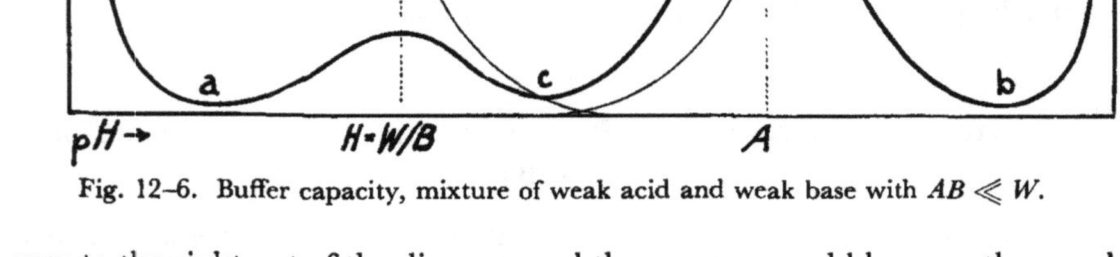

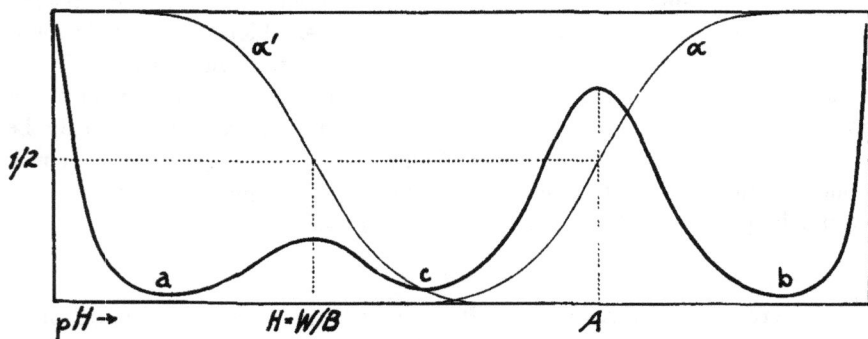

Fig. 12–6. Buffer capacity, mixture of weak acid and weak base with $AB \ll W$.

over to the right out of the diagram and the π curve would become the usual one for a weak acid, with a single maximum at $H = A$ and a minimum (minimum a and minimum b) on each side (Fig. 10–1). In such case, only the excess, $a - b$ (if $a > b$) of the weak acid over the base, can be titrated with strong base, with an inflection point at minimum b; the analysis may be completed however by titration of a separate sample with strong acid, to determine b, to minimum a.

If however, the base present is actually weak but such that $AB > W$, as in Fig. 12–4, there are two maxima and three minima in the π curve, the maxima as usual occurring at $H \cong A$ and $H \cong W/B$, with heights essentially proportional to the concentrations a and b respectively; the figure has been drawn for $a > b$.

If $a > b$, there are two inflection points in the titration of the original solution with strong base. The first, when $b_s = a - b$, occurs at minimum c, which is near $H = \sqrt{WA/B}$, the mid-point between pA and p(W/B), and which may therefore be acid or alkaline. To find H at this equivalent point, we set $b_s = a - b$ (and $a_s = 0$) in

$$H - OH = aA/(A + H) - bHB/(HB + W) + a_s - b_s, \tag{30}$$

and, neglecting $H - OH$, we obtain

$$H_e \cong - W(a - b)/2aB + \sqrt{(\quad)^2 + bAW/aB} \cong \sqrt{bAW/aB}; \tag{31}$$

whereas before titration, with $b_s = 0$, the same considerations give

$$H_{\text{orig}} \cong A(a - b)/2b + \sqrt{(\quad)^2 + aAW/bB} \cong A(a - b)/b. \tag{32}$$

The formulas of Eq. (31) are of course identical with Eqs. XI(15) and XI(20), if the weak base of the present problem is considered as a second weak acid with the conjugate constant W/B. Now from

$$\pi = 2.3(a\alpha\rho + b\alpha'\rho' + H + OH), \tag{33}$$

noting that $H_e < A$ and greater than W/B, in general, and neglecting H and OH, we have

$$\pi_e \cong (2.3)2\sqrt{abW/AB}, \tag{34}$$

and

$$F = (p/400)\sqrt{(a - b)^2AB/abW}, \tag{35}$$

with p relative to the excess, $(a - b)$.

For the second equivalent point, when $b_s = a$, the inflection occurs at minimum b, in alkaline solution. Here the solution is mathematically one of two bases: the "salt" M_sX, at concentration $b_s = a$, with basic ionization constant $B^* = W/A$, and the base b itself. Hence, from Eq. XI(26), $(OH)_e \cong \sqrt{bB + aW/A}$; and since this is a terminal equivalent point, then from Eq. (13),

$$F = pa/400\sqrt{bB + aW/A}, \tag{36}$$

with p relative to the total value of a. Less accurately, since $AB > W$, this means $F \simeq pa/400\sqrt{bB}$.

Hence, by comparison of this final approximation with Eq. (35), for $AB > W$ and for a fixed value of A, we see that the titration of the excess of acid, $(a - b)$, has a lower feasibility, while the titration of the total acid, a, is easier the lower is the value of B. As AB increases, however, the titration of the excess becomes easier and that of the total more difficult.

In fact, as B decreases and A and W/B approach each other, or as $AB \stackrel{\rightarrow}{=} W$, the two maxima of Fig. 12–4 will merge with each other and with minimum c at a certain critical relation of the concentrations and constants. By analogy with the problem of two weak acids, Eqs. XI(43) and XI(44), this is seen to occur when AB becomes smaller than either $27bW/a$ or $27aW/b$, whichever is larger. As AB decreases further, there will then be no minimum c, and only a single large maximum of the π curve, as in Fig. 12–5, until AB becomes smaller than $bW/27a$ or $aW/27b$, whichever is smaller, when minimum c reappears and the curve becomes that of Fig. 12–6. Between these limits, there will be no inflection point for the titration of the excess of acid, or $(a - b)$, but only a single inflection, at minimum b, for the titration of the total acid. The titration of the excess, $(a - b)$, however, remains out of the question even when $AB \ll W$ and the minimum c has reappeared in the π curve. When $AB > W$, the original solution, in which $a > b$, is distinctly acid compared to the equivalent point, $(b_s = a - b)$, corresponding to minimum c of Fig. 12–4, as seen in Eqs. (31) and (32). But when $AB < W$, the original solution, with $a > b$, is hardly acid compared to minimum c, and hence the titration is impossible. This may be understood by the following considerations: neglecting $H - OH$ in the general equation $H - OH = a\alpha - b\alpha'$, we have $a \simeq b\alpha'/\alpha$, whence $da/d\text{p}H \simeq -2.3b(\rho + \rho')\alpha'/\alpha$. Now if $AB > W$, then both ρ and $\rho' \simeq 0$ in minimum c; hence $|\, d\text{p}H/da \,|$ is high and a solution with $a > b$ is distinctly acid compared to minimum c. But when $AB < W$, then $\alpha'/\alpha \simeq 1$ and both ρ and ρ' are nearly equal to 1, in minimum c; in this case then $|\, d\text{p}H/da \,| \simeq 1/(4.6b)$, which is very low. An excess of the weak acid hardly brings the pH of the solution out of the minimum. This effect, which will be discussed further in Chapter XIII, Section B, is illustrated in Table 12–1, with values of pH calculated through Eq. XIII(19). It is seen that an increase of a above b causes a greater change in pH when $AB > W$ than when $AB < W$ despite the fact that with the same value of A the base is weaker in the second case. At the same time there is also a smaller increase in the buffer capacity π, or $db_s/d\text{p}H$, as a increases above b, with $AB < W$; π has been calculated through Eq. (34).

Finally, for the titration of the total acid, the feasibility of course increases as B decreases, and when $AB < W$ Eq. (36) becomes $F \simeq (p/400)\sqrt{aA/W}$, the usual function for titration of a weak acid with strong base.

Table 12–1. Relations at minimum c, Figs. 12–4 and 12–6.

A	B	b	a	pH	π
10^{-4}	10^{-8}	0.1	0.1	5.00	0.0380
			0.11	4.78	0.0434
			0.2	3.99	0.117
10^{-4}	10^{-12}	0.1	0.1	3.02	0.0404
			0.11	3.00	0.0422
			0.2	2.86	0.0566

2. Titration of the 1 : 1 Salt MX, or the Special Case in Which $a = b = c$. This case was referred to in Chapter v, Section F, in connection with the determination of the constants A and B. Here

$$H - OH = c[A/(A + H) - B/(B + OH)] + a_s - b_s. \tag{37}$$

When $a = b = c$ in Eq. (30), in other words, the solution to be titrated is that of the salt MX; the π curves will be the same as those of Figs. 12–4,5,6, except that with equal concentrations the two maxima, when separate, will be of equal height, and the combined maximum for Fig. 12–5, attaining its greatest height when $AB = W$, will be twice as high as the separate individual maxima, always with H and OH neglected in the expression for π. In the titration of such a salt solution then, the value of H to begin with is $H \cong \sqrt{WA/B}$, which is either in minimum c or at the combined maximum when there is no such minimum. In either case we may titrate the salt: with strong base, we titrate the weak acid of the salt to minimum b, with

$$F \cong (p/400)\sqrt{cA/(AB + W)}; \tag{38}$$

with strong acid, we titrate the weak base to minimum a, with

$$F \cong (p/400)\sqrt{cB/(AB + W)}. \tag{39}$$

The first titration (with b_s) is better than the second (with a_s) if $A > B$ whatever the relation between AB and W.

The equivalent point (b_s or $a_s = c$) is mathematically a solution of two weak bases (M_sX and M) or two weak acids (X and MX_s); with $b_s = c$,

$$OH_e \cong \sqrt{c(AB + W)/A}, \tag{40}$$

while with $a_s = c$,

$$H_e \cong \sqrt{c(AB + W)/B}. \tag{41}$$

With equal concentrations of a and b (or with $a = b = c$) the condition for the occurrence of the central minimum, at $H = \sqrt{WA/B}$, is $AB > 13.9W$ or $AB < W/13.9$, if H and OH are neglected in the expression for π; this follows by analogy from Eq. xi(39).

When $AB = W$, the π curve of the salt MX at concentration c, with a single maximum at $H = \sqrt{WA/B}$, can not be distinguished from that of a single acid or single base, with constant $A = W/B$, at concentration $2c$. This relation may be realized directly from Eq. (37) by setting $AB = W$ in it, and a_s (or b_s) $= c$, whereupon the equation reduces to that of a single acid (or single base), at concentration $2c$,

$$H - OH = 2cA/(A + H) = -2cB/(B + OH). \tag{42}$$

Thus, for $AB = W$, Eqs. (40) and (41) for the equivalent points become $OH_e \cong \sqrt{2cB}$ and $H_e \cong \sqrt{2cA}$ respectively; and the positions of the lateral minima of Fig. 12–5 may be written by analogy from the case of a weak acid (Chapter x): for minimum a, $H_i \cong \sqrt{2cA} - 2A$, and for minimum b, $H_i \cong \sqrt{AW/2c} + W/c$.

The buffer capacity of a salt solution of the type MX, with respect to strong acid or strong base, depends then not so much on the values of the constants A and B relative to each other as on the value of the product AB relative to W. As seen in Figs. 12–4,5,6, the buffer capacity is very low as long as AB is either much greater or much less than W. From Eq. (33), with $a = b = c$ and with $H \cong \sqrt{WA/B}$, the buffer capacity of the pure salt solution is

$$\pi \cong 2.3[2c\sqrt{W/AB}/(1 + \sqrt{W/AB})^2 + H + OH]. \tag{43}$$

With $H + OH$ neglected, then if $AB \gg W$, $\pi \cong (2.3)2c\sqrt{W/AB}$, and if $AB \ll W$, $\pi \cong (2.3)2c\sqrt{AB/W}$, in either case very small. If $A = B = K$, the solution is neutral for any value of c, and if $K \gg \sqrt{W}$, $\pi \cong (2.3)2c\sqrt{W}/K$, while if $K \ll \sqrt{W}$, $\pi \cong (2.3)2cK/\sqrt{W}$.

But if $AB = W$, then as seen in Fig. 12–5, π is very high, with the value $\pi \cong (2.3)c/2$. This is mathematically identical with $\pi = 2.3a/4$, the buffer capacity of a weak acid "half titrated" with strong base, for from the point of view of the hypothetical acid defined in Eq. (42), the pure salt at concentration c represents the "half titrated" acid at concentration $a = 2c$.

If the ionization constants, B and A, pertaining to the 1 : 1 salt MX, are equal, the salt solution is neutral whatever the value of AB relative to W. But the buffer capacity of the solution, for fixed concentration, starts at "zero" (rather, at the value for pure water) for the salt of strong acid and strong base, in which $AB = \infty$ relative to W; then as the acid and base become weaker, π rises, passes through a maximum, and falls again to "zero" when the acid and base have zero strength, the maximum occurring when $AB = W$. The solution has of course remained neutral through all these changes, provided $A = B$. The interesting question is: at what point or for what ratio of AB/W does the solution cease to be a "salt solution"? The solutions are all "equivalent points" and in that sense are all "salt solutions." But with $AB \ll W$, the solution will have little electrolytic character and it does not seem to warrant the name "salt solution." One is tempted to say that only if the salt solution has a buffer capacity which is in minimum c of the π curve of Fig. 12–4, with $AB > W$, should it be called a salt solution. Nevertheless there seems to be no necessary connection between the requirements for minimum c, involving A, B, and W, and the requirements, whatever they are, but certainly not involving W, for the formation of a solid substance of the specified stoichiometric composition, which may be called a salt. It does not necessarily follow that "salts" represented by minimum c of Fig. 12–4 are crystallizable salts whereas those of Fig. 12–5 or Fig. 12–6 can not exist as solid ionic compounds ("salts"). We shall return to this question in another connection, in Chapter XIII.

If $A \cong B \cong \infty$ then, or if $A \cong B \cong 0$, the "salt solution," with $a = b = c$, is mathematically equivalent to pure water and it may not be titrated either with strong acid or with strong base. If one of the ionization constants is infinite and the other zero, the solution is mathematically either a pure strong acid or a pure strong base and it may always be titrated. If one of the constants, such as A, is infinite and the other, B, is finite, the solution is mathematically a weak acid with ionization constant W/B; titration with b_s is now possible as long as $c > 27B$. If one constant (B) is zero and the other (A) is finite, the "salt" is mathematically a weak acid and may be titrated provided $cA > 27W$. In these two cases the pH titration curve passes through a horizontal inflection, at a maximum of π, making possible the usual estimate of the ionization constant involved.

If both A and B are finite, the "salt solution" may generally be titrated either with a_s or with b_s. If $AB > W$, titration with a_s may be used to determine the constant A since the process involves the π maximum at $H = A$, and the titration is possible (or it is marked by a vertical inflection point) if $c > 27A$. Similarly, titration with b_s, possible if $c > 27B$, allows determination of B. If $AB < W$, titration with a_s is possible if $cB > 27W$ and involves the π maximum at $H = W/B$, while titration with b_s, possible if $cA > 27W$, allows determination of A, at the π maximum at $H = A$. These relations are implied in the application of Eqs. v(129–132). But if $AB \cong W$, the salt may be titrated either with a_s or with b_s, but since π starts as a maximum there is no horizontal inflection of the pH titration curve in either case, so that these titrations are not useful for the determination of the constants. If $AB = W$, titration with a_s requires $2c > 27A$ or $2cB > 27W$, and titration with b_s requires $2c > 27B$ or $2cA > 27W$, for an end-point inflection.

3. Summary for Titration of Weak Acid in Presence of Weak Base.

(a) If $AB > W$, the original solution is distinctly acid relative to minimum c, if $a > b$, and it is distinctly basic relative to minimum c if $a < b$. In either case the total acid is determined by titration to minimum b and the total base by titration to minimum a; but the sharpest titration is that of the difference $(a - b)$ or $(b - a)$, to minimum c.

(b) If $AB \cong W$, the pH of the original solution lies somewhere between the minima of Fig. 12–5. Whatever the relative values of a and b, only their totals may be determined, a by titration with b_s to minimum b and b by titration with a_s to minimum a.

(c) If $AB < W$, the original solution is very near minimum c, with $a \gtrless b$. Hence it is again not possible to titrate the excess of either, such as $(a - b)$, but it is possible to titrate the total acid, with inflection at minimum b, and the total base with inflection at minimum a.

Practically, this means that for titration of weak acid in presence of strong base ($AB \gg W$), only the excess, $(a - b)$, may be determined (with minimum c); but for weak acid in presence of very weak base ($AB \leq W$), only the total acid may be determined, whether $a \gtrless b$ (with minimum b). In reverse, using strong acid for titration: for weak base in presence of strong acid ($AB \gg W$), only the excess, $(b - a)$, may be determined (with minimum c); but for weak base in presence of very weak acid ($AB \leq W$) only the total base may be determined, whether $b \gtrless a$ (with minimum a).

If ammonia then is caught in excess of HCl ($AB > W$), titration of the base with strong acid is impossible and in reverse only the excess of acid, $(a_s - b)$, may be titrated with good feasibility. Hence the ammonia is determined by difference: $b = a_s - b_s$, in which a_s is the total HCl taken and b_s the strong base used to titrate the excess, $(a_s - b)$; for titration of the excess of HCl,

$$F = (p/400)\sqrt{(a_s - b)^2 B/bW}. \tag{44}$$

But if ammonia is caught in excess of boric acid (AB approximately equal to or less than W), only the total weak base (even with $b > a$) may be determined, and the determination is direct; $b = a_s$ (used), and

$$F = pb/400\sqrt{aA + bW/B}. \tag{45}$$

Such titration, to minimum a, is possible whether $AB \gtrless W$.

D. Titration of Strong Acid in Presence of Two Weak Bases

The relations are shown graphically in Fig. 12–7. Given a solution containing strong acid at concentration a_s and the bases MOH and NOH at concentrations b_1 and b_2 (with $B_1 < B_2$) respectively, then in titration with b_s, the equivalent point when $b_s = a_s$ is a solution of two pure bases, and

$OH_e \cong \sqrt{b_2B_2 + b_1B_1}$; and if b_1 and b_2 are comparable in magnitude, $OH_e \cong \sqrt{b_2B_2}$. This point is near minimum b of Fig. 12–7, in alkaline solution, with $OH > B_2 > B_1$. Whatever the relative values of a_s and the b's, this inflection will always occur (provided the conditions for the existence of minimum b in the π curve are satisfied), with the feasibility function

$$F = pa_s/400\sqrt{b_2B_2 + b_1B_1} \cong pa_s/400\sqrt{b_2B_2}, \qquad (46)$$

since $\pi_e \cong (2.3)2\sqrt{b_2B_2 + b_1B_1} \cong (2.3)2\sqrt{b_2B_2}$. This value of F, Eq. (46), is the same as that in the titration of a_s in presence of two weak acids, with b

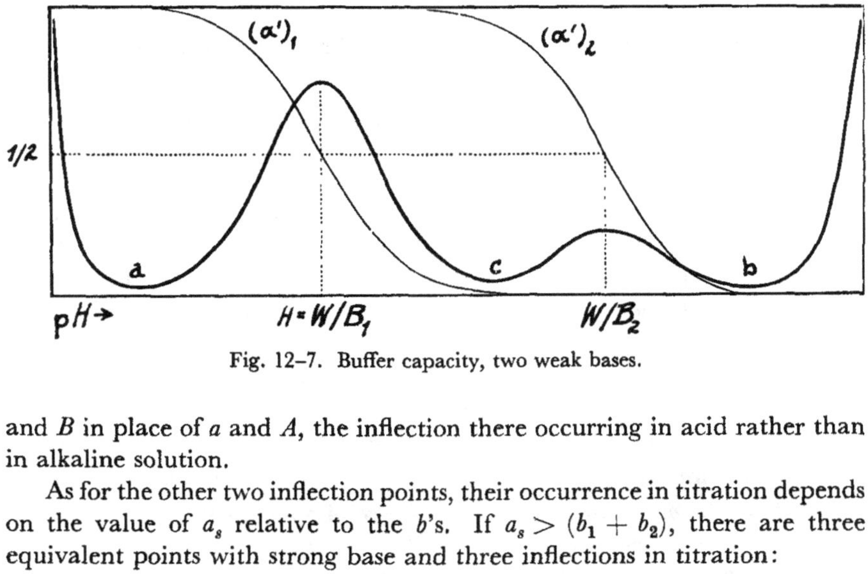

Fig. 12–7. Buffer capacity, two weak bases.

and B in place of a and A, the inflection there occurring in acid rather than in alkaline solution.

As for the other two inflection points, their occurrence in titration depends on the value of a_s relative to the b's. If $a_s > (b_1 + b_2)$, there are three equivalent points with strong base and three inflections in titration:

(1) At minimum a, in acid solution, when $b_s = a_s - (b_1 + b_2)$; this does not occur if $a_s < (b_1 + b_2)$. Here $H_e \cong \sqrt{b_1W/B_1 + b_2W/B_2} \cong \sqrt{b_1W/B_1}$, and (as a terminal equivalent point)

$$F = \frac{p[a_s - (b_1 + b_2)]}{400\sqrt{W(b_1/B_1 + b_2/B_2)}} \cong \frac{p[a_s - (b_1 + b_2)]}{400\sqrt{Wb_1/B_1}}, \qquad (47)$$

with p referring to the quantity $[a_s - (b_1 + b_2)]$. In other words, the "salts" MX_s and NX_s are treated as mathematical acids with constants $W/B_1 > W/B_2$ respectively.

(2) At minimum c, which may be acid or alkaline, when $b_s = a_s - b_2$; this does not occur if $a_s < b_2$. Here $OH_e \cong \sqrt{B_1B_2b_1/b_2}$, and

$$F = (p/400)\sqrt{b_1B_2/b_2B_1}, \qquad (48)$$

with p relative to b_1, which is here being determined, between minimum a and minimum c.

(3) At minimum b, in alkaline solution, when $b_s = a_s$, Eq. (46) gives F for the determination of the total acid; but if the titration is one from minimum c to minimum b, measuring b_2, then

$$F = pb_2/400\sqrt{b_2B_2 + b_1B_1} \cong (p/400)\sqrt{b_2/B_2}. \qquad (49)$$

Finally, minima or inflections not experienced in titration with strong base, because $a_s < b_2$ or $(b_1 + b_2)$, are traversed in titration with strong acid so that the combination of both titrations will give the complete analysis.

E. Titration of Strong Acid, a_s, in Presence of Both Weak Acid, a, and Weak Base, b

1. With AB either Much Greater or Much Smaller Than W.

(a) If $a_s > b$, the effective concentration of "strong acid" may be taken as $a_s - b$; and a and b represent concentrations of two weak acids, with constants A and W/B respectively; this makes use of the fact that the "salt," MX_s, here "present" at the concentration b, is mathematically an acid with the constant W/B. Three successive equivalent points may therefore be detected on titration with b_s, their order depending on the relation of AB to W.

(1) If $AB > W$, inflections occur successively when $b_s = a_s - b$, $a_s + a - b$, and $a_s + a$ at minimum a, minimum c, and minimum b respectively of Fig. 12–4. In this and in the three following cases, we have at minimum a, $H_e \cong \sqrt{aA + bW/B}$, and at minimum b, $H_e \cong W/\sqrt{bB + aW/A}$. In the present case, for minimum c, with $b_s = a_s + a - b$, Eq. (30), with D neglected, gives Eq. (31) for H_e.

(2) If $AB < W$, the successive inflections occur when $b_s = a_s - b$, a_s, and $a_s + a$ at minimum a, minimum c, and minimum b respectively of Fig. 12–6. At minimum c, Eq. (30) now gives

$$H_e \cong -A(b-a)/2b + \sqrt{(\quad)^2 + aAW/bB} \cong \sqrt{aAW/bB}. \qquad (50)$$

(b) If $a_s < b$, there are only two inflections on titration with b_s. The solution contains mathematically two weak acids and one weak base: weak acids at concentrations a and a_s (a_s being the concentration of the "salt" MX_s) with constants A and W/B respectively and a weak base (B) at concentration $(b - a_s)$. The pH of the solution is in this case on the alkaline side of minimum a of Fig. 12–4 or 12–6.

(1) If $AB > W$, titration with b_s gives inflections when $b_s = a$, at minimum c of Fig. 12–4, and when $b_s = a + a_s$ at minimum b; titration with a_s however will give an inflection when a_s' (used in titration) equals

$(b - a_s)$, thus yielding the complete analysis. In this case, for minimum c, Eq. (30) with D neglected gives

$$H_e = -\frac{bAB + aW - a_s(AB + W)}{2B(b + a - a_s)} + \sqrt{(\quad)^2 + \frac{a_sAW}{B(b + a - a_s)}}. \tag{51}$$

But since $AB > W$,

$$H_e \cong -A(b - a_s)/2(b + a - a_s) + \sqrt{(\quad)^2 + a_sAW/B(b + a - a_s)}. \tag{52}$$

(2) If $AB < W$, the inflections with b_s occur when $b_s = a_s$ and then $a_s + a$, at minimum c and minimum b respectively of Fig. 12–6; but again titration with a_s' to minimum a gives the quantity $(b - a_s)$. H_e at minimum c is given by Eq. (51).

In titration of a_s in presence of salt MX, or when $a = b = c$, all the above formulas for H_e will be simplified accordingly.

2. WITH $AB \cong W$. When $AB \cong W$, we have Fig. 12–5, and there will be no inflection corresponding to minimum c whatever the relative values of a_s, a, and b.

(a) If $a_s > b$, there is an inflection at minimum a, when $b_s = a_s - b$, and one at minimum b when $b_s = a_s + a$.

(b) If $a_s < b$, there will be an inflection only at minimum b, at $b_s = a_s + a$; reverse titration, with a_s', to minimum a, will measure the quantity $b - a_s$.

But in either case the set of titrations (with a_s' and with b_s) does not constitute an analysis since the results can not be used to determine the individual concentrations, a_s, a, and b.

The special case of titration of a_s in presence of a salt, at concentration c, and for which $AB \cong W$, will now be considered. This is like the titration of a strong acid at concentration $(a_s - c)$, in presence of weak acid with ionization constant A (or equal to W/B) at a concentration of $2c$. Titration with strong base gives an inflection at minimum a, when $b_s = (a_s - c)$, with $H_e \cong \sqrt{2cA}$ and

$$F = p(a_s - c)/400\sqrt{2cA}, \tag{53}$$

and a second inflection at minimum b, when $b_s = (a_s + c)$, with $H_e \cong \sqrt{AW/2c}$, and

$$F = p(a_s + c)/400\sqrt{2cW/A}. \tag{54}$$

In each case p is relative to the quantity being titrated, $(a_s - c)$ in Eq. (53), and $(a_s + c)$ in Eq. (54). The titration shows both inflections only if $a_s > c$, in which case the analysis is complete since the difference gives the quantity $2c$, from which a_s may then be calculated. On the other hand, if $a_s < c$, titration with b_s gives only the inflection at minimum b, when $b_s = a_s + c$; if the original solution is also titrated with a_s', however, an inflection occurs at minimum a, when a_s' (used) equals c minus the original a_s. Now the sum gives the quantity $2c$, etc.

But the important point to note is that in either case there is no inflection at the "equivalent point," $b_s = a_s$, for the titration with b_s, of a_s in presence of such a salt ($AB = W$). Such an equivalent point is mathematically a pure solution of the weak-weak salt MX, and $H_e \cong \sqrt{WA/B}$. The feasibility of the titration, however, is not determined merely by H_e or by H_e/a_s, where a_s is the concentration of the acid being titrated. It is a question, as discussed in Chapter IX, of the buffer capacity of the solution at the equivalent point. We see here, in other words, that H_e may have the value $\sqrt{W}$ (or the solution may be strictly neutral) if $A = B$, and yet if the buffer capacity is high the feasibility of the titration of the strong acid with strong base may be very low. While H_e depends on the ratio A/B, π_e depends on the ratio AB/W. If this ratio is either very small or very large, π_e is small and the feasibility high. But if $AB \cong W$, π_e is extremely high as pointed out under Eq. (43), there is no inflection of pH against b_s at $b_s = a_s$, and the titration is impossible. It is easier to titrate a strong (or even a weak) acid in the presence of a weaker acid than to titrate a strong acid in the presence of a salt with $AB \cong W$, even though such a salt be "neutral."

As an example, we shall consider the "titration" of a strong acid at concentration $a_s = 1$, in presence of a salt, MX, at concentration $c = 1$, with $A = B = 10^{-7}$. When $b_s = 0$, $H = 4.47 \times 10^{-4}$, and the total change in pH from $b_s = 0$ to $b_s = a_s$ (when $H = 10^{-7}$) is 3.35. When $b_s = 0.99$ (or 1 per cent short of the equivalent point, with $p = 2$), $H = 1.02 \times 10^{-7}$, while at $b_s = 1.01$, $H = 0.98 \times 10^{-7}$; hence Δ_p, from $-$ 1 to $+$ 1 per cent excess, is 0.0174. (These numerical values are calculated from the full Eq. (37) and are correct to the number of significant figures used.) For the region near the equivalent point one may safely neglect $H - OH$ in Eq. (37), which then gives

$$H \cong \frac{(a_s - b_s)(AB + W)}{2B(c - a_s + b_s)} + \sqrt{(\quad)^2 + \frac{AW(c + a_s - b_s)}{B(c - a_s + b_s)}}. \tag{55}$$

In contrast, the strong acid at $a_s = 1$ in presence of the weak acid at $a = 1$, with $A = 10^{-7}$, can be titrated relatively easily, with a feasibility of 0.630 pH units for one tenth of one per cent of titration through the equivalent point ($p = 0.1$), although the equivalent point is distinctly acid, with p$H = 3.5$.

We thus see that the titration of strong acid with strong base, the strong acid being at the high concentration of $a_s = 1$ at the equivalent point, is impossible in the presence of a salt with $AB = W$, at a concentration c equal to a_s, although the equivalent point is exactly neutral (unless c is also determined by combining titration with strong acid and titration with strong base, as explained above). Regardless of the ratio A/B and hence of the "reaction" of the solution at the equivalent point, the feasibility increases as AB diverges from W. The feasibility takes on its "full value," with

$F = pa_s/400\sqrt{W}$, when both A and B are "strong," which is the titration of a_s in presence of a salt like NaCl, or when both A and B are zero, which is the titration of a_s in presence of "nonelectrolytes."

F. TITRATION WITH STRONG BASE OF DIBASIC ACID STRONG IN K_1 IN PRESENCE OF WEAK BASE

This is mathematically the case of two acids, one strong and one weak (each at concentration a, of the dibasic acid) in presence of weak base.

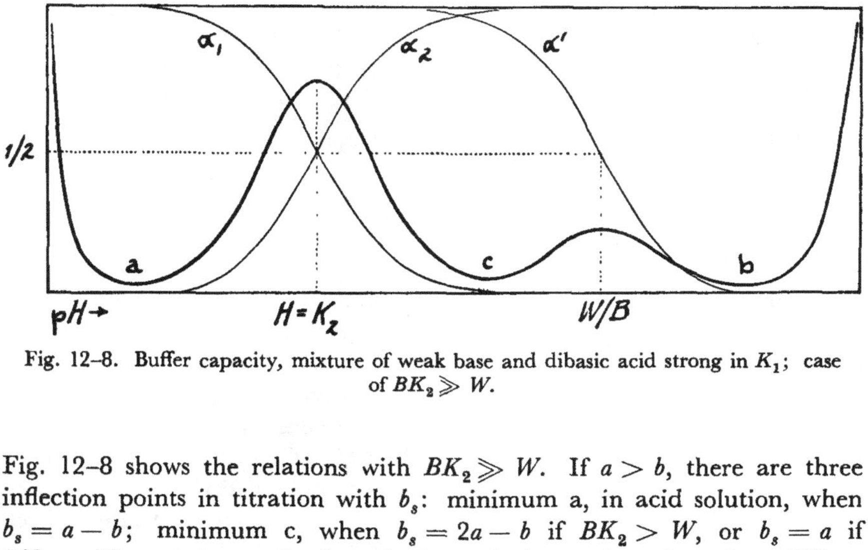

Fig. 12–8. Buffer capacity, mixture of weak base and dibasic acid strong in K_1; case of $BK_2 \gg W$.

Fig. 12–8 shows the relations with $BK_2 \gg W$. If $a > b$, there are three inflection points in titration with b_s: minimum a, in acid solution, when $b_s = a - b$; minimum c, when $b_s = 2a - b$ if $BK_2 > W$, or $b_s = a$ if $BK_2 < W$; minimum b, in alkaline solution, when $b_s = 2a$. (When ammonia is caught in sulfuric acid, with $BK_2 > W$, titration with strong base, to minimum c, measures the "excess," or $2a - b$, in equivalents.) If $a < b$, there are only two inflections with b_s: minimum c when $b_s = a$ and minimum b when $b_s = 2a$; but in this case, titration with strong acid to minimum a gives $b - a$. In either case $(a \gtrless b)$, the titration between minimum a and minimum c, in either direction, depends on K_2 if $BK_2 > W$, and on W/B if $BK_2 < W$; while that between minimum c and minimum b depends on W/B if $BK_2 > W$ and on K_2 if $BK_2 < W$. If $BK_2 \cong W$, minimum c disappears and the only inflections, in titration with b_s, occur at $b_s = a - b$ and $b_s = 2a$ if $a > b$ and at $b_s = 2a$ if $a < b$; in the latter case, titration with a_s to minimum a gives $b - a$.

The pertinent formulas for H_e at the various minima, for use in writing the feasibility function, may of course be written from those in Section E1 by writing a for a_s and K_2 for A in every case.

G. General Case of Titration of a Mixture of Many Acids and Bases

Given, for example, a solution containing strong acid, a_s, two weak acids, a_1 and a_2, with $A_1 > A_2$, and two weak bases, b_1 and b_2, with $B_2 > B_1$, the problem can always be simplified by considering the salt MX_s to be mathematically equivalent to an acid with $A^* = W/B$ (and the salt M_sX equivalent to a base with $B^* = W/A$). Thus if $a_s > (b_1 + b_2)$, the problem reduces to that of five acids since the two "salts," MX_s and NX_s, may be treated as two acids, at concentrations b_1 and b_2, and with constants W/B_1 and W/B_2 respectively; the solution is then considered as "containing" four weak acids and strong acid at concentration $a_s - (b_1 + b_2)$. Titration with b_s therefore shows five inflection points, the first when $b_s = a_s - (b_1 + b_2)$; but the order of the four remaining inflections depends on the relative magnitudes of the four constants, A_1, A_2, W/B_1 and W/B_2. If for example the largest of these constants are $W/B_1 > A_1$, then the second inflection occurs when $b_s = a_s - b_2$, with $H_e \cong \sqrt{a_1 A_1 W/b_1 B_1}$, and with p relative to b_1, which is being "titrated," $F = (p/400)\sqrt{b_1 W/a_1 A_1 B_1}$.

If $a_s < (b_1 + b_2)$ but greater than b_2, the solution "contains" mathematically four weak acids, at concentrations $(a_s - b_2)$, b_2, a_1, and a_2, with constants W/B_1, W/B_2, A_1, A_2 respectively, and one weak base, B_1, at concentration $(b_1 + b_2) - a_s$. But it does not necessarily follow in this case or in general that there will be four inflections in titration with strong base and one with strong acid. The total number will be five but their distribution will depend on the pH of the original solution, which is a complicated function of all the concentrations and all the constants. Those inflections occurring at values of pH greater than the pH of the original solution occur on titration with strong base; those in lower pH, on titration with strong acid. As another example, if $a_s < b_2$, then the solution "contains" three weak acids, at concentrations a_s, a_1, a_2 and with constants W/B_2, A_1, A_2 respectively, and two weak bases, B_1 at concentration b_1 and B_2 at concentration $(b_2 - a_s)$.

XIII

Titration with Weak Acid or Weak Base

A. Titration of Strong Base (or Acid) With Weak Acid (or Base)

1. General Relations. All the titrations so far discussed have been titrations with strong acid or strong base, and the feasibility then depended on the buffer capacity (π) of the solution, at the equivalent point, in respect to strong acid or strong base. In the present chapter we shall consider titrations with weak reagents. The feasibility of such titrations will always be relatively small, depending on the buffer capacity (π_a or π_b) with respect to the weak acid or weak base used for titration. Hence the first case to be treated, the titration of strong base with weak acid, is strictly not the reverse of the titration of weak acid with strong base since the two processes involve different buffer capacities. The usual "back titration" in the determination of a weak acid with strong base is not with the weak acid itself but with a strong acid, the effect of which is simply the negative of that of the strong base of the "forward titration," hence involving the same π curve. We are not concerned then merely with "back titration," but with determinations with weak titrating reagents.

For titration involving strong base and weak acid, in either direction, the titration curve through the equivalent point is best described by the H^2 sequence of Eq. x(22), simplifiable to

$$H \cong A(a - b_s)/2b_s + \sqrt{(\quad)^2 + AW/b_s}, \tag{1}$$

provided $aA \gg W$ and $b_s \gg A$. At the equivalent point ($b_s = a = c$), $(H_e)_* = \sqrt{AW/c}$ and $(OH_e)_* = \sqrt{cW/A}$, with c as the concentration of whichever of the two substances is being titrated.

2. Feasibility. From Eq. (1) the feasibility for the titration, with strong base, of the weak acid, at concentration c, is

$$\Delta_p \cong \log \frac{p/400(1 - p/200) + \sqrt{(\quad)^2 + W/cA(1 - p/200)}}{-p/400(1 + p/200) + \sqrt{(\quad)^2 + W/cA(1 + p/200)}}. \tag{2}$$

For titration, with the weak acid, on the other hand, of the strong base at the same concentration,

$$\Delta_p \cong \log \frac{-p/400 + \sqrt{(\quad)^2 + W/cA}}{p/400 + \sqrt{(\quad)^2 + W/cA}}. \tag{3}$$

The absolute numerical value of expression (2) is larger than that of (3), but the difference is clearly negligible as long as p is small and A not too small. If $A = 10^{-7}$, $c = 1$, and $p = 2$, Eq. (2) gives $\Delta_p = 3.005$, while Eq. (3) gives $\Delta_p = 3.001$.

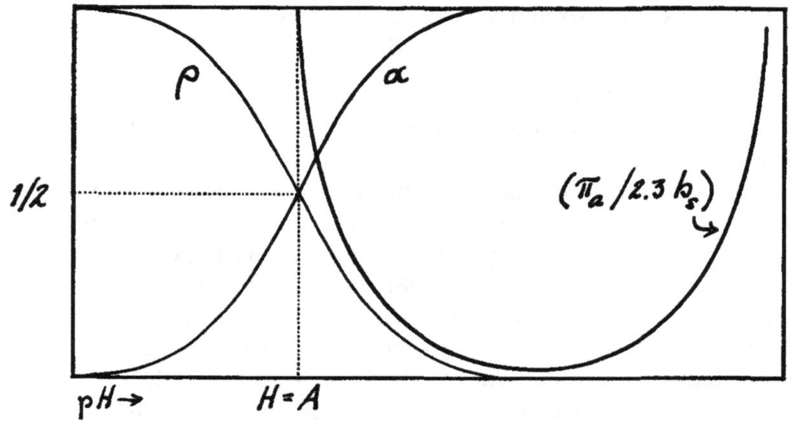

Fig. 13–1. Buffer capacity with respect to weak acid of solution containing (net) strong base.

The less accurate method (Procedure II of Chapter IX) for estimating the feasibility involves the buffer capacity at the equivalent point. From $D = a\alpha - b_s$,

$$\pi_a = |\, da/d\text{p}H \,| = 2.3(b_sH/A + OH + H + 2H^2/A). \tag{4}$$

Since $H_e \cong \sqrt{AW/c}$ and since $H_e < A$ and $H_e < OH_e$, then $(\pi_a)_e \cong 2(2.3)\sqrt{cW/A}$, and

$$F = (p/400)\sqrt{cA/W}. \tag{5}$$

While Procedure I, leading to Eqs. (2) and (3), brings out the small but real difference between the two titrations, this simpler procedure, operating through the approximate buffer capacity at the equivalent point, gives a feasibility function equal to that for the titration of weak acid with strong base, Eq. X(42), both being approximate formulas, however. With Eq. (5), the value of Δ_p is identical with that in Eq. (3) for either titration.

3. Buffer Capacity and Titration Curve. The relations between π_a and the fractions α and ρ are shown in Fig. 13–1, for unit value of b_s. The curve of π_a has a minimum at $OH \cong \sqrt{b_sW/A}$, near the equivalent point $a = b_s$, but no maximum since $d^2\pi_a/d\text{p}H^2$ is always positive. At high OH,

π_a approaches the value $2.3(OH)$; at $H = A$, $\pi_a \cong 2.3b_s$ and π_a then rises very steeply at higher H.

If strong acid is titrated with weak base, the curve of buffer capacity as $f(\mathrm{p}H)$ always lies above that for titration with strong base for any value of pH. If we distinguish π as $db_s/d\mathrm{p}H$ and π^B as $db/d\mathrm{p}H$ for these titrations, we have the various curves of Fig. 13–2 in which, with $B_1 > B_2$, π^B continually increases, at any value of H, as B decreases, approaching infinity as $B \overrightarrow{=} 0$.

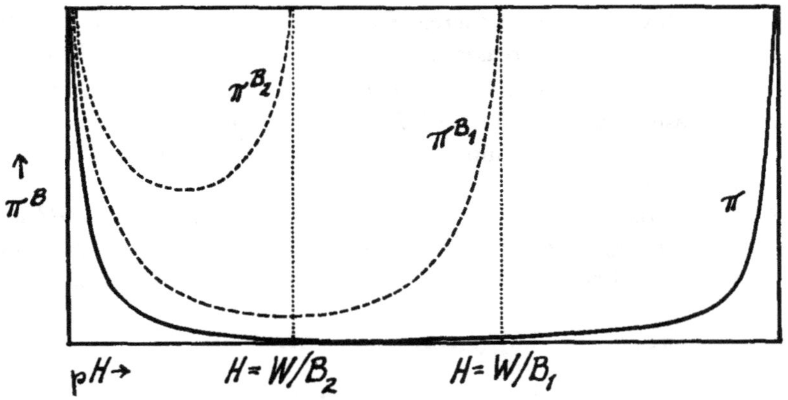

Fig. 13–2. Buffer capacity for titration of strong acid with weak base, π^B, for varying B; π is the buffer capacity with respect to strong base.

Moreover, whereas $\pi = 2.3(H + OH)$, with a value at any given H independent of the concentration of the strong acid being titrated, this is not the case for titration with weak base, in which

$$\pi^B = 2.3[a_sOH/B + H + OH + 2(OH)^2/B]. \quad (6)$$

In titration of a_s with b_s it is only the difference, $(a_s - b_s)$, which is significant in buffer action, whereas the actual concentration of weak base, b, and hence the actual concentration of the strong acid being titrated is in itself always significant in π^B. Thus while $\pi = 2.3(H + OH)$ may be written $\pi = 2.3(a_s - b_s + 2OH) = 2.3(2H + b_s - a_s)$, π^B, on the other hand, becomes $\pi^B = 2.3[b\rho' + (H + OH)/\alpha']$.

Finally, if we distinguish π_w as the buffer capacity in the addition of strong base to pure water and π_∞ as that in the addition of strong base to a solution of strong acid, we have $\pi_w = \pi_\infty = 2.3(H + OH)$. The only difference between the two curves is that π_w starts at $H = \sqrt{W}$ while π_∞ starts at a value of H fixed by the value of a_s. The π_w curve is merely the right half of Fig. 9–1. But in the addition of weak base to pure water the buffer capacity is

$$\pi_w^B = 2.3[H + OH + 2(OH)^2/B], \quad (7)$$

while in the addition of the weak base to a solution of strong acid $\pi_\infty^B = 2.3a_s OH/B + \pi_w^B$. Not only do these curves start at different values of H then, but the π_w^B curve is not merely part of the π_∞^B curve, the characteristics of which depend on the value of a_s. The π_∞^B curve is the sum of the two functions, π_w^B and $2.3a_s OH/B$. The latter rises continuously with OH, with neither maximum nor minimum. But π_w^B has a minimum at

$$H^3 - HW - 4W^2/B = 0. \tag{8}$$

For high B this minimum is therefore at $H \cong \sqrt{W}$, the curve approaching that of Fig. 9–1. As B decreases the minimum moves to higher H, occurring at $H \overset{\rightarrow}{=} \sqrt[3]{4W^2/B}$. When the function $2.3a_s OH/B$ is added to this curve, the effect, increasing with a_s and with $1/B$, is to move the minimum to higher H. The actual position of the minimum in this case will now be considered in connection with π_a, or π_∞^A.

4. Titration Error. To find more precisely the value of OH at the minimum of the π_a curve of Fig. 13–1 and hence at the inflection point in the pH curve for the titration of strong base with weak acid, we set $d\pi_a/d\text{p}H = 0$, or

$$d\pi_a/d\text{p}H = (2.3)^2(b_s H/A + H - OH + 4H^2/A) = 0; \tag{9}$$

hence

$$(OH)^3 A - (OH)W(b_s + A) - 4W^2 = 0. \tag{10}$$

Since $(OH)_e \cong \sqrt{b_s W/A}$, this may be rearranged to $(OH)^2 = W(b_s + A)/A + 4W^2/A(OH)$, whereupon

$$(OH)_i \cong \sqrt{W(b_s + A)/A} + 2W/(b_s + A), \tag{11}$$

$$H_i \cong \sqrt{AW/(b_s + A)} - 2AW/(b_s + A)^2. \tag{12}$$

Since $(OH)_i > (OH)_e$ the end-point (inflection point) occurs before the equivalent point. For the titration error we may introduce Eqs. (11) and (12) into Eq. x(59), and again neglecting only the second and third terms of H^2/A, we obtain

$$(b_s)_i \cong a_i + W/A + 3W/(\mathbf{c}_e + A), \tag{13}$$

in which, as usual, $\mathbf{c}_e = c_A c_B/(c_A + c_B)$ if the final volume is approximately $V_A + V_B$; cf. Eq. vii(28). The error is therefore negligible if A is large. But if the acid is weak, with $\mathbf{c}_e > A$, we have

$$(b_s)_i \cong a_i + W/A. \tag{14}$$

The relative percentage error for titration of strong base with weak acid (for $\mathbf{c}_e > A$) is then

$$\text{Error} \cong - 100W/\mathbf{c}_e A, \tag{15}$$

which is one third that for the titration of weak acid with strong base for comparable values of $\mathbf{c}_e A$; cf. Eq. x(62).

In the titration of a weak acid with a strong base then, titration back and forth through the inflection point, with strong acid and strong base, follows a single pH curve, that depending on π ($= dg/d\text{p}H$). The inflection then occurs at the same point, relative to the equivalent point, in either direction if dilution effects are negligible. The titration error is given by Eq. x(62) whether the titration is stopped in the forward direction, i.e. on addition of b_s up to the inflection point, or by back titration (to the same inflection point) with a solution of strong acid. But if the back titration is an alternation of strong base with the weak acid itself, through the equivalent point, the two paths are not identical. If dilution effects are negligible, the forward inflection (with b_s) occurs at $(OH)_i \cong \sqrt{aW/A} - 2W/A$, from Eq. x(54), with a negative error as in Eq. x(62); the reverse inflection, on addition of weak acid, occurs at $(OH)_i \cong \sqrt{aW/A} + 2W/a$, from Eq. (11), with now a positive error as in Eq. (15).

If a weak acid and a strong base are being titrated against each other, therefore, the inflection point experienced if the titration is finished with the weak acid is closer to the equivalent point than is that experienced if the titration is finished with the strong base. As seen in Eqs. (2) and (3), moreover, the sharpness of the titration is almost exactly the same in the two directions. Hence a more correct end-point is obtained in the determination of a weak reagent with a strong reagent if the process is finished by back titration with the weak reagent. If an acid with ionization constant 10^{-9} is titrated with strong base in such a way that the concentration at the inflection point is 10^{-2}, the error is $-$ 0.30 per cent by direct titration and only $+$ 0.10 per cent according to Eq. (15) if the inflection point is reached by back titration. The feasibility or sharpness of the titration would be only slightly higher in the first case. With H calculated from the full H^2 sequence of Eq. x(22), $\Delta_p = 0.141$ as against 0.137pH units respectively for two per cent of titration through the equivalent point, or for $p = 2$.

Similar considerations for titration of strong acid with weak base give, by symmetry, the value of H_i for the minimum of π_b and hence for the inflection in the titration curve, as $H_i \cong \sqrt{W(a_s + B)/B} + 2W/(a_s + B)$; and the titration error is $E = - W/\mathbf{c}_e B$.

B. Titration of Pure Weak Acid With Weak Base

1. Analytical Equations. The mathematical identity of this titration with the change in the relative concentrations (a_1, a_2) of two weak acids at their first equivalent point in titration with strong base, when $b_s = a_1$, was pointed out under Eq. xi(13); the equivalent point $a = b$, discussed under Eq. viii(8), is then identical with the first equivalent point for the strong base titration of two weak acids when $b_s = a_1 = a_2$, Eq. xi(21). However, since the problem is more familiar and practical in terms of acid and base

ionization constants, and of acid and base concentrations as the variables, it will be treated as the general problem of weak-weak titration. Here, then

$$D = H - OH = a\alpha - b\alpha' = aA/(A + H) - bB/(B + OH), \quad (16)$$

or

$$H^4B + H^3(bB + AB + W) + H^2[W(A - B) - AB(a - b)] - HW(aA + AB + W) - AW^2 = 0. \quad (17)$$

a. Approximate solutions of Eq. (17); *the titration curve.* The best approximation is the H^3 sequence. If $b < a$, the H^2 sequence is useless, and the H^4 is better than the H^3 only if $H^4B > HW(aA + AB + W)$ or roughly if $H > \sqrt[3]{aWA/B}$; but if $H \cong A(a - b)/b$, this means: if $b/a < \sqrt[3]{A^2B/aW}$. If $b > a$, the H^4 sequence is useless, and the H^2 is better than the H^3 only if $H^3(bB + AB + W) < AW^2$, or $H < \sqrt[3]{AW^2/bB}$; again, if $OH \cong B(b - a)/a$, this means: if $b/a > \sqrt[3]{AB^2/bW}$. It is obvious then that the H^3 sequence is much more widely applicable than the others. Therefore, in titration of the acid with the base, the start of the titration, when b/a $(= n)$ is small, is described by

$$H \cong -(b + A + W/B)/2 + \sqrt{(\quad)^2 + A(a - b) - W(A - B)/B}, \quad (18)$$

but for the major part of the titration and through the equivalent point,

$$H \cong \frac{AB(a - b) - W(A - B)}{2(bB + AB + W)} + \sqrt{(\quad)^2 + W\left(\frac{aA + AB + W}{bB + AB + W}\right)}. \quad (19)$$

If base is titrated with acid, the start is given by

$$H \cong \frac{W(aA + AB + W)}{2[W(A - B) - AB(a - b)]} + \sqrt{(\quad)^2 + \frac{AW^2}{W(A - B) - AB(a - b)}}, \quad (20)$$

but most of the curve, including the equivalent point, is again given by Eq. (19).

If the H^3 sequence, Eq. (19), is simplified to

$$H^3bB - H^2AB(a - b) - HaAW \cong 0, \quad (21)$$

then when $a > b$, H^3bB is generally greater than $HaAW$ since $H > \sqrt{WA/B}$, its value at the equivalent point; hence with $a > b$,

$$H \cong A(a - b)/b, \quad (22)$$

or the Henderson equation. Similarly, when $a < b$, $H < \sqrt{WA/B}$ and $H^3bB < HaAW$; then $H \cong aW/B(b - a)$, or $OH \cong B(b - a)/a$.

At the equivalent point, of course, with $a = b$, Eq. (17) becomes that for the pure salt MX already discussed as the principal example in Chapter VIII, Eq. VIII(8), while Eq. (19) reduces to Eq. VIII(15), etc.

b. Applicability of the Henderson equation, Eq. (22), *for weak acid and weak base.* By rearrangement of Eq. (16),

$$Hb = A(a - b) + [a(OH)A/B - D(A + H)(B + OH)/B]. \tag{23}$$

Following the procedure used in the case of titration of weak acid with strong base, under Eqs. X(28–32), we note that Eq. (23) will be valid approximately if $D \cong 0$ or if

$$(a - b) \gg | a(OH)/B - D(A + H)(B + OH)/AB |. \tag{24}$$

If $n = b/a$ as before, and $\eta = n/(1 - n)$, we now find that:

If the solution is acid, according to the condition of Eq. VII(21), the requirement is

$$a \gg A(1 + \eta W/AB)/n\eta(1 - n + \eta W/AB). \tag{25}$$

As long as $AB \gg W$, which, as we shall see in the following section, is the only significant possibility, this reduces to $n \gg \sqrt{A/a}$, so that the limit of applicability is roughly the same as that found for weak acid plus strong base or Eq. X(29), provided AB is large compared to W.

If the solution is alkaline, the condition is

$$a \gg W(1 + \eta W/AB)/A(1 - n)(1 - n - \eta W/AB). \tag{26}$$

Again then, provided $AB \gg W$, the condition is $n \ll (1 - \sqrt{W/aA})$, or the same as Eq. X(30) for weak acid plus strong base.

As long, then, as $AB \gg W$ or more precisely as long as η, or the ratio $b/(a - b)$, is small compared to AB/W, the range of applicability of the Henderson equation, or the ordinary "common ion effect," is wide and is approximately the same for titration with both strong and weak base.

2. Buffer Capacity and the Titration Curve.

a. General and graphical. For titration with weak base, the buffer capacity involved is db/dpH, or, from Eq. (16),

$$\pi_b = 2.3[a\alpha(\rho + \rho')/\alpha' + H + OH + 2(OH)^2/B], \tag{27}$$

with

$$d\pi_b/dpH = (2.3)^2[a\alpha(\rho^2 + 2\rho\rho' + \rho' - \alpha\rho)/\alpha' - H + OH + 4(OH^2)/B]. \tag{28}$$

For titration with weak acid, we consider da/dpH, or

$$\pi_a = 2.3[b\alpha'(\rho' + \rho)/\alpha + OH + H + 2H^2/A]. \tag{29}$$

The relations are shown graphically in Fig. 13–3 for $AB \gg W$; the figure refers schematically to unit concentrations of acid or base being titrated. The heavy solid curve (π_b) is the buffer capacity of the solution with respect to weak base and is therefore the significant derivative for titration with weak base, or from left to right. The heavy dotted curve (π_a) applies for the opposite case, titration of weak base with weak acid, or from right to left. In each case the curve starts from a definite point; π_b starts at the value of H in pure weak acid at concentration a, roughly $H \cong \sqrt{aA}$. Hence, with $a \gg A$, π_b starts at a low value of approximately $(2.3)2\sqrt{aA}$, according to Eq. (27). As the weak base is added, increasing the pH, the curve passes through a maximum when $b \cong a/2$, $H \cong A$, and $\pi_b \cong 2.3(a/4)$ and then through a minimum, which gives an inflection in the titration curve near the equivalent point $b = a$. When H reaches the value W/B, Eq. (27) gives $\pi_b \cong 2.3a$.

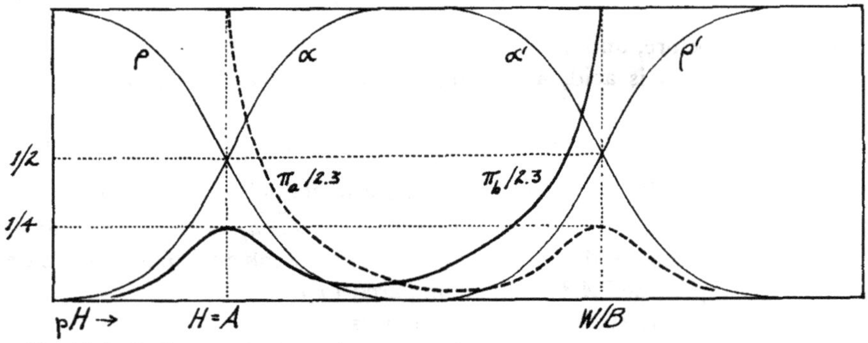

Fig. 13–3. Buffer capacity in weak-weak titration for unit concentrations with $AB \gg W$.

Similarly a solution of pure weak base starts at $OH \cong \sqrt{bB}$, with $\pi_a \cong (2.3)2\sqrt{bB}$. During titration with weak acid, the curve then traverses a maximum at $a \cong b/2$, $H \cong W/B$, and $\pi_a \cong 2.3(b/4)$ and then a minimum corresponding to an inflection in the pH curve near $a = b$. As H increases to the value $H = A$, π_a rises to approximately $2.3b$. In other words it is very difficult to make the solution of a base more acid than $H \cong A$ by addition of weak acid or to make an acid solution more alkaline than $H \cong W/B$ by addition of weak base. In fact, at $H = A$, $d\pi_a/d\text{p}H \cong (2.3)^2 b$, and at $H = W/B$, $d\pi_b/d\text{p}H \cong (2.3)^2 a$.

The two curves, π_b and π_a, however, are not symmetrically identical unless $A = B$. They cross at $H = \sqrt{WA/B}$ and hence at $H = \sqrt{W}$ only if $A = B$. The two minima therefore do not coincide; even with $A = B$, the minimum for π_b is at $H > \sqrt{WA/B}$ while that for π_a is at $H < \sqrt{WA/B}$, as we shall see later.

b. Feasibility of titration. At the equivalent point in either direction, $H_e \cong \sqrt{WA/B}$, and neglecting the terms in H and OH in Eqs. (27) and (29), we have for the approximate buffer capacity at the equivalent point in either direction (for $a = b = c$)

$$(\pi_b)_e \cong (\pi_a)_e \cong (2.3)2c\sqrt{W/AB}/(1 + \sqrt{W/AB}). \qquad (30)$$

Then if $AB > W$, $\pi_e \cong (2.3)2c\sqrt{W/AB}$; if $AB < W$, $\pi_e \cong (2.3)2c$, and if $AB = W$, $\pi_e \cong 2.3c$. We note therefore that π_e is low only if $AB \gg W$.

For the feasibility of the titration therefore in either direction, we have, through Eq. IX(27), provided $AB > W$,

$$F \cong (p/400)(1 + \sqrt{AB/W}). \tag{31}$$

The feasibility is fair then if $AB \gg W$, but otherwise poor. In fact if AB is not sufficiently greater than W, there is no practical meaning in a feasibility function since there will be no minimum in the π curve for an inflection point. Eq. (31) moreover shows no dependence upon the concentration because it has been assumed that $H_e \cong \sqrt{WA/B}$. The effect of the concentration would be introduced through the more accurate Procedure I of Chapter IX, using Eq. (19) for the titration curve through the equivalent point, simplified as the data may permit; but the effect is very small for the weak-weak titration.

c. Positions of maximum and minimum of π_b. The π_b curve in this titration is the sum of two functions, π_w^B of Eq. (7) and the contribution of the weak acid, or

$$\pi_j^B = 2.3a\alpha(\rho + \rho')/\alpha'. \tag{32}$$

The first, π_w^B, has already been discussed; this function, possessing a minimum, has in general a value much smaller than that of π_j^B for appreciable values of a except at very high H or very high OH. Except at the extremes, the shape of the total curve, or the sum $\pi_b = \pi_w^B + \pi_j^B$, therefore will be practically that of the specific curve π_j^B, which itself may have both a maximum and a minimum if AB is sufficiently large compared to W. From $d\pi_j^B/d\text{p}H = 0$, or from Eq. (28) with terms in H and OH neglected,

$$H^3B - H^2(AB - 4W) + 3HAW + A^2W = 0. \tag{33}$$

With $AB \gg W$ therefore, the maximum occurs at

$$H_{\text{max}} \cong A(1 - 8W/AB), \tag{34}$$

and the minimum at

$$H_{\text{min}} = H_i \cong \sqrt{AW/B} + 2W/B, \tag{35}$$

or

$$OH_{\text{min}} = OH_i \cong \sqrt{BW/A} - 2W/A. \tag{36}$$

If a is large enough so that the neglect of H and OH in Eq. (28) is justified, these are the positions of the maximum and minimum of the actual π_b curve of Fig. 13–3.

Neither of the minima in Fig. 13–3 therefore occurs at $H = \sqrt{WA/B}$. That for π_b occurs on the left and that for π_a on the right of $H = \sqrt{WA/B}$.

d. Titration error. Since, in titration with weak base, $H_i > H_e$, the inflection occurs early and the titration error is negative. To estimate a, from b_i (that is, from b at the inflection point), we may use Eqs. (35) and (36) in Eq. (16), obtaining

$$a_i = b_i(1 + 4W/AB) + \sqrt{WA/B} - \sqrt{WB/A}. \tag{37}$$

The percentage error is therefore about $100(4W/AB)$. The greater is the ratio AB/W, the smaller is the titration error, provided that the concentration is appreciable, so that $H_e \cong \sqrt{WA/B}$.

By analogy, for the titration of weak base with weak acid,

$$OH_{\max} \cong B(1 - 8W/AB), \tag{38}$$

$$OH_i \cong \sqrt{WB/A} + 2W/A, \tag{39}$$

$$H_i \cong \sqrt{WA/B} - 2W/B. \tag{40}$$

For the titration error

$$b_i \cong a_i(1 + 4W/AB) + \sqrt{WB/A} - \sqrt{WA/B}. \tag{41}$$

If a weak acid is measured by titration with a weak base, the error is negative if the titration is stopped (at the inflection point) in the forward direction; the error is positive if the titration is stopped by back titration to an inflection point with the weak acid itself. The forward titration gives an inflection at a ratio $b/a \cong 1/(1 + 4W/AB)$ while the reverse titration gives an inflection when $b/a \cong 1 + 4W/AB$.

e. Condition for the maximum and minimum in π_b (or for the inflection in titration curve). As the product AB decreases, the maximum and minimum move relatively closer to each other until they merge; thus Eq. (33) has no real positive root if $AB < 4W$. The maximum and minimum vanish therefore as AB decreases toward the value W. The combination of $d\pi_j^B/dpH = 0$ with $d^2\pi_j^B/dpH^2 = 0$ gives as the condition for merging of maximum and minimum

$$3H^3B - H^2AB - 5HAW - A^2W = 0, \tag{42}$$

which may be written as $H = (A/3)(1 + 5W/HB + AW/H^2B)$. As AB decreases toward the value W it may be seen from Fig. 13–3 that the merging must occur at $A > H > \sqrt{WA/B}$ and hence when

$$H \cong (A/3)(1 + 24W/B) \cong A/3. \tag{43}$$

With $H \cong A/3$ in $d\pi_j^B/dpH = 0$, or in Eq. (28) with the H and OH terms neglected, i.e. in Eq. (33), we then obtain

$$AB \cong 33W \tag{44}$$

as the condition for the maximum and minimum in the π_b curve during the titration of weak acid with weak base. Unless AB is greater than approximately $33W$ then, there is no end-point type of inflection in the pH titration curve, a result which by symmetry holds for titration either of a by b or of

b by a. This assumes always that the concentrations themselves are appreciable to justify the neglect of the H and OH terms in π_b or π_a; but the required ratio of AB/W would be still greater at low concentrations. For the effect of ionic strength, the condition of Eq. (44) becomes $\mathbf{AB} > 33\gamma^2\mathbf{W}$.

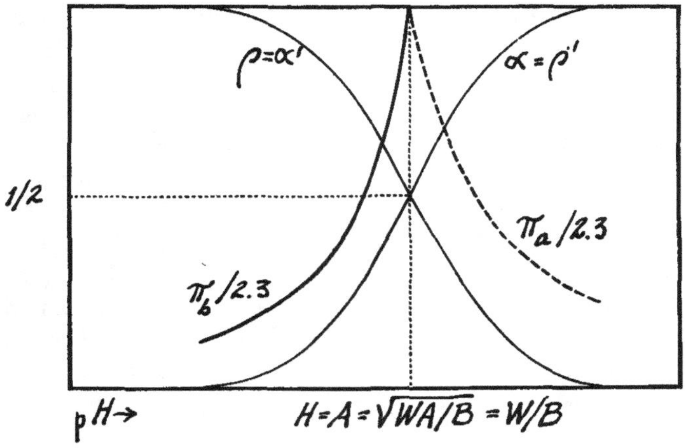

Fig. 13–4. The case of Fig. 13–3 with $AB = W$.

As AB decreases, therefore, the maximum and minimum of each curve vanishes and at $AB = W$ we would have curves as in Fig. 13–4 for $a = b = c = 1$. At $H = A = \sqrt{WA/B} = W/B$ we have $\pi_b \cong \pi_a \cong 2.3c$. With

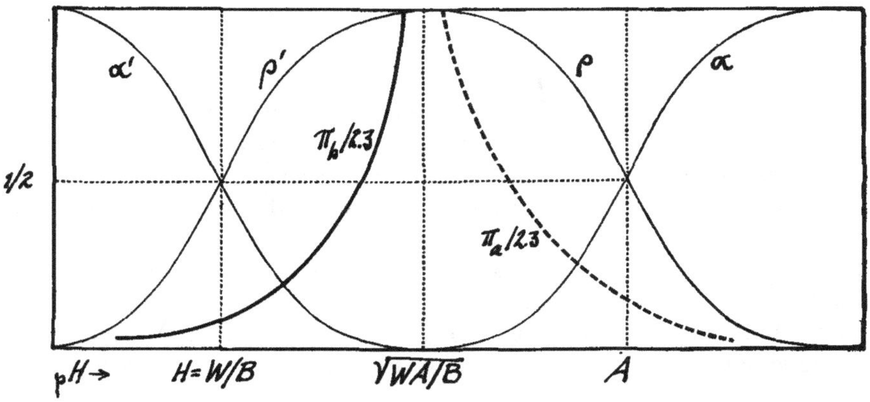

Fig. 13–5. The case of Fig. 13–3 with $AB \ll W$.

$AB < W$ the relations would be as in Fig. 13–5. At $H = \sqrt{WA/B}$ we have $\alpha = \alpha'$ and $\rho = \rho' = 1$ so that $\pi_b \cong \pi_a \cong (2.3)2c$.

If we equate π^B_∞, or π_b from Eq. (6), with π^B_A, or π_b from Eq. (27), for $a_s = a$, we find that the condition for their equality is $AB = W$. If $AB > W$,

$\pi_A^B > \pi_\infty^B$ at any value of H, and the curve π_A^B can have a maximum if $AB > 33W$; if $AB < W$, $\pi_A^B < \pi_\infty^B$ and π_A^B has no maximum. This means that the π_b curve of Fig. 13–4 is merely the right hand part of the curve for the titration of strong acid at the same concentration, with the same weak base, or of one of the particular curves of Fig. 13–2.

The maximum of the π_b curve then may vanish by merging with the minimum on its right as AB decreases, resulting in Fig. 13–5. The maximum may also vanish as A increases for a given value of B; in this case the maximum would move to the left until the curve takes on the character of Fig. 13–2. Since this disappearance of the maximum must occur at a high value of H, the condition for it will be quite independent of the value of B; as seen in Fig. 13–3 we may take $\alpha' \cong 1$ and $\rho' \cong 0$ in high H. The result, therefore, is the same (approximately) as the condition found in Chapter x, that the maximum vanishes when $A > a/27$.

f. Significance. Like the requirement noted in Chapter xii, Section C (that AB must be greater than $27Wb/a$ or $aW/27b$ for an inflection to occur in the titration, with strong base, of a weak acid in presence of weak base), the present requirement is also of interest in connection with the question of defining a "salt" in the water system. Unless $AB > 33W$, there is no way, "by titration," of forming a solution of the 1 : 1 salt, MX, given the separate acid and base. Only by mixing stoichiometrically equivalent quantities of the acid and base without recourse to an inflection point can such a solution be made. But again there is no necessary connection between this requirement ($AB > 33W$) for an inflection point and the possibility of the crystallization of the "salt" from aqueous solution. The evaporation of a pure salt solution causes an increase in the concentration and hence brings the solution toward the iso-electric point; when $AB \not> W$, the degrees of ionization of acid and base of a salt are both low at the iso-electric point and so it would appear that the crystallization of an *ionic* compound (a "salt") by evaporation of the aqueous solution becomes increasingly difficult, if not impossible, when $AB \not> W$. But this argument is incomplete since the values of α and α' reached upon evaporation depend on the solubility of the salt, which is not fixed solely by the quantities already considered. It remains nevertheless an interesting question whether there is a limiting ratio of AB/W for the stability of a "salt" as a solid in contact with aqueous solution. Such a limit may perhaps be used to give some quantitative meaning to the common but extremely vague kind of statement that certain acids and bases are "too weak" to form salts.

The relation, furthermore, between the ratio AB/W and the pH of the salt solution is interesting in a certain practical sense, in connection with what was pointed out under Eq. (30), that the buffer capacity with respect to either the weak acid or the weak base is very high if AB is not much greater than W. If $AB \gg W$, we may "titrate" the acid against the base, using the inflection point in general to arrive with some accuracy at a "pure" salt solution. The pH of the solution, however, is for this very reason very sensitive to slight impurity of the salt, "impurity" referring to inequality of a and b. Fortunately, such salts are usually crystallizable in the pure state and the exact pH of such a salt solution may be obtained by previous recrystallizations of the salt which is then handled as a physical solute. If $AB < 33W$, titration may not be used to prepare the pure "salt solution," but on the other hand the pH of the "salt solution" is hardly sensitive to impurity (or inequality of a and b). The pH of such a "salt solution" can then be determined by mixing the acid and the base in stoichiometric proportions with little error to be expected from inequality of a and b. Such "salts," however, can seldom be prepared and recrystallized as pure solids and it is questionable whether it is correct or useful to call their solutions salt solutions if these cannot be prepared by dissolving the pure salt as solute, but only by mixing the acid and the base. Although it is customary to speak of the relations in such a solution as the "hydrolysis" of the "salt,"(1) there is in this case, if the "salt" is not even obtainable as a crystal, no point whatever for the expression "hydrolysis."

C. Titration With Weak Base of Mixture of Strong Acid and Weak Acid

1. Analytical Formulas for H. With an added term in a_s in Eq. (16), Eq. (17) becomes

$$H^4B + H^3B(b - a_s + A + W/B) - H^2[W(a_s - A + B) + AB(a_s + a - b)] - HAW(a_s + a + B + W/A) - AW^2 = 0. \quad (45)$$

(1) See for example the paper (R. G. Bates and G. D. Pinching, *J. Am. Chem. Soc.*, **72,** 1939 [1950]) on the determination of the aqueous ionization constant of ammonia from the determination of pH in the solution of ammonia and potassium hydrogen *p*-phenolsulfonate ("KHPs") in equivalent proportions entitled "Dissociation Constants of Aqueous Ammonia at 0 to 50° from E.M.F. Studies of the Ammonium Salt of a Weak Acid," in which the authors speak of the "hydrolysis of potassium ammonium *p*-phenolsulfonate." The dibasic acid involved was treated as strong in respect to K_1. As pointed out under Eq. v(135) and, below, under Item (2) of Section E3, the "mixed salt" of type "$NaNH_4SO_4$," of a dibasic acid strong in respect to K_1 with strong and weak bases, is mathematically identical with the salt of monobasic weak acid (with ionization constant $A = K_2$) and weak base (B). The condition of Eq. (44) therefore here becomes $BK_2 > 33W$. With $B \cong 10^{-5}$, and $K_2 \cong 10^{-9}$, the "salt solution" in question was in fact prepared by the mixing of equivalent quantities of ammonia and KHPs. The general and exact equation for the calculation of B from the data in such a problem is given by Eq. v(135) modified only in respect to the activity coefficient.

The authors point out that the method is good only if pK_2 and p(W/B) "differ by less than two units." One reason for this practical limitation is the condition of Eq. (44); unless AB (or K_2B) is of the order of W or less, the dependability of the pH of the "salt solution" synthetically prepared is poor in respect to the inequality of a and b. In addition, as discussed in Chapter xii, Section C2 and Section E, unless $13.9W < BK_2 < W/13.9$ the pH of the solution, even with $a = b$, will be sensitive to traces of foreign strong acid or strong base.

The H^4 sequence applies from the start to somewhat past the first equivalent point, $b = a_s$, regardless of the ratio AB/W, so that

$$H \cong (a_s - b - A - W/B)/2 + \sqrt{(\quad)^2 + A(a_s + a - b) + (a_s - A + B)W/B}. \quad (46)$$

Hence, $H \cong a_s$ when $b = 0$ and $H \cong a_s - b$ up to near the equivalent point. At the equivalent point, when $b = a_s$,

$$H_e \cong \sqrt{aA + a_s W/B}. \quad (47)$$

This formula is expected since at this point the solution is mathematically one of two weak acids, the salt MX_s being mathematically a weak acid with ionization constant W/B.

(a) $AB > W$. If $AB > W$, then for comparable values of a_s and a Eq. (47) becomes $H_e \cong \sqrt{aA}$. Then for the second half of the titration, or in the range $a_s \ll b \ll a_s + a$, the terms in H^3 and H^2 give $H \cong A(a_s + a - b)/(b - a_s)$, the Henderson type of equation. Through the second equivalent point and up to great excess of the base, the H^3 sequence gives

$$H \cong \frac{W(a_s - A + B) + AB(a_s + a - b)}{2B(b - a_s + A + W/B)} + \sqrt{(\quad)^2 + \frac{AW(a_s + a + B + W/A)}{B(b - a_s + A + W/B)}}, \quad (48)$$

so that at the second equivalent point, when $b = a_s + a$,

$$H_e \cong \sqrt{WA(a_s + a)/aB}. \quad (49)$$

In considerable excess of the base, the terms in H^2 and H give $H \cong W(a_s + a)/B(b - a_s - a)$. The last term of Eq. (45) never becomes significant since the solution never becomes very alkaline.

(b) $AB < W$. Now Eq. (47) becomes $H_e \cong \sqrt{a_s W/B}$ and Eq. (48) has to be used for the titration from the first through the second equivalent point. For the second equivalent point itself however we have, from the terms in H^3 and H^2, $H_e \cong a_s W/aB$, and from the same terms, for large excess of base, $H \cong a_s W/(b - a_s)B$.

2. Buffer Capacity and Titration Curve. The buffer capacity, π_b, is now the sum of π_A^B $(= \pi_w^B + \pi_j^B)$ and the function $2.3a_s(OH)/B$; or, $\pi_b = \pi_\infty^B + \pi_j^B$:

$$\pi_b = 2.3[a_s OH/B + a\alpha(\rho + \rho')/\alpha' + H + OH + 2(OH)^2/B]. \quad (50)$$

The π_b curve is thus the sum of three curves or functions, π_w^B, π_j^B, and $2.3a_s(OH)/B$. With appreciable values of a_s and a, with $a \gg A$ and with $AB \gg W$, the curve of π_b against pH has a maximum near $H = A$, and two lateral minima. A particular curve such as Fig. 13–6 then applies for titration, with a particular weak base, of a solution with particular concentrations of strong acid and of a particular weak acid. It does not represent the curve for titration of the weak acid at the given concentration simply "in presence of strong acid" since the whole curve depends on the value of a_s. Minimum a determines the inflection for the equivalent point $b = a_s$ and minimum b that for the equivalent point $b = a_s + a$.

At the first equivalent point then combination with Eq. (47), with $H_e > A$, gives $(\pi_b)_e \cong (2.3)2\sqrt{aA + a_sW/B}$, and the feasibility function for the titration of the strong acid in presence of the weak acid is

$$F \cong (pa_s/200)/2\sqrt{aA + a_sW/B}, \tag{51}$$

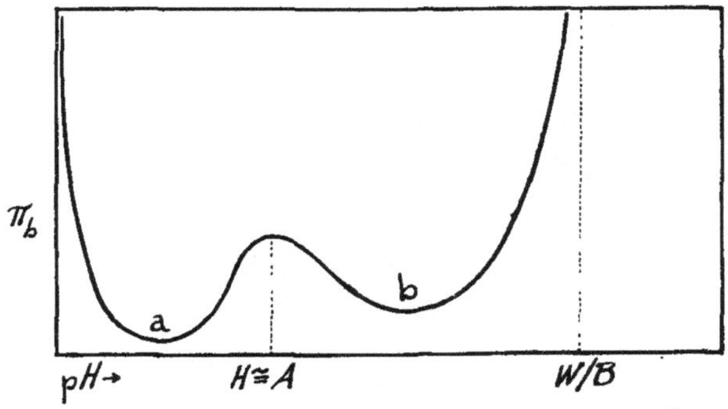

Fig. 13–6. Buffer capacity with respect to weak base of mixture of strong acid and weak acid.

which may be simplified according to whether $AB \gtrless W$. At the second equivalent point, provided $AB \gg W$, we have, with H_e from Eq. (49), $(\pi_b)_e \cong (2.3)2\sqrt{a(a_s + a)\,W/AB}$ if $\alpha/\alpha' \cong 1$; then for the measurement of the weak acid,

$$F \cong (p/200)\sqrt{ABa/(a_s + a)\,W}. \tag{52}$$

For the maximum and the minima of π_b in presence of a_s (with $AB \gg W$), we note that since the solution is never very alkaline, we may neglect the OH terms in the expression for π_b, so that when $d\pi_b/dpH = 0$,

$$a_sOH/B + a\alpha(\rho^2 + 2\rho\rho' + \rho' - \alpha\rho)/\alpha' - H \cong 0, \tag{53}$$

or

$$H^5B + 3H^4AB - H^3A(aB + a_sW/A - 3AB) + H^2A[a(AB - 4W) - 3a_sW + A^2B] - 3HA^2W(a + a_s) - A^3W(a + a_s) \cong 0. \tag{54}$$

The value of H at minimum a, near the equivalent point $b = a_s$, is given approximately by the first three terms, and more accurately as

$$H_{\text{min.a}} \cong \sqrt{aA} - 2A + a_sW/2B\sqrt{aA}, \tag{55}$$

so that the titration error at this inflection point may be either positive or negative. For the maximum of π_b, when $H \cong A$, the neglect of terms not involving the concentrations gives

$$H_{\max} \cong A[1 - 8W(a + a_s)/aAB]; \tag{56}$$

the principal terms are those in H^3 and H^2 of Eq. (54). For minimum b, near the equivalent point $b = a_s + a$, Eq. (54) gives, with the neglect of terms not involving a and a_s, and with those in H^2 and H^0 as principal terms,

$$H_{\min.b} \cong \sqrt{(AW/B)(a + a_s)/a} + 2(a + a_s)W/aB, \tag{57}$$

so that the titration error at this inflection point is negative.

For the conditions for the minima in the π_b curve, we may neglect the OH terms in π_b and combine the equations $d\pi_b/dpH = 0$ and $d^2\pi_b/dpH^2 = 0$; the result is

$$\begin{aligned} 2H^6B^2 &+ H^5B(7AB + 3W) + H^4AB(9AB + 11W) \\ &- H^3AB[3a(AB - W) - 5A(AB + 3W) - \mathbf{Y}/AB] \\ &+ H^2A^2B[a(AB - W) + A(AB + 9W) + 3\mathbf{Y}/AB] \\ &+ HA^2W[5a(AB - W) + 2A^2B + 3\mathbf{Y}/W] \\ &+ A^3W[a(AB - W) + \mathbf{Y}/W] \cong 0, \end{aligned} \tag{58}$$

in which $\mathbf{Y} = a_sW(AB - W)$. With $AB > W$, and with the terms not involving the concentrations neglected, this reduces to Eq. (42), and the condition for minimum b is approximately independent of a_s, namely $AB > 33W$. The condition for minimum a again remains approximately $a > 27A$ since at high H we may take $\alpha' \cong 1$ and $\rho' \cong 0$ and neglect the quantity a_sOH/B in Eq. (53). This amounts to neglecting the weakness of the base, and the stronger the base, the more correct is the condition $a > 27A$. That the condition is in fact quite independent of the value of B, for $AB > W$ and for an appreciable value of a_s, is verifiable by the following consideration. With W neglected relative to AB and with the terms in $\mathbf{Y}/AB$ also neglected, the first five terms of Eq. (58), which will be the significant ones in the acid region in question, give an expression independent of B:

$$2H^4 + 7H^3A + 9H^2A^2 - HA^2(3a - 5A) + A^3(a + A) \cong 0. \tag{59}$$

With $a = 27A$, this gives exactly $H = 2A$, which was the condition for this merging of minimum a and the maximum for titration with strong base, in connection with Eq. x(39). With $A = 10^{-2}$, $a = 0.27$, $H = 2 \times 10^{-2}$, and B as low as 10^{-8}, the neglect of the last two terms of Eq. (58) is still justified since the negative exponents of all the terms (to the base 10) are 26, 26, 26, 25, 26, 30, 31.

In general terms then the minimum a vanishes for given values of a_s, a, and B as A increases to leave, when $A > a/27$, a curve with only minimum

b. With fixed values of a_s and a, minimum b vanishes as the product AB becomes smaller than $33W$.

For fixed values of a_s and a, π^B, the buffer capacity with respect to weak base, is greater than π, that with respect to strong base, at any given value of H. Furthermore, for a fixed value of A, then in titration with two different weak bases, with $B_1 > B_2$, π^{B_2} is always greater than π^{B_1} for given pH, the curves never crossing. The schematic relations are shown in Fig. 13–7, in which it is assumed that $a > 27A$ to give a maximum in π. The minimum b disappears as AB falls to below $33W$.

For the effect of variation of the strength of the weak acid, with fixed values of a_s and a, in titration with a given weak base, the two buffer capacity

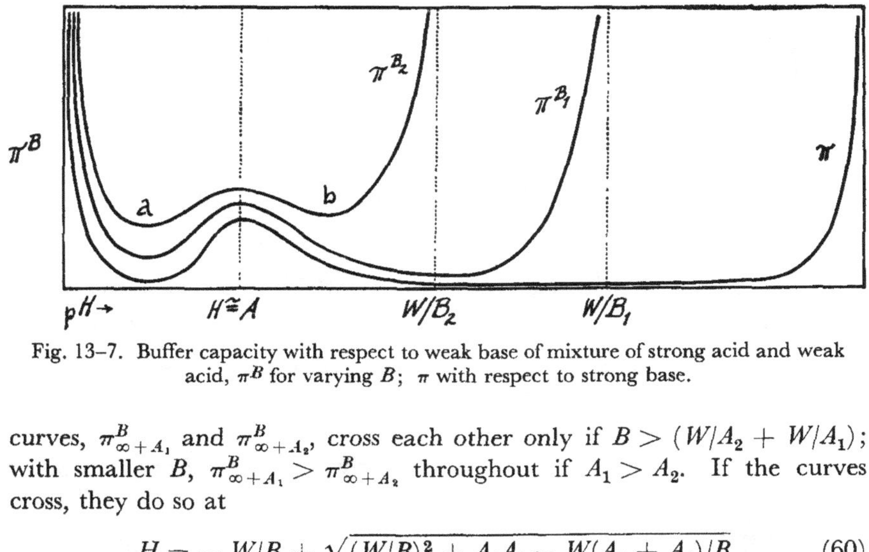

Fig. 13–7. Buffer capacity with respect to weak base of mixture of strong acid and weak acid, π^B for varying B; π with respect to strong base.

curves, $\pi^B_{\infty+A_1}$ and $\pi^B_{\infty+A_2}$, cross each other only if $B > (W/A_2 + W/A_1)$; with smaller B, $\pi^B_{\infty+A_1} > \pi^B_{\infty+A_2}$ throughout if $A_1 > A_2$. If the curves cross, they do so at

$$H = -W/B + \sqrt{(W/B)^2 + A_1A_2 - W(A_1 + A_2)/B}. \tag{60}$$

At higher H, $\pi^B_{\infty+A_1} > \pi^B_{\infty+A_2}$, and at lower H, $\pi^B_{\infty+A_1} < \pi^B_{\infty+A_2}$.

Graphically, for $AB \gg W$ and $B > (W/A_2 + W/A_1)$, we have Fig. 13–8 for fixed values of a_s and a.

The comparisons discussed under Figs. 13–7 and 13–8 apply for any fixed value of a_s, including $a_s = 0$. With $a_s = 0$, the curves start without minimum a, at the value of H in the pure weak acid solution.

D. Titration of Strong Base With Dibasic Acid Strong in K_1

This is mathematically the titration of strong base with an equimolar mixture of strong and weak monobasic acids. If a is the molarity of the dibasic acid,

$$H - OH = a(1 + \alpha_2) - b_s, \tag{61}$$

$$\pi_a = |\, da/dpH \,| = [2.3/(1 + \alpha_2)^2][b_s\alpha_2(1 - \alpha_2) + H(1 + 2\alpha_2 - \alpha_2^2) + OH(1 + \alpha_2^2)], \quad (62)$$

$$d\pi_a/dpH = [(2.3)^2/(1 + \alpha_2)^3][b_s\alpha_2(1 - \alpha_2)(1 - 3\alpha_2) - H(1 + 3\alpha_2 + 5\alpha_2^2 - 5\alpha_2^3) + OH(1 - \alpha_2 + 5\alpha_2^2 - \alpha_2^3)]. \quad (63)$$

The buffer capacity π_a may be considered to be the sum of two functions. One, represented by the terms in H and OH of Eq. (62), is a curve with a

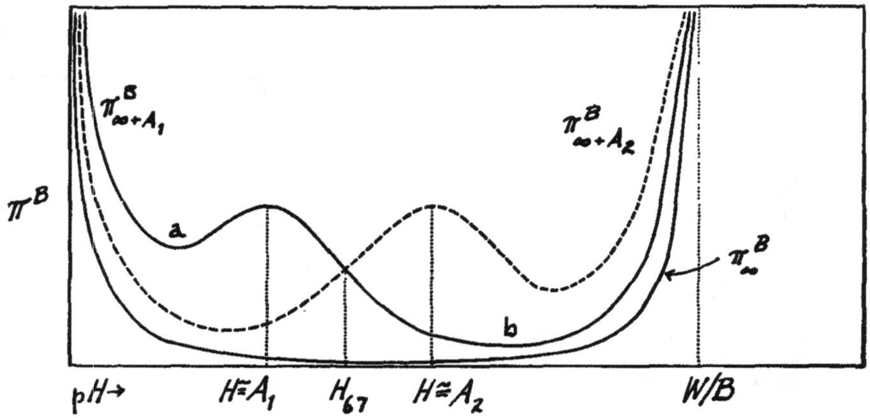

Fig. 13–8. Buffer capacity with respect to weak base of mixture of strong acid and weak acid for varying A.

single minimum and no maximum. Its derivative, taken from Eq. (63), is zero when

$$H^5 + 6H^4K_2 + H^3(14K_2^2 - W) + 2H^2K_2(2K_2^2 - W) - 6HK_2^2W - 4K_2^3W = 0, \quad (64)$$

so that the minimum of this function occurs very close to $H = \sqrt{W}$ whatever the value of K_2. The other contribution to π_a is the term in b_s in Eq. (62). From Eq. (63) this is seen to have a maximum but no minimum; its maximum occurs at $\alpha_2 = 1/3$ and $H = 2K_2$ and its contribution to π_a at this maximum is $2.3(b_s/8)$. (Its contribution to π_a at $H = K_2$, where $\alpha_2 = 1/2$, is $2.3b_s/9$.) For appreciable values of b_s then, the actual π_a curve has a maximum of $\sim 2.3(b_s/8)$ near $H = 2K_2$ and a minimum on each side of it. Minimum b, in alkaline solution, gives an end-point inflection in the titration curve corresponding to the equivalent point $a = b_s/2$ for the solution of the salt of type "Na_2SO_4"; minimum a, in acid solution, gives an inflection for the equivalent point $a = b_s$ for the salt of type "$NaHSO_4$." These relations are shown in Fig. 13–9.

For the positions of these minima we set $d\pi_a/dpH = 0$, obtaining

$$H^5 + 6H^4K_2 - H^3(b_sK_2 - 14K_2^2 + W) + 2H^2K_2(b_sK_2 + 2K_2^2 - W) - 6HK_2^2W - 4K_2^3W = 0. \quad (65)$$

The value of H at the maximum of π_a is given roughly by the terms in H^3 and H^2 as $H \cong 2K_2$.

(1) Minimum b. For minimum b, the last four terms of Eq. (65) give roughly, with $b_s \gg K_2$,

$$H_i \cong \sqrt{2K_2W/b_s} + 2W/b_s. \quad (66)$$

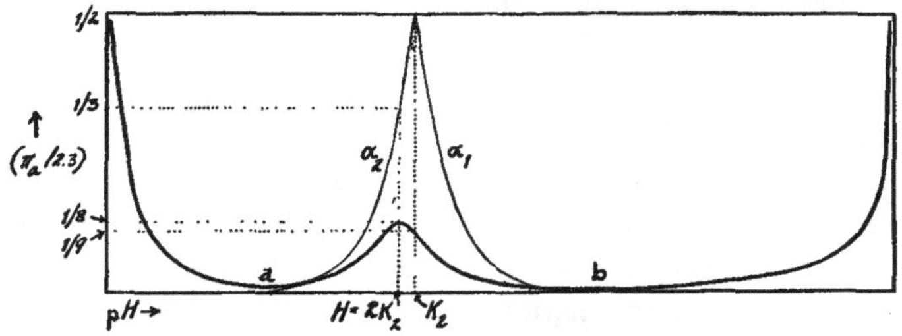

Fig. 13–9. Buffer capacity of solution of strong base with respect to dibasic acid strong in K_1; at unit value of b_s (schematic).

Hence, with $H_i > H_e$, this inflection occurs late. To find $(b_s)_i$ from a_i, we use Eq. (61), introducing into it H_i from Eq. (66) with $2a_i$ in place of $(b_s)_i$; roughly, since $(1 + \alpha_2) \cong 2 - H/K_2 + H^2/K_2$,

$$(b_s)_i \cong 2a_i - \sqrt{K_2W/a_i} - W/K_2. \quad (67)$$

For the feasibility of the titration at minimum b, $(\pi_a)_e$ is to be obtained from Eq. (62). With $H_e \cong \sqrt{2K_2W/b_s}$ at this equivalent point, $1 + \alpha_2 \cong 1 + \alpha_2^2 \cong 2$ and $\alpha_2(1 - \alpha_2) \cong H/K_2$; with the H term neglected, Eq. (62) then gives $(\pi_a)_e \cong 2.3\sqrt{b_sW/2K_2}$, so that

$$F \cong (p/400)\sqrt{2b_sK_2/W}, \quad (68)$$

with p as percentage of $(b_s/2)$; with p relative to b_s, F has twice this value. Eqs. (68) and (67) then give the feasibility and the titration error for the titration of strong base with acid of the type of "H_2SO_4," to the equivalent point "Na_2SO_4," where $a = b_s/2$; if K_2 is large, F is high and the error is negligible.

2. Minimum a. According to Fig. 13–9, there will be a second inflection in acid solution near $a = b_s$, the solution corresponding to aq. "$NaHSO_4$." Here $H_e \cong \sqrt{b_sK_2}$; hence with $1 + \alpha_2 \cong 1 + 2\alpha_2 - \alpha_2^2 \cong 1$, with $\alpha_2(1 - \alpha_2)$

$\cong K_2/H$, and with the OH term neglected, Eq. (62) gives the buffer capacity as $(\pi_a)_e \cong 2(2.3)\sqrt{b_s K_2}$, so that

$$F \cong (p/400)\sqrt{b_s/K_2}, \tag{69}$$

with p referring to b_s (total titration).

Again with $b_s \gg K_2$, the first four terms of Eq. (65) give

$$H_i \cong \sqrt{b_s K_2} - 4K_2, \tag{70}$$

so that this inflection occurs too early. The error is obviously very large if K_2 is large, as in H_2SO_4, and the feasibility, from Eq. (69), would be very small if indeed there is any inflection at all. From Eqs. (70) and (61), with $1 + \alpha_2 \cong 1 + K_2/H - K_2^2/H^2$, we have, for the titration error, roughly

$$(b_s)_i \cong a_i + 7K_2 + W/\sqrt{a_i K_2}. \tag{71}$$

For the requirements for the minima a and b, both will occur only if K_2 is neither too large nor too small. If K_2 is too large, there will be only minimum b, an inflection near $a = b_s/2$, for aq. "Na_2SO_4"; if too small, there will be only minimum a, an inflection near $a = b_s$, for aq. "$NaHSO_4$." The latter possibility is generally unlikely, however, for a dibasic acid "strong" in K_1. At any rate, differentiating $d\pi_a/d\text{p}H$, setting the result equal to zero and then introducing the equation $d\pi_a/d\text{p}H = 0$, we obtain exactly

$$H(1 - \alpha_2)(1 - 5\alpha_2) + 2\alpha_2 OH(1 - 2\alpha_2) = 0, \tag{72}$$

or

$$H^4 - 4H^3K_2 + 2HK_2W - 2K_2^2W = 0. \tag{73}$$

For large K_2 minimum a disappears in acid solution so that if we neglect the OH relative to the H term in Eq. (72), which is the same as using only the first two terms of Eq. (73), the condition is seen to be $\alpha_2 \cong 1/5$ or $H \cong 4K_2$. Then from $d\pi_a/d\text{p}H = 0$, again with OH neglected relative to H, the requirement for the occurrence of minimum a is

$$b_s > 110K_2. \tag{74}$$

With small K_2, minimum b disappears in basic solution so that the neglect of the H term of Eq. (72) gives $\alpha_2 \cong 1/2$ or $H \cong K_2$. Hence from $d\pi_a/d\text{p}H = 0$, with H neglected relative to OH, the requirement for this minimum is

$$b_s > 13W/K_2. \tag{75}$$

If $K_2 = \sqrt{W/2}$, Eq. (73) gives exactly $H = K_2 = \sqrt{W/2}$ so that the condition is exactly $b_s = 13W/K_2$ for this value of K_2. This means that if $K_2 > \sqrt{W/2}$, it is minimum a which vanishes as b_s decreases, and in titration with such an acid there will always be an inflection at minimum b near the equivalent point $a = b_s/2$ for the salt solution "Na_2SO_4," while that near

$a = b_s$ will occur only if $b_s > 110K_2$. On the other hand, if $K_2 < \sqrt{W/2}$ there will always be an inflection at minimum a near the equivalent point $a = b_s$ for the salt solution "$NaHSO_4$," while that near $a = b_s/2$ will occur only if $b_s > 13W/K_2$.

The general relations are therefore similar to those discussed in Chapter x for titration of monobasic weak acid with strong base, where the critical value of A is $A = \sqrt{W}$; see Eqs. x(36) and x(37).

Finally, although the *approximate* feasibility functions of Eqs. (68) and (69) are the same as those in the titration of the dibasic acid with strong base, the feasibility is strictly lower in the titration of the strong base with the dibasic acid since $|\pi_a|$ is always greater than $|\pi|$.

E. Titrations Involving Weak Base and a Dibasic Acid Strong in K_1

1. Titration of the Acid. The titration of a "strong" dibasic acid is the same as the titration of a mixture of strong and weak monobasic acids at equal concentrations. Hence this case is covered by the considerations in Section C, with the condition $a_s = a$. Provided $BK_2 > W$ there may be two inflections in the titration curve, one at $b = a$ and one at $b = 2a$, where a now represents the molarity of the dibasic acid; if BK_2 is not greater than W, there will be only one inflection, that at $b = a$. With "NH_3" and "H_2SO_4" as types, the first equivalent point is a solution of "NH_4HSO_4," the second, one of "$(NH_4)_2SO_4$." For the formulas for H in such salt solutions, the general equation is

$$H - OH = a(1 + \alpha_2) - b\alpha'. \qquad (76)$$

a. Acid salt, type "NH_4HSO_4"; with $b = a = c$, Eq. (76) gives

$$H^4 + H^3(K_2 + W/B) + H^2[K_2W/B - W - c(K_2 + W/B)] - H[2cK_2W/B + W(K_2 + W/B)] - K_2W^2/B = 0. \qquad (77)$$

This is identical with the equation for two monobasic acids at equal concentrations ($= c$), one with K_2 and one with W/B as ionization constants. This type of salt solution is always acid. The approximate solution of this equation must therefore be the same as that of Eq. xi(8) for two weak acids with the substitutions $a_1 = a_2 = c$ and K_2 and W/B for the constants A_1 and A_2. For appreciable values of c, the formula for H is that from the H^4 sequence then, or

$$H \cong -(K_2 + W/B)/2 + \sqrt{(\quad)^2 + c(K_2 + W/B) + W - K_2W/B}, \qquad (78)$$

$$\cong \sqrt{c(K_2 + W/B)}. \qquad (79)$$

b. Normal salt, type "$(NH_4)_2SO_4$"; with $a = c$ and $b = 2c$ in Eq. (76),

$$H^4B + H^3[B(c + K_2) + W] + H^2W(K_2 - B - c) - HW[K_2(2c + B) + W] - K_2W^2 = 0. \quad (80)$$

Although this too is only another special equivalent point in the problem of two weak acids, namely when $a_1 = 2c$, $a_2 = c$ and $(b_s - a_s) = c$, there will be greater practical significance in treating it as a case of acid and base, in terms of K_2 and B directly. A useful expression for H is obtained from the H^3 sequence since such a salt solution is usually neither very acid nor very alkaline, and the first and last terms of the equation will usually be the smallest. Because of signs the H^2 sequence gives no real answer unless $(c + B) < K_2$, and the H^4 sequence can be important only if the solution is very acid (which however would require B to be very small generally). Hence

$$H \cong \frac{W(c + B - K_2)}{2B(c + K_2) + 2W} + \sqrt{(\quad)^2 + W\left[\frac{K_2(2c + B) + W}{B(c + K_2) + W}\right]} = Q^3_{80}. \quad (81)$$

If c is large enough compared to B and K_2, H becomes independent of the concentration, or as in Eq. IV(9),

$$H \cong W/2B + \sqrt{(\quad)^2 + 2K_2W/B}. \quad (82)$$

Hence $H \cong \sqrt{2K_2W/B}$ if $K_2B \gg W$ and approximately equals W/B if $K_2B \ll W$. Eq. (82), which is H_* or H_{ie}, is obtainable from Eq. (76) by setting $H - OH = 0$, as explained in Chapter IV. Hence it gives the condition for neutrality of such a salt solution, which is, by setting $H = \sqrt{W}$ in Eq. (82), $B_n = \sqrt{W} + 2K_2$. If B is greater than (or less than) B_n, the solution is alkaline (or acid). Obviously Eq. (82) as a formula for H depends on having $B \cong B_n$ and c much greater than both B and K_2.

The application of Eq. (81) is illustrated in Table 13–1.

Table 13–1. Value of $[H^+]$ in solution of normal weak base salt of "strong" dibasic acid.

Constants K_2	Constants B	Molar concn. c	H Eq. (81)	H Exact, Eq. (80)
10^{-2}	10^{-5}	1	4.45×10^{-6}	4.45×10^{-6}
"$(NH_4)_2SO_4$"		10^{-2}	3.16×10^{-6}	3.16×10^{-6}
		10^{-4}	4.56×10^{-7}	4.56×10^{-7}
10^{-2}	10^{-10}	1	1.457×10^{-3}	1.457×10^{-3}
"Aniline sulfate"		10^{-2}	9.98×10^{-4}	9.75×10^{-4}
		10^{-4}	9.97×10^{-5}	9.93×10^{-5}

If there is some strong monobasic acid also present at the start of this titration, the successive equivalent points consist of the mixtures (typically): "$NH_4Cl + NH_4HSO_4$" when $b = a_s + a$ and "$NH_4Cl + (NH_4)_2SO_4$" when $b = a_s + 2a$. The first of these is mathematically a mixture of two weak monobasic acids, a_1 ($= a_s + a$) with W/B as ionization constant and a_2 ($= a$) with K_2. Hence

$$H_e \cong \sqrt{(a_s + a)W/B + aK_2}. \tag{83}$$

The second is similar to Eq. (48) with $a_s + a$ in place of a_s and with $b = a_s + 2a$, or:

$$H_e \cong W(a_s + a)/2aB + \sqrt{(\quad)^2 + K_2W(a_s + 2a)/aB}, \tag{84}$$

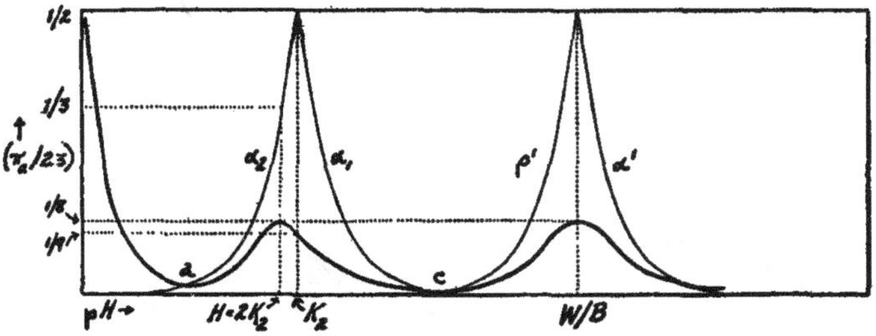

Fig. 13–10. Buffer capacity for titration with dibasic acid strong in K_1 of weak base at $b = 1$; case of $BK_2 \gg W$.

which may be obtained from Eq. (76) after adding a term in a_s, setting $b = (a_s + 2a)$ and assuming $H - OH = 0$.

2. Titration of the Base. The more important titration is that in the opposite direction: titration of a weak base such as NH_3 with a dibasic acid like sulfuric acid. For pure weak base and "strong" dibasic acid, Eq. (76) gives

$$\pi_a = |da/dpH| = [2.3/(1 + \alpha_2)^2][b\alpha'(\rho' + \rho'\alpha_2 + \alpha_1\alpha_2) + f_{62}(H,OH)], \tag{85}$$

$$d\pi_a/dpH = [(2.3)^2/(1 + \alpha_2)^3]\{\{b\alpha'\{\rho'(\alpha' - \rho') + \alpha_2[\alpha_1(\alpha_1 - 2\alpha_2) + 2\rho'(\alpha' - \rho' - \alpha_1)] + \alpha_2^2\rho'(\alpha' - \rho' - 2\alpha_1)\} + f_{63}(H,OH)\}\}. \tag{86}$$

In Eq. (85), $f_{62}(H,OH)$ means that terms in H and OH occur identical with those in Eq. (62); those in Eq. (86) are identical with those in Eq. (63). The relations are shown graphically in the usual way in Fig. 13–10 for $BK_2 > W$ and in Fig. 13–11 for $BK_2 < W$; $\pi_a/2.3b$ is plotted against pH. The curve starts at $OH \cong \sqrt{bB}$ for the solution of the pure weak base; in Fig. 13–11 it may start on either side of $H = K_2$.

With $BK_2 > W$ there are in general two inflections in the titration curve: the first at minimum c, near $a = b/2$, for a solution of a salt of the type "$(NH_4)_2SO_4$"; the second at minimum a in acid solution near $a = b$ for the salt of type "NH_4HSO_4." The π_a curve thus passes through two maxima in the course of titration: the first at $H \cong W/B$ and the second at $H \cong 2K_2$, with $\pi_a/2.3b \cong 1/8$ at both maxima; when $H = K_2$, $\pi_a/2.3b \cong 1/9$. The values of H_e at the two equivalent points involved are given by Eqs. (78) and (81).

Eq. (85) shows that the π_a curve is the sum of two functions. The function $f_{62}(H,OH)$ has already been discussed. Except at very high H or OH it is generally negligible compared to the contribution of the term in b of

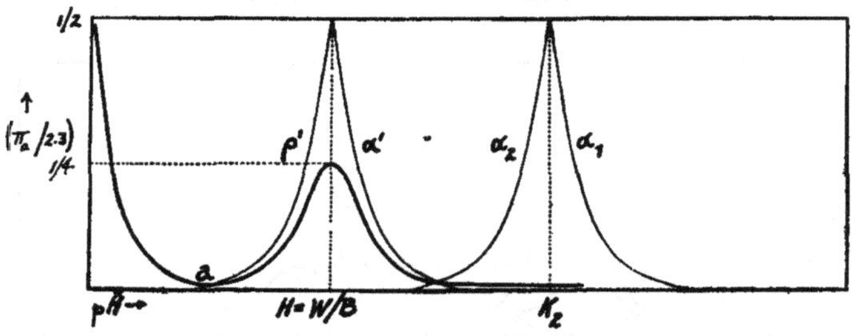

Fig. 13–11. The case of Fig. 13–10 with $BK_2 \ll W$.

Eq. (85). The shape of this weak base function depends on the value of BK_2 relative to W. With $BK_2 \gg W$ it has two maxima and a minimum; with $BK_2 < W$ it has a single maximum and no minima. This may be seen from its derivative, the b term of Eq. (86), which when set equal to zero gives

$$H^4B(BK_2 + W) - H^3[BK_2(2BK_2 - 5W) + W^2] - 6H^2K_2W^2 + 2HK_2^2W(2BK_2 - 7W) - 4K_2^3W^2 = 0. \quad (87)$$

With $BK_2 \ll W$ this becomes

$$H^4B/W - H^3 - 6H^2K_2 - 14HK_2^2 - 4K_2^3 \cong 0, \quad (88)$$

with only one positive root, at $H \cong W/B$, representing the single maximum, as shown in Fig. 13–11. With $BK_2 \gg W$, however, Eq. (87) has three positive roots. As BK_2 decreases toward the value W therefore, the two maxima of Fig. 13–10 merge with minimum c to leave a curve of the type of Fig. 13–11. (The critical condition could be investigated by setting both the first and the second derivatives of the b function of Eq. (85) equal to zero.)

With $BK_2 > W$, the first maximum of π_a during the titration occurs when $\alpha_2 \cong 1$ so that α_1 may be assumed negligible compared to the other fractions of the b term in Eq. (86), which then becomes $\alpha' - \rho' \cong 0$. Hence this maximum occurs at $\alpha' \cong \rho' \cong 1/2$, $H \cong W/B$, and $\pi_a \cong 2.3b/8$. For the second maximum, with $\rho' \cong 0$, the result is seen to be $\alpha_1 - 2\alpha_2 \cong 0$, or $\alpha_2 \cong 1/3$, $H \cong 2K_2$, and again $\pi_a \cong 2.3b/8$.

For minimum c we rearrange Eq. (87) to

$$H_i^2 = \frac{2K_2W}{B}\left(\frac{2BK_2 - 7W}{2BK_2 - 5W}\right) + \frac{H^4B^2K_2 - 4K_2^3W^2}{HBK_2(2BK_2 - 5W)} - \frac{HW^2(6K_2 - H^2B/W + H)}{BK_2(2BK_2 - 5W)}. \quad (89)$$

The second and third terms are small correction terms. If H is given the approximate value of $\sqrt{2K_2W/B}$, the second term vanishes, and since $K_2 \gg H$ at this inflection point, the result is

$$H_i \cong \sqrt{\frac{2K_2W}{B}\left(\frac{2BK_2 - 7W}{2BK_2 - 5W}\right)} - \frac{W^2}{B^2K_2}. \quad (90)$$

This is the inflection point for the titration of a weak base like NH_3 with a "strong" dibasic acid like sulfuric acid, to the point $a = b/2$, or to a solution of $(NH_4)_2SO_4$. Since $H_i < H_e$, from Eq. (81) or (82), in general, the end-point occurs early and the titration error is negative.

For minimum a we may simplify Eq. (86) by setting $\alpha' \cong 1$ and $\rho' \cong 0$, neglecting the OH term and neglecting α_2^2 and α_2^3 in the H term, obtaining

$$b\alpha_1\alpha_2(\alpha_1 - 2\alpha_2) - H(1 + 3\alpha_2) \cong 0, \quad (91)$$

or

$$H^3 + 6H^2K_2 - HK_2(b - 9K_2) + 2K_2^2(b + 2K_2) \cong 0. \quad (92)$$

Hence

$$H_i \cong \sqrt{bK_2} - 4K_2, \quad (93)$$

which is to be compared with $H_e \cong \sqrt{bK_2}$ for the equivalent point when $K_2B > W$, or Eq. (79). From Eqs. (79) and (85) moreover, the buffer capacity is $(\pi_a)_e \cong (2.3)\, 2\sqrt{b(K_2 + W/B)}$, and

$$F = (p/800)\sqrt{b/(K_2 + W/B)} \cong (p/800)\sqrt{b/K_2}, \quad (94)$$

in which p refers to percentage of $b/2$, which is the quantity being titrated between minimum c and minimum a, or from "$(NH_4)_2SO_4$" to "NH_4HSO_4." This titration is seen to be very poor if K_2 is large; compare Eqs. (69) and (70).

Finally, if $BK_2 < W$, as in Fig. 13–11, there is only one inflection point in the titration (minimum a) occurring near $a = b$, corresponding to the salt "NH_4HSO_4," with H_e now given from Eq. (79) as $H_e \cong \sqrt{bW/B}$. With $K_2 < W/B$, now the titration is like that of a weak base with monobasic strong acid, π_a passing through the usual maximum value of $\sim 2.3b/4$ near $H = W/B$.

3. Titration of Mixture of Strong and Weak Bases With Dibasic Acid Strong in K_1. The buffer capacity curve is now essentially but not

exactly the sum of Fig. 13–9 (strong base and "strong" dibasic acid) and either Fig. 13–10 or Fig. 13–11, depending on whether $BK_2 \gtrless W$. The relations are shown graphically in Figs. 13–12 and 13–13; the curve now starts at $OH \cong b_s$.

$$H - OH = a(1 + \alpha_2) - b_s - b\alpha'; \tag{95}$$

$$\pi_a = [2.3/(1 + \alpha_2)^2][f_{62}(b_s) + f_{85}(b) + f_{62}(H,OH)]; \tag{96}$$

$$d\pi_a/dpH = [(2.3)^2/(1 + \alpha_2)^3][f_{63}(b_s) + f_{86}(b) + f_{63}(H,OH)]. \tag{97}$$

In both cases the first minimum, minimum b, occurs near the equivalent point $a = b_s/2$ and the third, minimum a, occurs near $a = b_s + b$; the

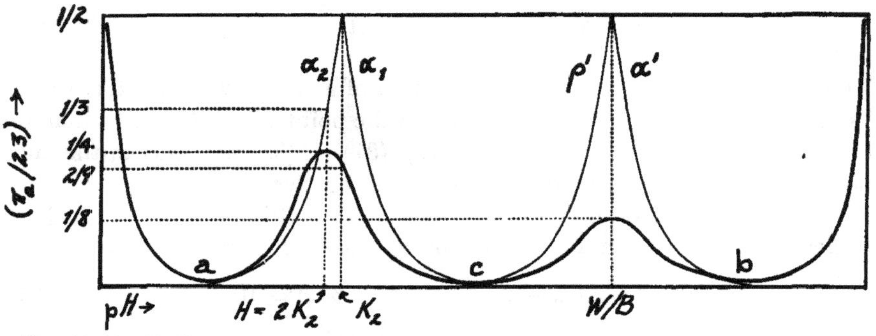

Fig. 13–12. Buffer capacity for titration with dibasic acid strong in K_1 of mixture of strong base and weak base at $b_s = b = 1$; case of $BK_2 \gg W$.

significance of the central minimum, minimum c, however, is not the same for the two cases. Minimum b is significant therefore for the titration of the strong base in presence of the weak base. For the value of OH at this minimum, we may assume $\alpha_2 \cong 1$ and $\alpha_1 \cong 0$ and neglect the H terms in Eq. (97) obtaining

$$b\alpha'\rho'(\alpha' - \rho') + OH \cong 0, \tag{98}$$

or

$$(OH)^3 + 3(OH)^2B - (OH)B(b - 3B) + B^2(b + B) \cong 0. \tag{99}$$

Approximately then,

$$(OH)_i \cong \sqrt{bB} - 2B. \tag{100}$$

In either case there are three minima of π_a. When $BK_2 > W$ there is a maximum at $H = W/B$ with the value $\pi_a/2.3 = 1/8$ for unit values of b_s and b and a second at $H = 2K_2$ with the value of 1/4; at $H = K_2$ the value is 2/9. With $BK_2 < W$ the first maximum $(= 1/8)$ is at $H = 2K_2$ with a value of 1/9 at $H = K_2$; the second $(= 1/4)$ is at $H = W/B$.

In this titration there are several possible equivalent points but since there are only three minima of π_a only three equivalent points will be marked by (vertical) inflection points in the curve of pH against a. The original solution, which we shall exemplify as "NaOH + NH_3," has a value of pH on the right of minimum b in either case, with $OH \cong b_s$. The various equivalent points and their approximate H_e (or OH_e) values are as follows:

(1) $a = b_s/2$: solution of "Na_2SO_4 + NH_3." Mathematically these are two weak bases, with

$$OH_e \cong \sqrt{b_s W/2K_2 + bB}. \qquad (101)$$

(2) $a = (b_s + b)/2$: "Na_2SO_4 + $(NH_4)_2SO_4$." Assuming $H - OH \cong 0$ in Eq. (95), we have

$$H_e \cong - W(b_s - b)/2B(b_s + b) + \sqrt{(\quad)^2 + (WK_2/B)2b/(b_s + b)}. \qquad (102)$$

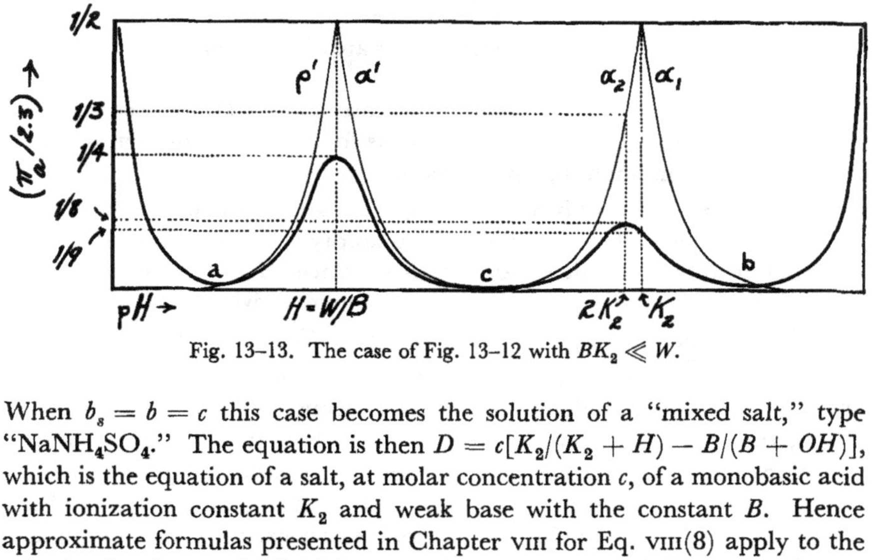

Fig. 13–13. The case of Fig. 13–12 with $BK_2 \ll W$.

When $b_s = b = c$ this case becomes the solution of a "mixed salt," type "$NaNH_4SO_4$." The equation is then $D = c[K_2/(K_2 + H) - B/(B + OH)]$, which is the equation of a salt, at molar concentration c, of a monobasic acid with ionization constant K_2 and weak base with the constant B. Hence approximate formulas presented in Chapter VIII for Eq. VIII(8) apply to the mixed salt of type $NaNH_4SO_4$ with the substitution of K_2 for A. In the problem of two acids this is again simply the special equivalent point in which $a_1 = a_2 = b_s$.

(3) $a = b_s$: "$NaHSO_4$ + NH_3." Again setting $H - OH = 0$ in Eq. (95) we have

$$H_e \cong K_2(b_s - b)/2b + \sqrt{(\quad)^2 + K_2 W b_s/bB}. \qquad (103)$$

This corresponds mathematically to a mixture of weak monobasic acid (K_2) at concentration b_s and weak base (B) at concentration b, for which the various approximations of Section B of this chapter may be used; thus Eq. (103) is a simplification of Eq. (19). Provided $BK_2 \gg W$, therefore, we may use the Henderson type of equation so that with $b_s > b$, $H_e \cong K_2(b_s - b)/b$, while for $b_s < b$, $OH_e \cong B(b - b_s)/b_s$. The preceding equivalent point, with $a = (b_s + b)/2$, or Eq. (102), is a special case of Eq. (103), the two being identical when $b_s = b$.

(4) $a = b_s + b/2$: "$NaHSO_4 + (NH_4)_2SO_4$."

$$H_e \cong (2b_sBK_2 + bW)/2bB + \sqrt{(\quad)^2 + 2K_2W(b_s + b)/bB}. \quad (104)$$

(5) $a = b + b_s/2$: "$Na_2SO_4 + NH_4HSO_4$."

$$H_e \cong [2bBK_2 - W(b_s - 2b)]/2b_sB + \sqrt{(\quad)^2 + 4bK_2W/b_sB}. \quad (105)$$

(6) $a = b_s + b$: "$NaHSO_4 + NH_4HSO_4$." This is mathematically a solution of two monobasic acids, acid K_2 at concentration $(b_s + b)$, and acid W/B at concentration b, as in Eq. (79) for "NH_4HSO_4"; hence

$$H_e \cong \sqrt{K_2(b_s + b) + bW/B}. \quad (106)$$

The equivalent points $a = b/2$ and $a = b$ are of no importance in presence of strong base.

With reference to Fig. 13–12 for the case $BK_2 > W$, the sequence of phenomena during titration will then be as follows. The solution starts at $OH \cong b_s$. At minimum b, an inflection occurs marking $a = b_s/2$, with $OH_e \cong \sqrt{bB}$, $\pi_e \cong 2.3\sqrt{bB}$, and $F = pb_s/400\sqrt{bB}$, for the determination of $b_s/2$; i.e. p is a percentage relative to the quantity $b_s/2$. A horizontal section of the pH titration curve then has its mid-point near $H = W/B$ and $a = b_s/2 + b/4$. A second inflection occurs at minimum c, when $a = (b_s + b)/2$, with $H_e \cong \sqrt{2bK_2W/B(b_s + b)}$; $\pi_e \cong (2.3)\sqrt{b(b_s + b)W/2BK_2}$, and $F = (p/200)\sqrt{bBK_2/2(b_s + b)W}$, with p referring to the quantity $b/2$, titrated between minimum b and minimum c. A second horizontal section now follows, traversing the equivalent points Nos. (4) and (5) as listed above, with $H_e \cong 2K_2b_s/b$ and $2K_2b/b_s$ respectively, their order simply depending on the relative magnitudes of b and b_s; both are near the maximum of π_a at $H \cong 2K_2$. Equivalent point 3 is traversed either before or after minimum c, depending on the relative magnitudes of b and b_s. Finally an inflection occurs at minimum a, when $a = b_s + b$ and $H_e \cong \sqrt{K_2(b_s + b)}$, with $\pi_e \cong (2.3)2\sqrt{K_2(b_s + b)}$ and $F = (p/800)\sqrt{(b_s + b)/K_2}$, p referring to $(b_s + b)/2$, titrated between minimum c and minimum a.

With $BK_2 < W$ (Fig. 13–13), the first inflection is again at minimum b, when $a = b_s/2$, but with $(OH)_e \cong \sqrt{b_sW/2K_2}$, $\pi_e \cong 2.3\sqrt{b_sW/2K_2}$ and $F = (p/400)\sqrt{2b_sK_2/W}$, p referring to $b_s/2$. Then π_a passes through the maximum at $H \cong 2K_2$, and the next inflection, at minimum c, marks the equivalent point $a = b_s$, with $H_e \cong \sqrt{b_sK_2W/bB}$, $\pi_e \cong (2.3)2\sqrt{b_sbBK_2/W}$, and $F = (p/800)\sqrt{b_sW/bBK_2}$, p again referring to $b_s/2$. The equivalent point $a = b_s + b/2$, with $H_e \cong W/B$, is traversed near the maximum of π_a at W/B, and the last inflection, $a = b_s + b$, occurs at minimum a, when $H_e \cong \sqrt{bW/B}$, $\pi_e \cong (2.3)2\sqrt{bW/B}$, and $F = (p/400)\sqrt{bB/W}$, p referring to the quantity b.

XIV

Problems Involving Two Interdependent Ionization Constants

(Dibasic Acid; Its Salts with Strong and Weak Bases; Ampholytes)

This chapter deals with equations involving two related ionization constants or two constants pertaining to a single solute.

A. Dibasic Acid and Strong Base (Formulas for H)

1. Fundamental Equation. The fundamental equation for this section is that of a solution of a general dibasic acid ("weak" in respect to both constants) together with *net* strong acid or strong base ($g = a_s - b_s$):

$$H - OH = a(\alpha_1 + 2\alpha_2) + g. \tag{1}$$

If g is now simply $- b_s$,

$$H - OH = a(1 + 2K_2/H)/(1 + H/K_1 + K_2/H) - b_s, \tag{2}$$

$$H^4 + H^3(b_s + K_1) + H^2[K_{12} - K_1(a - b_s) - W] - HK_1[K_2(2a - b_s) + W] - K_{12}W = 0. \tag{3}$$

It is to be noted that Eq. (3) is identical with that, namely Eq. XI(3), for two weak acids at equal concentrations, if A_2 is neglected in the sum $(A_1 + A_2)$. The approximate formulas derivable from Eq. (3) may therefore be written from those in Chapter XI with the condition $a_1 = a_2 = a$ and with neglect of A_2 relative to A_1. To avoid confusion, however, the important approximations for the present case will be taken directly from Eq. (3).

Assuming in general that $K_1 > K_2$, we note, using Eq. (2), the following values of the titration ratio n $(= b_s/a)$ for special values of H; for $H = \sqrt{W}$, see Eq. VII(53). Writing y for K_2/K_1, we have, for $H = K_1$,

$$n = (1 + 2y)/(2 + y) - (K_1 - W/K_1)/a \cong 1/2. \tag{4}$$

For $H = \sqrt{K_1K_2}$,

$$n = 1 - (\sqrt{K_1K_2} - W/\sqrt{K_1K_2})/a, \tag{5}$$

or $n = 1$ if $K_1K_2 = W$. We also note here for future reference that when $H = \sqrt{K_1K_2}$, the charge coefficient of a dibasic acid, or β, equals 1; hence $\alpha_1 + 2\alpha_2 = 1$ at this point.

For $H = K_2$,

$$n = 3/(2 + y) - (K_2 - W/K_2)/a \cong 3/2. \quad (6)$$

2. The Pure Acid ($b_s = 0$). The following are values of a for some special values of H:

For $H = K_2$,

$$a = (K_2 - W/K_2)(2 + y)/3; \quad (7)$$

impossible if $K_2 < \sqrt{W}$; otherwise $a \cong 2K_2/3$.

For $H = \sqrt{K_1K_2}$,

$$a = (\sqrt{K_1K_2} - W/\sqrt{K_1K_2}); \quad (8)$$

impossible if $K_1K_2 < W$; otherwise $a \cong \sqrt{K_1K_2}$.

For $H = K_1$,

$$a = (K_1 - W/K_1)(2 + y)/(1 + 2y); \quad (9)$$

impossible if $K_1 < \sqrt{W}$; otherwise $a \cong 2K_1$.

For approximate solutions of the equation for the pure acid, or

$$H^4 + H^3K_1 - H^2(aK_1 - K_{12} + W) - HK_1(2aK_2 + W) - K_{12}W = 0, \quad (10)$$

reference is made to the discussion in Chapter VIII, Section E, where it was pointed out that the principle of using a sequence of three terms for an approximate solution fails in this case if $K_1 \not\gg K_2$. Given $K_1 > K_2$ however, we find that the H^2 sequence is not important, requiring $a < K_2$, while of the other two the H^3 sequence is rarely necessary or applicable as long as $a \gg K_2$. The H^4 sequence is better as long as $H > \sqrt[3]{2aK_1K_2}$ and hence as long as $a \gg \sqrt{K_1K_2}$, as may be verified by consideration of Eqs. (2) and (8). Hence unless a is very small,

$$H \cong -K_1/2 + \sqrt{K_1^2/4 + K_1(a - K_2) + W} = Q_{10}^4. \quad (11)$$

When a is very large, this approaches $H_* = \sqrt{aK_1}$. But when a is very small, then from the H^3 sequence,

$$H \cong (a - K_2 + W/K_1)/2 + \sqrt{(\quad)^2 + 2aK_2 + W} = Q_{10}^3. \quad (12)$$

Finally, if $K_1 < \sqrt{W}$, the largest terms of Eq. (10), for any concentration, are the first and third, giving $H \cong \sqrt{aK_1 + W}$.

3. The First Equivalent Point ($b_s = a$). For the acid salt, type $NaHSO_3$, we have, from Eq. (3), with $b_s = a = c$, the molarity of the salt,

$$H^4 + H^3(c + K_1) + H^2(K_{12} - W) - HK_1(cK_2 + W) - K_{12}W = 0. \quad (13)$$

By transformation of constants ("A" $= K_2$ and $B^* = W/K_1$), this becomes

$$H^4B^* + H^3(cB^* + W) + H^2W(``A" - B^*) - HW(c``A" + W) - ``A"W^2 = 0. \quad (14)$$

Except for two minor terms in AB, this is the same as Eq. VIII(8), for the general 1 : 1 salt of weak acid and weak base. The approximations or simplifications of this equation then will not depend on whether "A"$B^* \gtrless W$ in Eq. (14) and hence on whether $K_1 \gtrless K_2$ in Eq. (13); the transformation involved makes the conditions $K_1 = K_2$ and "A"$B^* = W$ identical. Whereas therefore the general equation for dibasic acid and strong base, Eq. (3), "behaves" according to the expectations of the method of Chapter VIII, only if $K_1 > K_2$, the equation for the pure acid salt (and also later that for the normal salt) can be treated by that method whatever the relation between K_1 and K_2. Furthermore, since the quantities involving the product AB were hardly of any importance in determining the applicability of the approximate solutions of Eq. VIII(8), we shall simply reword those conditions here in terms of the constants of the dibasic acid for approximate solutions of Eq. (13). The H^4 sequence applies only if $c < \sqrt{K_1K_2}$ *and* $K_1K_2 < W$ (which compares to $A < B$). The H^2 sequence requires $c < W/\sqrt{K_1K_2}$ *and* $K_1K_2 > W$. Hence the H^3 is the approximation for general use, or

$$H \cong (W - K_1K_2)/2(c + K_1) + \sqrt{(\quad)^2 + (cK_1K_2 + K_1W)/(c + K_1)} = Q^3_{13}. \quad (15)$$

Now when $c \gg \sqrt{K_1K_2}$,

$$H \cong \sqrt{K_1K_2(c + W/K_2)/(c + K_1)}. \quad (16)$$

Hence if the solution is acid, or if $K_1K_2 > W$, further simplification means

$$H \cong \sqrt{cK_1K_2/(c + K_1)}, \quad (17)$$

while if it is alkaline, with $K_1K_2 < W$, then

$$H \cong \sqrt{K_1(K_2 + W/c)}. \quad (18)$$

Finally, in either case, for sufficiently high concentration, we have

$$H \cong \sqrt{K_1K_2}. \quad (19)$$

It is to be noted that if $K_1K_2 < W$, Eq. (17) is not better than Eq. (19) and may even be poorer, while if $K_1K_2 > W$, Eq. (17) is better than Eq. (19) but Eq. (18) is not better and may even be poorer than Eq. (19).

These formulas (15–19) are all approximations, valid only if the salt concentration is not too small. Eqs. (17–19) give absurd answers as $c \rightleftarrows 0$; and although Eq. (16) does reduce to $H = \sqrt{W}$ when $c = 0$, it begins to give wrong results when c is no longer greater than $\sqrt{K_1K_2}$. The simple formula, Eq. (19), $H = \sqrt{K_1K_2}$, is the value of H approached as c increases indefinitely and corresponds to H_* or H_{ie} (the iso-electric point), as discussed in Chapter IV.

4. THE SECOND EQUIVALENT POINT ($b_s = 2a$). For the normal salt, type Na_2SO_3, we have, with $a = c$ and $b_s = 2c$ in Eq. (3),

$$H^4 + H^3(2c + K_1) + H^2(cK_1 + K_{12} - W) - HK_1W - K_{12}W = 0. \quad (20)$$

Because of signs, the H^4 sequence is of no importance and the H^3 sequence is better than the H^2 only for very small values of c. For appreciable concentration then, the H^2 sequence gives the best approximation:

$$H \cong W/(2c + K_2 - W/K_1) + \sqrt{(\quad)^2 + K_2W/(c + K_2 - W/K_1)} = Q^2_{20}. \quad (21)$$

For high K_2, $H \cong \sqrt{K_2W/(c + K_2)}$, and for low K_2, $H \cong W/2c + \sqrt{(\quad)^2 + K_2W/c}$; in either case, for sufficiently high values of c, $H \cong \sqrt{K_2W/c}$.

5. TITRATION CURVE; EQ. (3). For appreciable values of the concentration a, the H^4 sequence applies when $b_s = 0$ (pure dibasic acid), the H^3 for the first equivalent point (salt NaHX) and the H^2 sequence for the second equivalent point (salt Na_2X). Hence, according to the principle explained in Chapter VIII, the very start of the titration of dibasic acid with strong base is best described by the H^4 sequence; or, with $a \gg b_s$,

$$H \cong -(b_s + K_1)/2 + \sqrt{(\quad)^2 + K_1(a - b_s) - K_{12} + W} = Q^4_3, \quad (22)$$

$$\cong -(b_s + K_1)/2 + \sqrt{(\quad)^2 + K_1(a - b_s)}. \quad (23)$$

In the middle range of the titration of the first equivalent, the H^4 and H terms are both small compared to those in H^3 and H^2; hence here

$$H \cong K_1(a - b_s - K_2 + W/K_1)/(b_s + K_1) \cong K_1(a - b_s)/b_s, \quad (24)$$

or the Henderson equation.

Through the first equivalent point ($b_s \cong a$), the H^3 sequence gives

$$H \cong \frac{K_1(a - b_s - K_2 + W/K_1)}{2(b_s + K_1)} + \sqrt{(\quad)^2 + \frac{K_1[K_2(2a - b_s) + W]}{b_s + K_1}} = Q^3_3, \quad (25)$$

$$\cong K_1(a - b_s)/2b_s + \sqrt{(\quad)^2 + K_1K_2(2a - b_s)/b_s}. \quad (26)$$

In the middle range of the second half of the titration ($a < b_s < 2a$), the largest terms are those in H^2 and H; hence

$$H \cong K_2(2a - b_s + W/K_2)/(b_s - a + K_2 - W/K_1)$$
$$\cong K_2(2a - b_s)/(b_s - a), \quad (27)$$

again the Henderson equation.

Through the second equivalent point ($b_s \cong 2a$), the H^2 sequence gives

$$H \cong \frac{K_2(2a - b_s) + W}{2(b_s - a + K_2 - W/K_1)} + \sqrt{(\quad)^2 + \frac{K_2 W}{b_s - a + K_2 - W/K_1}} = Q_3^2; \qquad (28)$$

or, with $aK_2 \gg W$,

$$H \cong K_2(2a - b_s)/2(b_s - a) + \sqrt{(\quad)^2 + K_2 W/(b_s - a)}. \qquad (29)$$

Beyond this equivalent point, with appreciable excess of b_s, the last two terms of Eq. (3) are the largest, and $H \cong W/(b_s - 2a - W/K_2) \cong W/(b_s - 2a)$; or merely $OH \cong b_s - 2a$.

Similarly then, if the solution contains strong acid plus dibasic acid, the first two terms of Eq. (3) with $- a_s$ in place of b_s, give $H \cong a_s - K_1 \cong a_s$, if a_s is large compared to both a and K_1. Otherwise it is better to use more of the H^4 sequence, or

$$H \cong (a_s - K_1)/2 + \sqrt{(\quad)^2 + (a_s + a)K_1}. \qquad (30)$$

Eqs. (25–29) may be rearranged for use in calculating H in buffers of the type $NaHX + Na_2X$, at concentrations c_1 and c_2 respectively. If the ratio c_1/c_2 is high, Eq. (25) is used:

$$H \cong -\frac{c_2K_1 + K_{12} - W}{2(c_1 + 2c_2 + K_1)} + \sqrt{(\quad)^2 + \frac{K_1(c_1K_2 + W)}{c_1 + 2c_2 + K_1}}, \qquad (31)$$

$$\cong -c_2K_1/2(c_1 + 2c_2) + \sqrt{(\quad)^2 + c_1K_1K_2/(c_1 + 2c_2)}. \qquad (32)$$

When $c_1/c_2 \cong 1$, Eq. (27) is used:

$$H \cong (c_1K_2 + W)/(c_2 + K_2 - W/K_1) \cong K_2c_1/c_2. \qquad (33)$$

When c_1/c_2 is small, Eq. (28) gives

$$H \cong \frac{(c_1K_2 + W)/2}{c_2 + K_2 - W/K_1} + \sqrt{(\quad)^2 + \frac{K_2 W}{c_2 + K_2 - W/K_1}}, \qquad (34)$$

$$\cong c_1K_2/2c_2 + \sqrt{(\quad)^2 + K_2W/c_2}. \qquad (35)$$

6. Titration of Diacid Base With Strong Acid. For diacid base with ionization constants $K_1' > K_2'$, titrated with strong acid, Eq. (3) becomes

$$(OH)^4 + (OH)^3(a_s + K_1') + (OH)^2[K_{12}' - K_1'(b - a_s) - W]$$
$$- (OH)K_1'[K_2'(2b - a_s) + W] - K_{12}'W = 0. \qquad (36)$$

All the foregoing considerations may now be applied, by analogy, by merely substituting OH for H, K_1' and K_2' for K_1 and K_2, b for the concentration of the diacid base in place of a, and a_s in place of b_s. In the use of Eq. (36) the various sequences of terms have the same application as in Eq. (3): the $(OH)^4$ sequence for he start of the titration, the $(OH)^3$ for the middle (basic salt), and the $(OH)^2$ sequence for the end (normal salt). For the basic salt then (MOHCl), $OH \cong \sqrt{K_1'K_2'}$, and for the normal salt (MCl_2), $OH \cong \sqrt{WK_2'/c}$. In terms of H and basic constants, Eq. (36) becomes

$$H^4K_{12}' + H^3K_1'[K_2'(2b - a_s) + W] - H^2W[K_{12}' - K_1'(b - a_s) - W] - HW^2(a_s + K_1') - W^3 = 0. \quad (37)$$

Now the application of the sequences is reversed, the H^4 sequence applying for the second equivalent point ($a_s = 2b$) and the H^2 sequence at the start ($a_s = 0$).

7. Titration of Alkali Salts of Dibasic Acid With Strong Acid, For the titration with strong acid (a_s), of the alkali normal salt of a dibasic acid, at concentration c, we may treat the salt as a diacid base, with base constants $K_1'^* = W/K_2$ and $K_2'^* = W/K_1$ and thereupon use Eq. (36), giving b the value c. Otherwise we may use Eq. (3) itself, with c in place of a and $(2c - a_s)$ in place of b_s. This follows from the fact that this titration is the addition of strong acid (a_s) to a solution in which $b_s = 2a = 2c$ for the normal salt. Hence Eq. (3) becomes

$$H^4 + H^3(2c - a_s + K_1) + H^2[K_{12} + K_1(c - a_s) - W] - HK_1[K_2a_s + W] - K_{12}W = 0. \quad (38)$$

But again, as in Eq. (37), the start of the titration ($a_s = 0$) is given by the H^2 sequence and the end ($a_s = 2c$) by the H^4 sequence.

B. Buffer Capacity and Feasibility of Titration

For dibasic acid plus strong base, Eq. (1) gives

$$db_s/dpH = \pi = 2.3[a(\alpha_1\rho + \alpha_1\alpha_2 + 4\alpha_2\rho) + H + OH], \quad (39)$$

$$\pi = 2.3[a(H/K_1 + K_2/H + 4K_2/K_1)/(1 + H/K_1 + K_2/H)^2 + H + OH], \quad (40)$$

$$d\pi/dpH = (2.3)^2[a(\alpha_2 - \rho)(\alpha_1^2 - \alpha_1\alpha_2 - \alpha_1\rho - 8\alpha_2\rho) - H + OH]. \quad (41)$$

With $K_1 \gg K_2$, the buffer capacity is practically that of a solution of two independent acids at equal concentration, so that the expression of Eq. (39) will exhibit two maxima and three minima, as shown in Fig. 14–1, in which $(\pi/2.3)$ is plotted for unit concentration of dibasic acid. The three minima correspond to three inflections (of the vertical or end-point type) in the titration curve of pH against b_s or a_s. In titration with b_s (left to right) the central minimum represents the inflection point for the first equivalent point ($b_s = a$) in titration of the pure dibasic acid, and the alkaline minimum, minimum b, the inflection for the second equivalent point ($b_s = 2a$). The acid minimum, minimum a, represents the inflection for the equivalent point for titration of strong acid in presence of the dibasic acid. In reverse, for titration with a_s, minimum b gives the inflection for titration of strong base in presence of the alkali normal salt of the dibasic acid, minimum c the inflection for titration of normal salt to acid salt, and minimum a the inflection for the complete titration of the normal salt, or when $a_s = 2c$.

1. MAXIMA AND CENTRAL MINIMUM OF π. For this region of the π curve we may neglect both H and OH in Eq. (41) and note that then $d\pi/dpH = 0$ either if $\alpha_2 - \rho \cong 0$ or if $\alpha_1^2 - \alpha_1(\alpha_2 + \rho) - 8\alpha_2\rho \cong 0$. The first condition, $\alpha_2 \cong \rho$, gives $H \cong \sqrt{K_1K_2}$ for the central minimum. The second condition must therefore give the two maxima. Since $\alpha_2 = \alpha_1 K_2/H$, $\rho = \alpha_1 H/K_1$, and $\alpha_1 + \alpha_2 + \rho = 1$, this condition becomes $\alpha_1^2 - \alpha_1(1 - \alpha_1) - 8\alpha_1^2 y \cong 0$, or $\alpha_1 \cong 1/(2 - 8y)$. Since $\alpha_1 = 1/(1 + H/K_1 + K_2/H)$, this determines two values of H for the two maxima of π:

$$H_{\pi_{\max}} \cong (K_1 - 8K_2)/2 \pm \sqrt{(\quad)^2 - K_1K_2}. \tag{42}$$

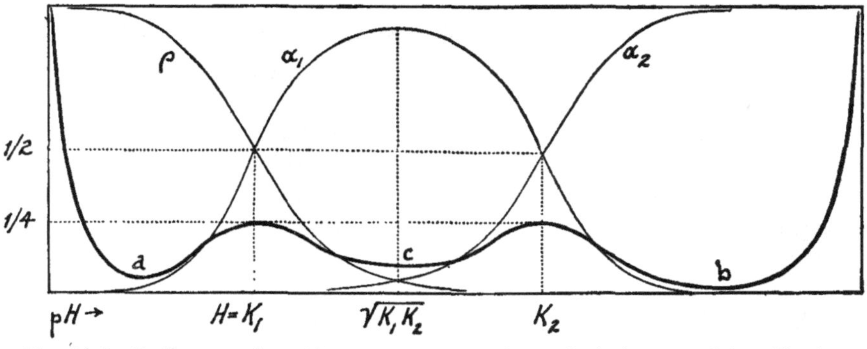

Fig. 14–1. Buffer capacity with respect to strong base of solution containing dibasic acid at unit concentration; case of $K_1 \gg K_2$.

If $K_1 \gg K_2$, these values are approximately $K_1(1 - 8y)$ and $K_2/(1 - 8y)$. The corresponding values of $\pi_{\max}$ may then be obtained from Eq. (40) with these values of H. These two maxima are very nearly equal; they appear to be equal only if H and OH are neglected in the expression for π; they would be exactly equal if $K_1K_2 = W$, for then the complete π curve of Fig. 14–1 would be symmetrical, with its center at $H = \sqrt{W}$. If it is assumed that $\pi_{\max} = \pi_{K_1} = \pi_{K_2}$, these maximum values are (approximately):

$$\pi_{\max} \cong 2.3a(1 + 4y)/4, \tag{43}$$

or $\pi_{\max} \cong 2.3a/4$ for small y.

Thus, in general, provided always that $K_1 \gg K_2$, the buffer capacity at the mid-point of each half of the titration of the dibasic acid, or near $H = K_1$ or $H = K_2$, is very close to a maximum value. The buffer capacity at equivalent points (salt solutions) on the other hand is very close to some minimum of π. At minimum c, the neglect of H and OH in Eq. (40) gives

$$\pi_{\min} \text{ or } \pi_{H = \sqrt{K_1K_2}} \cong (2.3)2a\sqrt{y}/(1 + 2\sqrt{y}), \tag{44}$$

or, for small y, $\pi_{\min} \cong (2.3)2a\sqrt{K_2/K_1}$.

2. Feasibility of Titration. As already mentioned, the titration curve of the dibasic acid is identical with that of two weak acids at equal concentrations, with neglect of A_2 in the sum $(A_1 + A_2)$. With $a_1 = a_2 = a$ therefore, we may use Eqs. IX(37) and IX(38) for the buffer capacity at minimum c of Fig. 14–1 and Eqs. IX(32–35) for the feasibility function for the titration to the first equivalent point since these expressions are based on the neglect of A_2 in $(A_1 + A_2)$. The final approximate feasibility function therefore becomes, from Eq. IX(35),

$$F = (p/400)\sqrt{K_1/K_2}, \tag{45}$$

so that $\Delta_p \cong (2/2.3) \sinh^{-1} (p\sqrt{K_1/K_2}/400)$. With the usual understanding about γ as the activity coefficient of a univalent ion, $F = (p\gamma/400)\sqrt{\mathbf{K}_1/\mathbf{K}_2}$, with $-\log \gamma \cong 0.505\sqrt{a}$, if $K_1 \gg K_2$, from Table 5–1. More accurately, μ is expressed as in Eqs. V(50) and V(51a), with the α's fixed by $H = \sqrt{K_1 K_2}$.

The slight effect of the concentration itself, which is ignored in Eq. (45), would be brought out by the use of the formulas of Eqs. IX(32–34), with K_1 and K_2 in place of A_1 and A_2 respectively, with $a_1 = a_2 = a$, and with $x = pa/200$. The most accurate estimate would come from the H^3 sequence of Eq. (3) used as in Eq. IX(31) for the case of two acids.

If we neglect ρ at minimum b and α_2 at minimum a, the approximate relations at these lateral minima are given by the considerations in Chapter X. For the total titration of the dibasic acid to its second equivalent point, or to the inflection corresponding to minimum b, the relative error is $-3W/2aK_2$, from Eq. X(88), and the feasibility function is $(p/200)\sqrt{aK_2/W}$, from Eq. X(87). From Eq. X(74) the error in the titration of strong acid in presence of the dibasic acid, with an inflection at minimum a, is given by the relation $(a_s)_i \cong (b_s)_i - 3K_1$, and from Eq. X(67) the feasibility function is $pa_s/400\sqrt{aK_1}$.

3. Reverse Titration With Strong Acid of $b_s + a$. If $b_s > 2a$, this is the titration of a solution containing normal alkali salt of the dibasic acid at concentration $c_2 = a$, and excess of strong base at concentration $b_s - 2a$. There will be three inflection points: the first at $a_s = b_s - 2a$, when the solution is mathematically simply the normal salt, with $H_e \cong \sqrt{WK_2/a}$ and $F = (p/400)(b_s - 2a)\sqrt{K_2/aW}$, for the titration of the excess of strong base; the second at $a_s = b_s - c_2$, with $H_e \cong \sqrt{K_1 K_2}$ and $F = (p/400)\sqrt{K_1/K_2}$, for the titration of the normal salt to the acid salt; the third at $a_s = b_s$, for the complete titration, forming a solution (mathematically) of pure dibasic acid, with $H_e \cong \sqrt{aK_1}$, and $F = (p/400)\sqrt{a/K_1}$ for titration of the acid salt to this point.

If $a < b_s < 2a$, the solution is one of NaHX at concentration c_1 + Na_2X at concentration c_2, so that $b_s = c_1 + 2c_2$ and $a = c_1 + c_2$. Then titration with strong base shows only one inflection, when b_s' (titrating strong base)

$= c_1$. Titration with strong acid gives two inflections, one at $a_s = c_2$ the second at $a_s = c_1 + 2c_2$.

If $b_s < a$, or a solution with $[H_2X] = c_0$ and $[NaHX] = c_1$, with $b_s = c_1$ and $a = c_0 + c_1$, titration with strong base gives two inflections, one at $b'_s = c_0$ and the second at $b'_s = 2c_0 + c_1$; titration with with strong acid gives only one inflection, at $a_s = c_1$.

If $b_s = a = c_1$, the solution, now NaHX, may be titrated with either strong base or strong acid; with strong base $F = (p/400)\sqrt{c_1K_2/W}$, while with strong acid, $F = (p/400)\sqrt{c_1/K_1}$. Hence if $K_1K_2 > W$, the salt, acid in "reaction," is titrated more efficiently with base than with acid.

4. Conditions for the Minima in π (Inflection Points in Titration). The existence of the central minimum in π and hence of an inflection corresponding to the equivalent point $b_s = a$, requires that the ratio y $(= K_2/K_1)$ be small enough. The critical ratio above which the minimum disappears, leaving a single maximum at $H \cong \sqrt{K_1K_2}$, is readily found, simply from the condition for a positive root in Eq. (42), which defines the lateral maxima of the π curve. From $(K_1 - 8K_2)^2/4 - K_1K_2 = 0$, then, with $(K_1 - 8K_2) > 0$, the critical ratio is

$$y = K_2/K_1 \cong 1/16. \tag{46}$$

This relation, Eq. (46), would be exact except for the neglect of the terms H and OH in Eqs. (39)–(41). If $K_1 < 16K_2$ then, there is no inflection in the titration curve at $H = \sqrt{K_1K_2}$ or $b_s = a$, and the dibasic acid *appears* to behave like a monobasic acid, with a single inflection at $b_s = 2a$; but the shape of the curve is not necessarily that of a monobasic acid. Similarly, of course, the titration with strong acid, of the normal salt of a dibasic acid, can show no inflection at the acid salt unless $K_1 > 16K_2$. (For the effect of ionic strength, the condition of Eq. (46) becomes $\mathbf{K}_2/\mathbf{K}_1 = \gamma^2/16$.)

Furthermore, the familiar relations, from Eqs. (4) and (6), that $H \cong K_1$ at $b_s = a/2$ and $H \cong K_2$ at $b_s = 3a/2$ hold only when $K_1 \gg K_2$. When $K_1 = 16K_2$, $H = K_1$ at $b_s = 18a/33 - (K_1 - W/K_1)$, and $H = K_2$ at $b_s = 48a/33 - (K_2 - W/K_2)$. But $H = \sqrt{K_1K_2}$ at $b_s = a - (\sqrt{K_1K_2} - W/\sqrt{K_1K_2})$ whatever the value of y.

Provided $K_1 \gg K_2$, the requirements for the lateral minima of π may be considered to remain approximately the same as for a single weak acid: at minimum a, that $a > 27K_1$ and at minimum b, that $aK_2 > 27W$.

5. The π Curve When K_1 Is Not Large Compared to K_2. Neglecting H and OH in Eq. (39), we shall consider simply the function $\pi' = \pi/2.3a = \alpha_1\rho + \alpha_1\alpha_2 + 4\alpha_2\rho$, which is symmetrical with respect to the point $H = \sqrt{K_1K_2}$ where $\rho = \alpha_2$. As K_1/K_2 decreases, the maxima of π', Fig. 14–1, approach each other and the central minimum rises until they merge when $K_1 = 16K_2$. From this point on, as K_1/K_2 decreases further, π' remains a maximum at $H = \sqrt{K_1K_2}$, whether $K_1 \gtrless K_2$. The function π',

moreover, is identical with α_1, for all values of H when $K_1 = 4K_2$; $\pi' < \alpha_1$ for $K_1 > 4K_2$, and $\pi' > \alpha_1$ for $K_1 < 4K_2$. Also, π' crosses the curve of ρ when $H^3 - 3HK_1K_2 - K_1^2K_2 = 0$, or at $H > \sqrt{K_1K_2}$; at a symmetrically corresponding value of $H < \sqrt{K_1K_2}$, π' crosses the α_2 curve. Finally, $\pi' = \rho + \alpha_2$ at $H = \sqrt{K_1K_2}$ for all relations of K_2 and K_1.

The interesting special cases are $K_1 = 16K_2$, $K_1 = 4K_2$, and $K_1 = K_2$, which are compared schematically in Figs. 14–2,3,4. The inflection for the first equivalent point vanishes at $K = 16K_2$, but the titration curve is identical with that of monobasic acid (with ionization constant $A = K_1/2$ and at concentration $2a$) only for the special relation $K_1 = 4K_2$; if $K_1 \neq 4K_2$ the two curves differ in shape whether or not there is an inflection at $H = \sqrt{K_1K_2}$. As K_1 decreases below $K_1 = 16K_2$, the maximum of π' rises, becoming narrower at the same time, since the area under the curve of Fig. 14–1 is constant for fixed concentration of dibasic acid being titrated from minimum a to minimum b. As K_1 becomes smaller than K_2, as in Fig. 14–5, for $K_1 = K_2/100$, $\pi'_{\max}$ increases toward a limit of 1, since with $\rho = \alpha_2$ at $H = \sqrt{K_1K_2}$ these fractions never exceed the value 1/2. We note, also, that $\rho = \alpha_1$ at $H = K_1$ and $\alpha_1 = \alpha_2$ at $H = K_2$, whether $K_1 \gtrless K_2$.

If $K_1 \not> 16K_2$ the alkali acid salt NaHX may still be titrated in either direction, but the separate titration curves would not show horizontal inflections and would not serve for the separate estimation of the constants, which have to be determined simultaneously from the total curve of titration between the pure acid H_2X and its normal salt Na_2X.

C. Shape of the Titration Curve (Comparison With Curves for Monobasic Acid and for Mixture of Two Monobasic Acids)

We shall here compare the shapes of the titration curves of:
(I) Monobasic acid,

$$H - OH = aA/(A + H) - b_s; \tag{47}$$

(II) Dibasic acid,

$$H - OH = a(1 + 2K_2/H)/(1 + H/K_1 + K_2/H) - b_s; \tag{48}$$

(III) Two independent monobasic acids at equal concentrations, $(A_1 > A_2)$,

$$H - OH = a[A_1/(A_1 + H) + A_2/(A_2 + H)] - b_s. \tag{49}$$

Complete titration requires $n\ (= b_s/a) = 1$ for (I), and $n = 2$ for (II) and (III). Provided always that n is not too close to zero or to either of these

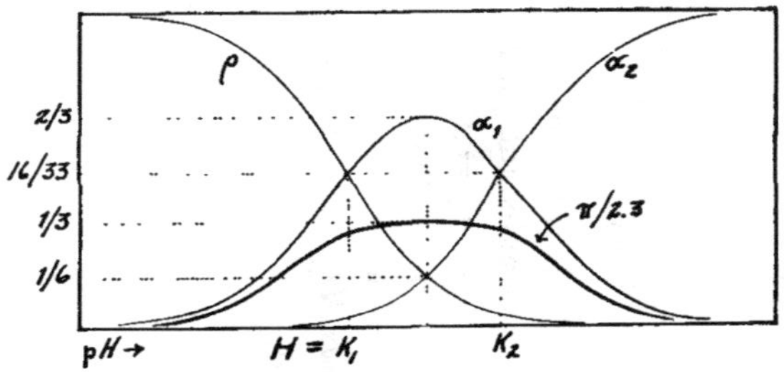

Fig. 14–2. The function $\alpha_1\rho + \alpha_1\alpha_2 + 4\alpha_2\rho$ with $K_1 = 16K_2$.

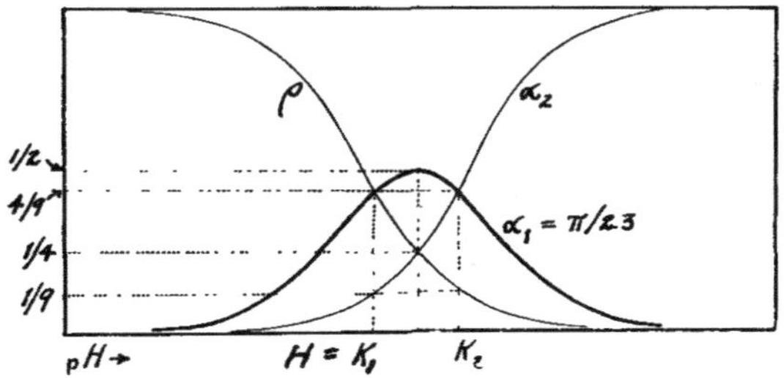

Fig. 14–3. The case of Fig. 14–2 with $K_1 = 4K_2$.

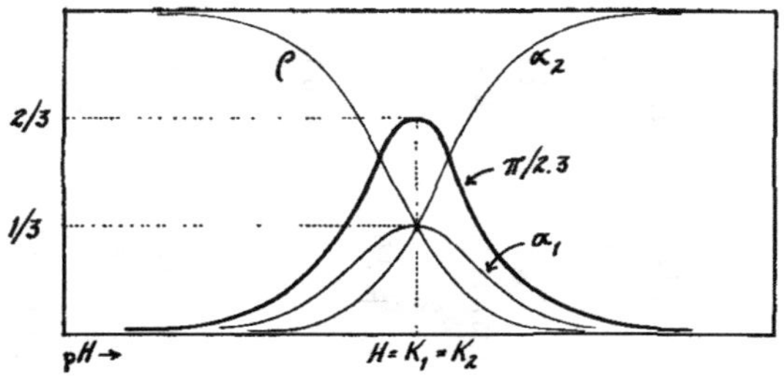

Fig. 14–4. The case of Fig. 14–2 with $K_1 = K_2$.

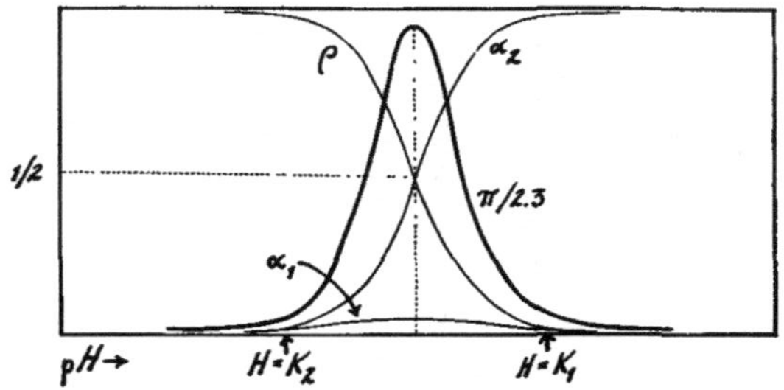

Fig. 14–5. The case of Fig. 14–2 with $K_1 \cong K_2/100$.

limits (1 and 2 respectively), we may neglect $(H - OH)$ and write (with y as the ratio K_2/K_1 or A_2/A_1),

$$\text{(I)}\quad H = A(1-n)/n, \tag{50}$$

$$\text{(II)}\quad H = K_1\left[(1-n)/2n + \sqrt{(\quad)^2 + y(2-n)/n}\right], \tag{51}$$

$$\text{(III)}\quad H = A_1\left[(1+y)(1-n)/2n + \sqrt{(\quad)^2 + y(2-n)/n}\right]. \tag{52}$$

The similarity between the titration curve of a dibasic acid and that of two independent acids (at equal concentrations) is marked and interesting. Thus Eq. (49) may be written as

$$H - OH = a[(1+y) + 2A_2/H]/[(1+y) + H/A_1 + A_2/H] - b_s. \tag{53}$$

Hence as long as $A_1 \gg A_2$, or as long as $y \ll 1$, the curve will be hardly distinguishable from that of a dibasic acid; as already noted, the curves are identical if we neglect A_2 in the sum $A_1 + A_2$.

All three of the curves, Eqs. (50), (51), and (52), are symmetrical with respect to the mid-point, the mid-point being at $n = 1/2$ in (I) and at $n = 1$ in (II) and (III). In other words, the ratio $H_n/H_{1/2}$ equals the ratio $H_{1/2}/H_{(1-n)}$ in (I), while $H_n/H_1 = H_1/H_{(2-n)}$ in (II) and (III). For the general ratio of H_{n_1} to H_{n_2}, the following formulas may be written:

$$\text{(I)}\quad H_{n_1}/H_{n_2} = \left(\frac{1-n_1}{1-n_2}\right)\left(\frac{n_2}{n_1}\right), \tag{54}$$

$$\text{(II)}\quad H_{n_1}/H_{n_2} = \sqrt{\frac{(2-n_1)n_2}{(2-n_2)n_1}}\left(\frac{-\mathbf{X}_1 + \sqrt{\mathbf{X}_1^2+1}}{-\mathbf{X}_2 + \sqrt{\mathbf{X}_2^2+1}}\right), \tag{55}$$

$$\text{(III)}\quad H_{n_1}/H_{n_2} = \sqrt{\frac{(2-n_1)n_2}{(2-n_2)n_1}}\left[\frac{-\mathbf{X}_1(1+y) + \sqrt{[\mathbf{X}_1(1+y)]^2+1}}{-\mathbf{X}_2(1+y) + \sqrt{[\mathbf{X}_2(1+y)]^2+1}}\right], \tag{56}$$

in which $\mathbf{X}_1 = (n_1 - 1)/2\sqrt{n_1 y(2-n_1)}$, while $\mathbf{X}_2$ is the same function of n_2.

The most interesting ratio is that for points equidistant (in respect to n) from the mid-point of the total titration; i.e. when $n_1 = 1 - n_2$ in (I) and $n_1 = 2 - n_2$ in (II) and (III). This will be called the "symmetrical ratio" with the symbol R_n. R_n is always understood to be less than 1, representing the ratio $H_n/H_{(1-n)}$ in (I), with $n > 1/2$, and $H_n/H_{(2-n)}$ in (II) and (III), with $n > 1$. Now, with $\mathbf{X} = (n-1)/2\sqrt{ny(2-n)}$,

$$\text{(I)}\quad R_n = \left(\frac{1-n}{n}\right)^2, \tag{57}$$

$$\text{(II)}\quad R_n = \left(\frac{2-n}{n}\right)\left(\frac{-\mathbf{X} + \sqrt{\mathbf{X}^2+1}}{+\mathbf{X} + \sqrt{\mathbf{X}^2+1}}\right), \tag{58}$$

$$\text{(III)}\quad R_n = \left(\frac{2-n}{n}\right)\left[\frac{-\mathbf{X}(1+y) + \sqrt{[\mathbf{X}(1+y)]^2+1}}{+\mathbf{X}(1+y) + \sqrt{[\mathbf{X}(1+y)]^2+1}}\right]. \tag{59}$$

For the difference in pH corresponding to the ratio R_n of the two values of H, or the increase in pH in titration from $1 - n$ to n in (I) and from $2-n$ to n in (II) and (III), Eqs. (57–59) become respectively

$$\text{(I)}\quad \Delta \mathrm{p}H = 2\log\,[n/(1 - n)], \tag{60}$$

$$\text{(II)}\quad \Delta \mathrm{p}H = \log\,[n/(2 - n)] + (2/2.3)\sinh^{-1}\mathbf{X}, \tag{61}$$

$$\text{(III)}\quad \Delta \mathrm{p}H = \log\,[n/(2 - n)] + (2/2.3)\sinh^{-1}[\mathbf{X}(1 + y)]. \tag{62}$$

The ratio of the constants, $y = K_2/K_1$, may then be calculated from the symmetrical ratio, $R_{(n)}$, at any value of n:

For dibasic acid, from Eq. (58),

$$y = (q - 1)(1 - n)^2/4n(2 - n), \tag{63}$$

with

$$\sqrt{q} = [(2 - n)/n + R_n]/[(2 - n)/n - R_n]. \tag{64}$$

When $n = 3/2$, so that $R_{3/2}$ is the ratio of the values of H at the mid-points of the two *halves* of the titration, then

$$y = (q' - 1)/12, \tag{65}$$

with

$$\sqrt{q'} = (1 + 3R_{3/2})/(1 - 3R_{3/2}). \tag{66}$$

Ordinarily the ratio y is taken as $R_{3/2}$ itself. But from Eqs. (65) and (66) it is seen that

$$y = R_{3/2}/(1 - 6R_{3/2} + 9R_{3/2}^2). \tag{67}$$

Hence $y \cong R_{3/2}$ only if $R_{3/2} \ll 1/6$. Correspondingly, from Eq. (58), the ratio $R_{3/2}$ is not exactly y, but

$$R_{3/2} = (1/3)(\sqrt{1 + 12y} - 1)/(\sqrt{1 + 12y} + 1), \tag{68}$$

so that $R_{3/2} \cong y$ only if $y \ll 1/12$. In pH, Eq. (68) corresponds to

$$(\Delta \mathrm{p}H)_{3/2} = \log 3 + (2/2.3)\sinh^{-1}(1/\sqrt{12y}), \tag{69}$$

which may also be written from Eq. (61) for $n = 3/2$.

We may therefore determine $K_1\sqrt{y}$ or $\sqrt{K_1K_2}$ from measurement of H at $n = 1$, or in aq. M_sHX, according to Eq. (51), and we may find the ratio y or K_2/K_1, according to Eq. (65) from the *difference* in pH between $n = 1/2$ and $n = 3/2$ without the actual values of pH involved. But the equations of this section are approximate only, involving the neglect of D in Eq. (48). The exact equations for the determination of K_1K_2 and K_2/K_1 were discussed in Chapter v.

For two independent acids (at equal concentrations), we have, from Eq. (59),

$$1/y = [2n(2 - n)/(q - 1)(1 - n)^2 - 1] + \sqrt{(\quad)^2 - 1}, \tag{70}$$

with q defined as in Eq. (64). When $n = 3/2$,

$$1/y = (7 - q')/(q' - 1) + \sqrt{(\quad)^2 - 1}, \tag{71}$$

with q' as in Eq. (66). Hence,

$$1/y = (1 - 8R_{3/2} + 9R_{3/2}^2)/2R_{3/2} + \sqrt{(\quad)^2 - 1}, \tag{72}$$

so that

$$y \cong R_{3/2}/(1 - 8R_{3/2} + 9R_{3/2}^2). \tag{73}$$

Therefore $y \cong R_{3/2}$ only if $R_{3/2} \ll 1/8$. Correspondingly, from Eq. (59)

$$R_{3/2} = (1/3)[\sqrt{1 + 12y/(1+y)^2} - 1]/[\sqrt{1 + 12y/(1+y)^2} + 1], \tag{74}$$

and therefore $R_{3/2} \cong y$ only if $y/(1+y)^2 \ll 1/12$, or if $y \ll 1/10$.

In terms of pH,

$$(\Delta pH)_{3/2} = \log 3 + (2/2.3) \sinh^{-1} [(1+y)/\sqrt{12y}]. \tag{75}$$

Some examples of values of the ratio $R_{3/2}$ corresponding to various values of y are given in Table 14–1 for the comparison of the curve of a dibasic acid with that of two independent acids at equal concentration. For the independent acids, with $A_1 > A_2$, the limiting value of $R_{3/2}$ is 1/9 (for $y = 1$) as for the case of a single monobasic acid at the same total concentration; note Eq. (57) with $n = 3/4$. For the dibasic acid, this value of $R_{3/2} = 1/9$ holds for $y = 1/4$ when, as explained in connection with Eq. VI(24), the dibasic acid is not distinguishable from a monobasic acid at concentration $2a$ or from two independent acids each at the concentration a. Finally, for both (II) and (III), the ratio of H at $n = 3/2$ to H at $n = 1$ (or the mid-point) is the square root of $R_{3/2}$ as tabulated.

Table 14–1. Symmetrical ratio $R_{3/2}$ as function of y.

$1/y$	$1/R_{3/2}$	
	(II)	(III)
1000	1006.3	1008.0
100	105.92	108.3
16	21.58	23.68
12	17.49	19.62
10	15.42	17.59
4	9	11.47
2	6.645	9.56
1	5.302	9
0.5	4.5	
0.1	3.6	
0.01	3.178	
0	3	

D. Simple Ampholyte

1. Analytical Equations.

$$H - OH = c(\alpha_- - \alpha_+) + (a_s - b_s). \tag{76}$$

If $a_s - b_s = 0$,

$$H^4K_b + H^3(cK_b + W) + H^2W(K_a - K_b) - HW(cK_a + W) - K_aW^2 = 0. \tag{77}$$

Also, from Eq. (76),

$$\pi = |\, dg/d\text{p}H \,| = 2.3[c(\alpha_0\alpha_- + \alpha_0\alpha_+ + 4\alpha_-\alpha_+) + H + OH], \tag{78}$$

$$= 2.3\left[cW\frac{(K_aOH + K_bH + 4K_aK_b)}{(K_aOH + K_bH + W)^2} + H + OH\right]. \tag{79}$$

As pointed out under Eqs. I(67–72) a simple ampholyte is mathematically identical in respect to equations defining H, with the alkali acid salt of a dibasic acid (the hydrochloride of the ampholyte), or with the basic salt of a diacid base with strong acid; the identity of Eq. (77) with Eq. (14) is obvious. Since the first of the two possible (and equivalent) analogies is the more familiar one, we shall write the approximate solutions of Eq. (77) directly from those of Eq. (13) by making the transformations

$$K_a = \text{“}K_2\text{”}, \qquad W/K_b = K_1^*, \tag{80}$$

in which K_1^* and "K_2" now represent the successive acid constants of the hydrochloride considered as a dibasic acid. If the concentration c, then, is not vanishingly small,

$$H \cong -W(K_a - K_b)/2(cK_b + W) + \sqrt{(\quad)^2 + W(cK_a + W)/(cK_b + W)} = Q_{77}^3; \tag{81}$$

and provided $c > \sqrt{WK_a/K_b}$ when $K_a > K_b$ and $c > \sqrt{WK_b/K_a}$ when $K_a < K_b$, then $H \cong \sqrt{W(cK_a + W)/(cK_b + W)}$, etc., to $H \cong \sqrt{WK_a/K_b}$, all by analogy from Eqs. (15–19). Furthermore, just as the approximations for the salt NaHX hold with $K_1 \gtrless K_2$, they similarly apply for the pure ampholyte whether $K_aK_b \lesseqgtr W$ since, according to Eq. (80), $K_1 = K_2$ when $K_aK_b = W$.

As in the case of the dibasic acid, the π curve will depend on whether $K_aK_b \lesseqgtr W$. When the product $K_aK_b \ll W$, then for the constants of the hydrochloride (as a dibasic acid), $K_1^* \gg$ "K_2," and the π curve is given by Fig. 14–1, if we identify α_+, α_0, and α_- of the ampholyte with ρ, α_1, and α_2 of the dibasic acid. Then there are two inflections in the titration of the hydrochloride with strong base; one at $b_s = c$ (the solution of the ampholyte plus neutral salt such as NaCl) and the second at $b_s = 2c$ (sodium salt of the ampholyte plus NaCl). But if $K_aK_b > W$, the π curve is that of Fig. 14–5, with no minimum at $b_s = c$. From Eq. (46), the central minimum of the

π curve disappears when K_aK_b becomes larger than $W/16$. It seems that it would be difficult to recognize as such an ampholyte with $K_aK_b > W/16$. Whereas many dibasic acids are known with $K_1 < 16K_2$, all known amino acids for example seem to have constants such that $K_aK_b < W/16$. For the simple ionization schemes discussed in Chapter VI, it will be recalled, K_aK_b could not be greater than $W/4$.

The point $H = \sqrt{WK_a/K_b}$ (corresponding to $H = \sqrt{K_1K_2}$ in Figs. 14–1 to 14–5) is the iso-electric point of the ampholyte, where α_0 (or "α_1" of the "dibasic acid") is a maximum, and $\alpha_- = \alpha_+$; this value of H_{ie} is reached, in a solution of the ampholyte, when a_s (or $-\ b_s$) $= \sqrt{WK_a/K_b} - \sqrt{WK_b/K_a}$.

The analytical equations for the curve of the titration of the hydrochloride with strong base would be Eqs. (22–30), with c for a and with the constants substituted according to Eq. (80). With $K_aK_b \ll W$, the titration of excess of strong acid in presence of the hydrochloride is marked by an inflection at minimum a of Fig. 14–1, with $H_e \cong \sqrt{cW/K_b}$ and F, the feasibility function, is $(pa_s/400)\sqrt{K_b/cW}$. For the first equivalent point in the titration of the pure hydrochloride itself, when $b_s = c$, $H_e \cong \sqrt{WK_a/K_b}$. Hence assuming no effect of ionic strength and hence of the NaCl present and neglecting H and OH in Eq. (79), we have, for the buffer capacity of the ampholyte solution at $H = \sqrt{WK_a/K_b}$,

$$\pi_e \cong \pi_{ie} \cong (2.3)2c\sqrt{K_aK_b/W}/(1 + 2\sqrt{K_aK_b/W}), \tag{82}$$

which follows directly from Eq. (44). With $K_aK_b \ll W$, π_e is low and F high: $\pi_e \cong (2.3)2c\sqrt{K_aK_b/W}$, $F \cong (p/400)\sqrt{W/K_aK_b}$. Otherwise π_e is high; if $K_aK_b = W$, $\pi_e \cong (2.3)2c/3$, and if $K_aK_b \gg W$, $\pi_e \cong (2.3)c$.

For the next equivalent point, $b_s = 2c$, when the solution is mathematically the sodium salt of the ampholyte, $H_e \cong \sqrt{WK_a/c}$ and $F = (p/400)\sqrt{cK_a/W}$.

The pure ampholyte itself, at concentration c, may be titrated either with strong acid or with strong base. With strong acid, the equivalent point is at $H_e \cong \sqrt{cW/K_b}$, with $F = (p/400)\sqrt{cK_b/W}$; with strong base, $H_e \cong \sqrt{WK_a/c}$, $F = (p/400)\sqrt{cK_a/W}$.

2. ON THE ANALOGY BETWEEN THE AMPHOLYTE AND THE SALT NaHX. The transformations involved in Eqs. VIII(1–4) and in writing those of this section from Eqs. (15–19), are purely mathematical in nature. The convenient treatment of the ampholyte problem as a special case of the general problem of a dibasic acid does not then depend upon any theoretical interpretation such as the "zwitter-ion theory," which in this connection is merely consistent with the mathematics. Some special points must be brought out, however, to avoid possible confusion when this theory is used in connection with the mathematical problem of the hydrogen ion concentration of solutions containing ampholytes.

The mathematical analogy, with its usefulness, is unquestionable, but the analogy is between the ampholyte and a salt, not between the ampholyte and a dibasic acid. Because of the emphasis laid upon the theory as preceding the mathematics, this is not always made clear and we read of "treating amino acids as dibasic acids." This may be misleading. The ampholyte itself is not a dibasic acid (any more than it is a diacid base) in any sense of the word, physically or mathematically, and does not possess two "acid constants"; in solution it is (mathematically but not physically) of the nature of a salt rather than of an acid. And it is physically neither a salt nor an acid but an ampholyte, which still remains a category of solutes.

We may also say, of course, reversing the analogy, that the acid salt, NaHX, is like an ampholyte, and more specifically that the ion, "HX^-," may be said to have mathematically both an acid constant (K_2) and a basic constant (W/K_1) corresponding to K_a and K_b of the ampholyte. We thus find that while the analogy is commonly regarded as useful in the treatment of the ampholyte problem, it is also helpful in explaining the properties of the solution of the salt NaHX. While it is obvious and self-evident for example that the solution of a pure ampholyte with $K_a = K_b$ should be neutral, it is not quite obvious that the salt NaHX of an acid with $K_1K_2 = W$ should also be neutral. But if we now merely think of the salt mathematically as an ampholyte with "K_a" $= K_2$ and $K_b^* = W/K_1$, it follows that the salt solution must be acid if $K_1K_2 > W$, neutral if $K_1K_2 = W$ and alkaline if $K_1K_2 < W$.

Nevertheless even this analogy with the salt must be applied with careful distinction. Modern discussion of ions emphasizes their "hydrolysis constants," so that W/K_1 for example is called a "hydrolysis constant" of the ion HX^-. As a result the transformation between acid and basic constants involved in the application of the zwitter-ion theory (the mathematical essence of the theory and of the analogy involved) is confused with the mathematically insignificant (conjugate) relation between acid (or basic) ionization constants and so-called "hydrolysis constants"; and the definition and description of the constants is not always clear, one or both of them being sometimes called "hydrolysis constants."[(1)]

(1) Thus while the zwitter-ion theory constants are correctly given in S. Glasstone, *Introduction to Electrochemistry*, D. Van Nostrand, N.Y., 1942, p. 425 (where, however, we read, p. 421, of "treating amino acids as dibasic acids"), as "K_2" $= K_a$ and $K_1^* = W/K_b$, Kolthoff and Rosenblum, on the other hand (Ref. VI–2, pp. 42–45) define K_a' ($= W/K_b$) and K_b' ($= W/K_a$) as "the true acidic and basic dissociation constants" on the basis of the zwitter-ion theory and go on to explain that "the true acidic dissociation constant of the amino acid is evidently identical with the hydrolysis constant of the basic group with an apparent dissociation constant K_b; the true basic dissociation constant corresponds to the value of the hydrolysis constant for the acid group on the older basis" (p. 44). Not only does the parallel transformation of both constants leave the problem unchanged, failing to relate it to that of a dibasic acid, but by developing the subject on the basis of the theory (explanation and mechanism) rather than on the basis of mathematical definitions, these authors, it seems, proceed then to draw a false conclusion in an important matter (p. 45), that "A complete analogy exists between the behavior of an amino acid towards acids and bases and that of a salt of the ammonium acetate type, with the restriction that hybrid ions do not

The distinction made in Chapter 1 between "salt solutions" as mathematical equivalent points and the concept of "salt" as a physical or crystallographic entity may be recalled at this point. The solution of an amino acid, in respect to "buffer properties," is mathematically identical with that of an acid salt of the type $NaHSO_3$, but the solid amino acid is not a salt. At the same time the conductivity and the osmotic properties of the solution, depending not on the charge coefficient, β, but on the conductivity coefficient, λ, and on the osmotic or colligative coefficient, ν, are altogether different from those of the solution of the *actual* salt, $NaHSO_3$. Similarly the simple salt MX_s (of weak base and strong acid) is identical, in buffering action, with a weak acid with ionization constant $A^* = W/B$. It is mathematically an acid because it has the same value of the coefficient β, but the two are altogether different physically and in respect to the coefficients λ and ν.

For the ampholyte and the acid salt, β is the same, equal to $(``K_2"/H - H/K_1^*)/(1 + H/K_1^* + ``K_2"/H)$, as related through Eq. (80). But the values of the coefficient λ are $(\alpha_+ + H/c)$ and $(1 + H/c)$ respectively. The term H/c is of course the same for both, but $\alpha_+ = (H/K_1^*)/(1 + H/K_1^* + ``K_2"/H)$ and hence $\alpha_+ \neq 1$. The values of the coefficient ν are $1 + (H + OH)/c$ and $2 + (H + OH)/c$ respectively and hence practically in the ratio 1/2.

Perhaps it would be better in order to avoid the kind of confusion discussed to put less emphasis on the interpretation and to make it clear that we are simply using a mathematical transformation or analogy because of its convenience in making two apparently unrelated problems (the ampholyte and the salt NaHX) mathematically identical. It must be remembered that an equally valid and useful analogy may be made with the problem of a diacid base, and even the zwitter-ion theory applies with equally satisfactory consistency and clarity on the basis of the sodium salt of the ampholyte as a diacid base.

E. Weak Base Salts of Dibasic Acids

In this section all approximations assume $K_1 > K_2$. The general equation is, with both strong and weak base present,

$$H - OH = a(1 + 2K_2/H)/(1 + H/K_1 + K_2/H) - bHB/(HB + W) - b_s, \quad (83)$$

conduct an electric current." It is true that conductivity has nothing to do with the "behavior towards acids and bases," since the latter depends solely on the charge coefficient (β) and the buffer capacity of the solution. But the buffer capacity of a solution such as that of ammonium acetate is distinctly different from that of an ampholyte. While the expression for π of an ampholyte, Eq. (79), is identical, through the transformations of Eq. (80), with that for π for the salt NaHX, obtainable from Eq. (40), it is not the same, however the constants are transformed or redefined, as that for a simple 1:1 salt, which would be $\pi/2.3 = c[HA/(H + A)^2 + HBW/(HB + W)^2] + H + OH$.

or

$$H^5B + H^4[B(b + b_s + K_1) + W] \\ + H^3[BK_1(K_2 - a + b + b_s) + W(b_s + K_1 - B)] \\ + H^2[K_1W(K_2 - B - a + b_s) - W^2 - BK_{12}(2a - b - b_s)] \\ - HK_1W[BK_2 + W + K_2(2a - bs)] - K_{12}W^2 = 0. \quad (84)$$

1. Acid Salt, Type NH_4HSO_3. With $b = a = c$ and $b_s = 0$,

$$H^5B + H^4[B(c + K_1) + W] + H^3[BK_{12} + W(K_1 - B)] \\ + H^2[W(K_{12} - BK_1 - cK_1 - W) - cBK_{12}] \\ - HK_1W(BK_2 + W + 2cK_2) - K_{12}W^2 = 0. \quad (85)$$

(a) For neutrality, with $H - OH = 0$ in Eq. (83), the requirement is $B_n = K_1(W + 2K_2\sqrt{W})/(W - K_{12})$. The solution therefore cannot be neutral or alkaline unless $K_1K_2 < W$. If $K_1K_2 < W$, then for neutrality, $B_n \cong K_1(1 + K_2/\sqrt{W})$.

(b) This solution usually is not likely to be very acid or very alkaline, and the best practical approximation for Eq. (85) will be its H^4 sequence. The H^5 sequence is possible only if $B > K_1$ and $K_1K_2 < W$ or, in general, if the solution is alkaline, whereupon the high degree terms become very small. The H^2 sequence is possible only if the concentration is extremely small, so that the H^2 term may become positive. Of the two remaining sequences, that in H^4 will tend to be the more important if the solution is acid because of the higher powers of H and even, in general, when the solution is alkaline since it can be alkaline only with large B ($> K_1$) and small acid constants ($K_1K_2 < W$). Therefore, from the H^4 sequence, $H = Q^4_{85}$, or

$$H \cong -\frac{BK_{12} + W(K_1 - B)}{2[B(c + K_1) + W]} \\ + \sqrt{(\quad)^2 + \frac{cBK_{12} - W(K_{12} - BK_1 - cK_1 - W)}{B(c + K_1) + W}}, \quad (86)$$

$$\cong \sqrt{cK_1(BK_2 + W)/B(c + K_1)}. \quad (87)$$

Then if $BK_2 > W$, $H \cong \sqrt{K_1K_2}$, and if $BK_2 < W$, $H \cong \sqrt{WK_1/B}$.

(c) If $K_1K_2 < W$ and $B \cong B_n$, then neglecting $H - OH$ in Eq. (83), we have

$$H^3B - HK_1(BK_2 + W) - 2K_{12}W = 0. \quad (88)$$

This equation, which is Eq. IV(13), strictly fixes a value of H (H_* or H_{ie}) at which β_s, or $\beta - \beta'$, equals 0 and hence the iso-electric point for this salt. As explained in Chapter IV, Eq. (88) should give a rough formula for H independent of the concentration and applicable at high concentrations. In the present and remaining cases (Chapter XVI included) the equation for H_* is of high degree and approximations will be written only for certain extreme or important relative values of the constants. For Eq. (88) to hold in this case, or for $H - OH$ to be negligible in Eq. (83), we have already noted that K_1K_2 should be less than W; hence $K_2 < \sqrt{W}$ and less than H if $H - OH \cong 0$; therefore we may neglect the last term in Eq. (88) as the smallest, obtaining

$$H \cong \sqrt{WK_1/B + K_1K_2}. \tag{89}$$

In any actual case, at high concentrations this would reduce, if $B \cong B_n$, to either $\sqrt{WK_1/B}$ or $\sqrt{K_1K_2}$, as under Eq. (87), depending on the relative magnitudes of the constants.

2. Normal Salt, Type $(NH_4)_2SO_3$. With $b = 2a = 2c$ and $b_s = 0$,

$$\begin{aligned} H^5B + H^4[B(2c + K_1) + W] + H^3[BK_1(c + K_2) + W(K_1 - B)] \\ + H^2W(K_{12} - BK_1 - cK_1 - W) - HK_1W(BK_2 + W + 2cK_2) \\ - K_{12}W^2 = 0. \end{aligned} \tag{90}$$

(a) Neutrality, from Eq. (83), with $H = \sqrt{W}$, requires $B_n = K_1(\sqrt{W} + 2K_2)/(K_1 + 2\sqrt{W})$.

(b) For approximations of Eq. (90), the sequence in H^5 is possible only if $B > K_1$ and $c < W/K_1$; that in H^2 only if $c < K_2$. On the whole then:

(1) If the solution is acid ($B < B_n$), we have, using the H^4 sequence, $H = Q^4_{90}$, or

$$H \cong -\frac{BK_1(c + K_2) + W(K_1 - B)}{2[B(2c + K_1) + W]} + \sqrt{(\quad)^2 + W\frac{K_1(c + B - K_2) + W}{B(2c + K_1) + W}}, \tag{91}$$

or $H \cong \sqrt{WK_1/2B}$.

(2). If alkaline ($B > B_n$), the H^3 sequence is better, and $H = Q^3_{90}$, or

$$H \cong W\frac{K_1(c + B - K_2) + W}{2[BK_1(c + K_2) + W(K_1 - B)]} + \sqrt{(\quad)^2 + K_1W\frac{2cK_2 + BK_2 + W}{BK_1(c + K_2) + W(K_1 - B)}}, \tag{92}$$

$$\cong W/2B + \sqrt{(\quad)^2 + 2K_2W/B}. \tag{93}$$

(c) If $B \cong B_n$, then neglecting $H - OH$ in Eq. (83), with $b = 2a$, or setting β_s $(= \beta - 2\beta') = 0$, we have

$$2H^3B + H^2BK_1 - HK_1W - 2K_1K_2W = 0. \tag{94}$$

This again defines the iso-electric point and leads to approximations independent of the concentration. With $K_1 > \sqrt{W}$ and $B_n \cong \sqrt{W} + 2K_2$, then $2H^3 < H^2BK_2$, and the result is Eq. (93). With $K_1 < \sqrt{W}$ and

$B_n \cong (K_1/2)(1 + 2K_2/\sqrt{W}) \cong K_1/2$, the largest terms are $2H^3B$ and HK_1W, giving $H \cong \sqrt{K_1W/2B}$ as under Eq. (91).

3. MIXED SALT, TYPE $NaNH_4SO_3$. With $a = b = b_s = c$, Eq. (84) gives

$$H^5B + H^4[B(2c + K_1) + W] + H^3[BK_1(c + K_2) + W(K_1 + c - B)] + H^2W(K_{12} - BK_1 - W) - HK_1W(BK_2 + W + cK_2) - K_{12}W^2 = 0. \quad (95)$$

(a) For neutrality, $B_n = (K_1K_2 - W)/(K_1 + 2\sqrt{W})$. The solution is therefore alkaline unless $K_1K_2 > W$, regardless of B. It can be neutral or acid only if $K_1K_2 > W$, whereupon $K_1 \gg \sqrt{W}$, and then $B_n \cong K_2$.

(b) If the solution is nearly neutral by such a test, we may use the equation for the iso-electric point by setting $\beta_s\ (= \beta - \beta' - 1) = 0$ in Eq. (83), or

$$2H^3B + H^2(BK_1 + W) - K_1K_2W = 0. \quad (96)$$

[The close interrelation of Eqs. (88), (94), and (96) is interesting. We may point out also that the condition for neutrality, or the formula for B_n, may in each case be obtained from the iso-electric equation by setting $H = \sqrt{W}$ in it.] Eq. (96) may now be used for the approximate value of H at high concentrations, provided $B \cong B_n$. With $K_1 > \sqrt{W}$ (and $H \cong \sqrt{W}$), we neglect $2H^3B$, and

$$H \cong \sqrt{K_1K_2W/(BK_1 + W)}. \quad (97)$$

Then if $BK_1 > W$, $H \cong \sqrt{WK_2/B}$, and if $BK_1 < W$, $H \cong \sqrt{K_1K_2}$.

(c) For the effect of concentration, an approximate formula must be taken from Eq. (95). Since these salt solutions are usually more alkaline than the normal weak base salts, the best sequence, on the whole, will be that in H^3, or $H = Q^3_{95}$,

$$H \cong \frac{W(BK_1 - K_{12} + W)/2}{BK_1(c + K_2) + W(c + K_1 - B)} + \sqrt{(\quad)^2 + \frac{W(cK_2 + BK_2 + W)}{B(c + K_2) + W(c + K_1 - B)/K_1}}, \quad (98)$$

$$\cong \sqrt{W(cK_2 + BK_2 + W)/[B(c + K_2) + W(c + K_1 - B)/K_1]}. \quad (99)$$

Unless K_2 is of the order of magnitude of W, there is obviously very little effect of concentration on H in solutions of this type of mixed salt. Eq. (99) will reduce to Eq. (97) with no effect of concentration as long as $cK_2 \gg (BK_2 + W)$.

4. TABLE OF VALUES OF H FOR TYPICAL CASES. Some values calculated for typical weak base salts of general dibasic acids are listed in Table 14–2, which shows the applicability of the equations developed by comparison with theoretical values obtained from the full Eq. (84). Good agreement is

Table 14–2. Weak base salts of dibasic acids.

Salt Type	K_1	K_2	B	c	Eq.	H Calcd.	H Exact
NH_4HSO_3	10^{-2}	10^{-7}	10^{-5}	1	86	3.16×10^{-5}	*
				10^{-2}		2.25×10^{-5}	*
aniline bisulfite	10^{-2}	10^{-7}	10^{-10}	1	86	9.95×10^{-4}	*
				10^{-2}		6.81×10^{-4}	*
NH_4HCO_3	10^{-7}	10^{-11}	10^{-5}	1	86	1.005×10^{-8}	*
				10^{-2}		1.006×10^{-8}	1.015×10^{-8}
				10^{-4}		1.059×10^{-8}	1.064×10^{-8}
NH_4HS	10^{-7}	10^{-15}	10^{-5}	1	86	1.000×10^{-8}	*
				10^{-2}		1.001×10^{-8}	*
				10^{-4}		1.054×10^{-8}	1.058×10^{-8}
$(NH_4)_2SO_3$	10^{-2}	10^{-7}	10^{-5}	1	92	1.465×10^{-8}	*
				10^{-2}		1.466×10^{-8}	*
				10^{-4}		1.50×10^{-8}	1.51×10^{-8}
aniline sulfite	10^{-2}	10^{-7}	10^{-10}	1	91	9.8×10^{-5}	9.82×10^{-5}
				10^{-2}		9.71×10^{-5}	9.73×10^{-5}
				10^{-4}		4.12×10^{-5}	4.15×10^{-5}
$(NH_4)_2CO_3$	10^{-7}	10^{-11}	10^{-5}	1	92	1.02×10^{-9}	1.00×10^{-9}
				10^{-2}		1.02×10^{-9}	1.003×10^{-9}
				10^{-4}		1.11×10^{-9}	1.18×10^{-9}
$(NH_4)_2S$	10^{-7}	10^{-15}	10^{-5}	1	92	1.00×10^{-9}	9.81×10^{-10}
				10^{-2}		1.00×10^{-9}	9.82×10^{-10}
				10^{-4}		1.185×10^{-9}	1.16×10^{-9}
				1	91	9.81×10^{-10}	
				10^{-2}		9.82×10^{-10}	
				10^{-4}		1.08×10^{-9}	
$NaNH_4SO_3$	10^{-2}	10^{-7}	10^{-5}	1	98	1.000×10^{-8}	*
				10^{-2}		1.001×10^{-8}	*
				10^{-4}		1.054×10^{-8}	*
sodium aniline sulfite	10^{-2}	10^{-7}	10^{-10}	1	98	3.15×10^{-6}	*
				10^{-2}		3.13×10^{-6}	*
				10^{-4}		2.21×10^{-6}	2.20×10^{-6}
$NaNH_4CO_3$	10^{-7}	10^{-11}	10^{-5}	1	98	9.96×10^{-11}	*
				10^{-2}		1.049×10^{-10}	1.053×10^{-10}
				10^{-4}		3.85×10^{-10}	3.88×10^{-10}
$NaNH_4S$	10^{-7}	10^{-15}	10^{-5}	1	98	3.31×10^{-12}	*
				10^{-2}		3.20×10^{-11}	*
				10^{-4}		3.69×10^{-10}	3.67×10^{-10}

found in every case when the approximation to be used is chosen according to the conditions of applicability indicated for each type of salt. [*Note:* an asterisk in the last column means that the exact value is the same as the "approximate," to the number of significant figures shown.] In the case of aq. "$(NH_4)_2S$," which is alkaline, the "indicated" approximation is Eq. (92). Although this gives answers within two per cent of the exact value, it is not on the whole the best approximation, as shown by the numerical test with orders of magnitude of the terms of Eq. (90). In this case it turns out that Eqs. (91) and (92) are about equally good, as applying to a *range* of concentration.

XV

Some Complex Cases

We shall here consider some special cases of acids in which, because of association or other unusual interactions, the solution is "complex" in the sense discussed in Chapters I and II. The equilibria in these solutions involve solute species other than those related by the fundamental ionization constants of the acid, or X^0, X^-, $X^=$, · · ·, X^{-z}, and the charge coefficient of the acid is a function not only of H and equilibrium constants but also of the concentration of the acid. The cases to be treated are those called "homopoly-acids," with only one acid involved in the complex or complexes.

At the end of the chapter, in Section F, we shall also consider the equations involved in some simple cases of the homogeneous hydrolysis of nonelectrolytes.

A. Monobasic Acid (Type HF) With Complex Univalent Anion

We refer here to the case already outlined in Chapter II regarding the supposition that the behavior of aqueous HF or hydrofluoric acid, which forms both "normal salts" and what may be called "acid salts," or types NaF and $NaHF_2$, is attributable to an ordinary ionization process with ionization constant A, coupled with a "complex" process giving rise to the complex anion HF_2^- by combination of F^- with HF^0. With K defined as in Eq. II(15) as $[HF_2^-]/[HF^0][F^-]$, we have, from Eqs. II(14–19), for a solution containing such a complex acid at concentration a, together with strong base,

$$D = \frac{a}{2} + \left(\frac{A^2 - H^2}{8HAK}\right)\left[\sqrt{\frac{8aHAK}{(A + H)^2} + 1} - 1\right] - b_s; \tag{1}$$

this is a rearrangement of Eq. II(19).

1. Analytical Concentrations for Specified Values of H. With $b_s = 0$,

$$a = (1/HK)[-(H - A - 4HKD)/2 + \sqrt{(\quad)^2 + HKD(H^2/A - A - 4HKD)}]. \tag{2}$$

For $H = A$, with $b_s = 0$, $a = 2(A - W/A)$, independent of K, and possible only if $A > \sqrt{W}$.

The value of b_s, for specified values of a and H, may of course be calculated directly from Eq. (1). The ratio b_s/a for $H = A$ is

$$(b_s/a)_A = 1/2 - (A - W/A)/a, \tag{3}$$

also independent of K.[(1)] The ratio for neutrality ($H = \sqrt{W}$) is

$$\left(\frac{b_s}{a}\right)_n = \frac{1}{2} + \left(\frac{A^2 - W}{8aAK\sqrt{W}}\right)\left[\sqrt{\frac{8aAK\sqrt{W}}{(A + \sqrt{W})^2} + 1} - 1\right]. \tag{4}$$

which varies with both K and a, since the solution is "complex." The ratio for neutrality is 1/2 only if $A = \sqrt{W}$, when it is also independent of a. If $A > \sqrt{W}$, $(b_s/a)_n$ is greater than 1/2 at any concentration, and furthermore increases with dilution; the salt of type "$NaHF_2$" is then acid at any concentration. If $A < \sqrt{W}$, $(b_s/a)_n$ is smaller than 1/2 at any concentration and decreases upon dilution; now the salt of type "$NaHF_2$" is alkaline at any concentration. Furthermore, as either a or K approaches zero, $(b_s/a)_n$ approaches the value for an ordinary monobasic acid, or $A/(A + \sqrt{W})$, which may be greater or less than 1/2. As either a or K approaches infinity, $(b_s/a)_n$ approaches the limit 1/2. With $K = \infty$, therefore, the acid behaves like a "strong" acid at half the concentration, and Eq. (1) becomes $D = a_s/2 - b_s$.

Examples of $(b_s/a)_n$, for neutrality, at $a = 1$, are listed in Table 15-1.

2. Behavior as Pseudodibasic Acid in Titration. As a polynomial in H, Eq. (1) becomes

$$\begin{aligned} &H^4(1 - 4AK) + H^3[b_s + 4(a - 2b_s)AK] \\ &\quad - H^2A[a + (a - 2b_s)^2K + A + W/A - 8KW] \\ &\quad + HA^2[(a - b_s) - 4(a - 2b_s)KW/A] + A^2W(1 - 4KW/A) = 0. \end{aligned} \tag{5}$$

(With $K = 0$, this reduces to Eq. x[22] multiplied by $[A - H]$.) While the equation in this form may be used to calculate H for specified values of a and b_s, the process involves tedious approximations if any precision is desired.

(1) Hence the ratio $(b_s/a)_A \cong 1/2$ if A is neither too large nor too small. In this sense this ratio is practically, though not exactly, independent of the concentration a. Such a relation has been observed experimentally, for germanic acid, by G. Carpéni [*Compt. rend.*, **226**, 807 (1948)]. Carpéni thereupon generalizes, as an expectation, that for certain "complex acids" involving this type of equilibrium between two species, there will be what he calls an "isohydric point," a ratio of b_s to a for which H is independent of a. It is possible, indeed, to show that relations similar to Eq. (3) may be derived even for other formulas of the complex. Thus, if the complex ion in the present case is $H_2F_3^-$, with $K = [H_2F_3^-]/[F^-][HF]^2$, then $H = 2A$ at $b_s/a = 1/3 - (2A - W/2A)/a$, and if the complex is $HF_3^=$, with $K = [HF_3^=]/[F^-]^2[HF]$, then $H = A/2$ at $b_s/a = 2/3 - (A/2 - 2W/A)/a$—in both cases independent of K and practically independent of a. There would then be, as Carpéni has in fact pointed out, the possibility of estimating A from the ratio b_s/a at his so-called "isohydric point."

In order to sketch a titration curve for the effect of variation of b_s for fixed a, it is much easier to calculate b_s for specified values of a and H directly through the equation in its original form of Eq. (1). This is in accordance with the general point made in Chapter VII that these high degree equations usually become extremely simple when H is taken not as the dependent but as the independent variable. In this way the titration curves shown in Fig. 15–1 to Fig. 15–6 have been calculated. Figs. 15–1 to 15–4 apply for unit concentration, or $a = 1$; Fig. 15–1 shows the effect of increasing K with A fixed at

Table 15–1. Values of $(b_s/a)_n$ for $a = 1$.

A	K	$(b_s/a)_n$
$A = 1$	10^{-5}	1.0000
	1	1.0000 –
	10^{5}	0.9904
	10^{10}	0.5111
$A = 10^{-5}$	10^{-5}	0.990
	1	0.981
	10^{5}	0.5111
	10^{10}	0.500035
$A = 10^{-10}$	10^{-5}	0.001
	1	0.002
	10^{5}	0.4657
	10^{10}	0.49989

the value of 1; Figs. 15–2,3,4 apply to $A = 10^{-3}$, 10^{-6}, and 10^{-12} respectively. Fig. 15–5 shows the effects at $a = 10^{-2}$, and Fig. 15–6 at $a = 10^{-4}$.

The general effect of increasing K upon a given acid at a given concentration is to make its behavior approach as a limit that of a strong acid at half the concentration of the actual "acid." With $K = \infty$, this means that the complex acid of type HF at concentration a would behave like HCl at concentration $a/2$. Thus the shape of the titration curve is modified from that typical of a weak acid at concentration a (when $K = 0$) to approach that of HCl at concentration $a/2$. In general, this means that the pH curve of the first half of the titration is lowered while the second half of the curve is raised, resulting in an inflection near $b_s = a/2$. As long as $a > |2(A - W/A)|$, the resulting family of curves (for fixed a and A) crosses, independently of K, at $H = A$, at b_s given by Eq. (3). Otherwise the "complex" curve lies entirely above the simple curve for $K = 0$ for high A and entirely below for low A.

As a result of this general effect, there is in each case a certain critical value of the constant K (depending both on a and on A) above which there

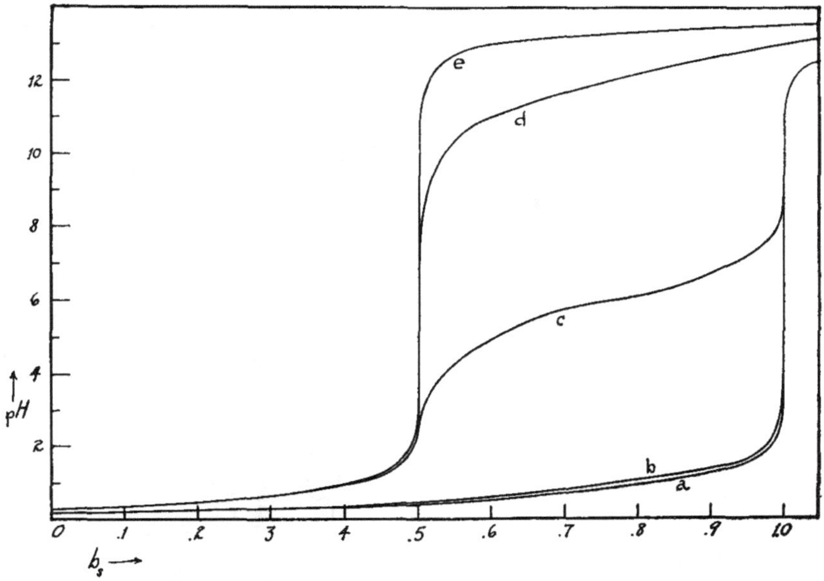

Fig. 15–1. Titration curves for $a = 1$ and $A = 1$: a, $K = 0$; b, $K = 1$; c, $K = 10^6$; d, $K = 10^{12}$; e, $K = \infty$.

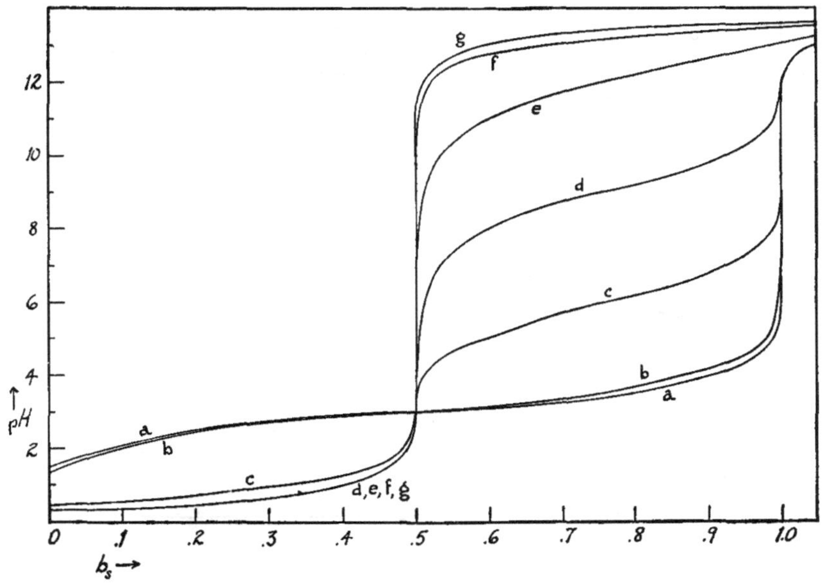

Fig. 15–2. Titration curves for $a = 1$ and $A = 10^{-3}$: a, $K = 0$; b, $K = 1$; c, $K = 10^3$; d, $K = 10^6$; e, $K = 10^9$; f, $K = 10^{12}$; g, $K = \infty$.

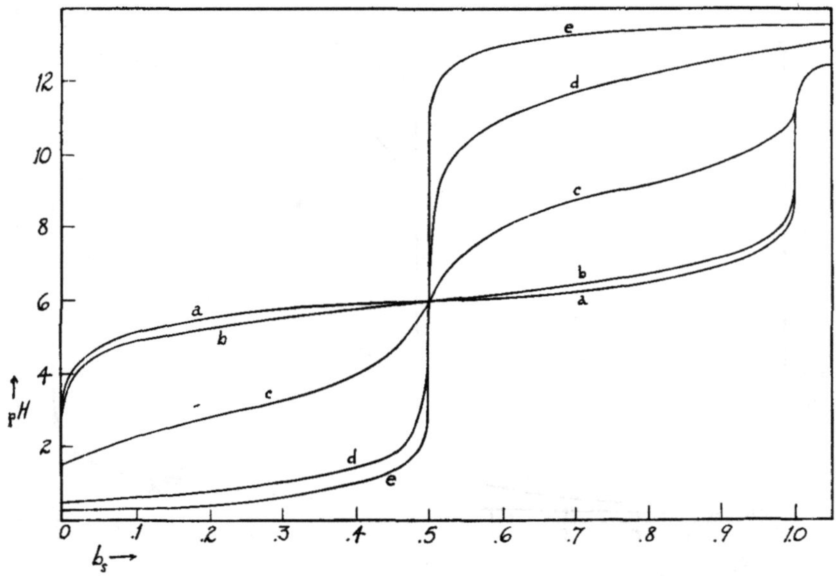

Fig. 15–3. Titration curves for $a = 1$ and $A = 10^{-6}$: a, $K = 0$; b, $K = 1$; c, $K = 10^3$; d, $K = 10^6$; e, $K = 10^{12}$ to $K = \infty$.

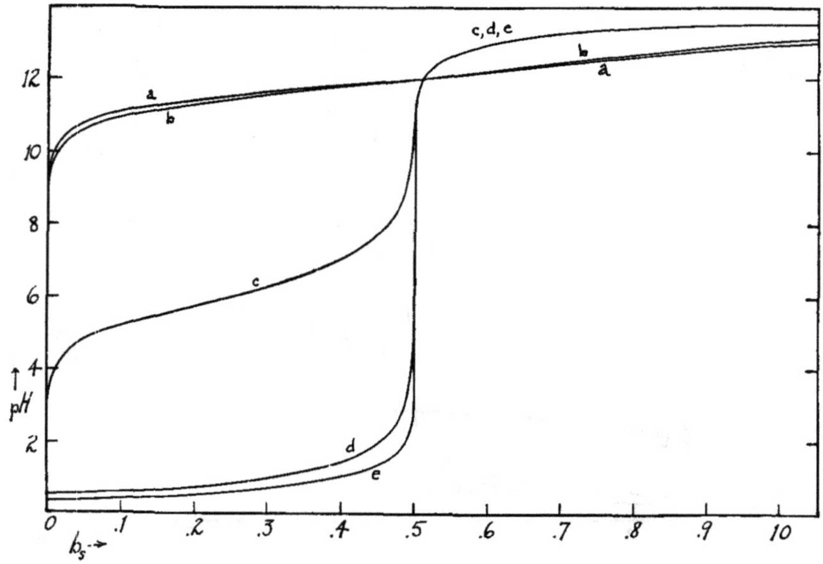

Fig. 15–4. Titration curves for $a = 1$ and $A = 10^{-12}$: a, $K = 0$; b, $K = 1$; c, $K = 10^6$; d, $K = 10^{12}$; e, $K = \infty$.

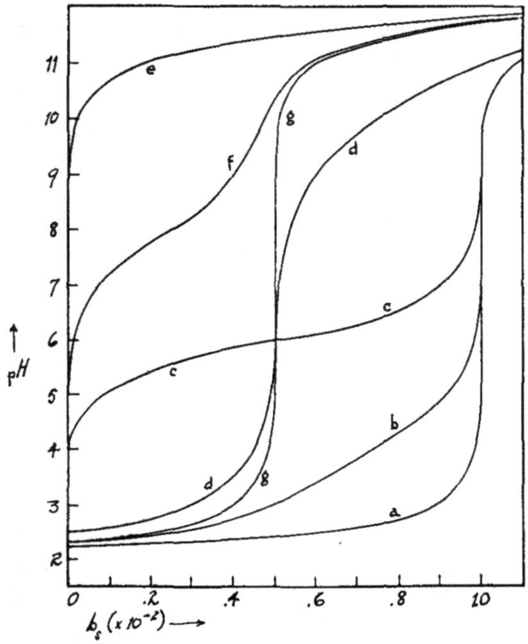

Fig. 15-5. Effect of variation of K at $a = 10^{-2}$; a: $A = 1, K = 0$; b: $A = 1, K = 10^6$; c: $A = 10^{-6}$, $K = 0$; d: $A = 10^{-6}$, $K = 10^6$; e: $A = 10^{-12}$, $K = 0$; f: $A = 10^{-12}$, $K = 10^6$; g: $K = \infty$.

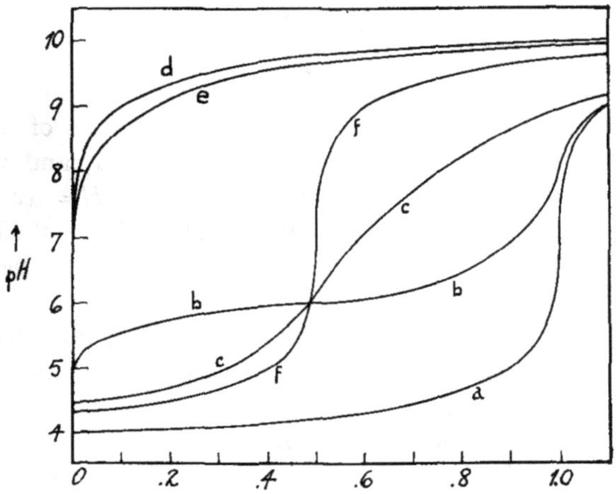

Fig. 15-6. Effect of variation of K at $a = 10^{-4}$; a: $A = 1$, $K = 0$ and $K = 10^6$; b: $A = 10^{-6}$, $K = 0$; c: $A = 10^{-6}$, $K = 10^6$; d: $A = 10^{-12}$, $K = 0$; e: $A = 10^{-12}$, $K = 10^6$; f: $K = \infty$.

appears a vertical inflection in the titration curve near $b_s = a/2$ corresponding to the equivalent point for the salt "$NaHF_2$." There is then a still higher critical value of K above which the original inflection at $b_s = a$ disappears, leaving only that at $b_s = a/2$. Between these limiting values of K, which are $f(a,A)$, the acid therefore shows two successive (vertical) inflections in its titration curve with strong base, somewhat like a dibasic acid. This *pseudodibasic acid* titration curve, however, does not have quite the shape of that of a true dibasic acid. The relation between the two types of curve will be seen below through consideration of possible simplifications of the exact equations.

The monobasic acid at molarity a, therefore, becomes mathematically a strong monobasic acid at molarity $a/2$, if K is very large, regardless of the value of A. But for intermediate values of K, such as 10^3 to 10^9, the acid behaves as a pseudodibasic acid at the molarity $a/2$; and now if A is very large this pseudodibasic acid is "strong in respect to K_1."

Examination of the pseudodibasic acid titration curves for high concentration shows that $H \cong A$ at $b_s = a/2$ and that the ratio of H at $b_s = 3a/4$ to H at $b_s = a/4$, $\cong 1/K^2$. Further consideration of Eq. (5) in comparison with that of a true dibasic acid, Eq. XIV(3), allows the "constants" of the pseudodibasic acid to be identified roughly as "K_1" $= aAK$ and "K_2" $= A/aK$, so that A corresponds to $\sqrt{\text{“}K_1K_2\text{”}}$ and aK corresponds to $\sqrt{\text{“}K_1/K_2\text{”}}$ or $1/\sqrt{\text{“}y\text{”}}$. The "constants" themselves, of course, depend on the concentration.

3. Calculation of H. An equation such as Eq. (5), involving the square of the difference of the concentrations, may have more than one positive root for a given set of values of the constants and concentrations. For the pseudodibasic behavior here described, which prevails only over certain restricted ranges of the values of the parameters, there will usually be little uncertainty regarding the choice of the correct root. In uncertain cases the value must be chosen with regard to the possible limits of H for the data given. For example, with $A = 10^{-6}$, $K = 10^9$, $a = 1$, and $b_s = 0.55$, the largest terms of Eq. (5) are the last three, giving $H = (2.225 \pm 1.025) \times 10^{-13}$. But even with $K = 0$ we would calculate $OH \cong 0.05$, so that with $K = 10^9$ the value of H must be greater than 2×10^{-13}. The true value is therefore 3.25×10^{-13}.

Approximate formulas for H for various ranges of the titration may be written from Eq. (5) only if we assume certain general ranges of values of the constants. The interesting case is that in which $AK \ll 1$ and $K \gg 1$, so that the behavior is that of a pseudodibasic acid weak in respect to both ionization constants, "K_1" and "K_2." Thereupon we may simplify Eq. (5) as follows:

$$H^4 + H^3[b_s + 4(a - 2b_s)AK] - H^2A[a + K(a - 2b_s)^2] + HA^2[(a - b_s) - 4(a - 2b_s)KW/A] + A^2W \cong 0. \quad (6)$$

For the start of the titration, near $b_s = 0$, the H^4 sequence would be used, which, for the pure acid, then gives very roughly $H \cong \sqrt{a^2AK}$. In the range near $b_s = a/4$, the second and third terms give $H \cong AK(a - 2b_s)^2/b_s$, and at $b_s = a/4$, $H \cong aAK$. Through the equivalent point $b_s = a/2$, the H^3 sequence is used, simplified to give

$$H \cong A[a + K(a - 2b_s)^2]/2b_s \pm \sqrt{(\quad)^2 - A^2(a - b_s)/b_s}. \tag{7}$$

The positive sign is used if $b_s < a/2$, the negative if $b_s > a/2$. For the "salt solution" of type "$NaHF_2$" then, when $b_s = a/2$, this equation gives $H \cong A$. For the second "middle range," near $b_s = 3a/4$, the H^2 and H terms of Eq. (6) are used to give $H \cong A(a - b_s)/K(a - 2b_s)^2$, so that when $b_s = 3a/4$, $H \cong A/aK$. Through the equivalent point, near $b_s = a$, the H^2 sequence gives the best approximation, so that for the salt "NaF," when $b_s = a = c$, $H \cong \sqrt{AW/c^2K}$ or $OH \cong \sqrt{c^2KW/A}$.

4. Species Concentrations. For the species concentrations, we have, starting with Eq. II(18),

$$[F^-] = -(H + A)/4HK + \sqrt{(\quad)^2 + aA/2HK}, \tag{8}$$

$$[HF] = -(H + A)/4AK + \sqrt{(\quad)^2 + aH/2AK}, \tag{9}$$

and

$$[HF_2^-] = a/2 - [F^-](H + A)/2A = a/2 - [HF](H + A)/2H. \tag{10}$$

If we define $[HF]/a$ as ρ, $[HF_2^-]/a$ as α_1 and $[F^-]/a$ as α_2, the analogy with dibasic acid may be brought out further by consideration of the derivatives of these quantities with respect to pH. It is then found that $d\rho/d\text{p}H$ is always negative, $d\alpha_2/d\text{p}H$ always positive, and $d\alpha_1/d\text{p}H = 0$ at $H = A$, so that α_1 passes through a maximum value at $H = A$, near the first equivalent point of titration with strong base, at $b_s = a/2$, corresponding to the salt solution "$NaHF_2$." This is similar to the maximum of α_1 for a dibasic acid at $H = \sqrt{K_1K_2}$, near the first equivalent point.

When $H = A$ (or roughly in pure "$NaHF_2$"), $[F^-] = [HF] = -1/2K + \sqrt{(\quad)^2 + a/2K}$, while $[HF_2^-] = a/2 - [F^-]$, which is the maximum for $[HF_2^-]$. For the mid-point of the first half of the titration, at $H = aAK$ (near $b_s = a/4$), we have, if $aK \gg 1$, $[F^-] \cong 1/2K$, an approximate value which is independent of a; and $[HF] \cong a/2$, $[HF_2^-] \cong a/4$. For $[HF_2^-] = [HF]/2$, the requirement is $H = A(aK - 1/2)$. At $H = A/aK$, near $b_s = 3a/4$, then again if $aK \gg 1$, $[HF] \cong 1/2K$, independent of a; and $[F^-] \cong a/2$, $[HF_2^-] \cong a/4$. Now $[HF_2^-] = [F^-]/2$ when $H = A/(aK - 1/2)$.

5. Shape of the Titration Curve. The approximations in Section (3) resemble those for a dibasic acid at concentration $a/2$, with "K_1" $= A(aK)$ and "K_2" $= A/(aK)$; but the effect of the concentration is not the same. For further comparison of the pseudodibasic and the true dibasic acid titration curves, we shall now consider the approximate equation obtainable

from Eq. (1) by neglecting D. This will be applicable for high a and for $0 < b_s < a$. If we furthermore define n^* as the titration ratio $b_s/(a/2)$—not b_s/a—the result is, from Eq. (7),

$$H = A\{[1 + aK(n^* - 1)^2]/n^* \pm \sqrt{(\quad)^2 - (2 - n^*)/n^*}\}, \tag{11}$$

the positive sign being used if $n^* < 1$, the negative for $n^* > 1$. Hence $H = A$ at $n^* = 1$, $H \cong aAK$ at $n^* = 1/2$ and $H \cong A/aK$ at $n^* = 3/2$. The corresponding expression for a true dibasic acid is Eq. XIV(51), in which, however, n has its usual meaning, b_s/a.

The pH curve described by Eq. (11), as in the case of Eqs. XIV(50–52), is symmetrical with respect to the point $n^* = 1$. The "symmetrical ratio" R_{n^*} (see Eqs. XIV[57–59]) is given by the formula

$$R_{n^*} = \left(\frac{2 - n^*}{n^*}\right)\left(\frac{\mathbf{Y} - \sqrt{\mathbf{Y}^2 - 1}}{\mathbf{Y} + \sqrt{\mathbf{Y}^2 - 1}}\right), \tag{12}$$

with

$$\mathbf{Y} = [1 + aK(n^* - 1)^2]/\sqrt{n^*(2 - n^*)}. \tag{13}$$

Hence

$$\Delta pH \text{ for } R_{n^*} = \log\,[n^*/(2 - n^*)] + (2/2.3)\,|\cosh^{-1}\mathbf{Y}\,|, \tag{14}$$

in contrast with the inverse sinh function of Eq. XIV(61) for a true dibasic acid. The ratio R_{n^*} is independent of A so long as the neglect of D in Eq. (1) is justified; and Eq. (12) may therefore be used to obtain the complex ion constant K from the value of some symmetrical ratio of the curve, or in other words from the *shape* of the curve, at some fixed value of the concentration, a. Thus,

$$K = [1/a(n^* - 1)^2][-1 + \sqrt{qn^*(2 - n^*)/(q - 1)}], \tag{15}$$

in which q is defined by Eq. XIV(64). At $R_{(3/2)^*}$,

$$K = (4/a)[-1 + \sqrt{3q'/4(q' - 1)}], \tag{16}$$

with q' defined by Eq. XIV(66). Now if $R_{(3/2)^*}$ is very small and $q' \cong 1$, then

$$K \cong (1/a)\sqrt{(1 + 6R_{(3/2)^*} + 9R^2_{(3/2)^*})/R_{(3/2)^*}}. \tag{17}$$

Hence the approximate value of A ($= \sqrt{\text{“}K_1K_2\text{”}}$) can be obtained at $n^* = 1$ (aq. "$NaHF_2$") and the approximate value of K ($= 1/a\sqrt{\text{“}y\text{”}}$) from $R_{(3/2)^*}$.

The titration curves of Eqs. (11) and XIV(51) are shown in Fig. 15–7. The solid curve is the true dibasic acid curve, and the dotted curve the pseudodibasic acid curve for comparable constants. It is assumed, in other words, that $aAK = K_1$ of the dibasic acid, and $A/aK = K_2$. The crossing of the curves at $n^* = 1$ (where $H = A = \sqrt{K_1K_2}$) is independent of the concentration and of K. The curves also cross near $n^* = 1/2$ and $n^* = 3/2$, where $H \cong$ "K_1" and "K_2" respectively. They otherwise differ in that the symmetrical ratio R_{n*} for the pseudodibasic acid is larger (or smaller) than that of the true dibasic acid when n^* is larger (or smaller) than 3/2. This may be seen by comparison of Eqs. (12) and XIV(58). The two expressions are equal when $\mathbf{Y}^2 = \mathbf{X}^2 + 1$. But since these shapes of curves apply only when

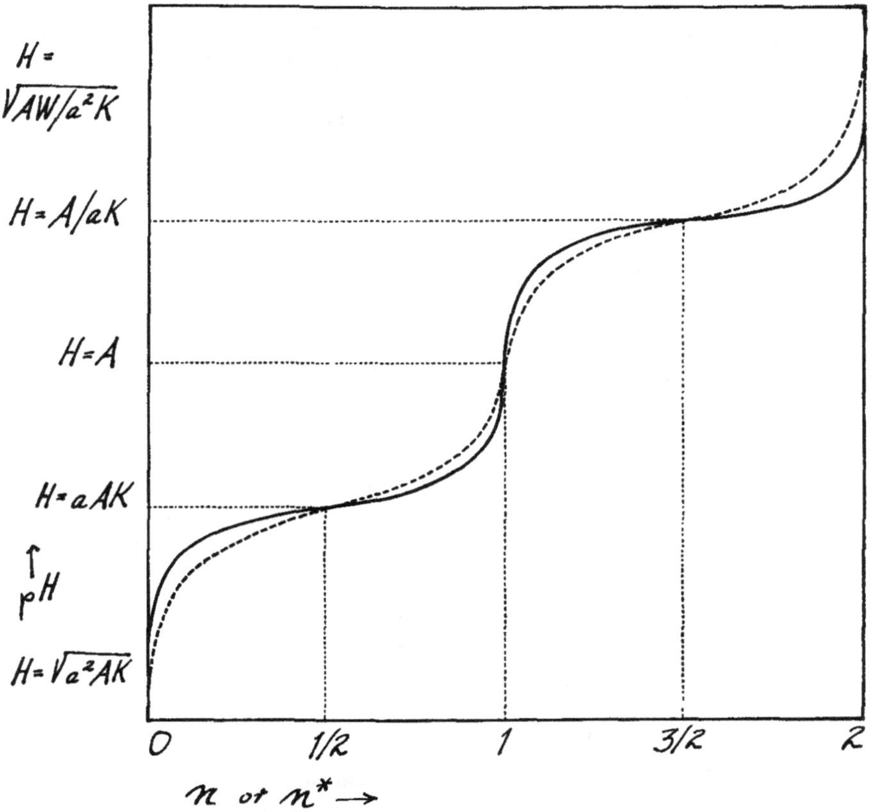

Fig. 15–7. Comparison of titration curves of pseudodibasic acid and true dibasic acid. Dashed curve represents Eq. (11) for pseudodibasic acid; full curve represents Eq. XIV(51) for true dibasic acid, with $K_1 = aAK$ and $K_2 = A/aK$. For true dibasic acid the abscissa is $n = b_s/a$; for pseudodibasic acid the abscissa is $n^* = 2b_s/a$.

the functions $\mathbf{X}$ and $\mathbf{Y}$ are quite large, we may neglect 1 compared to $\mathbf{X}$ and the term 1 in the numerator of $\mathbf{Y}$ in Eq. (13). Then since $aK = 1/\sqrt{y}$, it is seen that R_{n*} of Eq. (12) is greater (or less) than R_n of Eq. XIV(58) if n^* is greater (or less) than 3/2.

6. Buffer Capacity. For comparable constants then, as we have just seen, the steepness of the pH curve through the equivalent point at $n^* = 1$ is smaller for the pseudodibasic than for the true dibasic acid case. This steepness, furthermore, increases not only as K increases, just as it increases

with $1/\sqrt{y}$ for the true dibasic acid, but it also increases with the concentration itself. These effects may be seen from consideration of the buffer capacity of the solution.

For noncomplex solutions, the neglect of D in the full equation, such as Eq. XIV(1), leads to $\pi_j = db_s/d\text{p}H = a(d\beta/d\text{p}H)$, as in Eq. I(113); with $d\beta/d\text{p}H$ independent of a, the concentration of the acid being titrated. If then we are interested only in the shape of the curve, with D neglected, we may speak of $\pi_n = dn/d\text{p}H = d(b_s/a)/d\text{p}H$ or, dividing π_j by a, $\pi_n = d\beta/d\text{p}H$.

In the present complex case, however, π_{n*} is not independent of the concentration a, and with D neglected, Eq. (1) gives

$$\frac{\pi_{n*}}{2.3} = \left(\frac{H^2 + A^2}{4aHAK}\right)(\sqrt{\mathbf{J}} - 1) - \frac{(H - A)^2}{(H + A)^2\sqrt{\mathbf{J}}}, \tag{18}$$

in which $\mathbf{J}$ is the quantity under the radical sign in Eq. (1). Here n^* is still defined as $2b_s/a$. When $n^* = 1$, the mid-point of the curve of Fig. 15–7, at $H = A$, then

$$(\pi_{n*})_A = 2.3(-1 + \sqrt{1 + 2aK})/2aK \cong 2.3/\sqrt{2aK}, \tag{19}$$

$$d(\pi_{n*})_A/dK = 2.3(-1 - aK + \sqrt{1 + 2aK})/2aK^2\sqrt{1 + 2aK}. \tag{20}$$

With $aK \gg 1$ this is negative. Hence the steepness of the vertical inflection of the titration curve at the mid-point of Fig. 15–7, depending on a minimum value of π_{n*} at $H = A$, increases with K for a given acid at a given concentration. Furthermore,

$$d(\pi_{n*})_A/da = 2.3(-1 - aK + \sqrt{1 + 2aK})/2a^2K\sqrt{1 + 2aK}. \tag{21}$$

This is also negative for large aK. The buffer capacity at the mid-point therefore decreases with increase of the concentration. As the concentration increases then for fixed constants, the mid-point vertical inflection increases in steepness. For the corresponding true dibasic acid case, on the other hand, π_n at $H = \sqrt{K_1K_2}$, or π_{min}/a, is, according to Eq. XIV(44), independent of the concentration (if D is negligible, of course), and the value increases with y, just as it decreases with K in Eq. (20).

Finally, in *rough* analogy with the case of true dibasic acid, we may assume that the vertical mid-point inflection at $b_s = a/2$ will not occur unless $(aK)^2 \gg 16$, since the condition in the case of the dibasic acid is $K_1/K_2 > 16$.

7. Feasibility of Titration. Using Eq. (7) for the curve through the equivalent point $b_s = a/2$, and therefore assuming again that $A/K \ll AK \ll 1$, we may write, on the basis of Procedure I of Chapter IX,

$$\Delta_p = \log \frac{2[1 + aKx^2]/(1 - x) + \sqrt{(\quad)^2 - (1 + x)/(1 - x)}}{2[1 + aKx^2]/(1 + x) - \sqrt{(\quad)^2 - (1 - x)/(1 + x)}}, \tag{22}$$

in which $x = p/200$, p being percentage relative to $a/2$. For small p, this means $\Delta_p = \log[(G + \sqrt{G^2 - 1})/(G - \sqrt{G^2 - 1})]$, or

$$\Delta_p = (2/2.3) \mid \cosh^{-1} G \mid, \tag{23}$$

in which $G = 1 + aK(p/200)^2$.

The feasibility given by Eq. (23) is of course lower than that for a true dibasic acid with $\sqrt{K_1/K_2} = aK$, given by Eq. XIV(45). If $K = 0$, $\Delta p = 0$ according to Eq. (23), and the point $b_s = a/2$ is merely the mid-point of titration of a monobasic weak acid.

At the second equivalent point, $b_s = a$, there will be no inflection point unless the second constant, "K_2" (or A/aK) $\gg 27W/a$. If there is such an inflection, then the feasibility function at this point may be written by analogy with that for the true dibasic acid, which is $F = (p/200)\sqrt{aK_2/W}$, with p as percentage of titration relative to the total titration, or to twice the molarity a of the dibasic acid. In the present case the "molarity" of the pseudodibasic acid is $a/2$; with this substitution, and then with A/aK for "K_2," this expression becomes $F = (p/200)\sqrt{A/2KW}$.

In accordance with the correspondence indicated, the "complex" feasibility at the first equivalent point is dependent on the concentration and that at the second equivalent point is independent of the concentration, whereas the opposite is the case for the true dibasic acid.

8. Determination of the Constants. For the exact expressions defining the constants $\mathbf{A}$ and $\mathbf{K}$ in terms of measured values of $\mathbf{H}$, a, and b_s, Eq. (1) gives, with $x = a/2 - b_s - \mathbf{D}/\gamma$ and with the γ convention of Chapter V,

$$A^2 - 2AH\left(\frac{1 + 4Kx^2/a}{1 + 2x/a}\right) + H^2\left(\frac{1 - 2x/a}{1 + 2x/a}\right) = 0, \tag{24}$$

or, since the third term is always positive then as explained in Eqs. V(78–81),

$$-\log \mathbf{A} = \mathrm{p}\mathbf{H} + \frac{\tanh^{-1}(2x/a)}{2(2.3)} \pm \left(\frac{1}{2.3}\right)\left|\cosh^{-1}\left(\frac{1 + 4\mathbf{K}x^2/a}{\sqrt{1 - 4x^2/a^2}}\right)\right| - \log\gamma. \tag{25}$$

As expected from Eq. (3), $A = H$ when $x = 0$, or when $b_s = a/2 - D$. Hence since both dH/db_s and dx/db_s are negative, the positive sign is used in Eq. (25) when the experimental quantity x is positive, the negative sign when x is negative. When $b_s = a/2$, as for the salt "$NaHF_2$," we have simply $x = -\mathbf{D}/\gamma$, so that then $\mathbf{A} \cong \mathbf{H}\gamma$, provided that such salt solution is neither too acid nor too alkaline. With D, or $\mathbf{D}/\gamma$, neglected in x, and with $n^* = 2b_s/a$, Eq. (24) is identical with Eq. (11).

For $\mathbf{K}$, Eq. (1) gives

$$\mathbf{K} = \frac{a(\mathbf{H}\gamma - \mathbf{A})^2}{8\mathbf{H}\gamma\mathbf{A}x^2}\left[1 - \frac{2x}{a}\left(\frac{\mathbf{H}\gamma + \mathbf{A}}{\mathbf{H}\gamma - \mathbf{A}}\right)\right]. \tag{26}$$

Hence, with an approximate value of $\mathbf{A}$ obtained through Eq. (25) from measurements of the value of $\mathbf{H}$ in the region $b_s \cong a/2$, Eq. (26) may be used to evaluate $\mathbf{K}$ from values of $\mathbf{H}$ in the region $b_s = a/4$ when $\mathbf{K} \cong \mathbf{H}\gamma/a\mathbf{A}$, or in the region $b_s \cong 3a/4$ when $\mathbf{K} \cong \mathbf{A}/a\mathbf{H}\gamma$.

If a weak base such as ammonia with known ionization constant B and with concentration b is used, instead of the strong base b_s, these equations still hold for the determination of $\mathbf{A}$ and $\mathbf{K}$ from the measurement of $\mathbf{H}$ at known values of a and b. The only change required is that x now represents $a/2 - bHB/(HB + W) - D$ or $a/2 - b\mathbf{HB}/(\mathbf{HB} + \gamma\mathbf{W}) - \mathbf{D}/\gamma$. With $b=a/2$, this becomes $x = (a/2)W/(HB + W) - D = (a/2)\mathbf{W}/(\mathbf{HB}/\gamma + \mathbf{W}) - \mathbf{D}/\gamma$. We thus have, in Eqs. (25) and (26), with x so defined, the exact expressions defining the constants in terms of H or $\mathbf{H}$ for a solution of the salt "ammonium bifluoride," or NH_4HF_2, at the molar concentration $c = b = a/2$. If the solution contains a net concentration of strong acid, $g = a_s - b_s$, together with weak acid and weak base, Eqs. (25) and (26) still hold, but with $x = a/2 + g + a\alpha - b\alpha' - \mathbf{D}/\gamma$.

It is difficult to say to what extent the foregoing considerations actually apply to aqueous hydrofluoric acid. The following values of the constants have been reported(2): $A = 7.2 \times 10^{-4}$ and $K = 5.5$ at 25°, by Pick; $A = 7.4 \times 10^{-4}$ and $K = 4.7$ at 25°, by Davies and Hudleston; $A = 1.67 \times 10^{-4}$ at 25°, by Auméras; $A = 4.0 \times 10^{-4}$ and $K = 27$ at 15°, $A = 2.4 \times 10^{-4}$ and $K = 33$ at 25°, by Roth; $A = 6.71 \times 10^{-4}$ and $K = 3.96$ at 25°, by Broene and De Vries. Kolthoff and Rosenblum (Ref. VI–2, p. 381) list simply $A = 1.67 \times 10^{-5}$ at 25°. The discrepancies are serious. At any rate it would appear from one point of view that the complex constant must be rather small since aqueous "ammonium bifluoride," or NH_4HF_2, is a buffer. This buffering action indicates a small ratio of the pseudoconstants, "K_1/K_2," and hence a small value of the "complex" constant K.

B. Ionization of Dimeric Acid

For comparison we shall here also consider another "complex" monobasic acid, the case in which the uncharged acid, X^0, dimerizes in solution to form the species X_2^0, with A as the usual ionization constant of X^0, and $K = [X^0]^2/[X_2^0]$. The process here implied is either simply $X_2^0 \rightleftarrows 2X^0$ or, with the solvent involved, $X_2^0 + 2hH_2O \rightleftarrows 2X^0{\cdot}hH_2O$. The equilibrium constant is used as just defined, in either case, on the assumption of a constant activity of water.

1. Analytical and Graphical Relations. If the total analytical

(2) H. Pick, *Nernsts Festschrift*, Knapp, Halle, 1912, p. 360; C. W. Davies and L. J. Hudleston, *J. Chem. Soc.*, **125,** 260 (1924); M. Auméras, *Compt. rend.*, **184,** 1650 (1927); W. A. Roth, *Ann.*, **542,** 35 (1939); H. H. Broene and T. De Vries, *J. Am. Chem. Soc.*, **69,** 1644 (1947).

concentration of "X," or the analytical *normality* of the acid, is a, then $a = [X^0] + 2[X_2^0] + [X^-]$. Hence

$$[X^-] = -AK(H+A)/4H^2 + \sqrt{(\quad)^2 + aA^2K/2H^2}, \tag{27}$$

$$[X^0] = -K(H+A)/4H + \sqrt{(\quad)^2 + aK/2}, \tag{28}$$

$$[X_2^0] = a/2 - [X^0](H+A)/2H. \tag{29}$$

Differentiation shows $d[X^-]/dpH$ to be always positive and $d[X^0]/dpH$ and $d[X_2^0]/dpH$ to be always negative. With no maximum or minimum in any of these species concentrations, as $f(pH)$, there is now no "pseudodibasic" effect. Essentially the value of K merely controls the apparent strength of the monobasic acid. Nevertheless the solution is "complex" in that the charge coefficient is not independent of a, and the titration curve will not have the shape of that of a true simple monobasic acid. It is impossible, in other words, to express the relations by means of a single "apparent" ionization constant as was done in the noncomplex cases discussed in Chapter VI, and in this case the expression for H turns out to be actually quintic instead of cubic. For titration with strong base, we have

$$H - OH = [X^-] - b_s, \tag{30}$$

so that, with Eq. (27),

$$2H^5 + 4H^4b_s + H^3(2b_s^2 + AK - 4W) + H^2[AK(b_s + A) - 4b_sW] - H[A^2K(a - b_s) + W(AK - 2W)] - A^2KW = 0. \tag{31}$$

If either K or $A = 0$, Eq. (30) becomes simply $H - OH = -b_s$; if $K = \infty$, Eq. (31) becomes the cubic for titration of a simple monobasic acid, Eq. X(22); and if $A = \infty$, it becomes the equation for strong acid, or $H - OH = a - b_s$. For the determination of the constants $\mathbf{A}$ and $\mathbf{K}$ from measurements of $\mathbf{H}$, a, and b_s, the combination of Eqs. (27) and (30) gives

$$\mathbf{A} = \gamma\mathbf{H}[x/2(a-x) + \sqrt{(\quad)^2 + 2x^2/\mathbf{K}(a-x)}], \text{(2a)} \tag{32}$$

and

$$\mathbf{K} = 2(\gamma\mathbf{H}x)^2/\mathbf{A}[\mathbf{A}(a-x) - \gamma\mathbf{H}x], \tag{33}$$

with $x = (b_s + \mathbf{D}/\gamma)$.

For approximate formulas for H, with $AK \gg W$, the best approximation for the pure acid solution (with $b_s = 0$) is given by the terms in H^5, H^3, H^1 of Eq. (31),

$$H^2 \cong -(AK/4 - W) + \sqrt{(\quad)^2 + aA^2K/2 + W(AK/2 - W)}, \tag{34}$$

or $H^2 \cong -AK/4 + \sqrt{(\quad)^2 + aA^2K/2}$.

For the beginning of titration, the very start, with $b_s \cong 0$, requires consideration by approximation methods of all the first five terms of Eq. (31).

(2a) For application of an approximate form of this expression to acetic acid, see A. Katchalsky, H. Eisenberg and S. Lifson, *J. Am. Chem. Soc.*, **73**, 5889 (1951).

But when b_s becomes appreciable, of the order of magnitude of $a/2$, the H^3 sequence becomes a good approximation; for $b_s \gg A$ and $AK \gg W$, it gives

$$H \cong -AK/4b_s + \sqrt{(\quad)^2 + A^2K(a - b_s)/2b_s^2}. \tag{35}$$

Near $b_s = a$, the curve is best described by the last three terms of Eq. (31). Then as long as $b_s \neq a$,

$$H \cong A(a - b_s)/2b_s + \sqrt{(\quad)^2 + AW/b_s}. \tag{36}$$

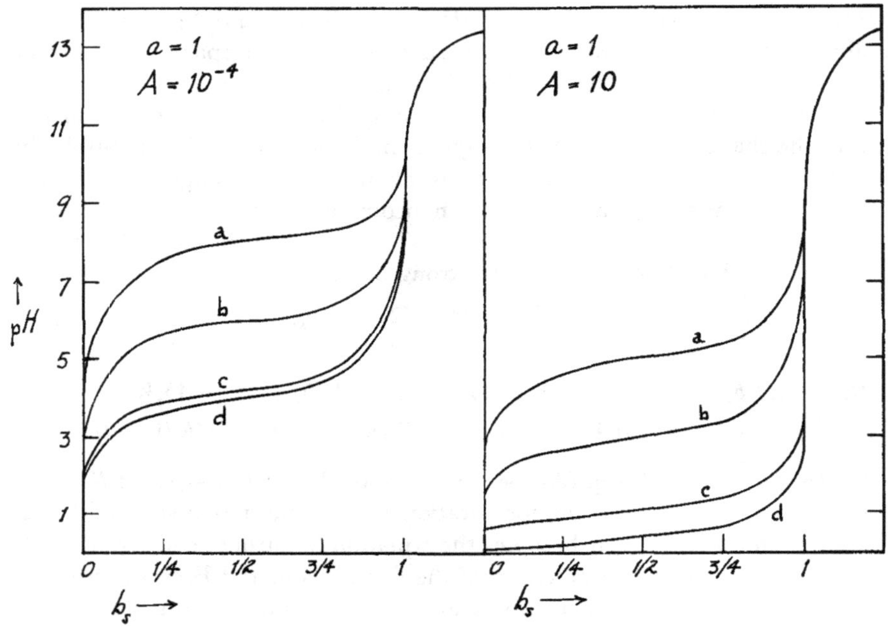

Fig. 15–8. Titration curves for "dimeric acid." For $A = 10^{-4}$: a, $K = 10^{-8}$; b, $K = 10^{-4}$; c, $K = 1$; d, $K = 10^4$ ($\rightarrow \infty$). For $A = 10$: a, $K = 10^{-12}$; b, $K = 10^{-8}$; c, $K = 10^{-4}$; d, $K = 1$ ($\rightarrow \infty$).

But if $a = b_s = c$ (the pure salt solution), then

$$H \cong \frac{W(AK - 2W)}{2[AK(c + A) - 4cW]} + \sqrt{(\quad)^2 + \frac{A^2KW}{AK(c + A) - 4cW}}. \tag{37}$$

Now if $AK > W, H \cong W/2(c + A) + \sqrt{(\quad)^2 + AW/(c + A)} \cong \sqrt{AW/(c + A)}$, as in Eq. x(14).

Two sets of titration curves have been sketched (Fig. 15–8) to illustrate the effect of K on an acid of given value of A, both for the same (unit) value of the concentration a; the pH values, calculated through the full Eq. (31), when necessary, are tabulated in Table 15–2. [Eq. (31) is necessary only for the calculation of pH values for round values of the ratio b_s/a; the titration curves may always be calculated directly very easily from the combination of Eqs. (27) and (30), giving b_s explicitly for specified values of a and H.]

2. Difficulty of Distinguishing Simple and Dimeric Acids. It is apparent that in certain cases it would be very difficult to distinguish such a "complex" monobasic titration curve from that of a true simple monobasic acid. With low K, the curve resembles that of a simple acid with apparent ionization constant $A\sqrt{K/a}$; with high K, the shape of the curve approaches that of a simple monomeric monobasic acid.

Table 15–2. Values of pH for $a = 1$.

	$A = 10^{-4}$				$A = 10$				
b_s \ K	10^{-8}	10^{-4}	1	$10^4 (\to \infty)$	10^{-12}	10^{-8}	10^{-4}	1	∞
0	4.08	3.06	2.15	2.00	2.58	1.58	0.60	0.043	0.038
1/4	7.61	5.61	3.78	3.52	4.61	2.61	0.84	0.171	0.164
1/2	8.00	6.00	4.21	4.00	5.00	3.00	1.10	0.345	0.341
3/4	8.33	6.33	4.60	4.48	5.33	3.33	1.39	0.644	0.644
1	10.10	9.04	9.00	9.00	8.10	7.06	7.02	7.02	7.02
5/4	13.40	13.40	13.40	13.40	13.40	13.40	13.40	13.40	13.40

If the acid is assumed to be simple, with $K = \infty$ and with the normality, a, representing the actual molarity of the acid, we would calculate A from the values of H, a, and b_s according to Eq. v(16). For the third acid in Table 15–2, with $A = 10^{-4}$ and $K = 1$, which at $a = 1$ and $b_s = 0.5$ gives $H = 6.17 \times 10^{-5}$, Eq. v(16) would give $A = 6.17 \times 10^{-5}$ (which is not far from the actual value for acetic acid, for example). The titration curve of a simple (monomeric) acid with $A = 6.17 \times 10^{-5}$, at the molarity $a = 1$, furthermore, would be hardly distinguishable from that of the dimeric acid with $A = 10^{-4}$ and $K = 1$ at the normality 1 or molarity 0.5. The pH values at the titration ratios listed in Table 15–2 would be 2.11, 3.73, 4.21, 4.69, 9.10, and 13.40.

At lower concentrations the difference between the titration curves of the two types of acids decreases still further. At the normality $a = 0.01$, for example, this same dimeric acid with $A = 10^{-4}$ and $K = 1$ gives a titration curve differing, at any point, by not more than 0.01 in pH from that of a simple monomeric acid with $A = 9.87 \times 10^{-5}$ at the molarity 0.01. The pH values are 3.02, 3.59, 4.02, 4.49, 8.00 at $b_s = 0$, 0.0025, 0.005, 0.0075, and 0.01 respectively. In such a case the close fit of a titration curve to that of a monomeric monobasic acid expected according to Eq. x(22) is not evidence that the solute is actually monomeric. In this connection it is to be recognized that it is extremely difficult to interpret any titration curve as a determination of more than an equivalent weight.

It should also be noted that with a as the analytical *normality* of the acid, the same Eq. (31) holds whether the "original" solute species is simply X, which then associates in solution to form X_2, or is actually X_2, which then dissociates in solution to form X. As in the problem of ampholytes (discussed in Chapter VI), this question of the nature of the original solute species is beyond resolution by means of the equilibrium relations of the aqueous solution. These relations can not tell us whether the ampholyte is, *per se*, of the nature of an "inner salt" or simply a "molecule" and whether a particular acid is *per se* "monomeric" or "dimeric" before being dissolved in water.

C. Chromic Acid; Dibasic Acid With Complex Divalent Anion

We shall here consider a dibasic acid, H_2X, having a univalent ion, HX^-, which associates to form the poly-acid (or condensed) ion $X_2^=$ called dichromate, $Cr_2O_7^=$, in the case of chromic acid, and pyrosulfate, $S_2O_7^=$, in the case of sulfuric acid. If the only species produced by the acid are H_2X, HX^-, $X^=$, and $X_2^=$, the equilibrium constants involved are the usual ionization constants K_1 $(= H[HX^-]/[H_2X])$ and K_2 $(= H[X^=]/[HX^-])$, and the complex constant K, or

$$K = [X_2^=]/[HX^-]^2. \tag{38}$$

We note at the start that the equilibrium relations do not depend on whether the "dichromate ion" is written as $Cr_2O_7^=$ or as $(HCrO_4^-)_2$, or on whether we consider it as being "formed" by any particular one of the following "reactions," which are in the final analysis indistinguishable from each other: $2HCrO_4^- \rightleftarrows (HCrO_4^-)_2$; $2HCrO_4^- \rightleftarrows Cr_2O_7^= + H_2O$; $2CrO_4^= + 2H^+ \rightleftarrows Cr_2O_7^= + H_2O$; $2CrO_4^= + H_2O \rightleftarrows Cr_2O_7^= + 2OH^-$; $2CrO_4^= + 2H^+ \rightleftarrows (HCrO_4^-)_2$; $2H_2CrO_4 \rightleftarrows Cr_2O_7^= + 2H^+ + H_2O$; $CrO_4^= + H_2CrO_4 \rightleftarrows Cr_2O_7^= + H_2O$; etc. The only requirement is that the ion must contain two moles of "CrO_3" and be doubly charged. Through the definitions of K_1 and K_2 the equilibrium constant for the formation of "$Cr_2O_7^=$" may always be put into the form of Eq. (38), according to which the ratio $[X_2^=]/[HX^-]^2$ is a constant independent of the hydrogen ion concentration of the solution and hence of the presence or absence of other acids and bases, weak or strong.

1. Concentrations of Species. As usual, however, the concentrations of the four species will vary, for a given concentration of the acid, with the hydrogen ion concentration, although since the relations are "complex" the distribution of the species for a given value of H varies with the analytical concentration. If a represents the number of moles of "CrO_3" per liter, and if

$$a = [H_2X] + [HX^-] + [X^=] + 2[X_2^=], \tag{39}$$

then, through the definitions of the constants, we have

$$[HX^-] = -(H^2 + HK_1 + K_1K_2)/4HK_1K + \sqrt{(\quad)^2 + a/2K}, \quad (40)$$

with $[H_2X] = [HX^-]H/K_1$, $[X^=] = [HX^-]K_2/H$, and $[X_2^=] = K\,[HX^-]^2$.

Differentiation then shows that $d[H_2X]/dH$ is always positive and $d[X^=]/dH$ always negative; but both $[HX^-]$ and $[X_2^=]$ have maximum values at $H = \sqrt{K_1K_2}$. Moreover, $[X_2^=]$ is greater than (or less than) $[HX^-]$ if $[HX^-]$ is greater than (or less than) $1/K$.

2. On the Formula of the Actual Solute. The ionizing molecular chromic acid in chromic acid solution is sometimes represented as H_2CrO_4 and sometimes as $H_2Cr_2O_7$. Concerning this question, we may note that still more complicated schemes may be considered, involving for example the seven species CrO_3, H_2CrO_4, $HCrO_4^-$, $CrO_4^=$, $Cr_2O_7^=$, $HCr_2O_7^-$, and $H_2Cr_2O_7$, which will be related by six independent equilibrium constants. It will be clear, however, that here too the ratio $[Cr_2O_7^=]/[HCrO_4^-]^2$ will be independent of H. Furthermore, $[CrO_4^=]$ increases with decreasing H without limit, and the sum of the concentrations of the three uncharged species, with a distribution independent of H, decreases with increasing H without limit. But $[HCrO_4^-]$ and $[Cr_2O_7^=]$ will have maximum values at the same value of H while $[HCr_2O_7^-]$ will have a maximum value at a different, and higher, value of H. As in the case discussed in Section B of this chapter, however, the formula of the uncharged species representing the "original" solute can not be determined from the equilibrium relations of the solution. The distribution of the seven species will be a function of a, H, and the constants, and the resulting mathematical equation relating H to these quantities and to the concentrations of other solutes which may be present will be independent of the formula chosen as that of the "chromic acid" itself. This question remains unanswerable and therefore meaningless from this point of view even if we study in addition the solubility of chromic acid or of any of its salts as a function of H or, in analytical terms, as a function of the concentrations of added solutes, including the chromates themselves.

If the empirical composition of the saturating solid phase, for example, is $CrO_3{\cdot}H_2O$, we may consider as possible "formulas" $CrO_3{\cdot}H_2O$, H_2CrO_4, $H(HCrO_4)$, $H_2Cr_2O_7{\cdot}H_2O$, with $P_1 = [CrO_3]$, $P_2 = [H^+]^2[CrO_4^=]$, $P_3 = [H^+][HCrO_4^-]$, $P_4 = [H^+]^2[Cr_2O_7^=]$, as their respective "solubility product" constants. But these are of course interrelated so that, for example, $P_1 = P_2/K_1K_2K'$ (with $K' = [H_2CrO_4]/[CrO_3]$), $P_3 = P_2/K_2$ and $P_4 = P_2^2K/K_2^2$. In any case, therefore, $[HCrO_4^-]$ is inversely proportional to H, at saturation. Moreover, the analytical concentration a of the solute, or the acid, is related to $[HCrO_4^-]$, or $[HX^-]$, according to some function similar to the inverse function of Eq. (40). Hence it follows that the variation of the solubility with respect to H can not distinguish the different "formulas" of the solid. This statement holds not only for the effect of foreign acids and bases on the solubility of the acid but also for the effect of the salts of the acid itself upon its solubility. This will be clear if we define the "solubility" as the total analytical concentration of "CrO_3"; the addition of an alkali chromate or dichromate is then seen to be merely the addition of strong base, the effect of which is to modify the hydrogen ion concentration.

A similar situation holds with regard to the "formula" or the constitution of a salt of the acid. If a solid has the empirical composition $Na_2O{\cdot}2CrO_3{\cdot}2H_2O$, two of the possible "formulas" are $Na_2Cr_2O_7{\cdot}2H_2O$ and $NaHCrO_4{\cdot}(1/2)H_2O$, with $P_1 = [Na^+]^2[Cr_2O_7^=]$, $P_2 = [Na^+][HCrO_4^-]$, $P_3 = [Na^+][H^+][CrO_4^=]$ as possible solubility products, expressions which are interrelated through the equilibrium constants. Again however, the variation of the solubility with respect to H or with respect to concentrations of added solutes can not distinguish the possibilities. In any case $[HCrO_4^-]$ is inversely proportional to $[Na^+]$ for saturation, and the relation between $[HCrO_4^-]$ and the analytical concentration of CrO_3 is always a function similar to that of Eq. (40). The solution of this solid, therefore, corresponds to the stoichiometric point $b_s = a$. The solution may be called either one of $NaHCrO_4$ or one of $Na_2Cr_2O_7$. Both ions, $HCrO_4^-$ and $Cr_2O_7^=$, are present (among others), and their ratio depends on K of Eq. (38) and on the concentration of the salt. Whether the solid salt is to be given one or the other of these formulas is not a question to be decided upon the basis of the relations in the aqueous solution.

3. Values of b_s for Special Values of H. We return now to the consideration of the ionization scheme involving the four species H_2CrO_4, $HCrO_4^-$, $CrO_4^=$, and $Cr_2O_7^=$ (in general symbols, H_2X, HX^-, $X^=$, and $X_2^=$) alone, and the three equilibrium constants K_1, K_2, and K already defined. The electroneutrality equation here becomes

$$D = H - OH = [HX^-] + 2[X^=] + 2[X_2^=] - b_s; \tag{41}$$

combined with Eq. (39), this means

$$H - OH = (a - b_s) + (K_2/H - H/K_1)[HX^-], \tag{42}$$

with $[HX^-]$ given by Eq. (40).

The general values of b_s for specified values of a, corresponding to special values of H such as $H = \sqrt{W}$, $H = K_1$, $H = K_2$, etc., may be expressed explicitly by means of Eq. (42) combined with Eq. (40). For the important case of $H = \sqrt{K_1K_2}$, we have, with no dependence upon K,

$$b_s = a - (\sqrt{K_1K_2} - W/\sqrt{K_1K_2}) \cong a. \tag{43}$$

4. Determination of the Constants. Eqs. (40) and (42) may be combined to give expressions for the evaluation of the constants from the

analytical concentrations and the value of H, as follows. With $x = (a - b_s - D)/a = (a - b_s - \mathbf{D}/\gamma)/a$,

$$K_1 = -(Hx + 2K_2)/2\mathbf{Y} \pm \sqrt{(\quad)^2 + H^2(1-x)/\mathbf{Y}}, \tag{44}$$

in which $\mathbf{Y} = 2ax^2K - xK_2/H - (1+x)K_2^2/H^2$. Since $(1-x)$ is always positive, then if $\mathbf{Y}$ is positive we have, according to Eq. v(75), and with the usual γ convention,

$$-\log \mathbf{K}_1 = \mathrm{p}\mathbf{H} - \tfrac{1}{2}\log\left(\frac{1-x}{\mathbf{Y}'}\right) + \left(\frac{1}{2.3}\right)\sinh^{-1}\left[\frac{x + 2\mathbf{K}_2/\mathbf{H}\gamma^3}{2\sqrt{(1-x)\mathbf{Y}'}}\right] - \log\gamma, \tag{45}$$

with $\mathbf{Y}' = 2ax^2\mathbf{K}/\gamma^2 - x\mathbf{K}_2/\mathbf{H}\gamma^3 - (1+x)\mathbf{K}_2^2/\mathbf{H}^2\gamma^6$. If $\mathbf{Y}$ is negative, then from Eq. v(81),

$$-\log \mathbf{K}_1 = \mathrm{p}\mathbf{H} - \tfrac{1}{2}\log\left|\frac{1-x}{\mathbf{Y}'}\right| \pm \left|\left(\frac{1}{2.3}\right)\right| \cosh^{-1}\left[\frac{x + 2\mathbf{K}_2/\mathbf{H}\gamma^3}{2\sqrt{(1-x)\mathbf{Y}'}}\right]\Bigg| - \log\gamma, \tag{46}$$

As in Eq. (25), the positive sign is used in Eq. (46) when x is positive, the negative when x is negative; this follows since, as expected from Eq. (43), $K_1 = H^2/K_2 = H\sqrt{\mathbf{Y}}$ when $x = 0$, while both dH/db_s and dx/db_s are negative.

Also,

$$K_2 = -H(x - 2H/K_1)/2(1+x) \pm \sqrt{(\quad)^2 + H^2\mathbf{J}/(1+x)}, \tag{47}$$

in which $\mathbf{J} = 2ax^2K + Hx/K_1 - (1-x)H^2/K_1^2$. Since $(1+x)$ is always positive, then if $\mathbf{J}$ is positive

$$-\log \mathbf{K}_2 = \mathrm{p}\mathbf{H} - \tfrac{1}{2}\log\left(\frac{\mathbf{J}'}{1+x}\right) + \left(\frac{1}{2.3}\right)\sinh^{-1}\left[\frac{x - 2\mathbf{H}\gamma/\mathbf{K}_1}{2\sqrt{(1+x)\mathbf{J}'}}\right] - 3\log\gamma, \tag{48}$$

in which $\mathbf{J}' = 2ax^2\mathbf{K}/\gamma^2 + x\mathbf{H}\gamma/\mathbf{K}_1 - (1-x)\mathbf{H}^2\gamma^2/\mathbf{K}_1^2$. If $\mathbf{J}'$ is negative,

$$-\log \mathbf{K}_2 = \mathrm{p}\mathbf{H} - \tfrac{1}{2}\log\left|\frac{\mathbf{J}'}{1+x}\right| \pm \left|\left(\frac{1}{2.3}\right)\right| \cosh^{-1}\left[\frac{x - 2\mathbf{H}\gamma/\mathbf{K}_1}{2\sqrt{(1+x)\mathbf{J}'}}\right]\Bigg| - 3\log\gamma; \tag{49}$$

and again, for the reasons given under Eq. (46), the sign used is the sign of the experimental quantity x.

Finally,

$$K = \frac{(K_1K_2 - H^2)^2}{2ax^2H^2K_1^2}\left[1 + \frac{x(K_1K_2 + HK_1 + H^2)}{K_1K_2 - H^2}\right]. \tag{50}$$

With activity coefficient and thermodynamic constants this becomes

$$\mathbf{K} = \frac{(\mathbf{K}_1\mathbf{K}_2 - \mathbf{H}^2\gamma^4)^2}{2ax^2\mathbf{H}^2\mathbf{K}_1^2\gamma^4}\left[1 + \frac{x(\mathbf{K}_1\mathbf{K}_2 + \mathbf{H}\mathbf{K}_1\gamma^3 + \mathbf{H}^2\gamma^4)}{\mathbf{K}_1\mathbf{K}_2 - \mathbf{H}^2\gamma^4}\right]. \tag{51}$$

If approximate values of two of the constants are available, the third may be estimated with some accuracy through one of these expressions. But there is no general region of values of a and b_s/a (or of a and x) in which the value of H is approximately dependent on only one constant, contrary to the usual case for ordinary dibasic or tribasic acids. Even the approximate determination of the constants must therefore always be simultaneous, for at least two at a time. For example, measurements in a region in which $H \gg K_2$ could be used in Eqs. (44) and (50) for the determination of K_1 and K from a series of values of H, x, a. In a region in which $H \ll K_1$, Eqs. (47) and (50) could be used for the determination of K_2 and K.

In the latter case, moreover, a certain degree of simplification of the exact equation is possible and useful, so that with $H \ll K_1$ Eq. (47) becomes

$$K_2 \cong \frac{H(b_s - a + D)}{2(2a - b_s - D)} + \sqrt{(\quad)^2 + \frac{2(b_s - a + D)^2 H^2 K}{(2a - b_s - D)}}, \tag{52}$$

while

$$K \cong K_2[K_2(2a - b_s - D) - H(b_s - a + D)]/2H^2(b_s - a + D)^2. \tag{53}$$

These equations may be expected to apply in the region $b_s \cong 3a/2$, or $a < b_s < 2a$, where the effect of K_1 will generally be negligible. Through them it becomes possible to determine, at least approximately, the constants K_2 and K from a series of measurements in this region, involving either variation of the ratio b_s/a for fixed a or variation of a for fixed ratio b_s/a, or both. The region in which $a < b_s < 2a$ corresponds to a solution containing various proportions of the salts K_2CrO_4 ($b_s = 2a$) and $K_2Cr_2O_7$ ($b_s = a$). If the molar concentration of K_2CrO_4 is c and that of $K_2Cr_2O_7$ is $c/\mathbf{r}$, then $a = c(2 + \mathbf{r})/\mathbf{r}$, $b_s = 2c(1 + \mathbf{r})/\mathbf{r}$ and $x = -[(c + D)/c][\mathbf{r}/(2 + \mathbf{r})]$, expressions to be used in Eqs. (47) and (50). If we neglect D relative to the analytical concentrations in the approximate Eqs. (52) and (53), we have, in terms of c and $\mathbf{r}$,

$$K_2 \cong \mathbf{r}H/4 + \sqrt{(\quad)^2 + \mathbf{r}cH^2K}, \tag{54}$$

$$K \cong K_2(2K_2 - \mathbf{r}H)/2\mathbf{r}cH^2. \tag{55}$$

With $\mathbf{r} = 2$ we have the mid-point of the second half of the titration of chromic acid, or $b_s = 3a/2$; again if c is the concentration of K_2CrO_4 at this point, $a = 2c$, $b_s = 3c$ and $x = -(c + D)/2c$, whereupon Eq. (47) becomes

$$K_2 = H(c + D + 4cH/K_1)/2(c - D) + \sqrt{(\quad)^2 + H^2\mathbf{J}/(c - D)}, \tag{56}$$

in which $\mathbf{J} = 2K(c + D)^2 - H(c + D)/K_1 - H^2(3c + D)/K_1^2$. In the approximation of Eq. (54) this means

$$K_2 \cong H/2 + \sqrt{H^2/4 + 2cH^2K}, \tag{57}$$

$$K \cong K_2(K_2 - H)/2cH^2. \tag{58}$$

Eq. (52) moreover may be rearranged to express the ratio $K_2/\sqrt{K}$, which we shall call $\mathbf{y}$, as follows:

$$\mathbf{y} = K_2/\sqrt{K} \cong H\sqrt{2(b_s - a + D)^2/(2a - b_s - D)}(\mathbf{Y} + \sqrt{\mathbf{Y}^2 + 1}), \quad (59)$$

with $\mathbf{Y} = 1/\sqrt{8K(2a - b_s - D)}$. Hence if K is fairly large, then $y \cong H\sqrt{a}$ at $b_s = 3a/2$, or $H \cong K_2/\sqrt{aK}$ at the mid-point of the second half of the titration.

Another function, $\mathbf{p} = K_1K_2$, may be estimated from a solution in which $b_s = a$, the first equivalent point of the titration. This is a pure solution of the salt $KHCrO_4$ or $K_2Cr_2O_7$. For $K_2Cr_2O_7$ at the molar concentration c, with $a = b_s = 2c$, we obtain, on multiplying Eq. (44) by K_2,

$$\mathbf{p} = K_1K_2 = H^2(2c - HD/2K_2)/\mathbf{Y} \pm \sqrt{(\quad)^2 - H^4(2c + D)/\mathbf{Y}}, \quad (60)$$

in which $\mathbf{Y} = 2c - D - HD/K_2 - 2H^2D^2K/K_2^2$. Hence, since $(2c + D)$ is positive, then if $\mathbf{Y}$ is negative,

$$-\log \mathbf{K}_1\mathbf{K}_2 = 2\mathrm{p}\mathbf{H} - \tfrac{1}{2}\log\left|\frac{2c + \mathbf{D}/\gamma}{\mathbf{Y}'}\right| - \frac{\sinh^{-1}}{2.3}\left[\frac{2c - \mathbf{HD}\gamma^2/2\mathbf{K}_2}{\sqrt{|(2c+\mathbf{D}/\gamma)\mathbf{Y}'|}}\right] - 4\log\gamma; \quad (61)$$

here $\mathbf{Y}'$ represents the expression $2c - \mathbf{D}/\gamma - \mathbf{HD}\gamma^2/\mathbf{K}_2 + 2\mathbf{H}^2\mathbf{D}^2\mathbf{K}\gamma^2/\mathbf{K}_2^2$. If $\mathbf{Y}$ is positive,

$$-\log \mathbf{K}_1\mathbf{K}_2 = 2\mathrm{p}\mathbf{H} - \tfrac{1}{2}\log\left(\frac{2c + \mathbf{D}/\gamma}{\mathbf{Y}'}\right) \pm \left|\frac{\cosh^{-1}}{2.3}\left[\frac{2c - \mathbf{HD}\gamma^2/2\mathbf{K}_2}{\sqrt{(2c+\mathbf{D}/\gamma)\mathbf{Y}'}}\right]\right| - 4\log\gamma; \quad (62)$$

the positive sign is used if the solution is alkaline, the negative if it is acid (positive D). This follows because, from Eq. (43), $H = \sqrt{K_1K_2}$ in this salt solution only if $K_1K_2 = W$.

Finally, K_1 may be estimated not only through Eq. (44), from approximate values of K_2 and K, but also as follows from approximate values of $\mathbf{p}$ $(= K_1K_2)$ and $\mathbf{y}$ $(= K_2/\sqrt{K})$:

$$K_1 = H\left(\frac{b_s + D}{a - b_s - D}\right) - \frac{\mathbf{p}}{H}\left(\frac{2a - b_s - D}{a - b_s - D}\right) - \frac{2H\mathbf{p}^2(a - b_s - D)}{\mathbf{y}^2(H^2 - \mathbf{p})}. \quad (63)$$

Eqs. (44–63) may be used, therefore, to determine the three constants from measurements at a variety of analytical relations, and the values first obtained as approximations may be refined by repetition of the calculation process until satisfactory constancy is reached. The introduction of activity coefficients on the basis of the convention assumed in Chapter v is simple, as already illustrated in these equations.

5. TITRATION CURVE. If the constants are known, then we may use

Eqs. (40) and (42), expanded, for an analytical expression for H in terms of constants and concentrations:

$$\begin{aligned} H^6 &+ H^5[b_s - K_1(2K_1K - 1)] + H^4[(a - b_s)K_1(4K_1K - 1) - W] \\ &- H^3K_1[2(a - b_s)^2K_1K + 2aK_2 + K_1K_2 - W(4K_1K - 1)] \\ &+ H^2K_1^2[(a - b_s)(K_2 - 4KW) - K_2^2] \\ &+ HK_1^2[(2a - b_s)K_2^2 + W(K_2 - 2KW)] + K_1^2K_2^2W = 0. \end{aligned} \tag{64}$$

[With $K = 0$, this reduces to Eq. XIV(3) multiplied by $(K_1K_2 - H^2)$.] For certain numerical values of data this equation gives two positive values of H, and again the correct value has to be chosen with reference to the possible limits of H. For example, with $a = b_s = 1$ and with $K_1 = 10^{-2}$, $K_2 = 10^{-4}$ and $K = 10^2$, the largest terms of the equation are those in H^5, H^3, and H, giving $H = 1.076 \times 10^{-3}$ and $H = 0.933 \times 10^{-3}$. But if $b_s = a$, then, from Eq. (43), H must be $< \sqrt{K_1K_2}$, so that the correct value is $H = 0.933 \times 10^{-3}$.

Eq. (64) has been used to calculate some points for the titration curve of an acid at unit concentration ($a = 1$) with $K_1 = 10^{-3}$ and $K_2 = 10^{-6}$, and with various values of the complex constant K; these are tabulated in Table 15–3 and plotted in Fig. 15–9. The shape of each curve has been fixed by calculation of b_s for specified a and H through Eq. (42) combined with Eq. (40).

With $K = 0$ the acid is of course simply a normal dibasic acid. With increasing K, the first half of the pH titration curve is lowered and the second half raised so that the effect of K may be said to increase the apparent magnitude of K_1 and to decrease that of K_2, if these constants are estimated with the neglect of K. The limiting behavior approached as K is increased to infinity is that of a strong monobasic acid at the concentration a; in this case $[X_2^{=}] \rightrightarrows a/2$ and $D \rightrightarrows 2[X_2^{=}] - b_s$ or $a - b_s$. For not too large values of K, in the example chosen, the value of H at $b_s = a$, or in a solution of "$KHCrO_4$" or "$K_2Cr_2O_7$," is $\sim \sqrt{K_1K_2}$ and then begins to fall off with very large K since its limit at $K = \infty$ is $\sqrt{W}$. For some additional values of K (besides those in Table 15–3), H at $b_s = a = 1$ has the values 30.6, 26.3, 13.5, 2.64, and 0.846($\times 10^{-6}$) for $K = 10^3, 10^5, 10^7, 10^{10}, 10^{12}$ respectively. At $b_s = 3a/2$, the mid-point of the second half of the titration, H starts at $\sim K_2$ and begins to decrease, having the approximate value of $K_2/\sqrt{aK}$ for a very wide range of values of K. The apparent value of "K_2" for such an acid, then, if read without further correction from the mid-point of this part of the titration curve, for appreciable values of a and large K, is not K_2 but, roughly, $K_2/\sqrt{aK}$. Additional values of the acid in Table 15–3, at $a = 1$ and $b_s = 1.5$, are $H = 311$, 31.6, 3.16, 0.1 and 0.01($\times 10^{-10}$) for $K = 10^3, 10^5, 10^7, 10^{10}, 10^{12}$ respectively.

6. Actual Values for Chromic Acid. For chromic acid the constants are reported as $K_1 \simeq 0.2$, $K_2 = 3.2 \times 10^{-7}$ and $K = 1/0.023$ or ~ 43. The

Table 15–3. Values of $[H^+]$ for $K_1 = 10^{-3}$, $K_2 = 10^{-6}$, $a = 1$.

b_s \ K	0	1	10	10^2	10^4	10^6	10^8	10^{20}
0	0.0311	0.0320	0.0385	0.0615	0.23	0.65	0.997	1.000
0.5	0.001	0.00162	0.00368	0.0102	0.079	0.30	0.495	0.500
1	3.16×10^{-5}	3.16×10^{-5}	3.16×10^{-5}	3.16×10^{-5}	2.97×10^{-5}	2.03×10^{-5}	8.13×10^{-6}	1.00×10^{-7}
1.5	10^{-6}	6.17×10^{-7}	2.70×10^{-7}	9.5×10^{-8}	9.95×10^{-9}	1.00×10^{-9}	1.00×10^{-10}	2.00×10^{-14}
2	10^{-10}	10^{-10}	10^{-10}	9.9×10^{-11}	6.57×10^{-11}	1.7×10^{-11}	3.7×10^{-12}	10^{-14}

latter two were calculated simultaneously from determinations of H in solutions of fixed ratio of b_s to a (actually of K_2CrO_4 to $K_2Cr_2O_7$) with variation of a, the total range of H being from 3×10^{-7} to 8×10^{-7}; K_1 was then estimated from some measurements at lower ratio of b_s to a, with H between 10^{-3} and 10^{-2}.[(3)] A value of K of ~ 50 is also reported[(4)], based on conductometric titration curves and measurements of electrolytic conductivities and specific gravities.

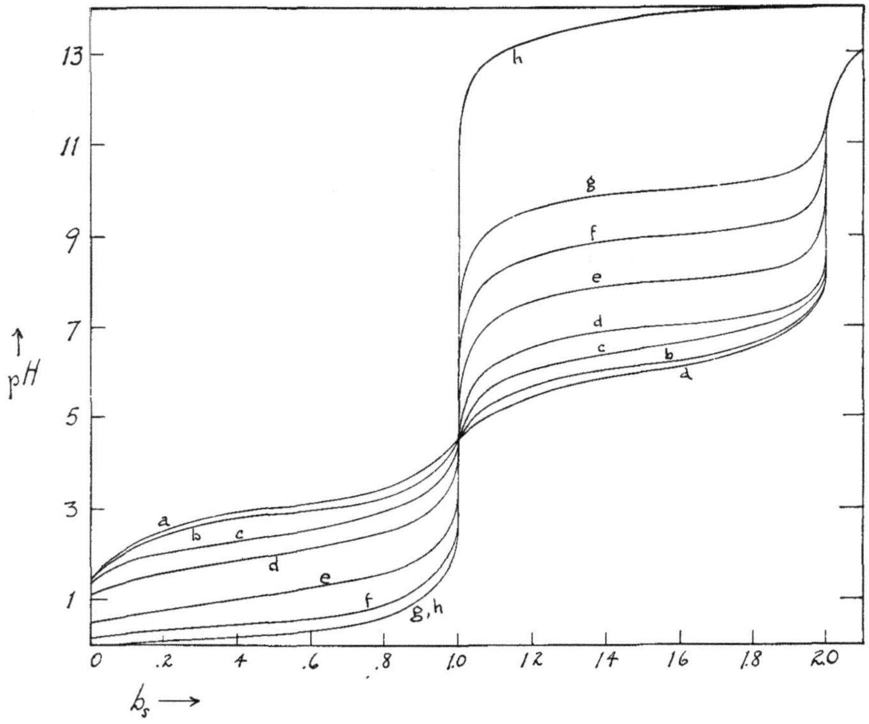

Fig. 15–9. Titration curves for acid of "chromic acid" type, with $a = 1$, $K_1 = 10^{-3}$ and $K_2 = 10^{-6}$: a, $K = 0$; b, $K = 1$; c, $K = 10$; d, $K = 10^2$; e, $K = 10^4$; f, $K = 10^6$; g, $K = 10^8$; h, $K = 10^{20}$ and $K = \infty$.

If the order of magnitude is 10^{-1} for K_1, 10^{-7} for K_2, and 10^2 for K, approximate values of H may be calculated as follows from Eq. (64):

(a) At $b_s = 0$, the first four terms are needed, to be solved by trial and error.

(b) At $b_s \cong a/2$, the three terms in H^5, H^4, H^3, as a simple quadratic, give

$$H^2(b_s - 2K_1^2K) + 4H(a - b_s)K_1^2K - 2(a - b_s)^2K_1^2K \cong 0, \qquad (65)$$

with H depending simply on the constant K_1^2K.

(3) J. D. Neuss and W. Rieman, *J. Am. Chem. Soc.*, **56,** 2238 (1934).
(4) E. Endrédy, *Math. naturw. Anz. ungar. Akad. Wiss.*, **54,** 459 (1936).

(c) At $b_s \cong a$ ($\gtrless a$), the five terms from H^5 to H must be solved by trial and error.

(d) At $b_s = a$ exactly (pure $K_2Cr_2O_7$ at concentration $c = a/2 = b_s/2$), the three terms in H^5, H^3, and H give a quadratic for H^2, so that

$$H^2 \cong -\frac{K_1K_2(4c + K_1)}{K_1(4K_1K - 2) - 4c} + \sqrt{(\quad)^2 + \frac{2cK_1^2K_2^2}{K_1(2K_1K - 1) - 2c}}. \quad (66)$$

(e) At $b_s \cong 3a/2$, the three terms in H^3, H^2, H give

$$2H^2(a - b_s)^2K - H(a - b_s)K_2 - (2a - b_s)K_2^2 \cong 0. \quad (67)$$

(f) At $b_s \cong 2a$, the last three terms of Eq. (64) give

$$H^2(b_s - a) - H(2a - b_s)K_2 - K_2W \cong 0, \quad (68)$$

independent of both K_1 and K.

Table 15–4 shows the value of H calculated for solutions of pure $K_2Cr_2O_7$ at various concentrations, based on the *rounded values* of $K_1 = 0.2$, $K_2 = 3 \times 10^{-7}$ and $K = 40$.

Table 15–4: $K_1 = 0.2$, $K_2 = 3 \times 10^{-7}$, $K = 40$.
$K_2Cr_2O_7$ at molarity $= c$.

c	H
1	1.61×10^{-4}
10^{-1}	1.04×10^{-4}
10^{-2}	5.53×10^{-5}
10^{-3}	2.28×10^{-5}
10^{-4}	7.5×10^{-6}

In Table 15–5 are listed values of H calculated for a solution with $[K_2CrO_4] = c$ and $[K_2Cr_2O_7] = c/2$, so that $a = 2c$ and $b_s = 3c = 3a/2$, for the same rounded values of the constants. The calculation of these values required the following portions of Eq. (64), all divided through by K_1^2, and the last also by K_2:

For $c = 1$ to $c = 10^{-3}$,

$$-H^3(2c^2K) - H^2(cK_2) + H(cK_2^2) \cong 0; \quad (69)$$

for $c = 10^{-4}$,

$$-H^3(2c^2K + K_2) - H^2(cK_2) + H(cK_2^2) \cong 0; \quad (70)$$

for $c = 10^{-5}$,

$$-H^3(2c^2K + K_2) - H^2(cK_2 + K_2^2) + H(cK_2^2 + K_2W) \cong 0; \quad (71)$$

for $c = 10^{-6}$ to 10^{-8},

$$-H^3 - H^2(c + K_2) + H(cK_2 + W) + K_2W \cong 0. \tag{72}$$

Table 15–5: $K_1 = 0.2$, $K_2 = 3 \times 10^{-7}$, $K = 40$.
$[K_2CrO_4] = c$; $[K_2Cr_2O_7] = c/2$.

c	H
1	3.17×10^{-8}
10^{-1}	8.90×10^{-8}
10^{-2}	1.97×10^{-7}
0.075	1.00×10^{-7}
10^{-3}	2.79×10^{-7}
10^{-4}	2.96×10^{-7}
10^{-5}	2.85×10^{-7}
10^{-6}	2.15×10^{-7}
10^{-8}	1.03×10^{-7}

The results are interesting in that they illustrate the effect of dilution upon the hydrogen ion concentration in a solution involving complex relations. The solution is alkaline in high concentration, passes through neutrality at $c = 0.075$, and becomes acid with further dilution. The value of H reaches a maximum near $c = 10^{-4}$ and then falls off with further dilution to approach $H = \sqrt{W}$ as $c \rightleftharpoons 0$. This effect was discussed in Chapter II under the Theorem of Isohydric Solutions. It must be noted moreover that the calculations used for the values in Table 15–5 are based on unity activity coefficients; not only however, are the concentrations so small that such an assumption is probably justified, but the kind of effect noted would persist also at some fixed ionic strength.

7. Special Case of Dibasic Acid Strong in Respect to K_1. If the dibasic acid is "strong" in respect to K_1, with $K_1 = \infty$, the foregoing equations may be considerably simplified. This treatment would apply approximately for chromic acid, often assumed to be strong in respect to its first ionization constant. In such a case, $[H_2X] = 0$, and

$$[HX^-] = -(H + K_2)/4HK + \sqrt{(\quad)^2 + a/2K}. \tag{73}$$

Now $d[X^=]/dH$ is negative while both $d[HX^-]/dH$ and $d[X_2^=]/dH$ are always positive. In place of Eq. (42) we have

$$D = a - b_s + [HX^-]K_2/H. \tag{74}$$

On expansion these equations combined give

$$2H^5K - 4H^4K(a - b_s) + H^3[2K(a - b_s)^2 + K_2 - 4KW] \\ - H^2[(a - b_s)(K_2 - 4KW) - K_2^2] \\ - H[(2a - b_s)K_2^2 - 2KW^2 + K_2W] - K_2^2W = 0, \quad (75)$$

obtainable also directly from Eq. (64) with $K_1 = \infty$. [With $K = 0$, this reduces to Eq. x(81).] For large K the pure acid solution would be described by the following approximate equation:

$$2H^4K - 4H^3aK + 2H^2a^2K - Ha(K_2 - 4KW) - 2aK_2^2 \cong 0. \quad (76)$$

For the salt of type "$K_2Cr_2O_7$" at molar concentration c, when $a = b_s = 2c$, the solution is of course acid. Hence we may take $D \cong H$ in Eq. (74) or drop all terms involving W in Eq. (75), obtaining, with $a = b_s = 2c$,

$$2H^4K + H^2(K_2 - 4KW) + HK_2^2 - 2cK_2^2 \cong 0. \quad (77)$$

For the salt of type "K_2CrO_4" at molar concentration c, when $a = c$ and $b_s = 2c$, the solution is alkaline. Hence with $D \cong -W/H$ Eq. (74) now gives

$$2H^3c^2K + H^2c(K_2 - 4KW) - HW(K_2 - 2KW) - K_2^2W \cong 0. \quad (78)$$

For a mixture of K_2CrO_4 at concentration c and $K_2Cr_2O_7$ at concentration $c/\mathbf{r}$, we may assume, if c is not too small and if $\mathbf{r}$ is of the order of 1, that $D \cong 0$ in Eq. (74). Then, with $a = c(1 + 2/\mathbf{r})$ and $b_s = 2c(1 + 1/\mathbf{r})$, we obtain the rough relation

$$2H^2cK + HK_2 - 2K_2^2/\mathbf{r} \cong 0. \quad (79)$$

This is identical with Eqs. (54) and (55).[5] The accuracy of this formula increases with the value of c, and it may be said to represent H_* or the value of H approached as c increases indefinitely. But at low c it may involve serious error since it approaches the limit $H \rightleftharpoons 2K_2/\mathbf{r}$ instead of $H \rightleftharpoons \sqrt{W}$ as c approaches zero.

D. Boric Acid

1. Ionization Scheme. If boric acid solution involves the species H_3BO_3, $H_2BO_3^-$, HBO_2, BO_2^-, $H_2B_4O_7$, $HB_4O_7^-$, and $B_4O_7^=$, the concentrations of the seven species are interrelated by six independent equilibrium constants; but the equilibrium relations determining the hydrogen ion concentration of the solution require only four over-all or apparent equilibrium constants, in addition to the ion-product constant of water, W. The ratio $[HBO_2]/[H_3BO_3]$ is constant ($= \kappa$) at equilibrium, and the same is true for the species BO_2^- and $H_2BO_3^-$, with $[BO_2^-]/[H_2BO_3^-] = \kappa'$. The

(5) This, in fact, is the relation used by Neuss and Rieman (*loc. cit.*, Ref. xv–3), for the determination of K_2 and K, for chromic acid, at $\mathbf{r} = 2$ and $c \cong 0.00057$ to 0.037.

symbol $[X^0]$ will be understood to represent the sum of the concentrations of the two monomeric uncharged species, or $[H_3BO_3](1 + \kappa)$, and $[X^-]$ will represent the quantity $[H_2BO_3^-](1 + \kappa')$, as in Section A of Chapter VI. Hence only five concentrations are to be related by four independent constants. Three of these may be taken as ordinary ionization constants and at least one as a complex constant for the association of the monomeric into the poly-acid or tetraborate forms:[6]

$$A = H[X^-]/[X^0] = 6 \times 10^{-10}; \tag{80}$$

$$K = [H_2B_4O_7]/[X^0]^4 = [H_2X_4]/[X^0]^4 = 3.6 \times 10^{-3}; \tag{81}$$

$$K_1 = H[HB_4O_7^-]/[H_2B_4O_7] = H[HX_4^-]/[H_2X_4] = 10^{-4}; \tag{82}$$

$$K_2 = H[B_4O_7^=]/[HB_4O_7^-] = H[X_4^=]/[HX_4^-] = 10^{-9}. \tag{83}$$

If a is the analytical molarity of boric acid ("H_3BO_3"), then

$$a = [X^-] + [X^0] + 4[H_2X_4] + 4[HX_4^-] + 4[X_4^=], \tag{84}$$

$$= [X^-](A + H)/A + (4H^2K/A^4)[X^-]^4(H^2 + HK_1 + K_1K_2). \tag{85}$$

For electroneutrality, with strong base present at concentration b_s,

$$D = H - OH = [X^-] + [HX_4^-] + 2[X_4^=] - b_s, \tag{86}$$

$$= [X^-] + (H^2KK_1/A^4)[X^-]^4(H + 2K_2) - b_s. \tag{87}$$

An explicit equation relating H, the constants, and the analytical concentrations a and b_s can be obtained through the combination of Eqs. (85) and (87) by elimination of $[X^-]$. This can be done by substituting $[X^-]^4$ from Eq. (85) into Eq. (87) and then solving for $[X^-]$ in terms of H, constants, and concentrations. The resulting expression for $[X^-]$ may then be used in either Eq. (85) or Eq. (87) to give the desired final relation, which, however, would be too involved for practical use. The calculation of H for a given set of data may also be done through the implicit Eqs. (85) and (87) as a pair, by testing values of H until the same value of $[X^-]$ is obtained from the solution of both equations by approximation methods.

For low values of a the degree of polymerization is relatively small, although it varies, for given a, with the hydrogen ion concentration. A solution of borax, or aq. $Na_2B_4O_7$, at concentration c (with $a = 4c$, $b_s = 2c$), is usually titrated with HCl to the methyl red end-point. At the equivalent point, which is mathematically a solution of boric acid at $a = 4c$, and which is therefore slightly acid, we may take $[X^0] \cong a$, $[X^-] \cong aA/H$, $[HX_4^-] \cong K_1Ka^4/H$, and $[X_4^=] \cong 0$. Then Eq. (86) becomes $H \cong \sqrt{aA + a^4K_1K}$. For $a = 0.06$ (or $c = 0.015$), this gives $H \cong 6.38 \times 10^{-6}$. Hence $[X^-]$

(6) The numerical values of A, K_1, and K_2 are those given in Latimer; K has been calculated from these and the constant $[\text{"}H_3BO_3\text{"}]^2[\text{"}H_2BO_3^-\text{"}]^2/[B_4O_7^=] = 10^{-3}$, also in Latimer (Ref. III-1, pp. 259–60).

$\cong 5.63 \times 10^{-6}$, $[H_2X_4] \cong 4.7 \times 10^{-8}$, $[HX_4^-] \cong 7.4 \times 10^{-7}$, $[X_4^=] \cong 1.2 \times 10^{-10}$, and $[X^0] \cong 0.06$, these values satisfying Eq. (86) with $H = 6.38 \times 10^{-6}$. For aq. $Na_2B_4O_7$ itself at the same concentration ($c = 0.015$, $a = 0.06$), the solution may be considered roughly as one of boric acid, as a simple monobasic acid, half titrated with strong base. The solution is alkaline, and if we assume $H \cong A = 6 \times 10^{-10}$, then $[X^-] \cong 0.028$, $[H_2X_4] \cong 2.2 \times 10^{-9}$, $[HX_4^-] \cong 3.7 \times 10^{-4}$, $[X_4^=] \cong 6.1 \times 10^{-4}$, and $[X^0] \cong 0.028$. (Consideration of the actual Eqs. (85) and (86) confirms this estimate of H, giving $H \cong 6.1 \times 10^{-10}$.)

On the basis of the number of charges per mole of original "H_3BO_3" it is evident that $d([H_3BO_3] + [HBO_2])/dH$ (or $d[X^0]/dH$) and $d[H_2B_4O_7]/dH$ must be always positive, while $d([H_2BO_3^-] + [BO_2^-])/dH$ (or $d[X^-]/dH$) must be always negative. Each of the concentrations of the other two species, however, $[HB_4O_7^-]$ and $[B_4O_7^=]$ must pass through a maximum with respect to change in H. Thus, for the same value of a ($= 0.06$) at $H = 10^{-14}$: $[X^-] \cong 0.06$, $[H_2X_4] \cong 3.6 \times 10^{-27}$, $[HX_4^-] \cong 3.6 \times 10^{-17}$, $[X_4^=] \cong 3.6 \times 10^{-12}$, and $[X^0] \cong 10^{-6}$.

Because the degree of polymerization is in general so low at low concentrations of boric acid, the feasibility functions given in Chapter x, Section E3, *for the titration of borax may be accepted as at least approximately valid.*

2. Effect of Poly-alcohols on Boric Acid. Boric acid will now be considered as a simple monobasic acid with ionization constant A defined as usual as $A = H[X^-]/[X^0]$; its polymerization will be neglected. The apparent strength is known to be increased by poly-hydroxyl compounds, expecially 1,2 and 1,3 diols. One interpretation assumes the formation of a complex with uncharged boric acid, with an ionization constant, A', presumably higher than that of the simple boric acid itself, or A. For the effect of glycerine we have, according to Kolthoff,(7) $K = [\mathbf{G}X^0]/[\mathbf{G}][X^0] = 0.9$, and $A' = H[\mathbf{G}X^-]/[\mathbf{G}X^0] = 3 \times 10^{-7}$, in which $\mathbf{G}$ represents glycerine.

If a is again the analytical concentration of boric acid and c that of the glycerine,

$$a = [X^-] + [X^0] + [\mathbf{G}X^0] + [\mathbf{G}X^-] = [X^-]\,(H + A)/A + [X^-][\mathbf{G}]\,(H + A')K/A; \quad (88)$$

$$c = [\mathbf{G}] + [\mathbf{G}X^0] + [\mathbf{G}X^-] = [\mathbf{G}] + [\mathbf{G}][X^-](H + A')K/A. \quad (89)$$

By elimination of $[X^-]$ we have

$$[\mathbf{G}] = [(c - a) - (H + A)/K(H + A')]/2 + \sqrt{(\quad)^2 + c(H + A)/K(H + A')}. \quad (90)$$

(7) I. M. Kolthoff, *Rec. Trav. Chim. Pays-Bas,* **44,** 975 (1925).

Finally, for titration with strong base in presence of glycerine, then,

$$D = [X^-] + [\mathbf{G}X^-] - b_s = \frac{aA(1 + KA'[\mathbf{G}]/A)}{H + A[1 + K(H + A')[\mathbf{G}]/A]} - b_s, \quad (91)$$

with $[\mathbf{G}]$ given by Eq. (90). If the concentration c of the glycerine is so large relative to a, the concentration of the boric acid being titrated, that the excess of glycerine, or the concentration of the species $[\mathbf{G}]$, is practically constant and approximately equal to c, Eq. (91) becomes

$$D \cong aA(1 + \tau c)/\{H + A[1 + \tau c(1 + \tau' H)]\} - b_s, \quad (92)$$

in which τ and τ' are constants. For a given value of the excess of glycerine then, the behavior of boric acid would not be quite the same as that of a simple monobasic acid of higher ionization constant, because of the term in $\tau' H$.

The effect has also been interpreted, at least in the case of mannitol, as involving the formation of charged complexes between the borate ion X^- and the poly-alcohol. This interpretation does not require the assumption that a stronger acid is being formed. If we assume that these complexes may have the formulas $\mathbf{M}X^-$, $\mathbf{M}_2X^-$, $\mathbf{M}_3X^-$, etc., with $\mathbf{M}$ representing mannitol, the "complex" equilibrium constants may be defined as follows: $\kappa_1 = [\mathbf{M}X^-]/[\mathbf{M}][X^-]$, $\kappa_2 = [\mathbf{M}_2X^-]/[\mathbf{M}]^2[X^-]$, $\kappa_3 = [\mathbf{M}_3X^-]/[\mathbf{M}]^3[X^-]$, etc.(8)

Since $a = [X^-] + [X^0] + [\mathbf{M}X^-] + [\mathbf{M}_2X^-] + \cdots$, then through the definitions of A and the κ's,

$$[X^-] = a/(1 + H/A + \kappa_1[\mathbf{M}] + \kappa_2[\mathbf{M}]^2 + \cdots). \quad (93)$$

For titration with strong base, therefore, of boric acid at concentration a in presence of mannitol, $D = [X^-] + [\mathbf{M}X^-] + [\mathbf{M}_2X^-] + \cdots - b_s$, or

$$D = aA(1 + \kappa_1[\mathbf{M}] + \kappa_2[\mathbf{M}]^2 + \cdots)/[H + A(1 + \kappa_1[\mathbf{M}] + \kappa_2[\mathbf{M}]^2 + \cdots)] - b_s. \quad (94)$$

Furthermore, if the analytical concentration of mannitol is c, so that $c = [\mathbf{M}] + [\mathbf{M}X^-] + 2[\mathbf{M}_2X^-] + 3[\mathbf{M}_3X^-] + \cdots$, then from the definitions of the κ's,

$$[X^-] = (c - [\mathbf{M}])/(\kappa_1[\mathbf{M}] + 2\kappa_2[\mathbf{M}]^2 + 3\kappa_3[\mathbf{M}]^3 + \cdots). \quad (95)$$

For the relation between H, a, c, b_s, and the constants, it would be necessary to eliminate $[X^-]$ and $[\mathbf{M}]$ from Eqs. (94) and (95). This is possible if all the κ's above κ_2 are neglected. For the case then in which only κ_1 and κ_2 are considered, with x defined as $(b_s + D)$, we have, from Eqs.

(8) According to A. Deutsch and S. Osoling, *J. Am. Chem. Soc.*, **71,** 1637 (1949), there is evidence for the formation of at least two of these complexes, with $\kappa_1 \cong 3.0 \times 10^2$ and $\kappa_2 \cong 5 \times 10^4$. The product $\kappa_2 A$, however, has been reported as 1.0×10^{-4} by S. D. Ross and A. J. Catotti, *ibid.*, **71,** 3563 (1949), and as 1.7×10^{-4}, by J. Böeseken, N. Vermaas and A. T. Küchlin, *Rec. Trav. Chim. Pays-Bas*, **49,** 711 (1930).

(93) and (94), $[X^-] = A(a-x)/H$, $\kappa_2[\mathbf{M}]^2 + \kappa_1[\mathbf{M}] + 1 = Hx/A(a-x)$, and $[\mathbf{M}] = -\kappa_1/2\kappa_2 + \sqrt{(\quad)^2 + [Hx - A(a-x)]/A\kappa_2(a-x)}$. Substitution of these expressions in Eq. (95) gives

$$\kappa_1 = \mathbf{J}/2 + \sqrt{(\mathbf{J}/2)^2 + (\kappa_2\mathbf{X} + \mathbf{Y})}, \quad (96)$$

$$\kappa_2 = (\kappa_1^2 - \kappa_1\mathbf{J} - \mathbf{Y})/\mathbf{X}, \quad (97)$$

in which

$$\mathbf{J} = H^2c/A(a-x)[A(a-x) + H(c-x)], \quad (98)$$

$$\mathbf{X} = \mathbf{J}\,[2A(a-x) + H(c-2x)]^2/H^2c, \quad (99)$$

$$\mathbf{Y} = \mathbf{J}\,[A(a-x) - Hx]/cA(a-x). \quad (100)$$

Eqs. (96) and (97) would then permit the simultaneous calculation of the constants κ_1 and κ_2 from the data (measurements of H over a range of known values of a, c, and b_s) if A is known and on the assumption that formation of tetraborate species is negligible. With regard to the assumption that the highest complex is $\mathbf{M}_2X^-$, it is possible that further information may be obtained through measurements of the effect of mannitol on the solubility of boric acid.

We shall not pursue these complex relations further. But again we shall consider the spɩcial condition in which the concentration c, of mannitol, is so large relative to a, the concentration of the boric acid being titrated, that the excess of mannitol, or the concentration of the species $[\mathbf{M}]$, is practically constant and equal to c. Then Eq. (94) becomes

$$D \cong aA(1 + \kappa_1 c + \kappa_2 c^2 + \kappa_3 c^3 + \cdots)/[H + A(1 + \kappa_1 c + \kappa_2 c^2 + \cdots)] - b_s. \quad (101)$$

Under these conditions, therefore, the behavior of the acid is like that of a monobasic acid with an apparent ionization constant $A' = A(1 + \kappa_1 c + \kappa_2 c^2 + \kappa_3 c^3 + \cdots)$. The apparent strength depends not only on the constants κ_1, κ_2, κ_3, etc., but also on the mannitol concentration c. For a given value of c, however, the behavior is simply that of a stronger monobasic acid. With $\kappa_1 = 3.0 \times 10^2$ and $\kappa_2 = 5 \times 10^4$ the apparent ionization constant of boric acid approaches the value $A' = (6 \times 10^{-10})\ (300c + 50{,}000c^2)$ or 1.24×10^{-6} in an excess of 0.2 mole mannitol per liter. With such an excess of mannitol, therefore, the equivalent point for titration of 0.05 molar boric acid with 0.05 molar strong base would be expected to have a pH of ~ 9.15 with a feasibility function of $4.4p$ (with p = per cent of titration) and a feasibility of ~ 1.9 pH units for one per cent of titration through the equivalent point.

E. Phosphoric Acid

Orthophosphoric acid solutions are usually treated as solutions of a simple noncomplex tribasic acid, as in fact we shall do later in Chapter XVI.

But if pyrophosphoric and metaphosphoric acids differ from orthophosphoric acid only in respect to the analytical ratio of water to "P_2O_5," then a solution of "orthophosphoric acid" *in complete equilibrium* must contain all three "acids" and their ions. For the sake of simplicity of illustration of the type of relations involved, we shall take metaphosphoric acid as simply "HPO_3," whereupon we have eleven species of the solute, related through ten independent equilibrium constants: three ionization constants of H_3PO_4, four of $H_4P_2O_7$, one of HPO_3, and two "complex" constants. One of these complex constants may be defined as a dimerization constant for the formation of $H_4P_2O_7$, of the form $K = [H_4P_2O_7][H_2O]/[X^0]^2$, with $[X^0] = [H_3PO_4] + [HPO_3]$, and the other as a hydration constant relating the ortho- and the meta-acids, or $K' = [HPO_3][H_2O]/[H_3PO_4]$. As usual, if the solution is sufficiently dilute, the factor $[H_2O]$ may be taken as constant. In a dilute solution *at equilibrium,* therefore, the ratio of the uncharged species $[HPO_3]/[H_3PO_4]$ is a constant independent of the analytical concentrations and independent of the value of H. This constancy of ratio holds for any two or more species of the same degree of association or polymerization and of the same charge per analytical mole of phosphorus. Thus the ratio $[H_2PO_4^-]/[PO_3^-]$ would similarly be constant as long as the activity of water may be regarded as constant.

The total concentration in the form of species of the same charge per mole of solute varies with H, but the distribution within the total varies with the analytical concentration though not with H. In boric acid the total analytical concentration of boron in uncharged species is $m(1 + \kappa) + 4K(1 + \kappa)^4m^4$ if m is the concentration of the species H_3BO_3; in phosphoric acid, this total is $m(1 + K') + 2K(1 + K')^2m^2$ if $m = [H_3PO_4]$. If metaphosphoric acid is not simply HPO_3 but something like $(HPO_3)_6$, with $k = [H_4P_2O_7][H_2O]/m^2$ and $k' = [(HPO_3)_6][H_2O]^6/m^6$, then the ratio $[(HPO_3)_6]/[H_3PO_4]$ would not be constant and the total would be $m + 2km^2 + 6k'm^6$, the distribution still being independent of H.

Furthermore each of the ratios $[H_2P_2O_7^=]/[H_2PO_4^-]^2$, $[P_2O_7^=]/[HPO_4^=]^2$ and $[H_2P_2O_7^=]/[PO_3^-]^2$ is also constant for any concentration and for any value of H. The total concentration of the uncharged species increases steadily with H without passing through a maximum; and the concentration of $[PO_4^{\equiv}]$, with the highest charge per mole of solute, increases steadily with decreasing H. All the other species pass through maxima with respect to change in H; these maxima occur, with decreasing H, in the order $H_3P_2O_7^-$, $H_2PO_4^-$ (together with $H_2P_2O_7^=$ and PO_3^-), $HP_2O_7^{\equiv}$, and $HPO_4^=$ (together with $P_2O_7^{\equiv\!\!\equiv}$).

F. Homogeneous Hydrolysis of Nonelectrolytes

1. Hydrolysis Producing Two Acids. The reaction of a substance such as the halide of a nonmetal, or an acyl halide, etc., with water leads to two

acids as hydrolysis products. With "YX" as the symbol for the formula of the reactant and if the two acids formed, HX and HY, are both weak, with ionization constants A_X and A_Y respectively, then $K_h = [HX][HY]/[YX][H_2O]$. "HX" and "HY" are symbols only, not formulas, for one of the acids of course will be a hydroxyacid. We also define $K = K_h[H_2O]$, so that with the same assumptions as were involved in the definition of the ionization constants themselves (Chapter 1, Eqs. 1[6–12]), K is a constant if the activity of water is a constant. In the following equations, therefore, the equilibrium constants, K, A_X, A_Y, and W, are each to be multiplied by the activity of water when the deviation of this activity from unity is to be considered. Such simple correction, however, assumes that the products of the reactions, HX and HY in K_h and the ions in the ionization reactions, are themselves not hydrated. Unless these degrees of hydration are known, therefore, the introduction of the activity of water as a correction in the equilibrium constants is of questionable value, and the equations in every case approach exactness only in dilute solution.

Furthermore, if c is the original analytical concentration of YX dissolved in water, then $c = [YX] + [X^-] + [HX] = [YX] + [Y^-] + [HY]$, while $H - OH = D = [X^-] + [Y^-] + g$, in which g, as usual, is the net concentration of foreign strong acid $(= a_s - b_s)$. The combination of these equations with the definitions of K_h, A_X, and A_Y, leads to

$$[X^-] = -KA_X(H + A_Y)/2H^2 + \sqrt{(\quad)^2 + cA_X^2K(H + A_Y)/H^2(H + A_X)}, \tag{102}$$

and

$$[Y^-] = [X^-]A_Y(H + A_X)/A_X(H + A_Y). \tag{103}$$

Combined with the equation of electroneutrality, these would yield an expression defining H as $f(c,K,A_X, A_Y, W,$ and $g)$, but it would be extremely involved and of the eighth degree in H. Similarly an expression for K may be written for its determination from the measurement of H at known values of c and g if the constants A_X, A_Y, and W are known. The expression turns out to be first degree in K:

$$K = H^2(D - g)^2\mathbf{X}/[c\mathbf{Y}^2 - (D - g)\mathbf{XY}], \tag{104}$$

with $\mathbf{X} = (H + A_X)(H + A_Y)$ and $\mathbf{Y} = A_X(H + A_Y) + A_Y(H + A_X)$.

If one of the two acids formed is strong, so that $A_X = \infty$, $[HX] = 0$, and $K = [H^+][X^-][HY]/[YX]$, then

$$[X^-] = -K(H + A_Y)/2H^2 + \sqrt{(\quad)^2 + cK(H + A_Y)/H^2}, \tag{105}$$

$$[Y^-] = [X^-]A_Y/(H + A_Y), \tag{106}$$

$$K = \frac{H^2(D - g)^2(H + A_Y)}{c(H + 2A_Y)^2 - (D - g)(H + A_Y)(H + 2A_Y)}. \tag{107}$$

Finally, if both acids are strong, with $K = [H^+]^2[X^-][Y^-]/[YX]$, then

$$[X^-] = [Y^-] = -K/2H^2 + \sqrt{(\quad)^2 + cK/H^2}, \tag{108}$$

$$K = H^2(D-g)^2/[4c - 2(D-g)]. \tag{109}$$

2. Hydrolysis Producing an Acid and a Nonelectrolyte. The reaction of an ester with water leads to an alcohol and an acid as hydrolysis products. With RX as the symbol for the formula of the ester, ROH for the alcohol, and HX for the weak acid produced (with ionization constant A), then $K = [ROH][HX]/[RX]$, $c = [RX] + [ROH] = [RX] + [X^-] + [HX]$, and $D = [X^-] + g$. Hence,

$$[X^-] = -KA/2H + \sqrt{(\quad)^2 + cKA^2/H(H+A)}, \tag{110}$$

$$H^5 + H^4(A-2g) + H^3(AK - 2W + g^2 - 2gA) + H^2[A(AK - 2W + g^2) - g(AK - 2W)] - H[cA^2K + W(AK - W) + gA(AK - 2W)] - AW(AK - W) = 0, \tag{111}$$

$$K = H(D-g)^2(H+A)/[cA^2 - A(D-g)(H+A)]. \tag{112}$$

If the activity coefficient of an uncharged species is taken as unity, so that $\mathbf{K} = K$, and on the assumption of the Debye-Hückel limiting law for the activity coefficient of a univalent ion, $\mathbf{K}$ is given by Eq. (112) with $\mathbf{H}/\gamma$ for H, $\mathbf{D}/\gamma$ for D, and $\mathbf{A}/\gamma^2$ for A. The activity coefficient for given observed values of $\mathbf{H}$, c, and g is calculated directly by successive approximations from $-\log \gamma = 0.505 \sqrt{\mathbf{H}/\gamma + b_s}$, so that $\mathbf{K}$ may then be calculated from observed values of $\mathbf{H}$, c, b_s, and g ($= a_s - b_s$), and known values of $\mathbf{A}$ and $\mathbf{W}$.

If the acid HX is strong, with $K = [ROH][H^+][X^-]/[RX]$, then

$$[X^-] = -K/2H + \sqrt{(\quad)^2 + cK/H}, \tag{113}$$

$$H^4 - 2H^3g + H^2(K - 2W + g^2) - H[cK + g(K - 2W)] - W(K - W) = 0, \tag{114}$$

$$K = H(D-g)^2/[c - (D-g)]. \tag{115}$$

For the calculation of $\mathbf{K}$ from $\mathbf{H}$, a_s, b_s, c, and $\mathbf{W}$, we have in this case $\mathbf{K} = K\gamma^2$, again with $-\log \gamma = 0.505 \sqrt{\mathbf{H}/\gamma + b_s}$.

XVI

Tribasic Acid; Salts with Strong and Weak Bases

A. With Strong Base

The general equation is

$$D = H - OH = a\beta - b_s = a(\alpha_1 + 2\alpha_2 + 3\alpha_3) - b_s; \quad (1)$$

$$H - OH = a \frac{1 + 2K_2/H + 3K_{23}/H^2}{1 + H/K_1 + K_2/H + K_{23}/H^2} - b_s; \quad (2)$$

$$\begin{aligned} &H^5 + H^4(b_s + K_1) + H^3[K_{12} - W - K_1(a - b_s)] \\ &\quad + H^2K_1[K_{23} - W - K_2(2a - b_s)] - HK_{12}[W + K_3(3a - b_s)] \\ &\quad - K_{123}W = 0. \end{aligned} \quad (3)$$

The general problem has already been discussed briefly in Chapter VIII in illustration of the method of using various sequences of three terms for approximate solutions of this type of equation (always for the condition $K_1 > K_2 > K_3$). With appreciable concentration of the acid, we may expect that the titration curve through the four stoichiometric points ($b_s = 0$, a, $2a$, $3a$) will be best described by the H^5, H^4, H^3, and H^2 series of three terms respectively; that these sequences will then also apply in this order to the solution of the pure acid and of the three successive alkali salts; and that the middle ranges of the titrations of the three successive equivalents (or when $H \cong K_1$, K_2, K_3) will be given by the three successive *pairs* of terms, $H^4 + H^3$, $H^3 + H^2$, and $H^2 + H^1$.

1. The Pure Acid.

$$\begin{aligned} H^5 + H^4K_1 + H^3(K_{12} - W - aK_1) + H^2K_1(K_{23} - W - 2aK_2) \\ - HK_{12}(3aK_3 + W) - K_{123}W = 0. \end{aligned} \quad (4)$$

By analogy with the case of dibasic acid, we may say that if $a \gg \sqrt{K_1K_2}$, H will be given fairly accurately by the first three terms, or as $H \cong Q_4^5$, as in Eq. VIII(28). (In this chapter the quadratic solution of such sequences of three terms will not be written out, but will always be abbreviated in this fashion, as "Q" functions.) Hence when a increases indefinitely, H approaches the value $\sqrt{aK_1}$ or H_*. For lower a the first four terms of Eq. (4) would have to be considered, and if $a \ll \sqrt{K_1K_2}$ but much greater than $\sqrt{K_2K_3}$, the H^4 sequence becomes somewhat dependable.

2. Primary Alkali Salt ($b_s = a = c$), or NaH_2X.

$$H^5 + H^4(c + K_1) + H^3(K_{12} - W) - H^2K_1(K_{23} - W - cK_2)$$
$$- HK_{12}(W + 2cK_3) - K_{123}W = 0. \qquad (5)$$

Neglecting $H - OH$ in Eq. (2), for H_* or H_{ie}, we have

$$H^3 - HK_1K_2 - 2K_{123} = 0. \qquad (6)$$

This may be used to find the condition of neutrality for such a salt solution. With $H = \sqrt{W}$ in Eq. (6), we find that with $(K_3)_n = \sqrt{W}(W - K_1K_2)/2K_1K_2$, the solution may be alkaline only if $K_3 < \sqrt{W}/2$, since $K_1 > K_2 > K_3$. Also, it is always acid if $K_1K_2 > W$, and if $K_1K_2 < W$ it is still acid unless $K_3 < (K_3)_n$.

If these conditions for neutrality are approximately satisfied, then Eq. (6) may be used as a rough formula for H independent of the concentration; and since, with $H \cong \sqrt{W}$, K_3 must be less than $\sqrt{W}$, then the neglect of the last term in Eq. (6) leaves $H \cong \sqrt{K_1K_2}$. For the effect of concentration, we must use Eq. (5), noting that of the two principal sequences (H^4 and H^3), the H^3 can be important only if roughly $H < \sqrt[3]{K_{123}}$ or less than $\sqrt[3]{K_{12}W/c}$. But such a solution is never very alkaline; or since $K_1 > H > K_2$, then $H \cong Q_5^4$, as in Eq. VIII(29). This may be simplified in various ways, depending on the numerical values; ultimately again $H \cong \sqrt{K_1K_2}$ for appreciably high values of c.

3. Secondary Alkali Salt, Na_2HX; $b_s = 2a = 2c$. For H_* or H_{ie},

$$2H^3 + H^2K_1 - K_{123} = 0. \qquad (7)$$

For neutrality, setting $H = \sqrt{W}$ in Eq. (7), we note that since $K_{23} = W(1 + 2\sqrt{W}/K_1)$, then K_2K_3 must be approximately equal to W if $K_1 > \sqrt{W}$, or approximately to $2W\sqrt{W}/K_1$ if $K_1 < \sqrt{W}$. Also, since $K_1 = 2W\sqrt{W}/(K_2K_3 - W)$, then the solution is always alkaline if $K_2K_3 < W$. It can be acid only if $K_2K_3 > W$.

Subject to these conditions, H can be approximately equal to $\sqrt{W}$ only if $K_1 \gg \sqrt{W}$; hence the rough formula for H, from Eq. (7), is $H \cong \sqrt{K_2K_3}$. For the effect of concentration, the full equation is

$$H^5 + H^4(2c + K_1) + H^3(cK_1 + K_{12} - W) + H^2K_1(K_{23} - W)$$
$$- HK_{12}(cK_3 + W) - K_{123}W = 0. \quad (8)$$

The best approximation here is the H^3 sequence; the H^4 sequence is not useful, giving a real value of H only if $K_{23} < W$, in which case the solution is alkaline and the high powers of H are very small. Hence, $H \cong Q_8^3$, as in Eq. VIII(30), or $H \cong \sqrt{K_2K_3}$ for sufficiently high values of c.

4. Tertiary Alkali Salt, NA_3X; $b_s = 3a = 3c$.

$$H^5 + H^4(3c + K_1) + H^3(2cK_1 + K_{12} - W) + H^2K_1(cK_2 + K_{23} - W) - HK_{12}W - K_{123}W = 0. \quad (9)$$

This solution is of course always alkaline; hence using the last three terms, we have $H \cong Q_9^2$, as in Eq. VIII(31). If $cK_2 > W$,

$$H \cong W/2c + \sqrt{(\quad)^2 + K_3W/c}, \quad (10)$$

or its equivalent, $OH \cong - W/2K_3 + \sqrt{(\quad)^2 + cW/K_3}$.

Table 16–1. Tribasic acid and strong base.

Example: $K_1 = 10^{-2}$, $K_2 = 10^{-7}$, $K_3 = 10^{-12}$

Type	c	Eq.	H
"H_3X"	1	Q_4^5	0.0951
	0.01		0.00618
NaH_2X	1	Q_5^4	3.146×10^{-5}
	0.01		2.234×10^{-5}
Na_2HX	1	Q_8^3	3.178×10^{-10}
	0.01		4.477×10^{-10}
Na_3X	1	Q_9^2	1.051×10^{-13}
	0.01		1.618×10^{-12}

Table 16–1 gives an example of the use of the formulas discussed for the case of $K_1 = 10^{-2}$, $K_2 = 10^{-7}$, $K_3 = 10^{-12}$ (close to phosphoric acid). The exact value in each case, obtained from Eqs. (4), (5), (8), and (9), is the same as that shown in the table carried to the same number of figures. A simple test through the estimation of the orders of magnitude of the terms of the full equations easily shows the high precision of the formulas at least for the example taken.

If M^{++} is the divalent cation of a diacid base which is considered completely ionized (as is often assumed for salts of calcium, barium, and magnesium, for example), then the above formulas hold similarly for salts such as $M(H_2X)_2$, MHX, and M_3X_2; but c now always represents the formal molarity of the "acid-radical," H_2X^-, $HX^=$, or $X^{\equiv}$, not the analytical molarity of the salt, which would be $c/2$, for example, for $M(H_2X)_2$.

B. Buffer Capacity and Feasibility of Titration

$$\pi = db_s/dpH = 2.3\{a[\alpha_1(\rho + \alpha_2 + 4\alpha_3) + \alpha_2(4\rho + \alpha_3) + 9\alpha_3\rho] + H + OH\}, \quad (11)$$

$$= 2.3\left[(a) \frac{\frac{H}{K_1} + \frac{K_2}{H}\left(1 + \frac{K_{23}}{H^2}\right) + \frac{K_2}{K_1}\left(4 + \frac{4K_{13}}{H^2} + \frac{9K_3}{H}\right)}{\left(1 + \frac{H}{K_1} + \frac{K_2}{H} + \frac{K_{23}}{H^2}\right)^2} + H + OH \right]. \quad (11a)$$

At the successive mid-points of titration then, or when $H \cong K_1, K_2, K_3$ respectively, the value of π is a maximum, approximately equal to $2.3a/4$

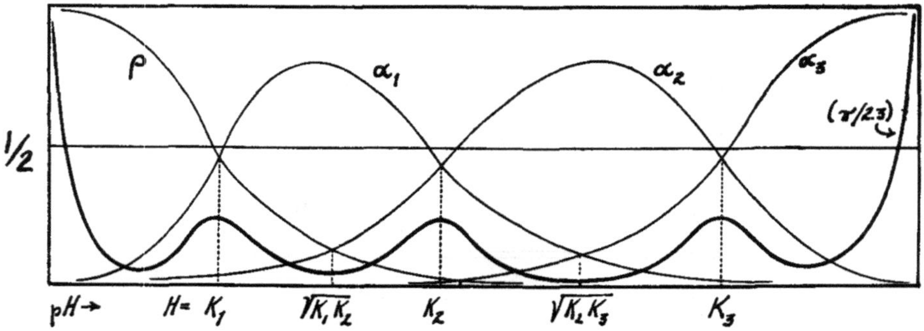

Fig. 16–1. Buffer capacity with respect to strong base of solution containing tribasic acid at unit concentration.

provided $(H + OH)$ is negligible. The relations are shown graphically in Fig. 16–1.

1. Feasibility (Approximate). From the approximate value of H_e at each of the four stoichiometric points ($b_s = 0, a, 2a, 3a$), the value of π at the four successive minima may be expressed and hence the feasibility function for titration at each of the four (vertical) inflections in the curve of pH against b_s.

Minimum (*o*): for titration of strong acid in presence of tribasic acid. If $H_e \cong \sqrt{aK_1}$, $\pi_e \cong (2.3)2\sqrt{aK_1}$, and

$$F = (p/400)\sqrt{a_s^2/aK_1}. \quad (12)$$

Minimum (*1*): titration of tribasic acid to the composition NaH_2X. If $H_e \cong \sqrt{K_1K_2}$, $\pi_e = (2.3)2a\sqrt{K_2/K_1}$, and

$$F = (p/400)\sqrt{K_1/K_2}. \quad (13)$$

Minimum (*2*): titration from $b_s = a$ to $b_s = 2a$ (NaH_2X to Na_2HX). If $H_e \cong \sqrt{K_2K_3}$, $\pi_e \cong (2.3)2a\sqrt{K_3/K_2}$, and

$$F = (p/400)\sqrt{K_2/K_3}; \quad (14)$$

here $b_s = a$ (from a to $2a$) is taken as 100 per cent; if the total titration is considered so that $b_s = 2a$ is required for 100 per cent, then F is doubled.

Minimum *(3)*: titration to the point Na_3X. If $H_e \cong \sqrt{K_3W/a}$, then $\pi_e \cong (2.3)2\sqrt{aW/K_3}$, and

$$F = (p/400)\sqrt{aK_3/W}. \tag{15}$$

Here again p refers to percentage of b_s used between minimum *(2)* and minimum *(3)*; if the total of three equivalents is considered as 100 per cent, this value of F must be tripled.

But it must also be noticed that there will be no minimum *(0)* unless roughly $a \gg 27K_1$, and no minimum *(3)* unless $aK_3 \gg 27W$. By analogy with the case of dibasic acid, we may also say that minimum *(1)* requires roughly $K_1 \gg 16K_2$ and minimum *(2)* requires $K_2 \gg 16K_3$.

2. Feasibility (More Accurate). The feasibility functions just given are quite approximate. As an example of the application of the more accurate procedure (Procedure I) of Chapter IX for the estimation of the feasibility of titration, we shall calculate the feasibility for the titration of tribasic acid with strong base from the first to the second equivalent points, or from minimum *(1)* to minimum *(2)*. The titration curve through the equivalent point ($b_s = 2a$) is given by the H^3 sequence of Eq. (3). The coefficients of this sequence, considered as $H^2 + vH + w = 0$, are

$$v = [K_2(2a - b_s - K_3) + W]/(a - b_s - K_2 + W/K_1),$$
$$w = K_2[(3a - b_s)K_3 + W]/(a - b_s - K_2 + W/K_1). \tag{16}$$

If the percentage $p = 100(b_s - a)/a$, then $(b_s)_{\pm p/2} = 2a \pm pa/200$. With the neglect of $pa/200$, $w_{-p/2} \cong w_{+p/2}$, satisfying the expectation and condition of Eq. IX(20). Hence again neglecting $pa/200$ in the denominator but not in the numerator of v, we have, from Eq. IX(22),

$$F_{\mp} = \frac{\pm pa/200 - K_3 + W/K_2}{2\sqrt{(a + K_2 - W/K_1)(a + W/K_3)K_3/K_2}}, \tag{17}$$

and $2.3\Delta_p = \sinh^{-1}F_- - \sinh^{-1}F_+$. Now neglecting K_3 and W/K_2 in the numerator and $(K_2 - W/K_1)$ in the denominator of Eq. (17), we have

$$F = F_- = -F_+ = (p/400)\sqrt{aK_2/(aK_3 + W)}. \tag{18}$$

If $F \gg 1$ then $\Delta_p \cong \log 4F^2$; but otherwise (which is more likely), Δ_p must be calculated as $(2\sinh^{-1}F)/2.3$. Further simplification of Eq. (18) depends on whether $aK_3 \gtrless W$. If $aK_3 \gg W$ then Eq. (14) follows; but it is not likely now that $F \gg 1$. At any rate, the feasibility is independent of the concentration only if $aK_3 \gg W$. If K_3 is so small that $aK_3 \ll W$ then $F = (p/400)\sqrt{aK_2/W}$. In titration the acid would in this case be effectively dibasic, with no inflection at the "third" equivalent point.

The effect of concentration of the acid on the feasibility of this titration may be estimated readily through Eq. (18), by means of hyperbolic functions. As an illustration, with $p = 1$ per cent, in the titration of a salt of the type of NaH_2PO_4 to Na_2HPO_4, we find the values shown in Table 16–2, assuming unity activity coefficients. H_e has been calculated from $H \cong Q_8^3$, or Eq. VIII(30), simplified to

$$H_e \cong \sqrt{K_2(cK_3 + W)/c}. \tag{19}$$

The use of Eq. (17) in place of Eq. (18) does not change the calculated values of Δ_p by more than 0.001.

Table 16–2. Second equivalent point ($b_s = 2a$) for unity activity coefficient.

	Constants					F	
	K_1	K_2	K_3	a	H_e	By Eq. (18)	Δ_p
(A)	10^{-2}	10^{-7}	10^{-12}	1	3.18×10^{-10}	0.787	0.627
				10^{-1}	3.32×10^{-10}	0.754	0.605
				10^{-2}	4.48×10^{-10}	0.559	0.463
				10^{-3}	1.05×10^{-9}	0.238	0.207
(B)	10^{-2}	10^{-7}	10^{-13}	1	1.05×10^{-10}	2.38	1.39
				10^{-1}	1.41×10^{-10}	1.77	1.16
				10^{-2}	3.32×10^{-10}	0.754	0.605
				10^{-3}	1.005×10^{-9}	0.249	0.215

With the usual convention for the introduction of activity coefficients (Chapter v), Eq. (18) becomes

$$F = (p\gamma/400)\sqrt{a\mathbf{K}_2/(a\mathbf{K}_3 + \mathbf{W}\gamma^4)}, \tag{20}$$

so that if the ionic strength is known, a correction (approximate) may be made for its effect on the constants. If the ionic strength is constant, the feasibility increases steadily with a. But if the solution being titrated is that of the pure salt "NaH_2PO_4," so that γ depends on a, then at the equivalent point, from Eq. v(51b), $-\log\gamma = 0.505\sqrt{H_e + a(2 + \alpha_2 + 3\alpha_3)}$ or, from Table 5–1, $-\log\gamma \cong 0.505\sqrt{3a}$. If this final expression applies, the feasibility now passes through a maximum as a varies, but the dependence is not simple. With $-\log\gamma = 0.505\sqrt{3a}$, we may obtain, from Eq. (20), that when $dF/da = 0$,

$$\log[\sqrt{3a} + 1/0.505(2.3)] - \log(a\sqrt{3a}) = 4(0.505)\sqrt{3a} + \log(\mathbf{K}_3/\mathbf{W}). \tag{21}$$

For $\mathbf{K}_3 = 10^{-12}$, this gives $a = 0.017$ for maximum feasibility; for $\mathbf{K}_3 = 10^{-13}$, $a_{max} = 0.052$. In Table 16–3 the values of H_e, F, and Δ_p for the titrations of Table 16–2 have been recalculated, together with $\gamma : \gamma$ from $-\log \gamma = 0.505\sqrt{3a}$ on the assumption that $\alpha_2 = 1$, H_e from Eq. (19) in the form $H_e \cong \sqrt{\mathbf{K}_2(c\mathbf{K}_3 + \gamma^4\mathbf{W})/c\gamma^{10}}$, and F from Eq. (20). Consideration of the full expression for μ from Eq. v(51b) shows in every case that the value of γ taken is sufficiently accurate for the purpose. The case of $a = 1$ has been omitted in Table 16–3 because of the inapplicability of the limiting law at such high ionic strength.

Table 16–3. Second equivalent point ($b_s = 2a$), on basis of limiting law.

	Constants $\mathbf{K}_1$	$\mathbf{K}_2$	$\mathbf{K}_3$	a	γ	H_e	F	Δ_ν
(1)	10^{-2}	10^{-7}	10^{-12}	10^{-1}	0.529	7.70×10^{-9}	0.416	0.351
				10^{-2}	0.818	1.04×10^{-9}	0.538	0.447
				10^{-3}	0.938	1.29×10^{-9}	0.248	0.212
(2)	10^{-2}	10^{-7}	10^{-13}	10^{-1}	0.529	2.51×10^{-9}	1.27	0.922
				10^{-2}	0.818	6.40×10^{-10}	0.875	0.686
				10^{-3}	0.938	1.22×10^{-9}	0.264	0.227

3. Buffer Capacity of "Buffer Mixtures." We here consider the buffer capacity of solutions near mid-points of titrations.

(a) For $[H_3X]/[NaH_2X] = c_0/c_1$, H is taken with some accuracy from the H^4 and H^3 terms of Eq. (3), as

$$H \cong K_1(c_0 - K_2)/(c_1 + K_1) \cong K_1c_0/c_1. \tag{22}$$

Neglecting OH in Eq. (11a), as also the terms in K_3, we have

$$\frac{\pi}{2.3} \cong (c_0 + c_1)\frac{H/K_1 + K_2/H + 4K_2/K_1}{(1 + H/K_1 + K_2/H)^2} + H. \tag{23}$$

With $H \cong K_1c_0/c_1$,

$$\frac{\pi}{2.3} \cong (c_0 + c_1)\frac{c_0/c_1 + (K_2/K_1)(4 + c_1/c_0)}{(1 + c_0/c_1 + c_1K_2/c_0K_1)^2} + K_1c_0/c_1; \tag{24}$$

then if $c_0 = c_1 = c$, $\pi \cong 2.3(c/2 + K_1)$. More accurate values follow by introduction of the full expression of Eq. (22) into Eq. (23).

(b) For $[NaH_2X]/[Na_2HX] = c_1/c_2$, H is taken from the H^3 and H^2 terms of Eq. (3):

$$H \cong [K_2(c_1 - K_3) + W]/(c_2 + K_2 - W/K_1) \cong K_2c_1/c_2. \tag{25}$$

Now with $H \cong K_2$, Eq. (11a) may be simplified to

$$\pi \cong 2.3(c_1 + c_2)\,\frac{(K_2/H)(1 + 4K_3/H + K_{23}/H^2) + H/K_1 + 4K_2/K_1}{(1 + K_2/H)^2}. \tag{26}$$

If $H \cong K_2c_1/c_2$,

$$\pi \cong 2.3(c_1 + c_2)\,\frac{c_2/c_1 + (K_2/K_1)(4 + c_1/c_2) + (K_3/K_2)(c_2/c_1)^2(4 + c_2/c_1)}{(1 + c_2/c_1)^2}, \tag{27}$$

and if $c_1 = c_2 = c$, $\pi \cong 2.3c/2$.

(c) For $[Na_2HX]/[Na_3X] = c_2/c_3$, H, from the H^2 and H terms of Eq. (3), is

$$H \cong (c_2K_3 + W)/(c_3 + K_3 - W/K_2) \cong K_3c_2/c_3. \tag{28}$$

In this region, with H of the order of K_3,

$$\frac{\pi}{2.3} \cong (c_2 + c_3)\,\frac{H/K_3 + (H^2/K_{23})(4 + H/K_3)}{(1 + H/K_3)^2} + \frac{W}{H}. \tag{29}$$

If $H \cong K_3c_2/c_3$,

$$\frac{\pi}{2.3} \cong (c_2 + c_3)\,\frac{c_3/c_2 + (K_3/K_2)(4 + c_2/c_3)}{(1 + c_2/c_3)^2} + \frac{c_3W}{c_2K_3}, \tag{30}$$

or, if $c_2 = c_3 = c$, $\pi \cong 2.3(c/2 + W/K_3)$.

C. "Higher Order" Ampholytes

Complex organic compounds possessing several "acidic" and "basic" groups or functions may be called "higher order ampholytes," and are in each case mathematically identical through the transformations of constants explained under Eqs. I(73) and I(74), with some particular alkali acid salt of a polybasic acid. As pointed out further in Chapter V, Section E, a substance like histidine, with two basic constants ($K_{b_1} > K_{b_2}$) and one acid constant, K_a, is mathematically equivalent to the secondary alkali salt of its dihydrochloride considered as a tribasic acid. With the transformations $K_1^* = W/K_{b_2}$, $K_2^* = W/K_{b_1}$, and "K_3" $= K_a$, then Eq. (4) describes a solution of the pure dihydrochloride at molar concentration a. Eq. (3) describes its titration with strong base, leading to three successive equivalent points as follows: Eq. (5) for the monohydrochloride; Eq. (8) for the histidine itself; and Eq. (9) for its sodium salt—all at molar concentration c.

Concerning the iso-electric point of such an ampholyte (discussed in Chapter IV), we observe that it is the equivalent of that of the secondary alkali salt of a tribasic acid. Hence H_{ie} is given by Eq. (7). As an iso-electric point, this equation, identical with Eq. III(36), corresponds to a maximum of ρ for the ampholyte (and hence of α_2 for a tribasic acid), as seen under Eq. III(36), where we also saw that the value of H at the isoelectric point then will be

$$H_{ie} \cong \sqrt{WK_a/K_{b_1}} - K_aK_{b_2}/K_{b_1}, \tag{31}$$

and therefore less than $\sqrt{WK_a/K_{b_1}}$.

D. Weak Base Salts of Tribasic Acid

In this case

$$D = a\beta - b\alpha' = a\beta - bHB/(HB + W), \tag{32}$$

$$\begin{aligned} H^6B &+ H^5[B(b + K_1) + W] + H^4[W(K_1 - B) - BK_1(a - b - K_2)] \\ &+ H^3[W(K_{12} - BK_1 - W) - K_1(aW - BK_{23}) - K_{12}B(2a - b)] \\ &+ H^2K_1[W(K_{23} - W - BK_2) - 2aK_2W - BK_{23}(3a - b)] \\ &- HK_{12}W[W + K_3(3a + B)] - K_{123}W^2 = 0. \end{aligned} \tag{33}$$

1. The Reaction ($H \gtrless OH$) of the Salts.

(a) Type NH_4H_2X; $b = a$; for neutrality,

$$B_n = (2K_2\sqrt{W} + W + 3K_{23})/(W/K_1 - K_2 - 2K_{23}/\sqrt{W}). \tag{34}$$

The solution is therefore always acid unless $K_1 < W/(K_2 + 2K_{23}/\sqrt{W})$.

(b) Type $(NH_4)_2HX$; $b = 2a$; for neutrality,

$$B_n = (2K_2\sqrt{W} + W + 3K_{23})/(\sqrt{W} + 2W/K_1 - K_{23}/\sqrt{W}). \tag{35}$$

This is also always acid unless $K_{23} < W(1 + 2\sqrt{W}/K_1)$.

(c) Type $(NH_4)_3X$; $b = 3a$; for neutrality,

$$B_n = (2K_2\sqrt{W} + W + 3K_{23})/(K_2 + 3W/K_1 + 2\sqrt{W}). \tag{36}$$

2. Primary Salt, Type NH_4H_2X. With $b = a = c$, Eq. (33) becomes

$$\begin{aligned} H^6B &+ H^5[B(c + K_1) + W] + H^4[W(K_1 - B) + BK_{12}] \\ &+ H^3[W(K_{12} - BK_1 - W) - cK_1(BK_2 + W) + BK_{123}] \\ &+ H^2K_1[W(K_{23} - W - BK_2) - 2cK_2(BK_3 + W)] \\ &- HK_{12}W[W + K_3(3c + B)] - K_{123}W^2 = 0. \end{aligned} \tag{37}$$

Since such a salt solution is usually acid the most important terms are generally those of the H^5 sequence, so that $H \cong Q^5_{37}$, or more approximately,

$$H \cong -K_1K_2/2c + \sqrt{(\quad)^2 + K_1K_2} \cong \sqrt{K_1K_2}. \tag{38}$$

With so many variable parameters it is obvious that no single sequence of terms will be the best approximation for all practical variations of the constants and concentration. The numerical test through the orders of magnitude of the terms of the general equation would settle the question in any specific example. But the approximation here chosen is the one which, by analogy with the preceding consideration of the alkali salts, promises to be the best one in general. Accordingly, as we go on to the secondary and tertiary salts, we shall take as the best approximations the various sequences in descending order of the degree of H.

3. Secondary Salt, Type $(NH_4)_2HX$. With $b = 2a = 2c$, Eq. (33) becomes

$$\begin{aligned} &H^6B + H^5[B(2c + K_1) + W] + H^4[W(K_1 - B) + BK_1(c + K_2)] \\ &\quad + H^3[W(K_{12} - BK_1 - W) - K_1(cW - BK_{23})] \\ &\quad + H^2K_1[W(K_{23} - W - BK_2) - cK_2(2W + BK_3)] \\ &\quad - HK_{12}W[W + K_3(3c + B)] - K_{123}W^2 = 0. \end{aligned} \tag{39}$$

Since such a solution is nearly neutral, we may use the H^4 sequence, so that $H \cong Q^4_{39}$. In certain cases, determined explicitly by the data tested in Eq. Q^4_{39},

$$H \cong W/2B + \sqrt{(\quad)^2 + 2WK_2/B}; \tag{40}$$

in others,

$$H \cong W/2B + \sqrt{(\quad)^2 + K_2K_3}. \tag{41}$$

4. Tertiary Salt, Type $(NH_4)_3X$. With $b = 3a = 3c$ in Eq. (33),

$$\begin{aligned} &H^6B + H^5[B(3c + K_1) + W] + H^4[W(K_1 - B) + BK_1(2c + K_1)] \\ &\quad + H^3[W(K_{12} - BK_1 - W) + cK_1(BK_2 - W) + BK_{123}] \\ &\quad + H^2K_1W[(K_{23} - W - BK_2) - 2cK_2] \\ &\quad - HK_{12}W[W + K_3(3c + B)] - K_{123}W^2 = 0. \end{aligned} \tag{42}$$

Such a solution is usually alkaline, so that $H \cong Q^3_{42}$; hence sometimes

$$H \cong W/B + \sqrt{(\quad)^2 + 3WK_3/B}. \tag{43}$$

If the solution is not highly alkaline, the H^4 sequence is better, so that $H \cong Q^4_{42}$; then, very roughly, $H \cong \sqrt{WK_2/B}$.

Table 16–4 compares some values for a typical example, calculated from the approximations indicated, with theoretical values obtained by numerical solution of Eqs. (37), (39), and (42). In the case of the tertiary salt ($b = 3a$) of the example chosen, the test, by orders of magnitudes, leaves little choice between Eqs. Q^3_{42} and Q^4_{42}, the magnitudes (again as negative exponents of 10) being, with $H = 10^{-9}$, 69, 50, 43, 41, 41, 44, 49, with Eq. Q^4_{42} slightly the better choice, as seen also in the table.

5. Approximations With No Dependence on Concentration. If Eqs. (34–36) show that the salt solution is nearly neutral, then "first approximations" independent of the concentration (provided the concentration is not too low) are obtainable from Eq. (32) by assuming $H - OH = 0$, or by

setting the charge coefficient of the salt equal to zero, which corresponds to the iso-electric point, or to the value of H_* approached at high concentration and therefore roughly independent of c.

For the primary salt ($b = a$),

$$H^4B - H^2K_1(BK_2 + W) - 2HK_{12}(BK_3 + W) - 3K_{123}W = 0. \quad (44)$$

Because of signs, an approximate solution of this equation must contain the term H^4B and since the largest of the remaining terms is generally that in H^2, as long as $B > \sqrt{W}$ and $K_3 < \sqrt{W}$, then

$$H_* \cong \sqrt{K_1(K_2 + W/B)}. \quad (45)$$

Table 16–4. Tribasic acid and weak base.

Example: $B = 10^{-5}$, $K_1 = 10^{-2}$, $K_2 = 10^{-7}$, $K_3 = 10^{-12}$

Type	c	Eq.	H (calcd.)	H (exact)
NH_4H_2X	1	Q^5_{37}	3.162×10^{-5}	3.162×10^{-5}
	0.01		2.252×10^{-5}	2.244×10^{-5}
$(NH_4)_2HX$	1	Q^4_{39}	1.465×10^{-8}	1.465×10^{-8}
	0.01		1.466×10^{-8}	1.466×10^{-8}
$(NH_4)_3X$	1	Q^3_{42}	2.02×10^{-9}	1.946×10^{-9}
	0.01		2.02×10^{-9}	1.947×10^{-9}
	1	Q^4_{42}	1.943×10^{-9}	
	0.01		1.943×10^{-9}	

Moreover, since B must be rather high for the solution to be approximately neutral, then $H \cong \sqrt{K_1K_2}$, as in Eq. (38); but this last formula, which may hold for high concentration, fails even when c drops to 0.01, as seen in Table 16–4. In this type of salt solution the dependence of H on c is still marked.

For the secondary salt ($b = 2a$),

$$2H^4B + H^3BK_1 - H^2K_1W - HK_{12}(BK_3 + 2W) - 3K_{123}W = 0. \quad (46)$$

Using the middle terms as the largest (as long as $K_1K_2 < \sqrt{W}$, $K_{123} < W\sqrt{W}$, and $B\sqrt{W} > K_{23}$), we have

$$H_* \cong W/2B + \sqrt{(\quad)^2 + K_2(K_3 + 2W/B)}, \quad (47)$$

which could reduce to either Eq. (40) or Eq. (41). Eq. (40) applies well for the example in Table 16–4.

For the tertiary salt ($b = 3a$),

$$3H^4B + 2H^3BK_1 + H^2K_1(BK_2 - W) - 2HK_{12}W - 3K_{123}W = 0. \quad (48)$$

Using the middle three terms (if $B\sqrt{W} > K_{23}$ and less than K_{12}), we have

$$H_* \cong -(BK_2 - W)/4B + \sqrt{(\quad)^2 + K_2W/B}, \quad (49)$$

reducing to $H \cong \sqrt{WK_2/B}$ as under Eq. Q^4_{42}. On the other hand, the last three terms of Eq. (48) give

$$H_* \cong W/(B - W/K_2) + \sqrt{(\quad)^2 + 3K_3W/(B - W/K_2)}, \quad (50)$$

which may reduce to Eq. (43). Eq. (43), but not $H \cong \sqrt{WK_2/B}$, applies fairly well to the example in Table 16–4.

E. Mixed Salts of Tribasic Acid

For salts of a tribasic acid with both strong base (at concentration b_s) and weak base (at concentration b),

$$H - OH = a\beta - b\alpha' - b_s; \quad (51)$$

$$\begin{aligned} &H^6B + H^5[B(b + b_s + K_1) + W] \\ &\quad + H^4[BK_1(b + b_s - a + K_2) + W(b_s - B + K_1)] \\ &\quad + H^3K_1[BK_2(b + b_s - 2a + K_3) + W(b_s - a - B + K_2 - W/K_1)] \\ &\quad + H^2K_{12}[BK_3(b + b_s - 3a) + W(b_s - 2a - B + K_3 - W/K_2)] \\ &\quad + HK_{123}W(b_s - 3a - B - W/K_3) - K_{123}W^2 = 0. \end{aligned} \quad (52)$$

We shall consider three cases:

(1) $b_s = a, \quad b = a = c;$ type $NaNH_4HX$;
(2) $b_s = 2a, \quad b = a = c;$ type Na_2NH_4X;
(3) $b_s = a, \quad b = 2a = 2c;$ type $Na(NH_4)_2X$.

1. Type $NaNH_4HX$. For a rough formula for H_* independent of concentration, we assume $H - OH = 0$ in Eq. (51) or $\beta - \alpha' - 1 = 0$, obtaining

$$2H^4B + H^3(BK_1 + W) - HK_{12}(BK_3 + W) - 2K_{123}W = 0. \quad (53)$$

Using terms in H^3 and H, we have

$$H \cong \sqrt{K_{12}\left(\frac{BK_3 + W}{BK_1 + W}\right)} = \sqrt{\frac{K_2(W + BK_3)}{B + W/K_1}}. \quad (54)$$

If $K_1 > W/B > K_3$, $H \cong \sqrt{WK_2/B}$. If $K_1 > K_3 > W/B$, $H \cong \sqrt{K_2K_3}$. If $K_1 < W/B$, $H \cong \sqrt{K_1K_2}$.

For neutrality, substituting $H = \sqrt{W}$, in Eq. (53), we have

$$B_n = (K_{12} + 2K_{123}/\sqrt{W} - W)/(K_1 + 2\sqrt{W} - K_{123}/W). \quad (55)$$

This requires roughly that $K_1K_2 > W$ and $K_3 \ll \sqrt{W}$.

For the dependence of H on concentration, Eq. (52), with $b_s = c$ and $b = a = c$, gives

$$\begin{aligned} &H^6B + H^5[B(2c + K_1) + W] + H^4[BK_1(c + K_2) + W(c - B + K_1)] \\ &\quad + H^3K_1[BK_{23} + W(K_2 - B - W/K_1)] \\ &\quad - H^2K_{12}[cBK_3 + W(c + B - K_3 + W/K_2)] \\ &\quad - HK_{123}W(2c + B + W/K_3) - K_{123}W^2 = 0. \end{aligned} \quad (56)$$

The largest terms will usually be those in the H^4 sequence, so that $H \cong Q_{56}^4$; therefore,

$$H \cong W(K_2 - B)/2B(c + K_2) + \sqrt{(\quad)^2 + cK_2(W + BK_3)/B(c + K_2)}, \quad (57)$$

$$\cong W(K_2 - B)/2cB + \sqrt{(\quad)^2 + K_2(W + BK_3)/B}, \quad (58)$$

$$\cong \sqrt{K_2(W + BK_3)/B}, \quad (59)$$

which may be compared with Eq. (54).

2. TYPE NA_2NH_4X. For the rough formula independent of concentration, Eq. (51), with $H - OH = 0$, or $\beta - \alpha' - 2 = 0$, gives

$$3H^4B + 2H^3(BK_1 + W) + H^2K_1(BK_2 + W) - K_{123}W = 0. \quad (60)$$

Since the solution is usually alkaline, neglecting the first two terms leaves

$$H \cong \sqrt{K_{23}W/(BK_2 + W)}. \quad (61)$$

Now if $BK_2 \gg W$, $H \cong \sqrt{WK_3/B}$; and if $BK_2 \ll W$, $H \cong \sqrt{K_2K_3}$.

For neutrality, Eq. (60) gives, with $H = \sqrt{W}$,

$$B_n = (K_{123}/\sqrt{W} - K_1\sqrt{W} - 2W)/(2K_1 + 3\sqrt{W} + K_{12}/\sqrt{W}). \quad (62)$$

Roughly, this is possible only if $K_{23} > W$; the solution is otherwise alkaline.

For an approximate formula showing the effect of concentration, Eq. (52), with $b_s = 2c$ and $b = a = c$, gives

$$\begin{aligned} &H^6B + H^5[B(3c + K_1) + W] + H^4[BK_1(2c + K_2) + W(2c - B + K_1)] \\ &\quad + H^3K_1[BK_2(c + K_3) + W(c - B + K_2 - W/K_1)] \\ &\quad + H^2K_{12}W(K_3 - B - W/K_2) - HK_{123}W(c + B + W/K_3) \\ &\quad - K_{123}W^2 = 0. \end{aligned} \quad (63)$$

The largest terms will usually be in the H^3 sequence, whence $H \cong Q_{63}^3$. Now provided c is not very small, then if $BK_2 > W$,

$$H \cong W/2c + \sqrt{(\quad)^2 + K_3W/B}, \quad (64)$$

and $H \cong \sqrt{K_3W/B}$, as under Eq. (61); if $BK_2 < W$,

$$H \cong W/2c + \sqrt{(\ \ )^2 + K_2K_3}, \tag{65}$$

or $H \cong \sqrt{K_2K_3}$, again as under Eq. (61).

3. TYPE $Na(NH_4)_2X$. A rough formula independent of concentration comes from H_* according to Eq. (51), with $(H - OH)$ neglected, or from $\beta - 2\alpha' - 1 = 0$:

$$3H^4B + H^3(2BK_1 + W) + H^2BK_{12} - HK_{12}W - 2K_{123}W = 0. \tag{66}$$

Using the middle three terms, we have

$$H \cong - BK_{12}/2(2BK_1 + W) + \sqrt{(\ \ )^2 + K_{12}W/(2BK_1 + W)}, \tag{67}$$

$$\cong - K_2/4 + \sqrt{(\ \ )^2 + K_2W/2B} \cong \sqrt{K_2W/2B}. \tag{68}$$

For neutrality, setting $H = \sqrt{W}$ in Eq. (66), we have

$$B_n = (K_{12} + 2K_{123}/\sqrt{W} - W)/(2K_1 + 3\sqrt{W} + K_{12}/\sqrt{W}). \tag{69}$$

Table 16–5. Mixed salts of tribasic acid.

Example: $B = 10^{-5}$, $K_1 = 10^{-2}$, $K_2 = 10^{-7}$, $K_3 = 10^{-12}$

Type	c	Eq.	H (calcd.)	H (exact)
$NaNH_4HX$	1	Q^4_{56}	1.0006×10^{-8}	1.0006×10^{-8}
	0.01		1.001×10^{-8}	1.001×10^{-8}
Na_2NH_4X	1	Q^3_{63}	3.163×10^{-11}	3.163×10^{-11}
	0.01		4.501×10^{-11}	4.52×10^{-11}
$Na(NH_4)_2X$	1	Q^4_{70}	9.81×10^{-10}	9.83×10^{-10}
	0.01		9.82×10^{-10}	9.85×10^{-10}

Roughly this is possible only if $K_{12} > W$; the solution is otherwise alkaline.

For the effect of concentration, with $b_s = c$ and $b = 2a = c$ in Eq. (52),

$$\begin{aligned} H^6B &+ H^5[B(3c + K_1) + W] + H^4[BK_1(2c + K_2) + W(c - B + K_1)] \\ &+ H^3K_1[BK_2(c + K_3) + W(K_2 - B - W/K_1)] \\ &- H^2K_{12}W(c + B - K_3 + W/K_2) \\ &- HK_{123}W(2c + B + W/K_3) - K_{123}W^2 = 0. \end{aligned} \tag{70}$$

This salt solution is not as alkaline as Type 2; therefore from the H^4 sequence, $H \cong Q_{70}^4$, which may reduce to Eq. (68) and finally to $H \cong \sqrt{WK_2/2B}$.

Table 16–5 lists some values for mixed salts of the same tribasic acid as that of table 16–4, and the values calculated through the approximations Q_{56}^4, Q_{63}^3, and Q_{70}^4 are compared with exact values calculated from the full Eqs. (56), (63), and (70). Again, if M^{++} is the divalent cation of a completely ionized diacid base, the foregoing equations apply equally to salts such as $Mg(NH_4HX)_2$, $MgNH_4X$, and $Mg\ [(NH_4)_2X]_2$, but with c as the analytical molarity of the acid radical $HX^{=}$ or $X^{\equiv}$ in each case.

XVII

Saturation with Respect to Acids and Bases

A. General Relations

We shall write P as the solubility product, in units of concentration, for any precipitate, and attach appropriate distinguishing subscripts in problems involving more than one precipitate.

Saturation with respect to an acid or a base means constancy of the activity (hence of the concentration, in ideal conditions) of the unionized form; if S is the solubility of an acid, then

$$S\rho = \mathbf{k}, \tag{1}$$

for saturation. But from Eqs. 1(20) and 1(21), the ionization fractions of an acid are so related that

$$\rho = \alpha_z H^z/K_{1\ldots z} = \alpha_{(z-1)} H^{(z-1)}/K_{1\ldots(z-1)} = \cdots = \alpha_1 H/K_1. \tag{2}$$

Hence, for saturation,

$$\begin{aligned} H^z[X^{-z}] &= H^z S\alpha_z = \mathbf{k}K_{1\ldots z};\\ H^{(z-1)}[X^{-(z-1)}] &= H^{(z-1)}S\alpha_{(z-1)} = \mathbf{k}K_{1\ldots(z-1)};\\ H[X^-] &= HS\alpha_1 = \mathbf{k}K_1. \end{aligned} \tag{3}$$

Any of these constant products may therefore be defined as the solubility product of the acid H_zX, whatever its basicity (z). We shall take as our definition consistently the first product listed, or

$$P = H^z[X^{-z}] = H^z S\alpha_z, \tag{4}$$

which holds whether or not the acid is "strong" so that $\rho = 0$; for a base, $M(OH)_{z'}$, $P = [M^{+z'}][OH^-]^{z'} = S\alpha'_{z'}[OH^-]^{z'}$.

From Eq. (4), for an acid, $\log S = \log P + zpH - \log \alpha_z$, or

$$\log S = \log P + zpH + \log\left(\frac{H^z}{K_{1\ldots z}} + \frac{H^{z-1}}{K_{2\ldots z}} + \cdots + \frac{H}{K_z} + 1\right). \tag{5}$$

A sufficient number of determinations then of the solubility as $f(pH)$ would theoretically allow the calculation of the various constants, both the K's and P. If the acid is completely ionized, $\alpha_z = 1$; the plot of $\log S$ against pH is then a straight line with slope z. If the acid is weak, the curve approaches a straight line at low H (i.e. as $\alpha_z \rightleftharpoons 1$), with a slope of z for the limiting straight line. The relations hold, of course, only so long as the solid phase remains the same; in real cases, the solid may change to a salt before the slope of z is reached. In very high H, the degrees of ionization approach zero, while ρ approaches 1, and then, from either Eq. (1) or Eq. (5), $S \rightleftharpoons P/K_{1\ldots z}$, or $\rightleftharpoons \mathbf{k}$, a constant, independent of pH, as for a nonelectrolyte. Moreover, since $S = [X^0] + [X^-] + [X^=] + \cdots + [X^{-z}] = \mathbf{k}(1 + K_1/H + K_{12}/H^2 + \cdots + K_{1\ldots z}/H^z)$, we may write

$$\log[(S - \mathbf{k})/\mathbf{k}] = pH - pK_1 + \log(1 + K_2/H + K_{23}/H^2 + \cdots + K_{2\ldots z}/H^{z-1}). \quad (5a)$$

Hence with $\mathbf{k}$ as the observed solubility at sufficiently high H, the measurement of S as $f(pH)$ allows the determination of pK_1 if K_2, K_3, etc. are negligibly small in this expression.[1]

For the solubility product of an acid in terms of activities, or $\mathbf{P}$, we have

$$\mathbf{P} = \mathbf{H}^z S\alpha_z \gamma^{z^2}, \quad (6a)$$

$$= H^z S\alpha_z \gamma^{z(z+1)}, \quad (6b)$$

$$= P\gamma^{z(z+1)}. \quad (6c)$$

Since the ionization fraction α_z is defined as a fraction of a concentration, it is to be calculated from mass ionization constants, so that in terms of H and K's we have simply the equations of Chapter I, or Eqs. I(20) and I(24). In terms of $\mathbf{H}$ and $\mathbf{K}$'s (activity of hydrogen ion and activity constants), α_3 for a tribasic acid, for example, becomes, with the γ convention explained in Chapter V,

$$\alpha_3 = \frac{\mathbf{K}_2\mathbf{K}_3}{\mathbf{H}^2\gamma^8} \Big/ \left(1 + \frac{\mathbf{H}\gamma}{\mathbf{K}_1} + \frac{\mathbf{K}_2}{\mathbf{H}\gamma^3} + \frac{\mathbf{K}_2\mathbf{K}_3}{\mathbf{H}^2\gamma^8}\right). \quad (7)$$

B. The Pure Saturated Aqueous Solution of an Acid or Base

1. General. With the quantity $(\ )_0$ always referring to the pure saturated solution,

$$D_0 = H_0 - OH_0 = S_0\beta_0, \quad (8)$$

or

$$\mathbf{D}_0 = D_0\gamma = S_0\beta_0\gamma. \quad (9)$$

Since β is a sum of ionization fractions, it is to be calculated from H and mass ionization constants as in Eq. I(26); or, in terms of $\mathbf{H}$ and activity constants, β for a polybasic acid becomes

$$\beta = \frac{1 + 2\mathbf{K}_2/\mathbf{H}\gamma^3 + 3\mathbf{K}_2\mathbf{K}_3/\mathbf{H}^2\gamma^8 + \cdots}{1 + \mathbf{H}\gamma/\mathbf{K}_1 + \mathbf{K}_2/\mathbf{H}\gamma^3 + \mathbf{K}_2\mathbf{K}_3/\mathbf{H}^2\gamma^8 + \cdots}. \quad (10)$$

(1) The case of monobasic acid was discussed by H. A. Krebs and J. C. Speakman, *J. Chem. Soc.*, **1945,** 593.

For the ionic strength, required for the estimation of γ, we have, according to Eqs. v(48) and v(49),

$$\mu = OH_0 + S_0(\alpha_1 + 3\alpha_2 + 6\alpha_3 + 10\alpha_4 + \cdot\cdot\cdot), \tag{11a}$$

$$= H_0 + S_0(\alpha_2 + 3\alpha_3 + 6\alpha_4 + \cdot\cdot\cdot). \tag{11b}$$

Eq. (4), defining P, and Eq. (8), representing electroneutrality, may always be combined so as to enable us to calculate two of the quantities S_0, H_0 (or $\mathbf{H}_0$), $\mathbf{P}$, $\mathbf{K}_1$, $\mathbf{K}_2$, $\cdot\cdot\cdot$ $\mathbf{K}_z$ if all but two of these are known. To find H_0 and $\mathbf{P}$, for example, from S_0 and the ionization constants, H_0 is first calculated from Eq. (8), with OH_0 neglected and with γ assumed unity. The preliminary value of H_0 is then used to find μ, through Eq. (11), and thus to estimate γ; thereupon H_0 may be refined through Eq. (8) with β_0 as in Eq. (10), etc. We may then calculate P through Eq. (4) and $\mathbf{P}$ from Eq. (6c), or $\mathbf{P}$ directly from either Eq. (6a) or Eq. (6b).

To find H_0 and S_0 from $\mathbf{P}$ and the $\mathbf{K}$'s, S_0 is eliminated from Eqs. (4) and (8), so that

$$D_0 H_0^{\,z} = P(\beta/\alpha_z)_0; \tag{12}$$

similarly, elimination of S_0 from Eqs. (6a) and (9) gives

$$\mathbf{D}_0 \mathbf{H}_0^z \gamma^{(z^2-1)} = \mathbf{P}(\beta/\alpha_z)_0. \tag{13}$$

With γ assumed as unity, and hence with $\mathbf{P} = P$, Eq. (12) is solved for H_0 in terms of constants; S_0 then follows from Eq. (4). These preliminary values are used in Eq. (11) to estimate μ and hence γ. Then H_0 is refined through Eq. (13) and S_0 is recalculated through Eq. (6a), etc.

Most of the subsequent considerations will be based on the ideal equations and their combinations, such as Eqs. (4), (8), and (12). Any resulting equation may be modified directly into one involving activity coefficients with activity constants through the relations $\mathbf{H} = H\gamma$, $\mathbf{D} = D\gamma$, Eq. (6c) for $\mathbf{P}$ and P, and Eq. v(43) for $\mathbf{K}$ and K. For very low solubility the ideal equations are of course sufficient.

Once $\mathbf{P}$ and the ionization constants are known, the constant $\mathbf{k}$ of Eq. (1), which is merely the concentration of the unionized form at saturation, is given, through Eq. (3), as $\mathbf{k} = \mathbf{P}/\mathbf{K}_{1\ldots z}$.

For the solubility, S_0, of a base such as $Fe(OH)_3$, from its solubility product, it would be incorrect to assume (according to what may be called the Arrhenius point of view, or as if we were dealing with a salt like AgCl) that $[Fe^{+++}] = S_0$ and $[OH^-] = 3[Fe^{+++}] = 3S_0$, whereupon $S_0 = \sqrt[4]{P/27}$[2] $= \sqrt[4]{10^{-37}/27} = 2.5 \times 10^{-10}$, so that H_0 then presumably has the value $1.3 \times 10^{-5} M$. The assumption that $[OH^-] = 3[Fe^{+++}]$ would be incorrect even if the base were *strong* or completely ionized to Fe^{+++}. It would thus be

(2) The value of the solubility product is 1×10^{-37}, according to A. B. Lamb and A. G. Jacques, *J. Amer. Chem. Soc.*, **60,** 967, 1215 (1938).

calculated that a pure saturated solution of $Fe(OH)_3$, which is by definition a base, is *acid*. Actually, if P is extremely small as in this case, we may safely assume even in ignorance of the ionization constants that $H_0 = OH_0 = \sqrt{W}$, as the solution of Eq. (12). Then we have $[Fe^{+++}] = S_0\alpha_3' = P/\sqrt{W^3}$. Without knowledge of the three ionization constants however, it is not possible to proceed beyond this point since $S_0 = P/\alpha_3'\sqrt{W^3}$, and α_3' can not be estimated. For $Fe(OH)_3$, we have $K_3' = 2.9 \times 10^{-12}$, $K_2' \cong 5 \times 10^{-10}$, and $K_1' \cong 2.5 \times 10^{-8}$(2a); with $H_0 = 10^{-7}$, then $\alpha_3' = 2.9 \times 10^{-8}$ and $S_0 = 3.5 \times 10^{-9}$.

2. Some Special Cases.

a. Dibasic acid. Eq. (12) now becomes

$$H_0^3 - H_0(W + P/K_2) - 2P = 0, \tag{14}$$

whether or not the acid is "strong in K_1," although K_1 is implicit in P, as seen in Eq. (3). If $W \ll P/K_2$, then $H_0 \cong \sqrt{P/K_2}$.

If H_0, S_0, and P are known, the two ionization constants are obtained from Eq. (14) as

$$K_2 = H_0P/(D_0H_0^2 - 2P), \tag{15}$$

and with this in Eq. (8), as

$$K_1 = H_0D_0(H_0^2D_0 - 2P)/[H_0^2D_0(S_0 - D_0) + D_0P]. \tag{16}$$

For a dibasic acid "strong in K_1," we may find P and K_2 from H_0 and S_0 through Eqs. (4) and (8):

$$K_2 = H_0(D_0 - S_0)/(2S_0 - D_0) \cong H_0(H_0 - S_0)/(2S_0 - H_0), \tag{17}$$

$$P = H_0^2(D_0 - S_0) \qquad \cong H_0^2(H_0 - S_0). \tag{18}$$

b. Monobasic weak acid. Now Eqs. (4) and (8) become $S_0 = P/H_0 + P/A$, and $D_0 = P/H_0$, so that

$$H_0 = \sqrt{P + W}, \text{ or } P = H_0^2 - W = H_0D_0. \tag{19}$$

Hence from any two of the quantities H_0, S_0, P, A, the other two are readily found. For monoacid weak base, Eq. (19) becomes

$$OH_0 = \sqrt{P + W}, \text{ or } P = (OH_0)^2 - W = (OH_0)(D_0'). \tag{20}$$

c. Strong acid. By definition, $\alpha_z = 1$, whatever the value of z. Hence

$$D_0 = zS_0, \tag{21}$$

$$P = H_0^zS_0. \tag{22}$$

(2a) From $[Fe(OH)^{++}][H^+]/[Fe^{+++}][H_2O] = 3.5 \times 10^{-3}$, $[Fe(OH)_2^+][H^+]/[Fe(OH)^{++}][H_2O] \cong 2 \times 10^{-5}$, and $[Fe(OH)_3][H^+]/[Fe(OH)_2^+][H_2O] \cong 4 \times 10^{-7}$; A. B. Lamb and A. G. Jacques (*loc. cit.*, Ref. XVII–2).

Now it is sufficient to know only one of the quantities H_0, S_0, or P (always if z is known) to find both of the others. Given S_0, H_0 is obtained from Eq. (21), and then P from Eq. (22). The value of H_0 may, moreover, be refined through an activity coefficient based on the ionic strength $\mu = H_0 + S_0z(z-1)/2 = (OH)_0 + S_0z(z+1)/2$, and introduced into Eq. (21) as $H_0 - \mathbf{W}/\gamma^2H_0 = zS_0$. Then Eq. (22) is used for $\mathbf{P}$, as $\mathbf{P} = H_0^zS_0\gamma^{z(z+1)} = \mathbf{H}_0^zS_0\gamma^{z^2}$. Given P, H_0 is first obtained by eliminating S_0 from Eqs. (21) and (22), as $H_0^{z+1} - H_0^{z-1}W - zP = 0$, and H_0 is then used in Eq. (22) for S_0. Given H_0, both S_0 and P are taken directly from Eqs. (21) and (22).

For monobasic strong acid, $H_0 = S_0/2 + \sqrt{(\quad)^2 + W} = \sqrt{P + W}$, which is the same as Eq. (19); also, $S_0 = D_0 = \sqrt{P/(1 + W/P)}$, and $P = H_0^2 - W = S_0^2(1/2 + \sqrt{1/4 + W/S_0^2})$.

C. Effect of Pressure on the Solubility of Volatile Acid or Base

1. Volatile Base, Type NH_3. Instead of Eq. (1), we have $S\rho' = \mathbf{k}\text{P}$, in which Henry's law is assumed to hold for the effect of the pressure, P, on the concentration of the unionized base. Then the solubility product is a function of the pressure, being equal to PP if P is the value at unit P. Since the solubility is not merely the concentration of the unionized solute, the effect of pressure on the solubility depends also on the ionization constant B and on other solutes affecting H. We shall let S be the solubility at unit P and S_P that at the pressure P, so that $S_\text{P}(OH)_\text{P}\alpha'_\text{P} = \text{P}P$. Then for the effect of pressure on the solubility in pure water, we have

$$(S_\text{P}/S)_0 = \text{P}OH\alpha'/(OH\alpha')_\text{P}. \tag{23}$$

But since $OH_\text{P} = \sqrt{\text{P}P + W}$, then if we may assume throughout this discussion that the ionization constants are independent of the pressure,

$$\left(\frac{S_\text{P}}{S}\right)_0 = \text{P}\left[\frac{\sqrt{P+W}\,(B + \sqrt{\text{P}P + W})}{\sqrt{\text{P}P + W}\,(B + \sqrt{P + W})}\right]. \tag{24}$$

Now if $P \gg W$,

$$(S_\text{P}/S)_0 \cong \sqrt{\text{P}}(B + \sqrt{\text{P}P})/(B + \sqrt{P}), \tag{25}$$

which equals $\sim \sqrt{\text{P}}$ if $B > \sqrt{P}$, or $\sim$ P if $B < \sqrt{P}$. But if $P \ll W$, then $(S_\text{P}/S)_0 \cong \text{P}$.

In the presence of strong base, $\rho' \cong 1$ and the effect is simply a direct proportionality with the pressure, or $(S_\text{P}/S)_0 \cong \text{P}$. On the other hand in the presence of strong acid, $H - OH = a_s - S\alpha' = a_s - P/OH$, so that $OH = -a_s/2 + \sqrt{(\quad)^2 + P + W}$. If $a_s > \sqrt{P}$, then $OH \cong (P + W)/a_s$.

With this, Eq. (23) becomes

$$\left(\frac{S_{\mathrm{P}}}{S}\right)_{a_s} = \mathrm{P}\left(\frac{P + W}{\mathrm{P}P + W}\right)\left[\frac{a_s + (\mathrm{P}P + W)/B}{a_s + (P + W)/B}\right]. \tag{26}$$

Hence if $P > W$, $(S_{\mathrm{P}}/S)_{a_s} \cong 1$ if $P/B < a_s$ and $\cong \mathrm{P}$ if $P/B > a_s$. If $P < W$, $(S_{\mathrm{P}}/S)_{a_s} \cong \mathrm{P}$.

2. Volatile Dibasic Acid, Type H_2S. With $S\rho = \mathbf{k}\mathrm{P}$ (if Henry's law applies), and with $S_{\mathrm{P}}H_{\mathrm{P}}^2(\alpha_2)_{\mathrm{P}} = \mathrm{P}P$, the effect of pressure on the solubility in pure water is $(S_{\mathrm{P}}/S)_0 = \mathrm{P}H^2\alpha_2/(H^2\alpha_2)_{\mathrm{P}}$, or

$$\left(\frac{S_{\mathrm{P}}}{S}\right)_0 = \mathrm{P}\left[\frac{H(1 + H_{\mathrm{P}}/K_1 + K_2/H_{\mathrm{P}})}{H_{\mathrm{P}}(1 + H/K_1 + K_2/H)}\right]. \tag{27}$$

Now if, from Eq. (14), we take $H_{\mathrm{P}} \cong \sqrt{\mathrm{P}P/K_2}$, and $H \cong \sqrt{P/K_2}$ at $\mathrm{P} = 1$, then, as long as $P > K_2^2/K_1$,

$$\left(\frac{S_{\mathrm{P}}}{S}\right)_0 \cong \sqrt{\mathrm{P}}\left(\frac{1 + \sqrt{\mathrm{P}P/K_2}/K_1}{1 + \sqrt{P/K_2}/K_1}\right). \tag{28}$$

But if H is greater than K_1, this means $(S_{\mathrm{P}}/S)_0 \cong \mathrm{P}$. In presence of a_s, with $\rho \cong 1$, $(S_{\mathrm{P}}/S)_{a_s} \cong \mathrm{P}$. In presence of strong base, $D = S(\alpha_1 + 2\alpha_2) - b_s$, or

$$H - OH = P(H + 2K_2)/H^2K_2 - b_s. \tag{29}$$

As long as $H > K_2$, therefore, $H \cong -b_s/2 + \sqrt{(\quad)^2 + (P + K_2W)/K_2}$, and with $b_s > \sqrt{P/K_2}$, $H \cong (P + K_2W)/b_sK_2$. Then, with $H > K_2$, Eq. (27) becomes

$$\left(\frac{S_{\mathrm{P}}}{S}\right)_{b_s} \cong \mathrm{P}\left(\frac{P + K_2W}{\mathrm{P}P + K_2W}\right)\left[\frac{b_s + (\mathrm{P}P + K_2W)/K_1K_2}{b_s + (P + K_2W)/K_1K_2}\right]. \tag{30}$$

Hence with $P > K_2W$, $(S_{\mathrm{P}}/S)_{b_s} \stackrel{\rightarrow}{=} 1$ if $P < K_1K_2$, and $\stackrel{\rightarrow}{=} \mathrm{P}$ if $P > K_1K_2$.

D. Fractional Precipitation of Bases

The corresponding equations for the simultaneous precipitation of acids may be written directly from those here given for bases by substitution of W/H for H and $-g$ for g.

The condition for saturation with respect to a univalent base, or the value of "H_{req}" for its precipitation, with $P = [M^+][OH^-] = b\alpha' OH$, is

$$H = bW/P - W/B, \tag{31}$$

or simply $H = bW/P$ if the base is strong. For a divalent base,

$$H = -W/2K_2' + \sqrt{(\quad)^2 + bW^2/P - W^2/K_{12}'}, \tag{32}$$

so that if the base is strong in respect to K_1' the quantity W^2/K_{12}' is dropped. For a trivalent base,

$$H^3K_{123}' + H^2K_{12}'W + HK_1'W^2 - W^3(bK_{123}'/P - 1) = 0, \tag{33}$$

which sets the form for still higher valences. If the base is strong in respect to K_1', Eq. (33) becomes

$$H^3K_{23}' + H^2K_2'W + HW^2 - bK_{23}'W^3/P = 0, \tag{34}$$

and if it is strong in respect to both K_1' and K_2',

$$H^3K_3' + H^2W - bK_3'W^3/P = 0. \tag{35}$$

(The usual condition for the applicability of the various approximate solutions of high degree equations defining H, as discussed in Chapter VIII, is that of an appreciably high value of the concentration of the solute in question. Such a condition is not in general to be expected to be satisfied in problems involving the solubility of the solute. Hence the equations in this and in the remaining chapters will as a rule be left in their full exact form, whether implicit or explicit, since simplifications or approximations will seldom be generally applicable.)

In a solution containing two bases (MOH and NOH, abbreviated as M and N) at the original concentrations $\mathbf{b}_M$ and $\mathbf{b}_N$ respectively, the equation of electroneutrality is

$$H - OH = -(b\beta')_M - (b\beta')_N + g. \tag{36}$$

If the first base (M) is the less soluble, then the range of H for a separation leaving p per cent of the first in solution without precipitating any of the second, is that between $H_{b_M'}$, or H_{req} for saturation with respect to the first base at the concentration b_M', defined as $p\mathbf{b}_M/100$, and $H_{\mathbf{b}_N}$, the value for saturation with the second base at the concentration $\mathbf{b}_N$. The corresponding range of g is therefore that between the value in

$$H - OH = -\left[\frac{P\beta'}{(OH)^{z'}\alpha_{z'}'}\right]_M - (\mathbf{b}\beta')_N + g, \tag{37}$$

with $H = H_{b_M'}$, and the value of g in

$$H - OH = -\left[\frac{P\beta'}{(OH)^{z'}\alpha_{z'}'}\right]_M - \left[\frac{P\beta'}{(OH)^{z'}\alpha_{z'}'}\right]_N + g, \tag{38}$$

with $H = H_{\mathbf{b}_N}$.

For better control of the value of H the adjustment may be made not with a strong base or acid but with a weak one. We shall suppose in such a case that the solution contains, besides the two bases to be separated, a certain value of g which may be considered as the algebraic value of [HCl] − [NaOH], a certain concentration a of a weak acid such as acetic acid, and a certain concentration b of a weak base such as ammonia. Then Eq. (36) becomes

$$H - OH = -(b\beta')_{M} - (b\beta')_{N} + g + a\alpha - (b\alpha')_{\text{“}NH_3\text{”}}. \quad (39)$$

If g and a are kept constant, the calculation of the value of $b_{\text{“}NH_3\text{”}}$ required to give the specified value of H, such as $H_{b'_M}$ or $H_{\mathbf{b}_N}$, is still simple and explicit. It must be remembered however that this "b" so calculated is the total analytical concentration of *ammonia* as a solute in the solution without regard to whether it is ionized or not or whether it is "present" as an "ammonium salt" or as "free ammonia," etc.

The relation between the concentrations of the two bases, b_M and b_N (to be distinguished from their initial concentrations $\mathbf{b}_M$ and $\mathbf{b}_N$) in a solution saturated with both, is obtainable by equating the two expressions for H_{req} for saturation with each base. Finally, the value of H in the solution saturated with both bases varies with the value of g in the solution; the expression for H as $f(g)$ is simply that given by Eq. (38). With H known, then the values of the two individual concentrations may be calculated directly from the solubility products.

We shall now write the equations for H as $f(g)$ and for the relation between b_M and b_N in a number of cases of saturation with two bases.

(1) Two univalent bases, MOH and NOH.

$$H^2(P_M + P_N + W) - HgW - W^2 = 0; \quad (40)$$

$$b_M = b_N P_M/P_N + P_M/B_M - P_M/B_N. \quad (41)$$

If either of the bases is strong, or even if both are strong, Eq. (40), involving the ionization constants only implicitly, remains unchanged. In Eq. (41) the term $P_M/B_M = 0$ if MOH is strong, and $P_M/B_N = 0$ if NOH is strong.

(2) Univalent base, MOH, and divalent base, $N(OH)_2$.

$$2H^3P_N/W + H^2[P_M + (P/K'_2)_N + W] - HgW - W^2 = 0; \quad (42)$$

$$b_M = H_{32N}P_M/W + P_M/B_M, \quad (43)$$

$$b_N = H^2_{31M}P_N/W^2 + H_{31M}(P/K'_2)_N/W + (P/K'_{12})_N. \quad (44)$$

Here H_{32N} means the expression for H in Eq. (32) for $N(OH)_2$, and H_{31M} means the expression for H in Eq. (31) for MOH.

(3) Two divalent bases, $M(OH)_2$ and $N(OH)_2$.

$$2H^3(P_M + P_N)/W + H^2[(P/K'_2)_M + (P/K'_2)_N + W] - HgW - W^2 = 0; \quad (45)$$

the (interchangeable) relation between the concentrations is given by Eq. (43).

(4) Univalent base, MOH, and trivalent base, $N(OH)_3$.

$$3H^4P_N/W^2 + 2H^3(P/K'_3)_N/W + H^2[P_M + (P/K'_{23})_N + W] - HgW - W^2 = 0; \quad (46)$$

$$b_N = H^3P/W^3 + H^2P/K_3'W^2 + HP/K_{23}'W + P/K_{123}'. \qquad (47)$$

In Eq. (47) the constants refer to $N(OH)_3$, and $H = H_{31M}$. The corresponding expression for b_M in terms of b_N would be Eq. (43) with $H = H_{33N}$, but Eq. (33) is cubic, so that the expression can not be made explicit.

(5) Divalent base, $M(OH)_2$, and trivalent base, $N(OH)_3$.

$$3H^4P_N/W^2 + 2H^3[P_M + (P/K_3')_N]/W + H^2[(P/K_2')_M + (P/K_{23}')_N + W] - HgW - W^2 = 0. \qquad (48)$$

For b_N as $f(b_M)$, Eq. (47) would apply, with $H = H_{32M}$. For b_M as $f(b_N)$ we would have Eq. (44) with M in place of N and with $H = H_{33N}$, so that again the expression could not be made explicit.

(6) Two trivalent bases, $M(OH)_3$ and $N(OH)_3$.

$$3H^4(P_M + P_N)/W^2 + 2H^3[(P/K_3')_M + (P/K_3')_N]/W + H^2[(P/K_{23}')_M + (P/K_{23}')_N + W] - HgW - W^2 = 0. \qquad (49)$$

The (interchangeable) relation between the concentrations is Eq. (47), with H from Eq. (33).

E. The System Acid–Base–Water

1. Schematic Diagrams. General solubility relations are best presented in connection with schematic phase diagrams. The system acid–base–water is a ternary system so that the isothermal relations of the condensed system, with vapor ignored, may be represented on a plane diagram. In connection with such diagrams we shall represent as **a** the number of moles of the acid whether or not completely dissolved, per liter of solution, and as a the actual concentration of the acid in the solution; similarly **b** will be the number of moles of the base, both solid and dissolved, per liter, and b its concentration in the solution. In Fig. 17–1, **b** is plotted vertically against **a** on rectangular coordinates. S_1 and S_2 represent the aqueous solubilities of base and acid respectively. As the acid is added to the saturated solution of the base in presence of excess of solid base, the solubility of the base is increased, following the curve S_1x; the solubility curve of the acid, as affected by the base, is S_2x, and the two curves meet at the point x, which is therefore a solution saturated with both solids. It may be that such a simple system is never realizable because of the precipitation of a solid salt as an intermediate phase. The hypothetical simple relations here assumed are nevertheless mathematically instructive and represent at least a metastable equilibrium state.

The region L, between the corner H_2O and the solubility curves, represents compositions of unsaturated liquid solution. If the total composition of the system falls in the region vertically above the curve S_1x, the line xu being vertical, the equilibrium state is a heterogeneous mixture of solid base and saturated liquid on the curve S_1x; the region L + Acid, below the horizontal line xv, is a similar mixture of solid acid and saturated liquid on the curve S_2x. If the composition falls in the region between the lines xu and xv, the system consists at equilibrium (which may be metastable) of a mixture of the liquid x and two solids.

2. Consideration of g as a Variable. The general equation for the solution is

$$H - OH = a\beta - b\beta' + g; \tag{50}$$

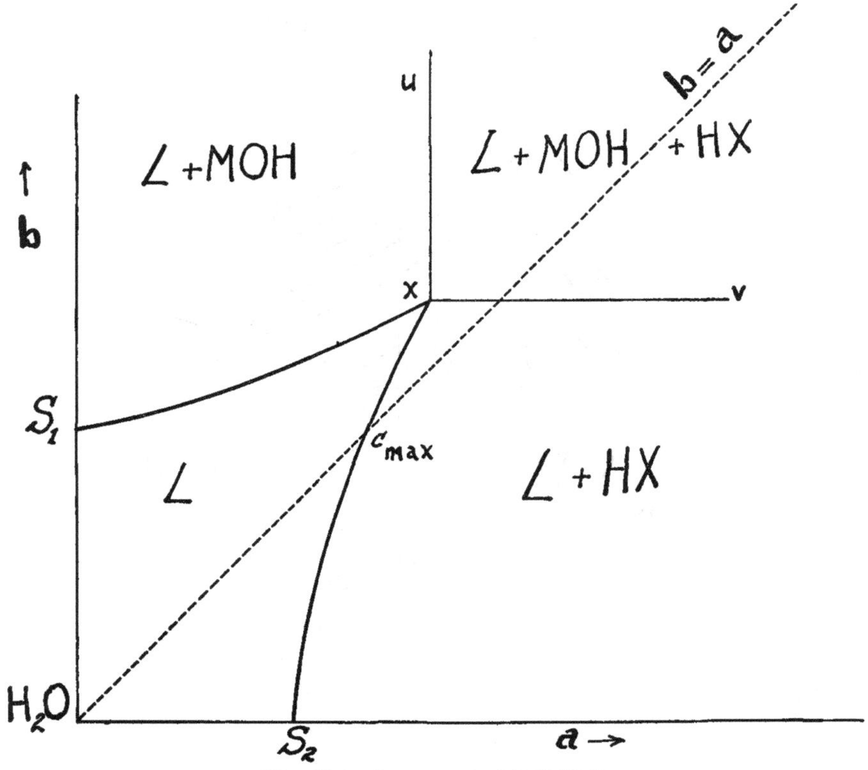

Fig. 17–1. System MOH–HX–H_2O.

here g, which may be positive or negative, represents the net concentration of foreign strong acid or strong base, or $g = a_s - b_s$. The plane of Fig. 17–1 represents the relations for a given value of g, specifically $g = 0$ if the diagram refers to the pure ternary system acid–base–water. There will be a similar plane for any given value of g, and if g is considered as an experimental variable similar to the experimental variables of temperature and pressure for the ternary system, we may think of each such plane diagram as a horizontal section of a solid model with g as the vertical coordinate. The general aspect of such a diagram will have certain resemblances to a three-dimensional temperature-composition phase diagram.

Fig. 17–2 shows schematically the system of Fig. 17–1 with *g* plotted on

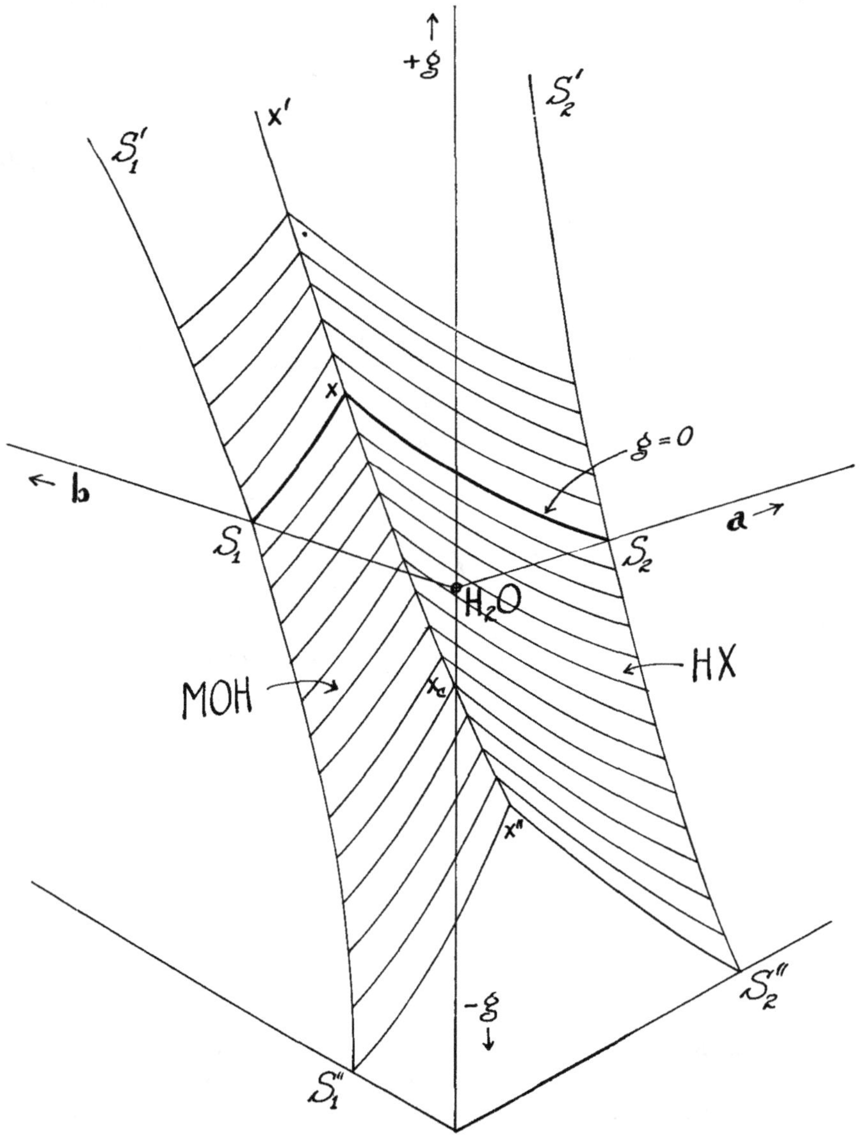

Fig. 17–2. Solubility relations of acid and base as function of *g*.

the vertical axis. It is drawn so that we look into the diagram from the H_2O corner, along the vertical diagonal plane $\mathbf{a} = \mathbf{b}$, and with a downward angle. In this way we look through the space representing unsaturated liquid, a space bounded by two solubility surfaces, one for base on the left and one for the acid on the right. The plane horizontal section of this figure at the value $g = 0$ represents the relations of "MOH" and "HX," the specific base and acid, in pure water. The part of the solid figure above this level represents relations in presence of excess of strong acid, with $g > 0$ or $a_s > b_s$; the part below, with negative g, shows the relations in excess of strong base, or $b_s > a_s$. Since we are dealing with ideal relations, the actual values of a_s and b_s, or the concentration of a salt such as NaCl, for example, will be considered to be of no significance; it is only the excess, $g = a_s - b_s$, which is here plotted. The diagram may be said to represent solubility relations at constant temperature, constant pressure, and constant ionic strength, as function of **a** and **b** as composition variables and of the special variable g.

For our purposes this variable "g" is neither a composition variable nor an intensive variable of the nature of temperature or pressure. The true intensive variable may be said to be H (or some function of H such as pH) and g bears a relation to H or pH similar to the relation of the heat content to temperature. The temperature of a system at constant pressure is controlled by addition or removal of heat, and the effect is subject to physical and chemical changes, both homogeneous and heterogeneous, caused by the change in heat content. The change in temperature caused by an increase of heat content for a system at equilibrium may be positive or zero but never negative. In the same way an increase or decrease of g, at constant temperature, pressure, and ionic strength, causes a change in H, and the effect is subject to the accompanying homogeneous and heterogeneous reactions. The homogeneous effects are those of buffering solutes, namely weak acids, weak bases, and ampholytes, while the heterogeneous effects are the precipitating and dissolving of solids (other phases, in general). The change in H caused by an increase of g in a system at equilibrium can be positive or zero but never negative. From the point of view of this comparison the buffer capacity of the solution ($\pi = dg/d\text{p}H$) is seen to be the analog of heat capacity, and a titration curve in analytical chemistry corresponds to a heating or cooling curve in thermal analysis. In both cases the curve is smooth unless there is a change in the number of phases during the process, a break corresponding to the appearance or disappearance of a phase, such as a precipitate.

In Fig. 17–2 then, the point labeled "H_2O" represents either pure water or aq. "NaCl"; the vertical axis directly above "H_2O" represents a solution of pure foreign strong acid, with or without "NaCl," etc. The left hand bounding plane of the figure represents compositions containing the specific base MOH and H_2O (but no HX) at various values of g; the right hand face is HX + H_2O at various values of g. The limiting horizontal plane used as the base of the figure has no particular significance. A change in **a** in the system, at constant **b**, means motion on a plane parallel to the face H_2O-HX;

change in **b** at fixed **a** occurs on a plane parallel to the face H_2O-MOH. Changes at fixed *g* occur on horizontal sections; and changes in *g* for fixed **a** and **b** occur on vertical lines in the figure.

The curve $S_1' - S_1 - S_1''$ is therefore the solubility curve of the base as a function of *g*; ideally the solubility of the base is increased in positive *g*, decreased in negative *g*. The opposite effect holds for the solubility of the acid, curve $S_2' - S_2 - S_2''$. The curve x′ — x — x″, the intersection of the two solubility surfaces, represents the liquid saturated with both solids.

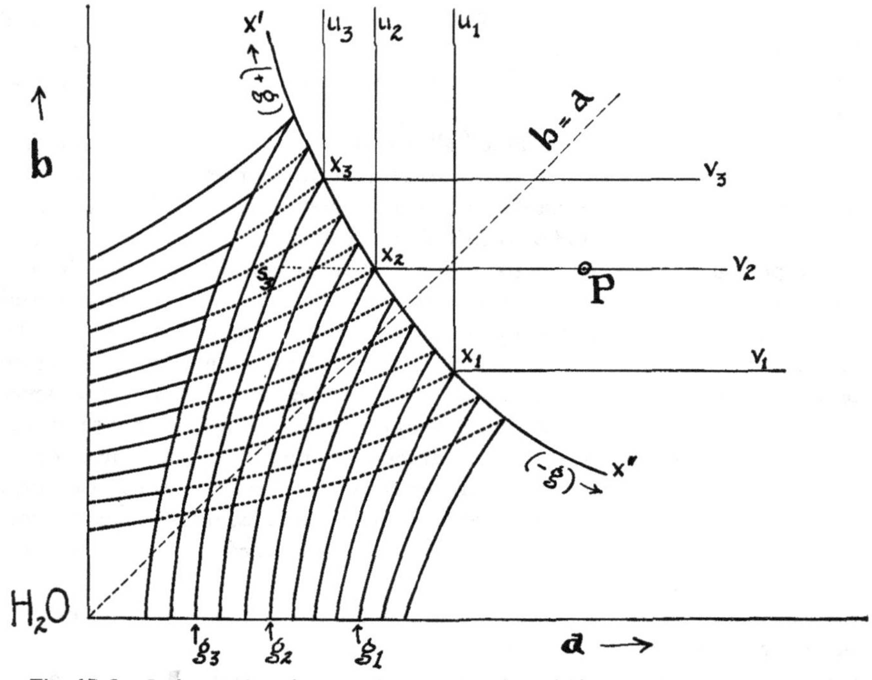

Fig. 17–3. Orthogonal projection of the curve of twofold saturation on a horizontal plane of Fig. 17–2.

This is a three-dimensional curve; the composition of this liquid (*a*,*b*,*H*, concentrations of individual solute species, etc.) is a function of *g*. The course of this curve could be shown by projection, for example, on the faces of the figure or, perhaps most usefully for the present, on a horizontal plane, as in Fig. 17–3.

For the position of the model as drawn in Fig. 17–2, the solubility surface of the acid falls toward the observer with increasing *g*, while that of the base recedes from the observer with increasing *g*. If an unsaturated solution, with known values of *a* and *b*, is treated with strong acid or strong base, saturation with one of the solids of the system will occur when its solubility surface is reached, obviously solid acid with increasing *g* and solid base with decreasing

g. Continued change of g increases the amount of solid, but because of the course of the curve x′x″ a second solid phase never appears.

Behind the curve x′x″ in Fig. 17–2 is a space in which the system, with given values of **a** and **b**, consists of three phases, liquid on the curve, solid base, and solid acid. A horizontal section of this space is shown between the lines xu and xv in Fig. 17–1. Fig. 17–3 shows three such sections, complete with their solubility curves, superposed upon the projection of the curve of twofold saturation, x′x″. In the order g_1, g_2, g_3, these represent sections for increasing values of g. Additional constant g contours are also shown to bring out the shape of the surfaces, that for HX lying above that for MOH when viewed from high g.

A system with values of the coordinates **a** and **b** represented by the point **P** in Fig. 17–3 consists, at g_1, of solid base, solid acid, and liquid of composition x_1. As g is increased, the composition of the liquid, still saturated with both solids, travels along the curve x′x″ toward x′. During the process the amount of solid base is diminishing. At g_2, when the liquid is at x_2 and when the boundary line x_2v_2 passes through **P**, the base is all dissolved and only the acid is left as solid phase. Continued increase of g now merely decreases the solubility of the acid, so that at $g = g_3$ the composition of the solution is at point s_3 on the solubility curve of the acid at g_3. This curve is simply the horizontal section of the solubility surface of the acid at g_3, shown in projection on Fig. 17–3.

These relations, it will be seen, are similar to the more familiar changes in phase diagrams occurring with change of temperature. The variable g is related to the changes in solubilities, at constant temperature, through the equilibrium constants and the condition of electroneutrality, somewhat as the heat content is related to the changes in solubilities at constant pressure, ideally, through the heats of fusion of the solids involved.

3. Analytical Equations. Limiting ourselves for the present to monobasic acid and monoacid base, we shall now consider the analytical equations for various kinds of sections of the solubility surfaces. A vertical section (parallel to a side of Fig. 17–2) on the surface of saturation for the base, for example, is one at constant b or constant a; a horizontal section is one at constant g.

a. Saturation with base. The condition for saturation with respect to the base is $P_1 = b\alpha' OH$, or as in Eq. (31),

$$H = W(b - P_1/B)/P_1; \quad OH = P_1/(b - P_1/B), \tag{51}$$

from which we see that saturation requires a minimum value of $b > P_1/B$. A strong base can always be precipitated whatever its concentration if the solution is made sufficiently alkaline, but a weak base can not be precipitated unless $b > P_1/B$. By using the value of H_{51} in Eq. (50), with $b\beta' = P_1/OH$, we may calculate, for specified g, the value of a corresponding to the specified value of b on the MOH solubility curve. The result may also be expressed as a relation between a and b, as a "solubility curve" at constant g. The equation is cubic in b. But since it is first degree in a it may be used to calculate a explicitly for any desired number of values of b in order to plot the curve:

$$a = \left[1 + \frac{W(b - P_1/B)}{AP_1}\right]\left[(b - P_1/B) + \frac{W(b - P_1/B)}{P_1} - \frac{P_1}{(b - P_1/B)} - g\right]. \quad (52)$$

If the acid HX is strong ($A = \infty$), the equation is quadratic in b:

$$b = (a + g)P_1/2(P_1 + W) + \sqrt{(\quad)^2 + P_1^2/(P_1 + W)} + P_1/B. \quad (53)$$

If $a = 0$ this gives the dependence of the solubility of the base upon g, and therefore describes the curve $S_1' - S_1''$ of Fig. 17–2.

According to Eq. (52), in which $a = f(b - P_1/B)$, the shape of the MOH saturation surface for given values of P_1 and A is independent of B. For specified values of a and g the effect of B appears explicitly simply as the added constant term P_1/B in the value of b, as in Eq. (53), the term vanishing for strong base. For given P_1 and A then, the surface, with fixed shape, simply moves farther from the HX–H_2O face of Fig. 17–2, as the base becomes weak, by the displacement P_1/B parallel to the MOH–H_2O face. With constant a and variable g, Eq. (53) describes a vertical section of the surface parallel to the MOH–H_2O face of Fig. 17–2.

For saturation with MOH alone, moreover, Eq. (50) becomes

$$H - OH = a\alpha - P_1/OH + g, \quad (54)$$

$$H^3(1 + P_1/W) + H^2[A(1 + P_1/W) - g] - H[A(a + g) + W] - AW = 0; \quad (55)$$

this gives H for known a and g, and the corresponding value of b follows from Eq. (51). Eq. (54) holds whether the saturating base is strong or weak; if the dissolved acid is strong, the equation becomes simply

$$H^2(1 + P_1/W) - H(a + g) - W = 0. \quad (56)$$

b. Saturation with acid. For saturation with the acid the expressions corresponding to Eqs. (51–56) are, with $P_2 = a\alpha H$:

$$OH = W(a - P_2/A)/P_2; \; H = P_2/(a - P_2/A); \quad (57)$$

$$b = \left[1 + \frac{W(a - P_2/A)}{BP_2}\right]\left[(a - P_2/A) + \frac{W(a - P_2/A)}{P_2} - \frac{P_2}{(a - P_2/A)} + g\right]. \quad (58)$$

If the base MOH is strong ($B = \infty$),

$$a = (b - g)P_2/2(P_2 + W) + \sqrt{(\quad)^2 + P_2^2/(P_2 + W)} + P_2/A; \quad (59)$$

with $b = 0$ this is the equation of the curve $S_2' - S_2''$ of Fig. 17–2. Again, the shape of the HX saturation surface is independent of A for fixed values of P_2 and B; the effect of variation in A is simply a displacement of the whole surface, equal to P_2/A, parallel to the HX–H_2O face of Fig. 17–2.

$$H - OH = P_2/H - b\alpha' + g; \tag{60}$$

$$H^3B + H^2[B(b-g) + W] - H[B(P_2 + W) + gW] - W(P_2 + W) = 0; \tag{61}$$

and if the dissolved base is strong,

$$H^2 + H(b - g) - (P_2 + W) = 0. \tag{62}$$

If the solubility of the acid, as a', is studied in presence of its alkali salt at concentration c, these equations still hold, with $a = a' + c$, $b = 0$ and $g = -c$. Similarly if the solubility of the base, as b', is studied in presence of its chloride or nitrate at concentration c, the relations are still those in Eqs. (51–56) with $b = b' + c$, $a = 0$, and $g = c$.

*b*1. *Suspension of strong acid titrated with strong base.* With solid acid present the titration curve is given by Eq. (62) as

$$H = -b_s/2 + \sqrt{(\quad)^2 + P_2 + W}, \tag{63a}$$

$$OH = b_sW/2(P_2 + W) + \sqrt{(\quad)^2 + W^2/(P_2 + W)}. \tag{63b}$$

Hence, also, with g representing the net concentration of foreign strong acid, or $a_s - b_s$,

$$P_2 = H(D - g), \tag{63c}$$

an expression to be compared with Eq. (19). We may thus calculate P_2 from the value of H in the saturated solution at a specified value of g. For the concentration of the dissolved acid at a given value of b_s, Eq. (59) gives

$$a = b_sP_2/2(P_2 + W) + \sqrt{(\quad)^2 + P_2^2/(P_2 + W)}. \tag{64}$$

The titration (increase of b_s) is represented by a vertical rise in Fig. 17–1. If $\mathbf{a}$ is not greater than a_x or if the quantity of acid in the system is such that when completely dissolved its concentration is not greater than a_x, then the acid will dissolve (to "dissolve" will here always mean to dissolve completely) when the total composition reaches the curve S_2x. The concentration of the base required to cause it to dissolve is, from Eq. (58), with A and $B = \infty$,

$$(b_s)_{\text{req}} = \mathbf{a} - (P_2^2 - \mathbf{a}^2W)/\mathbf{a}P_2. \tag{65}$$

If $(b_s)_{\text{req}} < \mathbf{a}$, i.e. if $\mathbf{a} < P_2/\sqrt{W}$, then the solid dissolves before the equivalent point ($b_s = \mathbf{a}$) is reached and the rest of the titration is normal as for unsaturated solution. If $(b_s)_{\text{req}} > \mathbf{a}$, or if $\mathbf{a} > P_2/\sqrt{W}$, the solid dissolves only in excess of the base. With solid always present the solution becomes neutral when $(b_s)_{\mathbf{n}} = a_{\mathbf{n}} = P_2/\sqrt{W}$, and at the equivalent point ($b_s = \mathbf{a}$)

the solution is alkaline, H being given by Eq. (63a) with $b_s = \mathbf{a}$. The equivalent point is of course neutral if the acid has already dissolved.

The equivalent point (pure salt plus water) is clear then or is free of precipitate, only if $\mathbf{a}$ and hence if c, the concentration of the salt, is not greater than $P_2/\sqrt{W}$. This critical value of $\mathbf{a}$ (or c) as $P_2/\sqrt{W}$ is that of the intersection of the solubility curve S_2x with the diagonal $\mathbf{b} = \mathbf{a}$, as may be seen by setting $a = b_s = c$ in Eq. (64), which then gives $c = P_2/\sqrt{W}$. It is for this reason, as will be discussed further in Section F of this chapter, that the intersection has been labeled c_{max} in Fig. 17–1.

If $\mathbf{a} > a_x$, addition of the base can not possibly dissolve all the solid. In this case the solubility curve is followed to the point x and then excess of the base remains as a second solid phase.

b2. Insoluble weak acid plus strong base. If a suspension of the weak acid is treated with strong base, the titration curve is still Eq. (63a,b) and again $P_2 = H(D - g)$ as in Eq. (63c), regardless of the value of the ionization constant. [This equation was recently used by Back and Steenberg, Ref. v–4a.] When $H = A$, the solution being saturated, $(b_s)_A = P_2/A - (A^2 - W)/A$; for neutrality, $(b_s)_n = P_2/\sqrt{W}$ as before, but, still for saturation, $a_n = P_2/A + P_2/\sqrt{W}$.

The suspension of the acid dissolves to give the concentration $\mathbf{a}$ when, from Eq. (58),

$$(b_s)_{req} = (\mathbf{a} - P_2/A) + W(\mathbf{a} - P_2/A)/P_2 - P_2/(\mathbf{a} - P_2/A). \qquad (66)$$

The concentration of the acid at a given value of b_s is the same as in Eq. (64), with an added term, $+ P_2/A$. The solid dissolves before or after the equivalent point ($b_s = \mathbf{a}$), depending on whether $(b_s)_{req} \lesseqgtr \mathbf{a}$ and hence on whether $\mathbf{a} \lesseqgtr c_{max}$, the intersection of the solubility curve of the acid with the diagonal $b = a$ in Fig. 17–1. In this case Eq. (59), with $a = b = c$ and $g = 0$, gives

$$c_{max} = P_2^2/2AW + \sqrt{(\quad)^2 + P_2^2/W} + P_2/A. \qquad (67)$$

If the alkali salt NaX of the insoluble weak acid at concentration c ($= \mathbf{a}$) is titrated with strong acid, the acid HX may precipitate during the addition of foreign strong acid. It does so when H has the value in Eq. (57) and when the concentration of strong acid added reaches the value $\mathbf{a} - (b_s)_{66}$. Now the precipitate appears before the equivalent point if $c > c_{max}$ of Eq. (67).

[If the salt MCl_2 of weak insoluble divalent base, $M(OH)_2$, is treated with foreign strong base, the base $M(OH)_2$ may precipitate in the process. If P is the solubility product of $M(OH)_2$ and K_1', K_2' its ionization constants and if c is the concentration of the salt, the precipitate $M(OH)_2$ appears when $P = c\alpha_2(OH)^2$, or when $OH = P/2K_2'(c - P/K_{12}') + \sqrt{(\quad)^2 + P/(c - P/K_{12}')}$. No precipitate ever appears therefore unless

$c > P/K_1'K_2'$. With this value of OH the value of b_s required for the precipitation may be obtained through the equation $H - OH = c(2 - \alpha_1' - 2\alpha_2') - b_s$. The precipitate appears before the equivalent point if $(b_s)_{\text{req}} < 2c$.]

b3. Determination of constants from solubility measurements. The constants $\mathbf{A}$ and $\mathbf{P}_2$ of the acid may be determined through measurements of $\mathbf{H}$ and S, the solubility, in presence of a strong base. For $\mathbf{A}$, the equations would be the same as Eqs. v(16) and v(31) but with S in place of a. For $\mathbf{P}_2$, the definition $P_2 = S\alpha H$ gives

$$\log \mathbf{P}_2 = \log S + \log \left(\frac{\mathbf{A}}{\mathbf{A} + \mathbf{H}\gamma}\right) - p\mathbf{H} + \log \gamma, \tag{68}$$

in which $\mathbf{A}$ and $\mathbf{P}_2$ are the thermodynamic constants.

It is theoretically also possible to obtain the two constants from measurements of the solubility S with varying b_s, or from the solubility curve, without the measurement of H. This involves the use of Eq. (66) solved for A or for P_2 and with S in place of $\mathbf{a}$:

$$A = \frac{P_2}{S - [b_sP_2/2(P_2 + W) + \sqrt{(\quad)^2 + P_2^2/(P_2 + W)}]}; \tag{69}$$

$$P_2^3 + P_2^2A(b_s - 2S - A + W/A) - P_2SA^2(b_s - S + 2W/A) + S^2A^2W = 0. \tag{70}$$

With a rough estimate of A available, the second equation may be used to calculate P_2 and then the first may be used to reestimate A from different data, etc., until the constants fit the data.

c. Saturation with both base and acid. The conditions $P_1 = b\alpha'OH$, $P_2 = a\alpha H$, and the equation of electroneutrality, or Eq. (50), which is now simply $H - OH = a\alpha - b\alpha' + g$, fix the relations of the four quantities H, a, b, and g for a solution saturated with both MOH and HX. These relations, which may then be used in various combinations in connection with problems of precipitation, are first Eqs. (51) and (57) for the relations between H and b and between H and a respectively. If these two expressions for H are equated, we obtain the interrelation of a and b:

$$(a - P_2/A)(b - P_1/B) = P_1P_2/W. \tag{71}$$

This is the equation of the hyperbola representing the projection of the curve x′x″ shown in Fig. 17–3. Its asymptotes are $a = P_2/A$ at $b = \infty$ and $b = P_1/B$ at $a = \infty$. It crosses the diagonal of Fig. 17–3 when $a = b = c$, or when

$$c = (P_1/B + P_2/A)/2 + \sqrt{(\quad)^2 + (P_1P_2/W)(1 - W/AB)}. \tag{72}$$

Only the position of this curve in its a/b projection and not its shape changes with a change in the strength of either the acid or the base or both, for fixed values of P_1 and P_2. If the acid is strong the curve shifts to the left by the constant amount P_2/A and if the base is strong its position drops by the constant amount P_1/B. If both are strong the curve is symmetrically placed, as $ab = P_1P_2/W$, and crosses the diagonal at $c = \sqrt{P_1P_2/W}$.

The substitution of H_{57} and P_1/OH for $b\alpha'$ in the electroneutrality equation gives the relation between a and g:

$$a = -\,gP_2/2(P_2 + W) + \sqrt{(\quad)^2 + P_2^2(P_1 + W)/W(P_2 + W)} + P_2/A; \quad (73)$$

similarly, for the relation of b and g:

$$b = gP_1/2(P_1 + W) + \sqrt{(\quad)^2 + P_1^2(P_2 + W)/W(P_1 + W)} + P_1/B. \quad (74)$$

These may also be rearranged to give the first degree equations $g = f(a)$ or $g = f(b)$. With $a = b = c$ from Eq. (72) the value of g at the point $\mathbf{x}_c$, the crossing of the curve x′x″ through the vertical diagonal plane $\mathbf{b} = \mathbf{a}$ as shown in Figs. 17–2 and 3 may be calculated explicitly in terms of the five constants A, B, P_1, P_2, W; the value of H at $\mathbf{x}_c$ follows from either Eq. (51) or Eq. (57) with $a = b = c_{72}$. From Eqs. (73) and (74) moreover we have, for $g = 0$,

$$a(\text{at } g = 0) = P_2\sqrt{(P_1 + W)/W(P_2 + W)} + P_2/A; \quad (75)$$

$$b(\text{at } g = 0) = P_1\sqrt{(P_2 + W)/W(P_1 + W)} + P_1/B. \quad (76)$$

Finally, the relation between H and g follows from

$$H - OH = P_2/H - P_1/OH + g, \quad (77)$$

as

$$H = \frac{gW/2}{P_1 + W} + \sqrt{(\quad)^2 + W\left(\frac{P_2 + W}{P_1 + W}\right)} \underset{(g=0)}{=} \sqrt{W\left(\frac{P_2 + W}{P_1 + W}\right)}. \quad (78)$$

Eq. (77) also gives the buffer capacity of the double suspension as

$$\pi = |\,dg/dpH\,| = 2.3(P_2/H + P_1/OH + H + OH), \quad (79)$$

with a minimum value, at $g = 0$, of $(2.3)2\sqrt{(P_1 + W)(P_2 + W)/W}$. For a pure suspension of the acid, when $P_1/OH = 0$ in Eq. (79), the buffer capacity is also a minimum at $g = 0$, with the value $(2.3)2\sqrt{(P_2 + W)}$; for a pure suspension of base, the minimum, again at $g = 0$ is $(2.3)2\sqrt{(P_1 + W)}$.

In general, unless both the acid and the base are strong, the product ab is not constant on the curve x′x″. If the base alone is weak, ab increases continuously with g, and if only the acid is weak, ab increases continuously with $-\,g$. If both are weak, then since $a\alpha b\alpha' = P_1P_2/W$, the product ab is a minimum when $\alpha\alpha'$ is a maximum, or at $H = \sqrt{WA/B}$. From Eq. IV(49), therefore,

$$(ab)_{\min} = (P_1P_2/W)(1 + \sqrt{W/AB})^2, \quad (80)$$

so that the minimum of ab is the intersection of this symmetrical hyperbola with curve x′x″. The minimum of ab occurs, moreover, from Eq. (77) at $g = \sqrt{WA/B} - \sqrt{WB/A} - \sqrt{AB/W}(P_2/A - P_1/B)$, and, from Eqs. (51) and (57), with $a > b$ if $P_2/A > P_1/B$.

The concentration of the solution on this curve may also be expressed as the sum of a and b, or the "solubility" of the suspension of the two solids, as $f(g)$. This "solubility" passes through a minimum as $f(H)$ and therefore as $f(g)$. This follows because what may be called the "charge coefficient" of the mixed suspension, or the quantity $P_2/H - P_2/OH$ in Eq. (77), is "compound" in the sense defined in Chapter IV. Its value is zero at $H = \sqrt{WP_2/P_1}$, which may therefore be called the "iso-electric point" of the mixed suspension. (It must be noted that this holds regardless of the strengths of the acid and base, HX and MOH, involved.) The minimum solubility of the mixture therefore occurs at $g = (H - OH)_{ie} = (P_2 - P_1)\sqrt{W/P_1P_2}$: in excess of strong acid if $P_2 > P_1$ or of strong base if $P_2 < P_1$. If the curve x′x″ is projected orthogonally upon the vertical mid-plane $\mathbf{b} = \mathbf{a}$, the central vertical plane normal to the paper in Fig. 17–2, the projection appears as in Fig. 17–4, in which g is plotted vertically against $(\mathbf{a} + \mathbf{b})/\sqrt{2}$. At $g = 0$, or point x, the sum, $a + b$, is given directly by Eqs. (75) and (76). At the minimum point, x_{min}, the sum is

$$(a + b)_{min} = 2\sqrt{P_1P_2/W} + P_2/A + P_1/B; \tag{81}$$

the minimum occurs with $a > b$ if $P_2/A > P_1/B$. The minimum of ab and that of $a + b$ coincide only if $P_2/A = P_1/B$. If $P_2/A > P_1/B$, the value of a is greater (and g therefore lower) at ab_{min} than at $(a + b)_{min}$.

F. Maximum Concentration of a Salt in Pure Water Without "Hydrolytic" Precipitation

1. Meaning of "Hydrolytic" Precipitation (Congruent and Incongruent Solubility of Salts). If there is a limited solubility of the acid or base of a salt, then there is a limiting concentration of the pure salt in water beyond which there will be precipitation of either the acid or the base. We shall refer to this concentration as c_{max}—the maximum concentration of pure salt stable in water.

In the following discussion we shall consider the solid salt to be a compound of the acid and the base, a 1 : 1 compound in the case of the salt of monobasic acid and monoacid base. This is the original Berzelius point of view of a salt and it is still very clear if the acid is an oxygen acid. The view had to be discarded of course in connection with HCl and a salt like NaCl. From the point of view of the Phase Rule, however, the system "NaOH–HCl–H_2O" is still ternary even though NaCl is not an *additive* compound of the components. It is classed as a "reciprocal ternary system" and is handled by the device of using both positive and negative quantities of the components in describing the compositions of the possible phases. The phase NaCl is thus a "compound" containing + 1 mole of NaOH, + 1 mole of HCl, and − 1 mole of H_2O. For our present purpose we therefore find it convenient and not inconsistent to speak of a *salt solution* as one containing the acid and the base in stoichiometric proportions and of a *solid salt* as a "compound" of the acid and base containing in addition either positive or negative quantities of water. If the number of moles of 1 : 1 salt precipitated

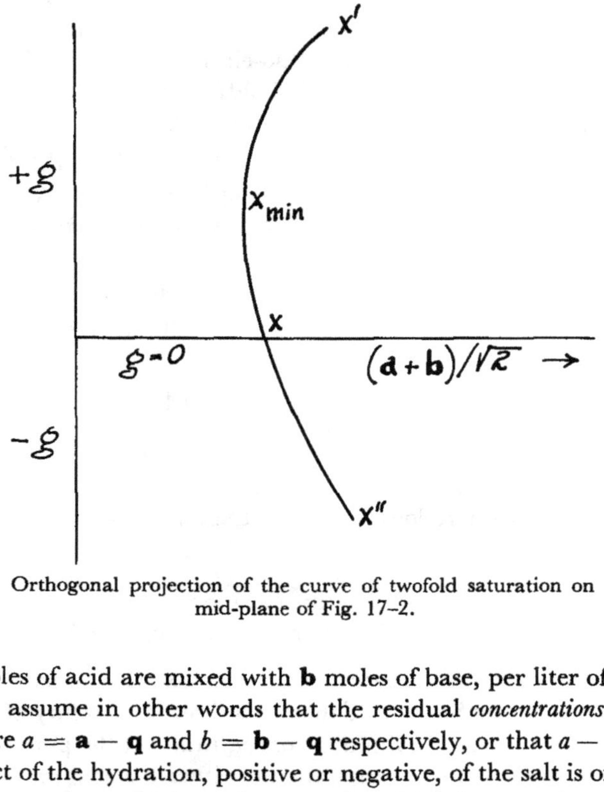

Fig. 17–4. Orthogonal projection of the curve of twofold saturation on the vertical mid-plane of Fig. 17–2.

when **a** moles of acid are mixed with **b** moles of base, per liter of solution, is **q**, we shall assume in other words that the residual *concentrations* of the acid and base are $a = \mathbf{a} - \mathbf{q}$ and $b = \mathbf{b} - \mathbf{q}$ respectively, or that $a - b = \mathbf{a} - \mathbf{b}$. This neglect of the hydration, positive or negative, of the salt is of the nature of other assumptions which we have made, which increase in validity with the degree of dilution of the system. It is related to the assumption originally made in connection with ionization constants that the activity of the water is constant.

If Fig. 17–5 then pertains to the system monobasic acid–monoacid base–water, or HX–MOH–H_2O, the composition of the salt is represented by the diagonal $\mathbf{b} = \mathbf{a}$. The curves S_1x and S_2x again represent the solubility curves of base and acid respectively; point x, the solution saturated with both base and acid, has the composition given by Eqs. (73–78). At this point moreover, as on any point of the curve x′x″ of Fig. 17–2, since both solubility products, P_1 for the base and P_2 for the acid, are satisfied simultaneously, there will be a constant ion-product for saturation with the mixture so that

$$[M^+][X^-] = P_1P_2/W = P_{mix}. \tag{82}$$

A "salt solution" lies on the diagonal line and the concentration of the salt, c, is measured either vertically or horizontally since $b = a = c$ for the

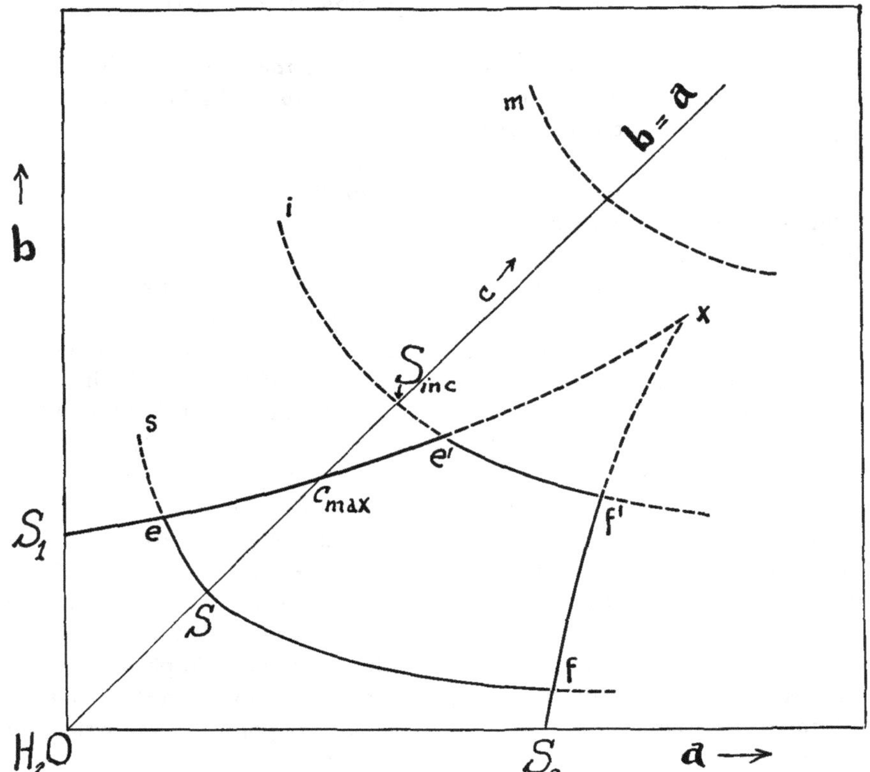

Fig. 17–5. Congruent and incongruent solubility relations of simple 1 : 1 salt; "hydrolytic" relations.

salt. For the specific relations assumed in Fig. 17–5, the maximum stable concentration of such a salt solution is given by point c_{max}. A higher concentration of the salt (pure in aqueous solution) is not possible at equilibrium since at this point the solution begins to precipitate the base, on the curve S_1x.

If a solid salt exists as a solid phase in the particular system HX–MOH–H_2O, it will have its own solubility curve and solubility product, P. The value of P may, at least hypothetically or in general, be either greater or smaller than P_{mix}, the product defined in Eq. (82) for a solution saturated with the acid and the base. If $P > P_1P_2/W$, a solution saturated with the salt would be supersaturated with respect to the mixture and the salt is never a stable solid phase in contact with the aqueous solution of the system. The solubility curve of the salt would then be entirely metastable, such as curve m, lying above and to the right of point x. The solubility curve of the salt can cut the curves S_1x and S_2x only if $P < P_1P_2/W$. The salt then has a region of stability as a solid phase in contact with aqueous solution and the range of its stability as a saturating phase, in respect to the values of a and b in the solution, will be given by that portion of this curve between the curves S_1x and S_2x.

If the solubility of the salt is low enough so that the solubility curve, curve s, crosses the diagonal below c_{max}, then a saturated solution of the salt is possible in pure water without excess of either acid or base. The solubility of the salt is now S; its solubility in excess of base lies on the branch Se of the curve and its solubility in excess of acid on the branch Sf. Point e is the solution saturated with both base and salt, point f the solution saturated with both salt and acid.

If the solubility curve of the salt, curve i, cuts the diagonal above c_{max} at point S_{inc}, the salt may exist as a stable saturating phase only along the limited section e′f′ in solutions containing an excess of the acid over the stoichiometric ratio $a = b$. The solution saturated with base and salt is now point e′ and the solution saturated with salt and acid is point f′. Only dilute unsaturated solutions of the salt are stable in pure water, namely those with concentrations below c_{max}. If the concentration of the salt in pure water is brought up to c_{max}, the solution at equilibrium there begins to precipitate the base. If the process of concentration is continued, as by evaporation of the solution, base continues to precipitate while the composition of the solution moves along the curve c_{max}e′ toward point e′, at which point the solid salt first appears. Or, if excess of the solid salt is added to pure water, equilibrium is established not with solid salt as the sole solid phase but with a mixture of base and salt, the solution having the composition e′. The pure saturated aqueous solution of the salt would be point S_{inc}, a system in metastable equilibrium.

In the first case, with $c_{max} > S$, the salt is said to be *congruently soluble in water*. Its stable solubility curve is cut by the line $\mathbf{b} = \mathbf{a}$, representing the composition of the salt. A saturated solution of the salt as sole solid phase may be prepared from salt and water without excess of either the acid or the base. Evaporation of the pure salt solution leads to precipitation of the pure salt as sole solid phase. The solid salt may be washed with water without decomposition.

In the second case, with $c_{max} < S_{inc}$ (S_{inc} being the metastable solubility of the salt in pure water), the salt is said to be *incongruently soluble in water*. With the relations assumed in Fig. 17–5, the salt is incongruently soluble *with respect to the base* (but similar considerations would apply to incongruence with respect to the acid if point x lies on the opposite side of the diagonal).

The stable solubility curve is not cut by the diagonal representing the proportions of acid and base in the salt. A saturated solution of the salt is possible only in excess of the acid, along the curve e′f′ Evaporation of the pure "salt solution" leads to precipitation of the base rather than of the salt. The pure salt solution can not be concentrated and the solid salt can not be washed with water without heterogeneous decomposition resulting in precipitation of the solid base.

This heterogeneous reaction is usually called "hydrolytic precipitation." Although the expression may be convenient, it is to be noted that the only equilibrium constants involved are solubility products and ionization constants already defined. No new equilibrium constant such as a "hydrolysis constant" has to be introduced, nor is there in fact any such constant definable with usefulness or with mathematical consistency, as already pointed out in Chapters IV, VII, VIII, etc. Whether or not a particular salt will suffer "hydrolytic precipitation" of its base depends on whether, in terms of Fig. 17–5, c_{max} is smaller or larger than the solubility of the salt in pure water or on whether the composition of the solution saturated with base and salt is on the right of the diagonal as at point e′ or on the left as at point e. In Chapters XIX and XX the composition of this solution will be calculated on the basis of the relations of the equilibrium constants involved, which for the simple uni-univalent case are A, B, W, and the solubility products of the two solids, the base and the salt. The expression "hydrolytic precipitation" may or may not be felt to be more convenient than "incongruence of solubility," but the mathematical relations do not involve any concept of "hydrolysis" at all. "Hydrolysis" is commonly understood as the reverse of ionization, but the heterogeneous decomposition involved in this precipitation of the base or of the acid is seen to be primarily a question of solubilities and secondarily (or implicitly) also a matter of the strengths of the acid and base. Even a salt of strong acid and strong base is subject to such "hydrolysis" or decomposition with precipitation of its acid or base if either of these is sufficiently insoluble. The considerations so far presented regarding Fig. 17–5 apply for any relations of the strengths of the acid and base, including even their possible ampholytic character. In Chapter XX we shall consider, moreover, the possible incongruence of solubility of an acid salt with respect to the normal salt; in such relations the acid salt $M(HX)_2$ is decomposed by water to precipitate the salt MX involving therefore the ion $X^{=}$, one formed by the further ionization (the opposite of the usual meaning of "hydrolysis") of the ion HX^- of the original acid salt.

2. CALCULATION OF c_{max}. To find c_{max} for any particular salt we simply combine the equation of electroneutrality for the pure salt solution at the concentration c, or $(H - OH) = c(\beta - \beta')$, with the requirement for saturation with respect to either its acid or its base, to eliminate H. Only a few general and typical cases will be considered.

a. Salt of strong acid and strong insoluble base; $AgNO_3$. From $(H - OH) = 0$

we have, if the base is univalent, as already indicated under Eq. (65), $c_{max} = P/\sqrt{W} = \mathbf{P}/\gamma\sqrt{\mathbf{W}}$, with P as the solubility product of the base. (Subscripts will be used only in Subsection d to distinguish between base and acid solubility products.) If the base is divalent, $c_{max} = P/W = \mathbf{P}/\gamma^4\mathbf{W}$. In each case the second expression introduces the activity coefficient on the basis of the convention explained in Chapter v.

For $AgNO_3$ therefore, a salt for which both the acid and the base are considered strong, the base however having $\mathbf{P} = 2 \times 10^{-8}$, we may write, if $\log \gamma = -0.505\sqrt{c}$, $\log c_{max} - 0.505\sqrt{c_{max}} = \log \mathbf{P} + 7$, so that c_{max} is only $0.427M$. Since this is apparently incompatible with the high presumably congruent solubility of the salt, it would seem at first hand that there might be something wrong either with the definition of the solubility product of the base or with the estimate of its strength. But the quantity of insoluble base precipitated may be calculated and it is found to be negligibly small unless P is itself extremely small. If c is the original concentration of the salt and hence still the concentration of its acid after precipitation of the base, we have $H - OH = c - [M^+] = c - P/OH$, and H is given by Eq. (56) with $a + g = c$. The number of moles of precipitate formed per liter of solution is then simply $c - [M^+] = (H - OH)_{56}$. In the case of $10M$ $AgNO_3$ solution, Eq. (56) gives $H = 2.5 \times 10^{-6} + (6.25 \times 10^{-12} + 10^{-20}/2\gamma^2)^{1/2}$, if the activity coefficients of the ions H^+, OH^-, and Ag^+ are assumed to have the common value γ. Hence $H \cong 5 \times 10^{-6}$, the number of moles of precipitate is only 4.998×10^{-6} and the concentration of (foreign) soluble strong acid a_s, which could be some slight excess of nitric acid itself, required to prevent the precipitation is also only 4.998×10^{-6} since for the clear solution of the salt, $H - OH = a_s$.

b. Normal salt of weak acid and strong insoluble base. If we combine the solubility product expression and the electroneutrality equation with elimination of H, we obtain an expression for c_{max} in terms of constants. We shall consider two variations, salts of univalent base and acids of varying basicity (z) and salts in which the valence of the base equals the basicity of the acid.

b1. Univalent base; $AgC_2H_3O_2$, Ag_2CO_3, Ag_3PO_4. In this case $P = zc(OH)$.

For *monobasic acid* (type silver acetate), Eq. x(13) with $P = c(OH)$ gives

$$c^3W(P + W) + c^2APW - cP^2W - AP^3 = 0. \tag{83}$$

For example then, if $A = 1.8 \times 10^{-5}$ and $P = 2 \times 10^{-8}$, the maximum concentration of such a salt without "hydrolytic" precipitation of the *strong base* is $c_{max} \cong 9.0 \times 10^{-3}$ ($\cong \sqrt[3]{AP^2/W}$). The situation is similar to that for silver nitrate. For a silver acetate solution at $c = 0.05$ ($> c_{max}$), Eq. (55), with $a + g = c$, gives $H = 2.4965 \times 10^{-8}$; the number of moles of precipitated AgOH per liter is then $c - P/OH = 7 \times 10^{-5}$. With H_{55} and with $a = c$, Eq. (54) may be used to find the concentration of foreign strong acid to prevent the precipitation, as $(a_s)_{req} = 6.90 \times 10^{-5}$. From Eq. (52) moreover, we may calculate directly, with $b = c = 0.05$, the concentration of excess of acetic acid itself required to prevent the precipitation, as $a_{52} - c = 6.95 \times 10^{-5}$.

For *dibasic acid* (type Ag_2CO_3), Eq. XIV(20) with $P = 2c(OH)$ gives

$$8c^4W(P + W) + 2c^3K_1P(P + 2W) + 2c^2P^2(K_{12} - W) - cK_1P^3 - K_{12}P^4/2W = 0. \quad (84)$$

With P for AgOH $= 2 \times 10^{-8}$, and $K_1 = 4 \times 10^{-7}$, $K_2 = 5 \times 10^{-11}$ for CO_2, this gives c_{max} for $Ag_2CO_3 = 1.19 \times 10^{-4}$ (at $H = 1.19 \times 10^{-10}$, from $P = 2cOH$). For such a concentration and such a value of H however, the ion-product for silver carbonate, or $4c^3\alpha_2$, would be 2.0×10^{-12}. Since the reported value of the solubility product of Ag_2CO_3 is somewhat larger, 8.2×10^{-12}, this salt would therefore be congruently soluble if the constants are correct.

For *tribasic acid* (type Ag_3PO_4), c_{max} is given by Eq. XVI(9) with $P = 3c(OH)$ as

$$81c^5W^2(P + W) + 9c^4K_1PW(2P + 3W) + 9c^3P^2[W(K_{12} - W) + K_{12}P/3] + 3c^2K_1P^3(K_{23} - W) - cK_{12}P^4 - K_{123}P^5/3W = 0. \quad (85)$$

b2. Base with valence z; CaS; divalent carbonates. With $P = c(OH)^z$, the monobasic case is again described by Eq. (83). The *dibasic case* becomes, with $P = c(OH)^2$ in Eq. XIV(20),

$$2x^5P^{1/2}W^2 + x^4W(K_1P + W^2) + x^3K_1P^{1/2}W^2 + x^2PW(K_{12} - W) - xK_1P^{3/2}W - K_{12}P^2 = 0, \quad (86)$$

in which $x = \sqrt{c}$. For the *tribasic case* we have, with $x = \sqrt[3]{c}$, through Eq. XVI(9) with $P = c(OH)^3$,

$$3x^7P^{1/3}W^3 + 2x^6P^{2/3}K_1W^2 + x^5W(K_{12}P + W^3) + x^4K_1P^{1/3}W^3 + x^3P^{2/3}W^2(K_{12} - W) + x^2K_1PW(K_{23} - W) - xK_{12}P^{4/3}W - K_{123}P^{5/3} = 0. \quad (87)$$

For CaS, with $P = 8 \times 10^{-6}$ for $Ca(OH)_2$ and $K_1 = 10^{-7}$, $K_2 = 10^{-15}$ for H_2S, Eq. (86) gives $c_{max} \cong \sqrt[3]{P} = 0.02M$. It is often stated that CaS is decomposed by water into $Ca(OH)_2$ and $Ca(HS)_2$. The calculated value of c_{max} ($\sim$ 1.4g/liter) however is close to the reported solubility of the salt so that it seems questionable that its decomposition would be caused by precipitation of the hydroxide. It may be caused by the precipitation of the acid sulfide, but we have no information on the solubility of this compound, which is apparently fairly soluble.

Some calculated relations for carbonates of divalent metals are shown in

Table 17–1. The solubility of the carbonate, or S_0, listed in the table, is calculated from the reported solubility product in each case, as explained in Chapter XVIII, Section B2. It is seen that $CaCO_3$, with $c_{max} > S_0$, is congruently soluble, while both $MgCO_3$ and $FeCO_3$ appear to be incongruently soluble on the basis of the reported constants. It is to be noted that all the calculations involved in Table 17–1 assume the hydroxides to be strong bases. The value of "H at c_{max}" is calculated from $P = c(OH)^2$.

If the base is not completely ionized, c_{max} would be larger. We shall estimate the effect for $Mg(OH)_2$, assuming it to be "strong in K_1'," with

Table 17–1. Relations for c_{max} for divalent carbonates.

	$CaCO_3$	$MgCO_3$	$FeCO_3$
P for $M(OH)_2$	8×10^{-6}	10^{-11}	1.7×10^{-15}
c_{max}	0.203	3.28×10^{-4}	1.2×10^{-5}
S_0	1.28×10^{-4}	7.67×10^{-3}	1.70×10^{-5}
H at c_{max}	1.6×10^{-12}	5.71×10^{-11}	8.4×10^{-10}

$K_2' = 2.6 \times 10^{-3}$(3). The critical value of H, or H at c_{max}, may be found by elimination of c from $P = c\alpha_2'(OH)^2$ and the electroneutrality equation,

$$H - OH = c(\alpha_1 + 2\alpha_2 - \alpha_1' - 2\alpha_2'). \tag{88}$$

Since the solution at c_{max} will be slightly alkaline, as we have already seen, we shall for simplification neglect H compared to OH in Eq. (88) and also assume that $\alpha_2 \cong 1 - \alpha_1 \cong K_2/(H + K_2)$. The result is

$$(OH)^4 + (OH)^3 W/K_2 + (OH)^2 P/K_2' - PW/K_2 \cong 0. \tag{89}$$

For $MgCO_3$ this gives $OH \cong 1.72 \times 10^{-4}$ (or $H \cong 5.81 \times 10^{-11}$, practically as in Table 17–1); but with $\alpha_2' \cong 0.938$ instead of 1, $c_{max} = 3.60 \times 10^{-4}$, a little larger than the value tabulated.

c. Normal salt of strong acid and weak insoluble base. Now with

$$P = c\alpha_{z'}'(OH)^{z'}, \tag{90}$$

it is not possible or feasible in general to eliminate OH from the equation of electroneutrality for an expression relating c_{max} and constants except in the univalent case. In the general case we shall therefore proceed by eliminating c to calculate the critical value of "H at c_{max}," with which c_{max} is then obtained from Eq. (90).

(3) D. I. Stock and C. W. Davies, *Trans. Far. Soc.*, **44**, 856 (1948).

c1. Monoacid base (salt MCl). For saturation with the base H_{req} is given by Eq. (51) with $b = c$; with this expression for H, the electroneutrality equation gives

$$c_{max} = P^2/2BW + \sqrt{(\ )^2 + P^2/W} + P/B. \tag{91}$$

(The acid analog of this equation is Eq. [67].) In terms of thermodynamic constants and activity coefficients, Eq. (91) becomes $c_{max} = \mathbf{P}^2/2\mathbf{BW} + \sqrt{(\ )^2 + \mathbf{P}^2/\gamma^2\mathbf{W}} + \mathbf{P}/\mathbf{B}$.

A concentration c, greater than c_{max}, without precipitation of the base, requires an excess of foreign strong acid equal to

$$(a_s)_{req} = W(c - P/B)/P - P/(c - P/B) - P/B, \tag{92}$$

which follows from $(H - OH)_{51} = c(1 - \alpha') + a_s$, or directly from Eq. (52). If the pure salt concentration, c, is less than c_{max}, the base will precipitate on the addition of this same quantity of foreign strong base. If the salt solution is treated instead with more of the weak base MOH itself, the solution becomes saturated with MOH when the concentration of the excess of the weak base added is $(b)_{53} - c$, Eq. (53) being used with $(a + g) = c$.

If excess of foreign strong acid at concentration a_s is being titrated with strong base in the presence of the salt MCl (MOH being weak and insoluble) at concentration c, the base MOH precipitates and thus interferes with the titration when $H = H_{51}$, requiring $c > P/B$, and $H - OH = a_s + c(1 - \alpha') - b_s$. Hence the precipitation occurs when $b_s = a_s - (a_s)_{92}$. The precipitation occurs before the equivalent point if $(a_s)_{92}$ is positive—roughly if c is greater than both P/B and P^2/BW.

c2. Higher valence types, especially $FeCl_3$. We first note that for saturation with the base at concentration c, Eq. (90) expanded becomes

$$(OH)^{z'}(1 - cK'_{1\ldots z'}/P) + (OH)^{z'-1}K'_1 + (OH)^{z'-2}K'_{12} + (OH)^{z'-3}K'_{123} + \cdots + K'_{1\ldots z'} = 0, \tag{93}$$

which reduces to Eq. (51) for $z' = 1$; for $z' = 3$, this is Eq. (33) in terms of H. Saturation requires in every case $c > P/K'_{1\ldots z'}$.

The general electroneutrality equation for these salts is

$$H - OH = c(z' - \alpha'_1 - 2\alpha'_2 - 3\alpha'_3 \cdots - z'\alpha'_{z'}). \tag{94}$$

When c is eliminated from Eqs. (90) and (94) we obtain an expression for the critical value of OH to be used in Eq. (90) to calculate c_{max}. For the monobasic case ($z' = 1$), this expression is $(OH)^2 + (OH)\ P/K'_1 - W = 0$; for $z' = 2$, $(OH)^2 + 2(OH)\ P/K'_{12} - (W - P/K'_2) = 0$; for $z' = 3$,

$$(OH)^3 + 3(OH)^2P/K'_{123} - OH(W - 2P/K'_{23}) + P/K'_3 = 0; \tag{95}$$

in general, for $z' > 2$,

$$(OH)^{z'} + z'(OH)^{z'-1}P/K'_{1\ldots z'} - (OH)^{z'-2}[W - (z'-1)P/K'_{2\ldots z'}] + (z'-2)(OH)^{z'-3}P/K'_{3\ldots z'} + (z'-3)(OH)^{z'-4}P/K'_{4\ldots z'} + \cdots + P/K'_{z'} = 0. \quad (96)$$

(Reference to the divalent case was made at the end of Section E3*b*2.)

These equations are interesting because of the difference between the cases with $z' > 2$ and those for $z' = 1$ or 2. For the univalent and divalent cases there is only a single value of *OH* for saturation with the base in a pure salt solution; this single value of *OH* then determines c_{max}. In these cases then, the salt solution is stable and clear only below c_{max}, and continued addition of the salt, to make $c > c_{max}$, never causes the precipitate to redissolve.

If $z' > 2$ however, we note first that a positive value of *OH* requires $P < K'_{2\ldots z'}W/(z'-1)$. If *P* is larger than this value there is no limit except for the solubility of the salt itself to the concentration of the pure salt in water without precipitation of the base. For such a system the solubility curve of the base, S_1x of Fig. 17–5, never crosses the line representing the composition of the salt, in this case, $a = 3b$ for the salt "MCl_3"; point x, that is, must lie above this line regardless of the solubility of the strong acid of the salt. But if $P < K'_{2\ldots z'}W/(z'-1)$ there are two values of *OH* for saturation with base. The solubility curve of the base in this case then crosses the salt line twice, and the point x, for saturation with both base and acid, may again be on either side of the line, depending on the relative solubilities of base and acid. As such a salt is added to pure water, the base comes to be precipitated when *OH* falls to the higher of the two values given by Eq. (96), or by Eq. (95) for the trivalent case. Continued addition of the salt, if not interrupted by saturation with the salt itself, causes the base to redissolve when *OH* reaches the lower of the two values given by the equation. The situation remains the same even if the base is strong in respect to all but its last ionization constant, provided that it is at least tribasic and that the acid of the salt is strong. This may be seen by setting K_1 and K_2 both equal to infinity in Eq. (95) for example. If the base is strong in respect to all its ionization constants, then $OH_{crit} = \sqrt{W}$, a single value again.

These considerations may be applied to $FeCl_3$, at least as a hypothetical case. With the values of the constants used in Section B1, Eq. (95) gives $OH_1 \cong 9.5 \times 10^{-8}$ and $OH_2 \cong 3.6 \times 10^{-12}$. With the first value, Eq. (90) gives $c_1 = 3.4 \times 10^{-9}$ and with the second value $c_2 = 0.0048$. (At such low concentrations the correction for the activity coefficient can not be serious. Possibly we may be justified even in our neglect of complex ion formation.) If the interpretation is correct then and if the equilibrium constants do have the meanings and the numerical values attributed to them, this means that even a practically infinitesimal concentration of $FeCl_3$ in pure water is impossible with "hydrolytic" precipitation of $Fe(OH)_3$. This precipitation must occur for concentrations of $FeCl_3$ (pure) between 3.4×10^{-9} and 0.0048. For higher concentrations however, the solution remains clear, being unsaturated with respect to the hydroxide. This may be checked as follows: For a solution of pure $FeCl_3$ at concentration c, OH is given by the base analog of Eq. XVI(9), with OH in place of H, and with K_1', K_2', K_3' of $Fe(OH)_3$ as the ionization constants. For the numerical values of the constants as assumed, we find, at the concentration $c = 0.06$, $OH \cong 7.8 \times 10^{-13}$, whence $\alpha_3' \cong 0.79$. Hence the solution is unsaturated with respect to $Fe(OH)_3$ since $[Fe^{+++}][OH^-]^3 = 0.06(0.79)(7.8 \times 10^{-13})^3 = 3.7 \times 10^{-40}$, or less than P. On the other hand, at $c = 10^{-4}$, $OH \cong 9.0 \times 10^{-11}$ and $\alpha_3' \cong 0.027$. Then $[Fe^{+++}][OH^-]^3 = (10^{-4})(0.027)(9.0 \times 10^{-11})^3 = 2.0 \times 10^{-36}$, so that the solution would be supersaturated with respect to $Fe(OH)_3$.

We have just seen that the solubility curve of the base of such a salt crosses the salt composition line at two points, such as c_1 and c_2 as calculated for $FeCl_3$. If the solubility of the salt in pure water, or S_0, is less than c_1, it is definitely congruently soluble with respect to the base. If the solubility is between c_1 and c_2, it is incongruently soluble, and a pure saturated solution of solid salt as sole solid phase is impossible in stable equilibrium. If $S_0 > c_2$, the salt is again congruently soluble, but a very slight addition of the salt to pure water causes precipitation of the base, which then redissolves upon further addition of the salt to give a concentration greater than c_2.

d. Simple 1 : 1 *salt of weak acid and weak base.* In this case MX is the salt of weak monobasic acid and weak monoacid base, both insoluble. For the clear salt solution, without precipitation, $D = c\beta = c(\alpha - \alpha')$. Elimination of H from this and Eq. (57) gives, for the maximum concentration without precipitation of the acid HX (with solubility product P_2),

$$\begin{aligned} &c^3A^3W(P_2 + W) - c^2A^2P_2W[3(P_2 + W) - AB] \\ &\quad + cAP_2^2[3W(P_2 + W) - A(BP_2 + 2BW + AW)] \\ &\quad - P_2^3(AB - W)(A^2 - P_2 - W) = 0. \end{aligned} \tag{97}$$

For the maximum concentration without precipitation of the base, we obtain, from Eqs. (51) and $D = c\beta$, an expression, Eq. ("97a"), identical with Eq. (97) with A and B interchanged and with P_1, the solubility product of the base, in place of P_2. Both expressions are solved for c and the smaller answer (which however must be greater than P_2/A or greater than P_1/B) is c_{max} for the salt. If we eliminate c from Eqs. (57) and $D = c\beta$, the critical value of H is given by the expression

$$H^3B + H^2(W + BP_2/A) - HBW - W(P_2 + W) = 0. \tag{98}$$

This equation gives only one value of H; hence, although each of the Eqs. (97) and ("97a") may give more than one positive value for c, only one value from each is significant. Example: if $A = 10^{-6}$, $B = 10^{-8}$, P_1 (for the base) $= 10^{-8}$ and P_2 (for the acid) $= 10^{-10}$, then Eq. (97) gives $c = (1.5 \pm 0.51) \times 10^{-4}$ and Eq. ("97a") gives $c = 1.5 \pm 0.5$. Since c_{max} must be greater than P_2/A, the value sought is $c_{max} = 2.01 \times 10^{-4}$, above which the acid would be precipitated.

Instead of solving both Eqs. (97) and ("97a"), we may first use Eqs. (75) and (76) to find the composition of point x in Fig. 17–5. In the present numerical example, these equations give $a_x = 0.0101$ and $b_x = 1.01$; hence the diagonal (for the salt composition) is crossed not by the solubility curve of the base but by that of the acid. In this case "hydrolytic" precipitation of the base is impossible at equilibrium since the salt is incongruently soluble with respect to the acid, and hence only Eq. (97) need be used to find c_{max}.

XVIII

Pure Saturated Aqueous Solution of a Salt

In this chapter, dealing with the pure saturated solution of a salt, the simple symbols H and S will be used throughout in place of H_0 and S_0. Again as in the case of the pure saturated solution of an acid or a base, two of the quantities H, S, P, K's, and (K')'s may be found, given all the others, by combination of the electroneutrality equation of the solution of the salt, $H - OH = c\beta$, with the definition of its solubility product, P.

A. Simple 1 : 1 Salt

The two conditions involving the five quantities H, S, A, B, P are

$$H - OH = S(\alpha - \alpha') = S[A/(A + H) - HB/(HB + W)], \qquad (1)$$

and

$$P = S^2\alpha\alpha' = S^2HAB/(H + A)(HB + W). \qquad (2)$$

Hence if both acid and base are strong, we have simply $H = \sqrt{W}$ and $S = \sqrt{P}$.

For the salt of weak acid and strong base,

$$H - OH = - S\rho = - SH/(A + H), \qquad (3)$$

$$P = S^2\alpha = S^2A/(A + H); \qquad (4)$$

and these equations may be used to obtain two of the four quantities if the other two are known. Given P and A, for example, elimination of S gives

$$H^5 + H^4(A - P/A) - 2H^3W - 2H^2AW + HW^2 + AW^2 = 0. \qquad (5)$$

Then with H_5, S is calculated from either of the Eqs. (3) and (4). For the salt of strong acid and weak base, the equations become $D = S\rho' = SW/(HB + W)$ and $P = S^2\alpha' = S^2HB/(HB + W)$.

For the weak-weak salt, Eqs. (1) and (2) are used, so that from S, A, and B for example, H is calculated from Eq. (1) and P from Eq. (2), while from S, H, and A, B is found from Eq. (1) and P from Eq. (2). From P and the ionization constants, H is obtained by elimination of S giving

$$H^6B + H^5(AB + W - BP/A) + H^4W(A - 2B) + H^32W(P - AB - W) - H^2W^2(2A - B) - HW^2(AP/B - AB - W) + AW^3 = 0. \qquad (6)$$

B. Salts of Dibasic Acid and Strong Base

We shall discuss these salts through the specific examples of the sulfides and the carbonates.

1. Solubility of Sulfides of Strong Bases.[1] The simple relations $P = S^2$ for the type MX and $P = 4S^3$ for the type M_2X would hold only if both the acid and the base were strong, whereupon all the degrees of ionization involved would equal 1. Otherwise, for the type MX, involving strong divalent base,

$$P = S^2\alpha_2; \qquad S = \sqrt{P/\alpha_2}; \tag{7}$$

Table 18–1. Corresponding values of H $(= x \times 10^{-7})$, S, and P for sulfides of type MX or M_2X.

x $(= H \times 10^7)$	$S(\times 10^7)$	pS	pP for MX	pP for M_2X
0.95	0.0690	8.161	24.612	32.171
0.9	0.143	7.844	23.921	31.163
0.8	0.312	7.506	23.170	30.074
0.7	0.516	7.287	22.649	29.334
0.6	0.776	7.110	22.202	28.710
0.5	1.125	6.949	21.773	28.120
0.4	1.633	6.787	21.322	27.507
0.3	2.465	6.608	20.807	26.813
0.2	4.114	6.386	20.152	25.936
0.1	9.075	6.042	19.125	24.565
0.05	19.04	5.720	18.160	23.278
0.01	99.0	5.004	16.012	20.414

and for the type M_2X, with strong univalent base,

$$P = 4S^3\alpha_2; \qquad S = \sqrt[3]{P/4\alpha_2}. \tag{8}$$

Also, $\alpha_2 = (K_2/H)/(1 + H/K_1 + K_2/H)$; but with $K_1 \cong 10^{-7}$ and $K_2 \cong 10^{-15}$ for H_2S, then as long as $H \cong \sqrt{W}$, or of the order of magnitude of K_1, the third term of the denominator in this expression for α_2 may be neglected, so that

$$\alpha_2 \cong K_1K_2/H(K_1 + H). \tag{9}$$

The calculation of P from S or that of S from P therefore requires the calculation of H, either from S or from P, for the evaluation of α_2. If S is known, H is obtained from Eq. xiv(20), for the solution of the normal salt of H_2S with a strong base, the equation being the same for both MX and M_2X, at the concentration S moles per liter:

$$H^4 + H^3(2S + K_1) + H^2(SK_1 + K_1K_2 - W) - HK_1W - K_1K_2W = 0. \tag{10}$$

(1) This treatment is substantially similar to that of P. van Rysselberghe and A. H. Ropp, *J. Chem. Educ.*, **21**, 96 (1944).

For small S (less than $K_1/10$, or 10^{-8}) this clearly gives $H \cong \sqrt{W}$. For large S (greater than $10K_1$, or 10^{-6}) Eq. (10) becomes $2H^3S + H^2SK_1 - HK_1W \cong 0$, or $H \cong -K_1/4 + \sqrt{(\quad)^2 + K_1W/2S}$; and if $S \gg K_1$, then $H \cong W/S$. But for intermediate values of S, of the order of K_1 or between 10^{-8} and 10^{-6}, Eq. (10) becomes

$$H^4 + H^3(2S + K_1) + H^2(SK_1 - W) - HK_1W \cong 0. \tag{11}$$

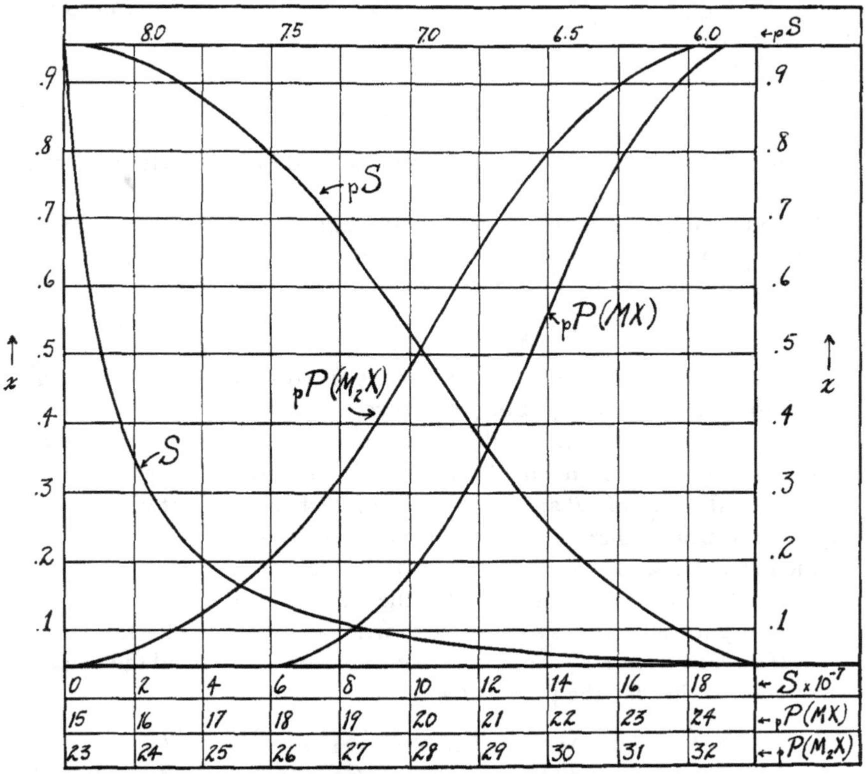

Fig. 18–1. Relation between [H^+], S, and P for saturation in pure water for sulfides of strong bases; [H^+] $= x \times 10^{-7}$.

There is no simple generally useful approximate solution of this expression even for the specific case of the sulfides. But a chart or curve of corresponding values of S and H for any specified value of K_1 may easily be prepared, with almost any desired degree of accuracy, if S is taken as the dependent variable, according to the suggestion made in Chapter VII. With $H = x \times 10^{-7}$, Eq. (11) gives $S = (1 + x)(1 - x^2)(10^{-7})/x(1 + 2x)$, if $K_1 = \sqrt{W} = 10^{-7}$. With this expression we calculate the values listed in the first three columns of Table 18–1 and plotted as two of the curves of Fig. 18–1; p$S = -\log S$.

The fact that x must be less than 1 for the formula to apply corresponds to the expected alkalinity of the solution of such sulfides. With H thus determined (as $\sim \sqrt{W}$ for low S, as $\sim W/S$ for high S, and from the curve of Fig. 18–1 for S of the order of 10^{-8} to 10^{-6}), α_2 may be evaluated and then used in the relation between P and S, or Eqs. (7) and (8).

a. Type MX. For low S, $\alpha_2 \cong K_2/2\sqrt{W}$, and

$$P \cong S^2K_2/2\sqrt{W} \cong S^2(10^{-8}/2). \tag{12}$$

Since this applies for $S < 10^{-8}$, it may also be used for calculating S from P for $P < 10^{-24}$.

For high S, $\alpha_2 \cong SK_2/W$, and

$$P \cong S^3K_2/W \cong 10S^3. \tag{13}$$

Since this applies for $S > 10^{-6}$, it may also be used for calculating S from P for $P > 10^{-17}$.

For intermediate values (S between 10^{-8} and 10^{-6}, P between 10^{-24} and 10^{-17}), Eq. (9) gives $\alpha_2 \cong 10^{-8}/x(1 + x)$, in which x is obtained from Fig. 18–1 for a given value of S. Then

$$P \cong S^2(10^{-8})/x(1 + x) \cong (10^{-22})(1 + x)(1 - x^2)^2/x^3(1 + 2x)^2. \tag{14}$$

The corresponding values of x and S already listed in Table 18–1 may now be used to calculate, through Eq. (14), corresponding values of P and S or of P and x in this region. The resulting values, as p$P = -\log P$, are listed in column 4 of the table; in Fig. 18–1 they are plotted as pP against x, and Fig. 18–2 is the plot of pP against pS. The direct relation between P and S may be read from either figure and the value of H, with $H = x \times 10^{-7}$, for specified P or S may be read from Fig. 18–1.

In Eq. (10) we have a relation for the calculation of H from S for use then in computing P. It is also possible to write an expression for the calculation of H directly from P by the elimination of S from Eq. (7) and the equation of electroneutrality for the pure salt solution. The result is

$$H^6(1 - 4P/K_1K_2) + H^5(K_1 - 4P/K_2) + H^4(K_1K_2 - 2W - PK_1/K_2) \\ - 2H^3K_1W + H^2W(W - 2K_1K_2) + HK_1W^2 + K_1K_2W^2 = 0. \tag{15}$$

The practical problem of the sulfides, however, is better handled through the procedure just outlined.

b. Type M_2X. For a given value of S and hence of H, the solubility product constant for the salt M_2X equals that for the salt MX with the same solubility, multiplied by $4S$, as seen in Eqs. (7) and (8). Hence, from Eqs. (12–14), we have:

For low S,

$$P \cong 2S^3K_2/\sqrt{W} \cong S^3(2 \times 10^{-8}), \tag{16}$$

applying for $S < 10^{-8}$ or $P < 10^{-32}$.

For high S,

$$P \cong 4S^4K_2/W \cong 40S^4, \tag{17}$$

applying for $S > 10^{-6}$ or $P > 10^{-25}$.

For intermediate values (S between 10^{-8} and 10^{-6}, or P between 10^{-32} and 10^{-25}),

$$P \cong 4S^3(10^{-8})/x(1 + x) \cong 4(10^{-29})(1 + x)^2(1 - x^2)^3/x^4(1 + 2x)^3. \tag{18}$$

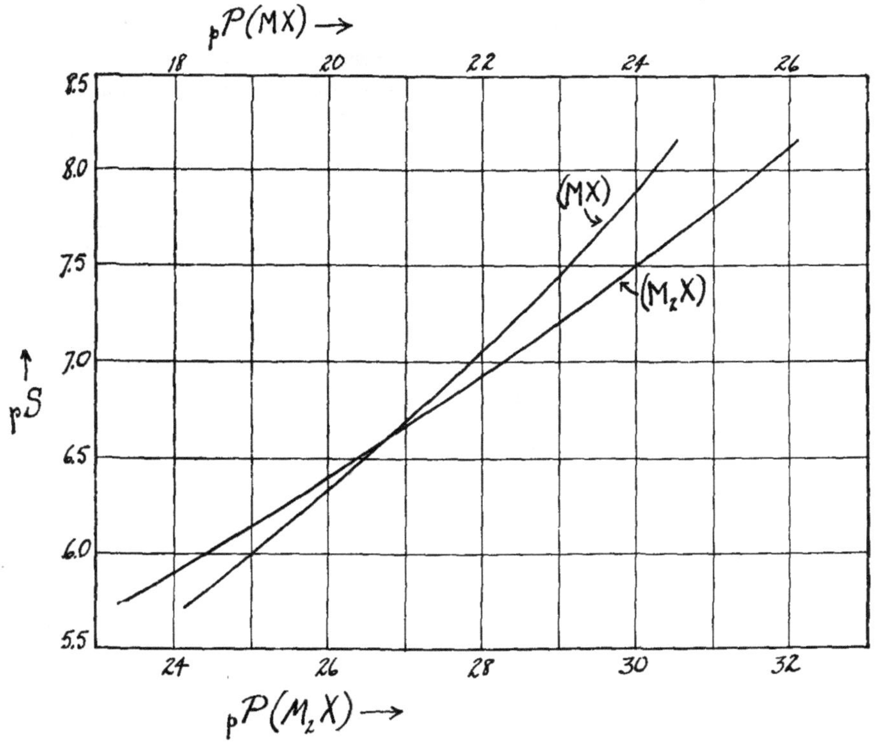

Fig. 18–2. Relation between pS and pP for sulfides of strong bases.

The corresponding values of pP for M_2X in this intermediate region are therefore obtained from the values for MX, as $\mathrm{p}P_{M_2X} = \mathrm{p}P_{MX} + \mathrm{p}S - \log 4$. These are listed in the last column of Table 18–1 and are plotted as additional curves in both Fig. 18–1 and Fig. 18–2. Again we may also write an expression relating H and P directly, through the elimination of S from Eq. (8) and the equation of electroneutrality, but the result is even more involved than Eq. (15), being of the tenth degree in H.

[*Note:* The formulas for the extremes, at high or low S or P, have been given by van Rysselberghe and Ropp (*loc.cit.*, Ref. XVIII–1). Their treatment appears different because it is presented in the language of "hydrolysis constants." Their formulas for the intermediate range, however, are less accurate than Eqs. (14) and (18) since they assumed a single value of H ($= 5 \times 10^{-8}$) in this range instead of the relation of Eq. (11) fixing H as $f(S)$.]

In summary then, the curves of Figs. 18–1 and 18–2 serve to determine P from S or S from P in the intermediate region of solubility of $\sim 10^{-7}$, while the analytical formulas of Eqs. (12), (13), (16), and (17) may be used for either higher or lower values.

2. CARBONATES OF STRONG BASES. For the insoluble sulfides the direct measurement is usually that of P rather than that of S. The carbonates are generally more soluble, and the direct measurement may be either S or P. It must be kept in mind, however, that it is not possible to calculate one of these quantities from the other without the intermediate calculation of H.

Table 18–2. Carbonates of divalent metals.

	P	H	$S (= \sqrt{P/\alpha_2})$	"S" $(= \sqrt{P})$	c_{max}
$MgCO_3$	5×10^{-5}	8.76×10^{-12}	7.67×10^{-3}	7.07×10^{-3}	3.28×10^{-4}
$CaCO_3$	5×10^{-9}	1.13×10^{-10}	1.28×10^{-4}	7.07×10^{-5}	0.203
$FeCO_3$	2.1×10^{-11}	6.4×10^{-10}	1.70×10^{-5}	4.58×10^{-6}	1.2×10^{-5}

We shall proceed on the assumption that the P values listed by Latimer (Ref. III–1) are the true values, i.e. that if they are derived from direct solubility measurements they were calculated not as S^2 (for type MCO_3) or $4S^3$ (for type M_2CO_3) but as $S^2\alpha_2$ and $4S^3\alpha_2$ respectively.

We shall illustrate the effect of assuming α_2 to be 1, in the case of the carbonates of the divalent metals Mg, Ca, Fe, Table 18–2; H is calculated from Eq. (15).

The last column of the table lists again c_{max}, the maximum possible concentration of the carbonate in pure water without "hydrolytic" precipitation of the hydroxide, as discussed and calculated under Table 17–1.

It was there pointed out that one of the assumptions made in these calculations is that these bases are completely "strong." We there found that with $K_1' = \infty$ and $K_2' = 2.6 \times 10^{-3}$ for $Mg(OH)_2$, c_{max} for $MgCO_3$ $= 3.6 \times 10^{-4}$. We would similarly find S, the solubility of $MgCO_3$, to be a little higher if the incomplete ionization of $Mg(OH)_2$ is taken into account. In this case the calculation of H for the saturated solution of $MgCO_3$ from its solubility product requires the combination of the equation of electroneutrality, Eq. XVII(88), with $P = S^2\alpha_2\alpha_2'$. If we again assume, as under Eq. XVII(89), that in the slightly alkaline solution $\alpha_2 \cong 1 - \alpha_1 \cong K_2/(H+K_2)$ for CO_2 the result is an (approximate) equation identical with Eq. (6), with K_2 for A and K_2' for B, and with P as the solubility product of $MgCO_3$. We thereupon find $H = 1.00 \times 10^{-11}$ and $S = 9.12 \times 10^{-3}$, a little higher than the value tabulated.

The order of magnitude of the difference between S and c_{max} for $MgCO_3$ remains unchanged. Although this salt is commonly considered to be congruently soluble, these calculations show it to be incongruently soluble if the constants are correct. If the solubility product of $MgCO_3$ ($P = 5 \times 10^{-5}$) is based on an actual aqueous solubility, the discrepancy appears serious. If for simplicity we consider $Mg(OH)_2$ strong so that the concentration of Mg^{++} in the saturated solution of $MgCO_3 = S = 7.67 \times 10^{-3}$, the hydroxyl ion concentration must be less than $\sqrt{P_1/S}$, in which P_1 is the solubility product of $Mg(OH)_2$ in order to have no precipitation of the hydroxide. This requires a concentration of foreign strong acid equal to 6.46×10^{-3}, as calculated from the relation, $H - OH = S(\alpha_1 + 2\alpha_2 - 2) + a_s$.

C. The Thermodynamic Solubility Product of $MgNH_4PO_4 \cdot 6H_2O$

We shall here use as an illustration of the application of the various methods and formulas discussed in this and preceding chapters the calculation of the solubility product of a mixed salt like $MgNH_4PO_4$ from the following data: $S = 2.1 \times 10^{-3}M$; $\mathbf{B}$ (for ammonia) $= 1.8 \times 10^{-5}$; and for phosphoric acid, $\mathbf{K}_1 = 7 \times 10^{-3}$, $\mathbf{K}_2 = 6 \times 10^{-8}$, $\mathbf{K}_3 = 5 \times 10^{-13}$. The ionization constants are assumed to be in terms of activities. The diacid base, $Mg(OH)_2$, is assumed completely ionized and the salt itself completely dissociated. All these considerations have significance of course only provided that S refers to a solubility (equilibrium) with respect to $MgNH_4PO_4 \cdot 6H_2O$ as solid phase.[2]

By definition, if the hydration of the salt is neglected, $\mathbf{P} = [Mg^{++}][NH_4^+][PO_4^{\equiv}]\gamma_{Mg^{++}}\gamma_{NH_4^+}\gamma_{PO_4^{\equiv}}$. With the usual convention about γ, and since $[Mg^{++}] = S$, $[NH_4^+] = S\alpha'$, and $[PO_4^{\equiv}] = S\alpha_3$, then $\mathbf{P} = S^3\alpha'\alpha_3\gamma^{14}$; or, since $\alpha_3 = \alpha_2\mathbf{K}_3/\gamma^6 H$,

$$\mathbf{P} = S^3\alpha'\alpha_2\mathbf{K}_3\gamma^8/H. \tag{19}$$

The α's require H, given by Eq. xvi(63), for the salt of type Na_2NH_4X at concentration $c = S$. Eq. xvi(62) shows that the solution will be alkaline since $\mathbf{K}_2\mathbf{K}_3 < \mathbf{W}$. A preliminary value of H is obtained as H_{ie}, from Eq. xvi(61), since $\mathbf{BK}_2 \gg \mathbf{W}$, as $H \simeq \sqrt{\mathbf{WK}_3/\mathbf{B}} = 1.67 \times 10^{-11}$. It seems likely then that the H^3 sequence of Eq. xvi(63), which reduces to $H \simeq \sqrt{WK_3/B}$ when $BK_2 \gg W$, will give the accurate value of H, after introduction of the activity coefficient γ. With the introduction of the

(2) From the values of H and γ calculated below, the ion activity product $[Mg^{++}][OH^-]^2$ in the saturated solution is found to be 2.6×10^{-11}. With 10^{-11} as the solubility product of $Mg(OH)_2$, an uncertain value, however, the salt magnesium ammonium phosphate would seem to be incongruently soluble, like $MgCO_3$.

numerical values of S and the activity constants, Eq. xvi(63) becomes, when divided through by $\mathbf{K}_1$,

$$H^6(2.57 \times 10^{-3}) + H^5(10^{-5})(1.62 + 1.80/\gamma^2) + H^4(7.56 \times 10^{-8})/\gamma^2 + H^3(2.27 \times 10^{-15})/\gamma^6 - H^2(6 \times 10^{-27})(1.8 + 1.7 \times 10^{-3}\gamma^4)/\gamma^8 - H[(6 \times 10^{-36})/\gamma^8 + (0.63 \times 10^{-36})/\gamma^{12}] - (3 \times 10^{-48})/\gamma^{14} = 0. \quad (20)$$

For the ionic strength, $2\mu = [H^+] + [OH^-] + 4[Mg^{++}] + [NH_4^+] + [H_2PO_4^-] + 4[HPO_4^=] + 9[PO_4^{\equiv}] = H + OH + S(4 + \alpha' + \alpha_1 + 4\alpha_2 + 9\alpha_3)$. But since $H - OH = S(\alpha_1 + 2\alpha_2 + 3\alpha_3 - 2 - \alpha')$, then $\mu = H + S(3 + \alpha' + \alpha_2 + 3\alpha_3)$, in which H is certainly negligible, so that

$$-\log \gamma \cong 0.5\sqrt{S(3 + \alpha' + \alpha_2 + 3\alpha_3)}. \quad (21)$$

Using $H_{ie} = 1.67 \times 10^{-11}$, we may write

$$\alpha' = \mathbf{B}/(\mathbf{B} + \mathbf{W}/H) = 0.0291(\cong \mathbf{B}/OH); \quad (22)$$

$$\alpha_2 = \frac{H\mathbf{K}_{12}}{H^3 + H^2\mathbf{K}_1 + H\mathbf{K}_{12} + \mathbf{K}_{123}} = 0.971 \left(\cong \frac{H}{H + \mathbf{K}_3}\right). \quad (23)$$

This gives $\mu = 0.00858$, $\log \gamma = -0.0463$ and $\gamma = 0.899$. With this value of γ, the H^3 sequence of Eq. (20) becomes $H^2 - H(5.89 \times 10^{-12}) - 3.80 \times 10^{-21} = 0$, giving $H \cong 6.47 \times 10^{-11}$. This value is fairly accurate since the magnitudes of the terms neglected in Eq. (20) are much smaller than those for the sequence used; correcting nevertheless for the adjacent terms, we have $H = 6.50 \times 10^{-11}$. Since this value of H is quite different from the preliminary value ($H_{ie} = 1.67 \times 10^{-11}$), we now calculate more accurate values of the α's and of γ, for use in Eq. (19), and if necessary for refining H itself. Now, with activity coefficients introduced,

$$\alpha' = \mathbf{B}/(\mathbf{B} + \mathbf{W}/H) = 0.105, \quad (24)$$

$$\alpha_2 = \frac{H\mathbf{K}_2/\gamma^4}{H^2 + H\mathbf{K}_2/\gamma^4 + \mathbf{K}_{23}/\gamma^{10}} = 0.985 \left(\cong \frac{H}{H + \mathbf{K}_3/\gamma^6}\right); \quad (25)$$

in Eq. (25) the H^3 term of the denominator of Eq. (23) has been neglected. Although these values differ somewhat from those of Eqs. (22) and (23), the new value of μ from Eq. (21) is 0.00868, with $\log \gamma = -0.0466$ and $\gamma = 0.898$; the values of Eqs. (24) and (25) may therefore be considered as final.

Hence, from Eq. (19),

$$\mathbf{P} = \frac{(2.1 \times 10^{-3})^3(0.105)(0.985)(5 \times 10^{-13})(0.898)^8}{(6.50 \times 10^{-11})} = 3.12 \times 10^{-12}.\text{(3)}$$

(3) The data used for this calculation, including the factor 0.5 in Eq. (21), are taken from *Analytical Processes* by T. B. Smith (Ref. viii–3, pp. 139–42), where the same problem is treated, but differently, on the basis of the "hydrolysis" of the salt. Smith's calculation gave $\mathbf{P} = 3.3 \times 10^{-12}$.

[If we take into consideration the incomplete ionization of $Mg(OH)_2$, with $\mathbf{K}_2' = 2.6 \times 10^{-3}$, we introduce the factor α_2' into Eq. (19), and the term $-S(1 + \alpha_2')$ into the equation of electroneutrality. The result is: $OH = 1.56 \times 10^{-4}$, $\alpha' = 0.124$, $\alpha_2' = 0.962$, $\alpha_2 = 0.988$, $\gamma = 0.899$, and $P = 2.94 \times 10^{-12}$.]

If a suspension of this salt in pure water is treated with foreign acid, the solubility rises, approaching the relation $\log S \xrightarrow{=} (1/3)(\log P - \log K_{123} - 3\text{p}H)$. If it is treated with foreign base, the solubility falls, passes through a minimum at the iso-electric point, and then rises again, approaching the relation $\log S \xrightarrow{=} (1/3)(\log P - \log B + \text{p}H - \text{p}W)$.

To find S_{min}, we use the value $\gamma = 0.90$ (since we have seen that this value remains practically constant in the region under discussion) and calculate H_{ie} from Eq. XVI(61) as $H_{\text{ie}} \cong \sqrt{\mathbf{K}_2\mathbf{K}_3\mathbf{W}/(\gamma^6\mathbf{B}\mathbf{K}_2 + \gamma^{10}\mathbf{W})} = 2.28 \times 10^{-11}$. This point is reached, moreover, when there has been introduced a concentration of foreign strong base, according to Eq. IV(65), equal to $(b_s)_{\text{ie}} = (OH - H)_{\text{ie}}$. If necessary then, the activity coefficient could be recalculated as $\log \gamma = -0.5\sqrt{S(3 + \alpha' + \alpha_2 + 3\alpha_3) + (b_s)_{\text{ie}}}$. Now with $H = 2.28 \times 10^{-11}$ in Eqs. (24) and (25), α' and α_2 are calculated ($\alpha' = 0.0394$ and $\alpha_2 = 0.960$) for use in Eq. (19) to calculate S_{min}, found to be 2.06×10^{-3}, negligibly lower than S_0. Since $(b_s)_{\text{ie}}$ is only $\sim 4.4 \times 10^{-4}$, γ as just defined remains 0.897, giving a corrected value of 2.07×10^{-3} for S_{min}, which may therefore be accepted as final. With ammonia instead of strong base, the solubility is a minimum, according to the constants here assumed, at an analytical concentration of ammonia equal to $[(OH - H)/\alpha'_{\text{NH3}}]_{\text{ie}}$ or 0.0137 M. This concentration is much lower than that recommended on the basis of practical experience for the washing of the precipitate in analysis. According to direct measurements,[4] the solubility is lower in 0.1 than in 0.05 M ammonia.

(4) R. F. Uncles and G. B. L. Smith, *Anal. Chem.*, **18,** 699 (1946).

XIX

Saturation with Salts of Monobasic Acid

A. Graphical Relations

1. At Constant g. The solubility relations in pure water ($g = 0$) for the system MOH–HX–H_2O, forming the salt MX, are shown schematically in Fig. 19–1, in which it is assumed that the salt is congruently soluble in pure water. S_1e is the solubility curve of the base, eSf that of the salt, S_2f that of the acid. Point e is the solution saturated with both MOH and MX, point f that saturated with both MX and HX. Compositions falling in the region between the H_2O corner and the three curves give unsaturated solutions. The region vertically above the curve S_1e gives saturated liquid on the curve + solid MOH; the region bounded by the curve eSf and the lines ek and fn, both parallel to the diagonal $\mathbf{b} = \mathbf{a}$, give saturated liquid on the curve + solid MX; the region horizontally to the right of curve S_2f gives saturated liquid + solid HX. Compositions with coordinates ($\mathbf{a}$ and $\mathbf{b}$, moles of acid and base per liter of system, to be distinguished from a and b, concentrations in the liquid solution) falling in the region between the vertical line ej and the line ek give three phases, liquid e + MOH + MX; those in the region between the line fn and the horizontal line fm give the three phases, liquid f + MX + HX.

If HX is added to the unsaturated solution of MOH with concentration $\mathbf{b}'$, solid salt precipitates when the solubility curve of the salt is reached at point u. Continued addition of HX increases the amount of solid salt while the composition of the liquid travels along the curve ef toward point f. When for example the total composition reaches point v, the liquid has the composition w, fixed by the line vw with slope 1, because the precipitation of the salt consumes MOH and HX in the ratio 1 : 1. When the total composition reaches point x, on the line fn, the liquid reaches point f and solid HX appears mixed with MX. The value of $\mathbf{a}$ required to reach point x is $a_f + (\mathbf{b}' - b_f)$ because the slope of the line fn is 1.

The line wvv′ has important properties. If we define $\mathbf{h}$ as $\mathbf{a} - \mathbf{b}$, we note that since the precipitated salt is 1 : 1 in composition, then $\mathbf{a} - \mathbf{b} = \mathbf{h} = a - b$. All mixtures with the same value of $\mathbf{h}$ give at equilibrium (if salt is precipitated) the same saturated solution, or the same point on the solubility curve eSf. The line wvv′ is a line of constant $\mathbf{h}$. The greater the actual values of $\mathbf{a}$ and $\mathbf{b}$ on this line are, the greater is the quantity of solid salt precipitated, but for the given value of **h** represented by this line, the solution is always at point w. It follows then that the composition of a solution saturated with solid salt as sole solid phase may be expressed, as we shall see later, in terms of **h**, *g*, and constants in an equation not involving the actual values of **a** and **b**.

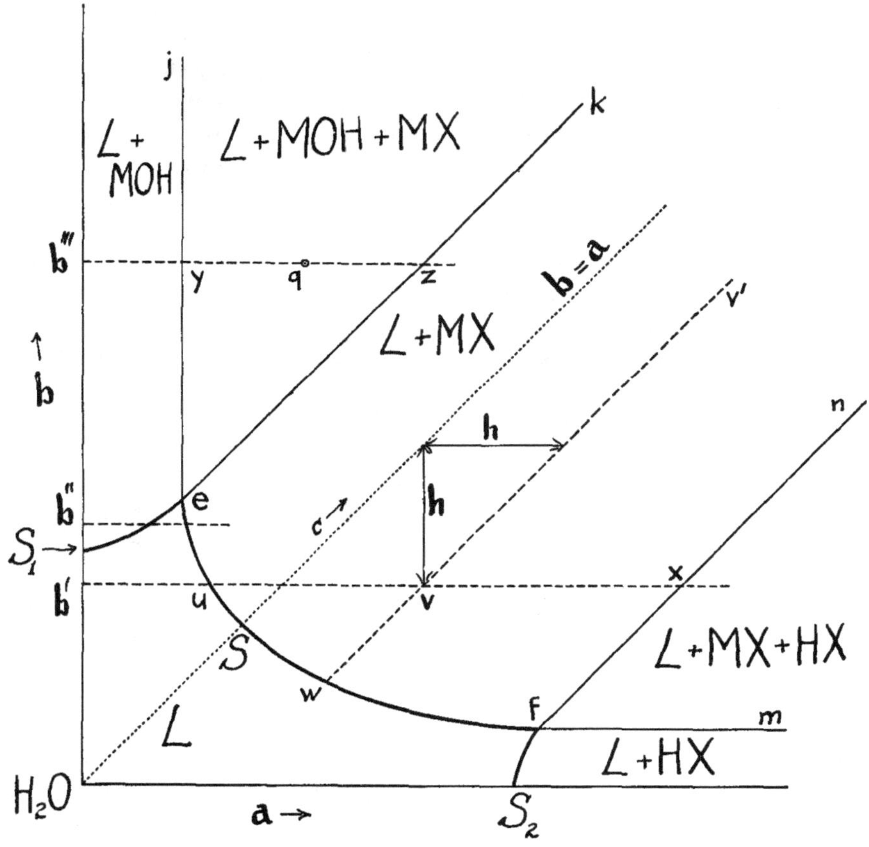

Fig. 19–1. System MOH–HX–H_2O involving the stable salt MX.

If HX is added to a suspension of the base with **b**″ moles of MOH per liter (including dissolved and solid MOH), the solid will have dissolved when the curve S_1e is reached; this is followed by precipitation of salt when the curve ef is reached. If the suspension of MOH has the composition **b**‴ (greater than b_e) addition of HX increases the solubility of the base until the liquid has the composition of point e, which is reached of course when the total composition reaches the vertical ej at point y, or when $\mathbf{a} = a_e$. At this point MX appears mixed with the remaining solid MOH. Continued addition of HX now causes solid MOH to be consumed at the expense of the formation of solid MX; the ratio of the amounts of the two solids changes therefore while the liquid remains fixed in composition at point e. The solid MOH is all consumed when the total composition reaches point z on line ek. Beyond this point, with MX as sole solid phase, the liquid travels along curve ef, the solubility curve of the salt. The value of **a** required to reach point z is $a_e + (\mathbf{b}''' - b_e)$. When the total composition is at some intermediate point q, the ratio of the number of moles of solid MX to the number of moles of solid MOH = yq/qz; the number of moles of solid MX $= a_q - a_y = a_q - a_e$, and the number of moles of solid MOH $= a_z - a_q$.

2. WITH VARIABLE g. Fig. 19–1, applying for $g = 0$, is a particular horizontal section of the three dimensional diagram, similar to Fig. 17–2, for the effect of g (net foreign strong acid or base) upon the solubility relations. The relation between the solubility product P of the salt and that of the mixture of acid and base, P_1P_2/W, defined in Eq. XVII(82), is a constant independent not only of H but also of the ionic strength since the "mass products" are $\mathbf{P}/\gamma^2$ and $\mathbf{P}_1\mathbf{P}_2/\gamma^2\mathbf{W}$ respectively. Hence a change in g can not make the salt disappear as a stable solid phase. Each horizontal section of the diagram, such as Fig. 19–1, for relations at constant g remains the same in having three stable solubility curves. But a change in g causes all five *points* of the horizontal section (S_1, e, S, f, S_2) to move in such a way as possibly to affect ultimately the congruence of solubility of the salt. At high g the salt will be congruently soluble with respect to the base, incongruently soluble with respect to the acid; at low g it will be incongruently soluble with respect to the base, congruently soluble with respect to the acid. This effect is shown in the picture of the solid model, Fig. 19–2. The relations at $g = 0$ (pure water) are shown as H_2O–S_1–e–S–f–S_2. At this value of g the salt is congruently soluble, its aqueous solubility being S. The curve for saturation with both MOH and MX, curve e′e″, is seen to swerve so as to cross the mid-plane at the value $g = g_{e_c}$ so that below this critical value of g the salt is incongruently soluble with respect to the base, as shown for the relations at the lowest value of g in the figure. Similarly, with increasing g the salt becomes incongruently soluble with respect to the acid when at $g = g_{f_c}$ the curve for saturation with respect to MX and HX crosses the mid-plane at f_c.

B. ANALYTICAL EQUATIONS

1. MINIMUM REQUIREMENT FOR SATURATION. The surfaces for saturation with MOH or HX alone have already been considered in Chapter XVII. The general electroneutrality is again

$$H - OH = a\alpha - b\alpha' + g. \tag{1}$$

The condition for saturation with respect to the salt is

$$P = a\alpha b\alpha', \tag{2}$$

or

$$H = (abA/P - A - W/B)/2 \pm \sqrt{(\quad)^2 - WA/B}, \tag{3}$$

$$OH = (abB/P - B - W/A)/2 \pm \sqrt{(\quad)^2 - WB/A}. \tag{3a}$$

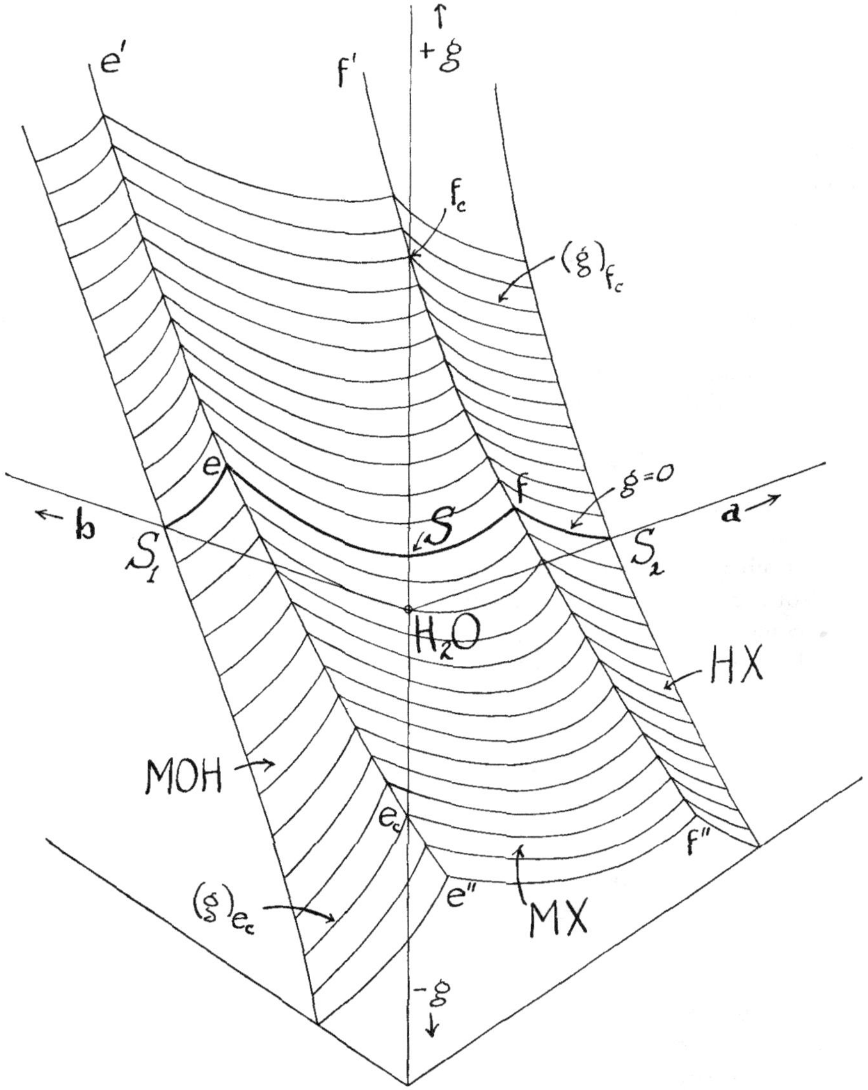

Fig. 19–2. Solubility relations of the salt MX as function of g.

This gives a single value of H and hence of g in Eq. (1) for saturation at specified values of a and b unless both the acid and the base are weak. If both are strong, with $\alpha = \alpha' = 1$, there is no effect of g on this solubility surface, the equation of which is then simply $ab = P$. In this case the solubility surface of the salt is strictly vertical, always in the *ideal* sense, in Fig. 19–2, although the intersection curves such as e′e″ for saturation with two solids will still swerve as g changes. If only the base is weak, Eq. (3) becomes

$$H = PW/B(ab - P), \tag{4}$$

while if only the acid is weak,

$$H = A(ab - P)/P. \tag{5}$$

Unless both are weak therefore, the minimum requirement for saturation is simply $ab > P$.

For the minimum requirement in the weak-weak case, the condition is, as under Eq. XVII(80), $H = \sqrt{WA/B}$ or the iso-electric point of the salt, so that

$$(ab)_{\min} = P(1 + \sqrt{W/AB})^2. \tag{6}$$

Since this still holds when a and b are equal, the minimum solubility of the suspension of the pure salt is

$$S_{\min} = \sqrt{(ab)_{\min}} = \sqrt{P}(1 + \sqrt{W/AB}). \tag{7}$$

These minimum requirements must be satisfied for saturation regardless of the composition of the solution, i.e. of whether or not $a = b$ and whether or not other acids and bases, strong or weak, are present.

For a given suspension of the salt moreover, not only the product ab but also a and b individually and their sum have minimum values at H_{ie}. Since $\mathbf{a} - \mathbf{b} = \mathbf{h} = a - b$, we have

$$P = a(a - \mathbf{h})\alpha\alpha' = b(b + \mathbf{h})\alpha\alpha', \tag{8}$$

or

$$a = \mathbf{h}/2 + \sqrt{(\quad)^2 + P/\alpha\alpha'}, \tag{9a}$$

$$b = -\mathbf{h}/2 + \sqrt{(\quad)^2 + P/\alpha\alpha'}. \tag{9b}$$

At constant $\mathbf{h}$ then, both $da/d\mathrm{p}H$ and $db/d\mathrm{p}H$ separately as also their sum are zero at the iso-electric point, when

$$(a + b)_{\min} = 2\sqrt{(\mathbf{h}/2)^2 + P(1 + \sqrt{W/AB})^2}. \tag{10}$$

But a and b are not equal of course unless $\mathbf{h} = 0$, when $a = b = c$, the concentration of the pure salt, for all values of H; nor does $(a + b)_{\min} = (ab)_{\min}$ except for the suspension of the pure salt, with $\mathbf{h} = 0$.

[Eqs. (6) and (10) should be compared with Eqs. XVII(80) and XVII(81) for the suspension of the mixture of the acid and base of the system.]

Fig. 19–3 is a vertical constant $\mathbf{h}$ section of Fig. 19–2, not to be compared with Fig. 17–4, which is a projection on the vertical mid-plane. In Fig. 17–4 the abscissa represents $(\mathbf{a} + \mathbf{b})/\sqrt{2}$; in the present figure the abscissa is

$(\mathbf{a} + \mathbf{b} - |\,\mathbf{h}\,|)/\sqrt{2}$ and hence $\mathbf{b}\sqrt{2}$ if $\mathbf{h}$ is positive and $\mathbf{a}\sqrt{2}$ if $\mathbf{h}$ is negative. In such a section the curve of twofold saturation e′e″ is crossed at low g and curve f′f″ at high g. These crossings would be at the points e_c and f_c of Fig. 19–2 if the section shown in Fig. 19–3 were the mid-plane section of Fig. 19–2, the section at $\mathbf{h} = 0$. Between these crossings there is a minimum on the section of the solubility surface of the salt. This minimum occurs at the same value of H $(= H_{ie})$ for any value of $\mathbf{h}$ and therefore for any such vertical section but not at the same value of g. It occurs at $g = \sqrt{WA/B} - \sqrt{WB/A}$

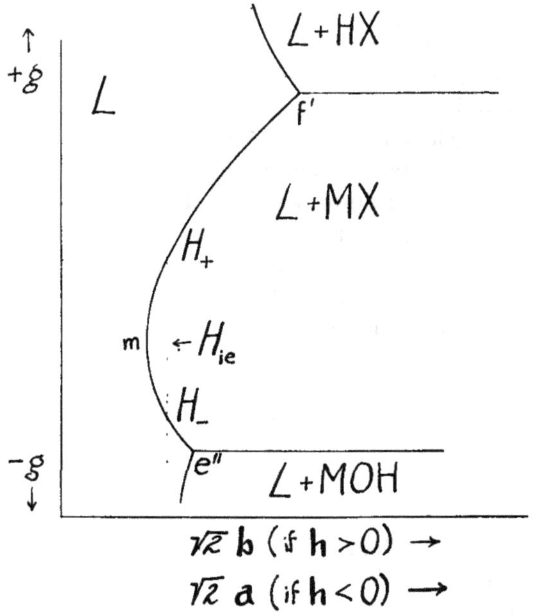

Fig. 19–3. A vertical section of Fig. 19–2 at constant $\mathbf{h}$ $(= \mathbf{a} - \mathbf{b})$ or parallel to the vertical mid-plane.

for the suspension of the pure salt ($a = b = c$; $\mathbf{h} = 0$), but otherwise, as may be seen from Eqs. (1) and (9), with H_{ie}, at $g = \sqrt{WA/B} - \sqrt{WB/A} - \mathbf{h}/(1 + \sqrt{W/AB})$. If a suspension of the salt (a liquid-solid sample with a fixed value of $\mathbf{h}$ in its composition) is subjected to change of g, the composition of the saturated solution moves up or down on the curve e″mf′ of Fig. 19–3 since this is the intersection of the vertical $\mathbf{h}$ plane with the solubility surface of the salt. If this motion of the composition of the saturated solution is projected orthogonally on a horizontal plane, it corresponds to motion back and forth on a line of constant $\mathbf{h}$ such as line wvv′ in Fig. 19–1.

2. Saturation With Salt Alone.

a. Calculation of solubility relations.

For a given value of $ab > (ab)_{min}$, or of $c > S_{min}$, there are two values of H (and hence of g) for saturation, given by Eq. (3), used with c^2 in place of *ab* for the pure salt. The upper value is shown as H_+, the value of H from Eq. (3) used with its positive sign, the lower as H_-, in Fig. 19–3. The solution is supersaturated if H lies between these limits. These points are not equidistant in their values of g either from $g = 0$ or from $g = g_{ie}$ but they are equidistant from H_{ie} in respect to pH. This may be seen since the product H_+H_- from Eq. (3) $= WA/B = (H_{ie})^2$. From this it follows also that α/α' at $H_+ = \alpha'/\alpha$ at H_-, for saturation at the two sides of the iso-electric point.

Even on a horizontal section at constant g such as Fig. 19–1 there may be points on the solubility curve on one side of the diagonal with the same value of ab as points on the other side. Such corresponding points are not at the same distance from the diagonal in the section and hence not at the same value of $|\mathbf{h}|$ unless the curve is symmetrical with respect to the diagonal. The curve is symmetrical only if $A = B$ (as for example for the salt of strong acid and strong base) when, however, there is no effect of H and hence of g upon the relations for saturation with salt alone.

If a suspension of the pure salt in water sufficient to give the concentration c when dissolved is treated with strong acid, the solubility increases from the start if $A < B$; otherwise it first falls to a minimum at H_{ie} and then rises again. With strong base, the solubility increases from the start if $A > B$ and otherwise passes through a minimum at H_{ie} before rising. To dissolve the salt, H_{req} is calculated from Eq. (3), with $ab = c^2$. The solid dissolves if H is either raised to the higher (H_+) of the two values given by the equation, by means of added foreign acid, or lowered to the lower value (H_-) by foreign base.

The value of g required for saturation at known values of a and b (or of c, when $a = b = c$) is found in every case from Eq. (1) through H_{req} from Eqs. (3), (4), and (5).

For saturation at known g, for specified a (or b), Eq. (1) is combined with Eq. (2) to find H_{req} and then Eq. (2) is used again to find b_{req} (or a_{req}). For known a we have

$$H - OH = a\alpha - P/a\alpha + g, \tag{11}$$

$$H^3(1 + P/aA) + H^2(A + 2P/a - g) - H[A(a + g) - AP/a + W] - AW = 0. \tag{12}$$

If the acid is strong, this becomes

$$H^2 - H(a + g - P/a) - W = 0. \tag{13}$$

Eqs. (11–13) do not involve explicitly the ionization constant of the base. On the other hand, for known b, for either strong or weak acid,

$$H - OH = P/b\alpha' - b\alpha' + g, \tag{14}$$

$$H^3B + H^2[B(b - g) - BP/b + W] - HW(B + 2P/b + g) - W^2(1 + P/bB) = 0; \tag{15}$$

and if the base is strong,

$$H^2 + H(b - g - P/b) - W = 0. \tag{16}$$

In each case the required value of the other concentration (b or a) may be found, with H_{req}, from Eqs. (2), (11), or (14).

A direct relation between a and b at saturation, for known g, requires elimination of H from Eq. (3) and either Eq. (12) or Eq. (15), in the general case. At constant g, a is continually decreased by increase of b, and b by increase of a. From Eq. (2),

$$db/da = [-P/(a\alpha\alpha')^2][\alpha'(da\alpha/da) + a\alpha(d\alpha'/da)]; \tag{17}$$

and from Eq. (11), $da\alpha/da = (dD/da)/(1 + P/a^2\alpha^2)$. Hence since both dD/da and $d\alpha'/da$ (like $d\alpha'/dH$) are positive, db/da is always negative. The same applies by symmetry to the effect of b on a.

For strong acid and weak base, Eqs. (4) and (13) give an equation which is cubic in a but quadratic in b so that

$$b = -P(a + g - P/a)/2aB + \sqrt{(\quad)^2 + P^2W/a^2B^2} + P/a. \tag{18}$$

Similarly for weak acid and strong base, Eqs. (5) and (16) give

$$a = P(g - b + P/b)/2bA + \sqrt{(\quad)^2 + P^2W/b^2A^2} + P/b. \tag{19}$$

These equations may also be rearranged for the direct calculation of the value of g required for saturation at given values of a and b without the intermediate calculation of H. For the salt of weak acid and strong base, for example, Eq. (18) becomes

$$g = b - P/b + A(ab - P)/P - PW/A(ab - P). \tag{20}$$

For the value of g required for a specified solubility c of the pure salt, such an equation is used with $a = b = c$; in this way the effect of g on the solubility may be plotted. On rearrangement the equation may then be used for the calculation of the solubility c of the pure salt as function of g:

$$c^5A/P + c^4 - c^3(g + 2A) - 2c^2P + cP(g + A - W/A) + P^2 = 0. \tag{21}$$

For the value of H in the saturated solution at specified g, the combination of $P = c^2\alpha$ and $H - OH = c(\alpha - 1) + g$ gives

$$\begin{aligned} H^5 + H^4(A - P/A - 2g) - H^3[2W - g(g - 2A)] \\ - H^2[2AW - g(gA + 2W)] + HW(W + 2gA) + AW^2 = 0, \end{aligned} \tag{22}$$

which reduces to Eq. XVIII(5) when $g = 0$. If a suspension of the pure salt sufficient to give the concentration $\mathbf{c}$ when completely dissolved is treated with strong acid, the solid may or may not dissolve before the equivalent point ($a_s = g = \mathbf{c}$); this is determined through Eq. (20). With solid always present the solution is neutral when $c_\mathbf{n} = \sqrt{P(A + \sqrt{W})/A}$ and $(a_s)_\mathbf{n}$

$= \sqrt{PW/A(A + \sqrt{W})}$, and at the equivalent point the solution is acid, H being given by Eq. (22) with $g = \mathbf{c}$.

Finally, if the solubility of the salt, as c, is studied in the presence of excess of its own specific acid, at concentration c', then all the preceding equations still apply, with $a = c + c'$ and $b = c$; if the solubility of the salt, as c, is studied in the presence of excess of its own specific base at concentration c'', then the equations are used with $a = c$ and $b = c + c''$.

b. Determination of equilibrium constants from solubility data. With S (the solubility of the salt) $= a = b$ in Eqs. (1) and (2), the dependence of log S on pH is given approximately as $2 \log S \cong \log P + \log [(A + H)/A]$ at high H and as $2 \log S \cong \log P + \log [(B + OH)/B]$ at low H.

The constants **A, B, P** may be calculated from measurements of **H** and S in presence of strong acid and of strong base. From Eq. (1) we have

$$-\log \mathbf{A} = \mathrm{p}\mathbf{A} = \mathrm{p}\mathbf{H} - \log \left[\frac{S\alpha' - g + D}{g + S(1 - \alpha') - D}\right] - \log \gamma. \qquad (23)$$

in which $D = \mathbf{D}/\gamma$. Then if g is a_s and H is high enough so that $\alpha' \cong 1$, this reduces to

$$\mathrm{p}\mathbf{A} \cong \mathrm{p}\mathbf{H} - \log [(S - a_s + D)/(a_s - D)] - \log \gamma; \qquad (24)$$

but if an approximate value of **B** is available, the calculation may be refined by evaluation of α' as $\mathbf{B}/(\mathbf{B} + \gamma\mathbf{W}/\mathbf{H})$. Similarly, from Eq. (1),

$$\mathrm{p}\mathbf{B} = \mathrm{p}\mathbf{W} - \mathrm{p}\mathbf{H} - \log \left[\frac{S\alpha + g - D}{-g + S(1 - \alpha) + D}\right] - \log \gamma. \qquad (25)$$

Now if g is $-b_s$ and OH is high enough, we may simplify this by taking $\alpha \cong 1$; but an approximate value of **A** may be used to evaluate α as $\mathbf{A}/(\mathbf{A} + \gamma\mathbf{H})$.

Eqs. (1) and (2) may also be combined to give

$$\log \mathbf{P} = 2 \log S + 2 \log \alpha' + \log [1 - (g - D)/S\alpha'] + 2 \log \gamma, \qquad (26)$$

and

$$\log \mathbf{P} = 2 \log S + 2 \log \alpha + \log [1 + (g - D)/S\alpha] + 2 \log \gamma. \qquad (27)$$

With g as a_s and in high H, α' in Eq. (26) approaches 1 and the expression reduces to

$$\log \mathbf{P} \cong 2 \log S + \log (1 - a_s/S + D/S) + 2 \log \gamma; \qquad (28)$$

with g as $-b_s$ and in high OH, α in Eq. (27) approaches 1 and a similar simplification becomes possible.

Eqs. (24) and (28) are exact for the salt of weak acid and strong base. For this case, moreover, Eq. (20), with $a = b = S$ and $g = a_s$, may be rearranged to give

$$\mathbf{A} = \frac{\mathbf{P}}{2S}\left(\frac{a_sS - S^2 + \mathbf{P}/\gamma^2}{S^2 - \mathbf{P}/\gamma^2}\right) + \sqrt{(\quad)^2 + \frac{\mathbf{P}^2\mathbf{W}}{\gamma^2(S^2 - \mathbf{P}/\gamma^2)^2}}, \qquad (29)$$

in which **A** and **P** are thermodynamic constants. This is useful in the calculation of **A** from S_0 (solubility in pure water) and S, the solubility in presence of a_s. Strictly, the solubility product **P** of the salt is not known unless **A** is also known. But provided that $\mathbf{A} > \sqrt{\mathbf{W}}$ and that S_0 is not too high, α_0 may be assumed approximately 1 and $\mathbf{P} \sim S_0^2$. A preliminary value of **A** from Eq. (29) may then be used, through Eq. XVIII(1), which expanded is Eq. VIII(8), to correct H_0 and thence both α_0 and **P**, thus leading finally to a precise value of **A** through Eq. (29). For the activity coefficient, the ionic strength may be taken as $\mu = a_s + S\alpha + OH \cong a_s + \mathbf{P}/S$.

Note: In all these equations P, the mass solubility product of the salt, always signifies not S_μ^2 but $(S^2\alpha\alpha')_\mu$; and **P**, the thermodynamic constant, is $(S^2\alpha\alpha')_{\mu=0}$, not simply $(S^2)_{\mu=0}$, since $(\alpha)_{\mu=0}$ for example is in general not 1 (although very close to 1 if $\mathbf{A} \gg \sqrt{\mathbf{W}}$).

c. Equations involving the composition of the total system. In terms of **h** (= **a** − **b**) and g as variables, the equation defining H for saturation with salt as sole solid phase is obtained by combination of Eqs. (9a) and (11) or identically of Eqs. (9b) and (14). With H thus calculated, either a or b may be calculated from Eqs. (9a) and (9b), the other following from $\mathbf{h} = a - b$.

If the acid and base are both strong ($\alpha = \alpha' = 1$), the expression for H is

$$H = (\mathbf{h} + g)/2 + \sqrt{(\quad)^2 + W}. \qquad (30)$$

If only the base is strong ($\alpha' = 1$),

$$\begin{aligned} H^5 - H^4(2g + \mathbf{h} - A + P/A) + H^3[(g + \mathbf{h})(g - 2A) - 2W] \\ + H^2[A(g + \mathbf{h})^2 + W(2g + \mathbf{h} - 2A)] \\ + HW[2A(g + \mathbf{h}) + W] + AW^2 = 0; \end{aligned} \qquad (31)$$

a similar expression may be written by symmetry for the salt of strong acid and weak base. For the general case, acid and base both weak,

$$\begin{aligned} H^3B + H^2[-(2g + \mathbf{h})B - BP/A + (AB + W)] \\ + H[(g + \mathbf{h})B(g - 2A) - W(2g - A + 2B)] \\ + [(g + \mathbf{h})^2AB + g^2W - (2g + \mathbf{h})W(A - B) + 2W(P - AB - W)] \\ + OH[(g + \mathbf{h})A(g + 2B) - W(-2g - B + 2A)] \\ + (OH)^2[(2g + \mathbf{h})A - AP/B + (AB + W)] + (OH)^3A = 0. \end{aligned} \qquad (32)$$

At constant g these equations give H for the solubility curve eSf of Fig. 19–1 in terms of **h**. Hence when the total composition, upon addition of HX to **b**′, there reaches the point v, with $\mathbf{h} = \mathbf{a}_v - \mathbf{b}'$, the composition of the solution may be calculated as just explained. At constant **h** on the other hand, these equations give H for the vertical section of the salt solubility surface shown in Fig. 19–3.

Eqs. (30–32) therefore apply to the general problem of saturation with respect to the salt as sole solid phase. With $\mathbf{h} = 0$ and $g \neq 0$, they describe the effect of foreign strong acid or base on the solubility of the pure salt in water or in aq. "$NaNO_3$" so that Eq. (31) for example reduces to Eq. (22). If both $g = 0$ and $\mathbf{h} = 0$, the equation reduces in each case to that for the saturated solution of the pure salt in pure water or in aq. $NaNO_3$; namely, $H = \sqrt{W}$ for strong-strong salt, Eq. XVIII(5) for salt of weak acid and strong base, and Eq. XVIII(6) for the weak-weak salt. With $g = 0$ and $\mathbf{h} \neq 0$, the equations describe a suspension of the salt in excess of the acid of the salt if $\mathbf{h} > 0$, or in excess of the specific base if $\mathbf{h} < 0$. A system made up of the alkali salt NaX, the salt MCl or MNO_3, and water alone, has $g = -\mathbf{h}$ for any value of $\mathbf{h}$, including 0. But when $g = \mathbf{h} = 0$, the result is again, as before, the equation for the pure suspension in water or in aq. $NaNO_3$. Finally, if $g \neq \mathbf{h} \neq 0$, the equations give the effect of foreign strong acid or strong base on the solubility of the salt in presence of excess of one of the components (MOH or HX) of the salt itself, both experimental variables being included in the equations.

d. Titrations involving precipitation of the salt M_sX. An expression such as Eq. (31) may be used in connection with certain familiar types of titration.

d1. Titration of weak acid after precipitation as salt. We here consider the titration, with strong base, b_s, of a weak acid after its precipitation by means of excess of a salt of strong acid. If the salt M_sNO_3, at total concentration $\mathbf{b}$, is added to a solution containing the weak acid HX at concentration $\mathbf{a}$, with $\mathbf{b} > \mathbf{a}$, precipitating the salt M_sX, with solubility product P, then the titration of the saturated solution is described by Eq. (31), with $g = \mathbf{b} - b_s$ and $\mathbf{h} = \mathbf{a} - \mathbf{b}$.

If $P \ll A$, the titration curve resembles that for strong acid and strong base; for the example shown in Table 19–1 ($A = 10^{-6}$ and $P = 10^{-10}$) the H^5 sequence (three terms) of Eq. (31) applies before the equivalent point, the H^4 sequence at the equivalent point ($b_s = \mathbf{a}$), and the H^2 sequence beyond the equivalent point.

If $A \ll P$, the important part of the titration curve (through the equivalent point) is described by the H^4 sequence, which may be simplified to

$$H^2P/A - H[(\mathbf{a} - b_s)(\mathbf{b} - b_s) - 2W] - W[\mathbf{a} + \mathbf{b} - 2b_s + (\mathbf{a} - b_s)^2A/W] \cong 0, \quad (33)$$

so that at the equivalent point, when $b_s = \mathbf{a}$, $H_e \cong \sqrt{(\mathbf{b} - \mathbf{a})AW/P}$. Applying Procedure I of Chapter IX to Eq. (33), we obtain the feasibility function for the titration for the following conditions: A small enough so that the term $(\mathbf{a} - b_s)^2A/W$ may be neglected near the equivalent point, and $A < P$; then $F \cong (p\mathbf{a}/400)\sqrt{(\mathbf{b} - \mathbf{a})A/PW}$.

Some examples of titration curves calculated from Eq. (31) are given in Table 19–1.

d2. Titration with "sodium palmitate" solution.

(*a*) "$AgNO_3$" (at $\mathbf{b}$) with NaX ($\mathbf{a}$). Here X^- is the anion of a weak acid, such as palmitic, giving an insoluble salt with the cation being titrated, here Ag^+. With P as the solubility product of AgX, the titration, with solid AgX present, is again described by Eq. (31), with $g = \mathbf{b} - \mathbf{a}$ and $\mathbf{h} = \mathbf{a} - \mathbf{b}$.

Table 19–1. Titration of weak acid after precipitation as salt; pH at various values of n ($= b_s/\mathbf{a}$).

	n	I	II(a)	II(b)
a		1	1	10^{-2}
b		2	2	1
A		10^{-6}	10^{-12}	10^{-12}
P		10^{-10}	10^{-8}	10^{-8}
	0	0	3.70	6.00
	0.5	0.3	4.12	6.30
	0.99	2	6.00	8.00
	0.999	3	7.00	8.79
	0.9999	4	8.00	
	1	7	9.00	9.00
	1.0001	10	10.00	
	1.001	11	11.00	9.21
	1.01	12	12.00	10.00

Table 19–2. Titration of M_sNO_3 with NaX, precipitating M_sX.

n	pH $\mathbf{b} = 1$	pH $\mathbf{b} = 10^{-2}$
0.9	7.02	7.52
0.99	7.15	7.90
0.999	7.52	
1	8.00	8.00
1.001	8.50	
1.01	9.00	8.11
1.1	9.50	8.50

If we assume values of 10^{-6} for A and 10^{-8} for P, the significant terms of this expression are those in H^4, H^2, and H^0, a quadratic in H^2; from the pH values at various titration ratios, both for $\mathbf{b} = 1$ and for $\mathbf{b} = 10^{-2}$, given in Table 19–2, the feasibility of this titration is seen to be low, especially at low concentration.

(b) "$CaCl_2$" (at **b**) with NaX (at **a**). Now P is the solubility product of CaX_2, "calcium palmitate," as in the determination of the hardness of water. With $\mathbf{h}^* = (2\mathbf{b} - \mathbf{a})$ we have, corresponding to Eq. (31),

$$H^7 - H^6(\mathbf{h}^* - A - 2P/A^2) - 3H^5W + H^4W(2\mathbf{h}^* - 3A) + 3H^3W^2 - H^2W^2(\mathbf{h}^* - 3A) - HW^3 - AW^3 = 0. \quad (34)$$

With $A = 10^{-6}$ and $P = 10^{-12}$, the feasibility is again poor at low concentration, as seen in Table 19–3 (with $n = \mathbf{a}/2\mathbf{b}$). With lower P ($= 10^{-18}$) the solution is practically neutral up to the equivalent point, when the pH begins to rise sharply even at the concentration $\mathbf{b} = 0.005$.

Table 19–3. Titration of $CaCl_2$(**b**) with NaX, involving precipitation of CaX_2; $A = 10^{-6}$.

	pH		
	$P = 10^{-12}$		$P = 10^{-18}$
n \ b	0.5	0.005	0.005
0.9	7.37	7.82	7.00 (+)
0.99	7.59	8.00	7.03
0.999	7.82		
1	8.05	8.05	7.15
1.001	8.50		
1.01	9.00	8.11	8.00
1.1	9.50	8.50	8.50

3. Saturation With Both Salt and Base (or Acid). For the solution saturated with both MX and MOH, we may write the following constant ion-products by combination of the two solubility products involved, P for the salt and P_1 for the base: $[M^+]^2[OH^-][X^-] = PP_1$; $[H^+][X^-] = PW/P_1$.

a. Interrelations of H, a, b, and g. For the interrelations of H, a, b, and g in the solution saturated with MOH and MX, the conditions are Eqs. (1) and (2) and $P_1 = b\alpha'OH$. The relation between H and b is Eq. XVII(51), $H = W(b - P_1/B)/P_1$; this value of H, to be called H_b, is H_{req} for saturation with MOH at concentration b, in any solution. The relation between H and a is

$$H = PW/P_1(a - PW/AP_1). \quad (35)$$

When this is combined with Eq. XVII(51) we obtain the relation between a and b:

$$(a - PW/AP_1)(b - P_1/B) = P. \quad (36)$$

This is the equation of the horizontal projection of the curve e′e″ of Fig. 19–2 on the a/b plane, shown later in Fig. 19–4. It is a hyperbola with the asymptotic limits of $a = PW/AP_1$ and $b = P_1/B$. The curve crosses the diagonal when $a = b = c$, with

$$c = (P_1/B + PW/AP_1)/2 + \sqrt{(\ \ )^2 + P(1 - W/AB)}. \tag{37}$$

This is the value of c $(= a = b)$ at the point e_c in Fig. 19–2.

If only the base is strong, Eqs. (36) and (37) become

$$(a - PW/AP_1)b = P; \qquad c = PW/2AP_1 + \sqrt{(\ \ )^2 + P}. \tag{38}$$

If only the acid is strong,

$$a(b - P_1/B) = P; \qquad c = P_1/2B + \sqrt{(\ \ )^2 + P}. \tag{39}$$

If both are strong, we have the symmetrical hyperbola $ab = P$, and $c = \sqrt{P}$.

Combination of Eqs. (1) and (35) with $P_1 = b\alpha'OH$ gives the relation between a and g:

$$a = -gP/2(P + P_1) + \sqrt{(\ \ )^2 + P^2(P_1 + W)/P_1(P + P_1)} + PW/AP_1, \tag{40}$$

$$a_{(g=0)} = P\sqrt{(P_1 + W)/P_1(P + P_1)} + PW/AP_1. \tag{41}$$

Similarly for the relation between b and g:

$$b = gP_1/2(P_1 + W) + \sqrt{(\ \ )^2 + P_1(P + P_1)/(P_1 + W)} + P_1/B, \tag{42}$$

$$b_{(g=0)} = \sqrt{P_1(P + P_1)/(P_1 + W)} + P_1/B. \tag{43}$$

Eqs. (41) and (43) give the coordinates of point e in Fig. 19–1, at $g = 0$. These expressions of course also give $g = f(a)$ and $g = f(b)$, and either one may be used to find g_c, or g at point e_c (Fig. 19–2), as f(constants) corresponding to the value of c $(= a = b)$ in Eq. (37). With $g > g_c$ the salt is congruently soluble, with $g < g_c$ it is incrongruently soluble, with respect to the base. If $g_c > 0$, the salt is incongruently soluble in pure water and this value of g $(= g_c)$ is required to make it congruently soluble. If $g_c < 0$, the salt is congruently soluble with respect to the base in pure water and this value of g makes it incongruently soluble. The combination of Eq. (37) with either Eq. (40) or (42) therefore gives an explicit test for the congruence of solubility of the salt with respect to its base, involving simply the constants A, B, P, P_1, and W. By setting $g = 0$ in the combination of the equations we obtain an expression involving only the constants, which may then be used to find the critical value of any one of them in terms of the other four: $P_1(P + P_1)(P_1 + W)(AP_1^2 - BPW)^2 = (ABP_1)^2(PW - P_1^2)^2$.

For the relation between H and g, we have

$$H - OH = PW/HP_1 - HP_1/W + g, \tag{44}$$

or

$$H = \frac{gW/2}{P_1 + W} + \sqrt{(\quad)^2 + \frac{W^2}{P_1}\left(\frac{P + P_1}{P_1 + W}\right)} \underset{(g=0)}{=} W\sqrt{\frac{P + P_1}{P_1(P_1 + W)}}. \tag{45}$$

The "solubility of the mixture," or the total concentration, $a + b$, in the solution saturated with both solids, passes through a minimum as g is changed. From Eq. (44) the "iso-electric point" of the mixture is at $H = W\sqrt{P}/P_1$; when $(a + b)_{\min} = 2\sqrt{P} + P_1/B + PW/AP_1$. This minimum occurs with $b > a$ therefore if $AP_1^2 > BPW$ and the value of g at the minimum is $g_{\min} = (PW - P_1^2)/P_1\sqrt{P}$ so that it occurs in positive g if $PW > P_1^2$.

We shall now write without discussion the corresponding equations for the curve f′f″ for saturation with respect to salt and acid, with P_2 as the solubility product of the acid. For the ion-product constants of the mixture: $[M^+][H^+][X^-]^2 = PP_2$; $[M^+][OH^-] = PW/P_2$. For the relation of H and a we have Eq. XVII(57) or $H = P_2/(a - P_2/A)$, to be called H_a; for H and b,

$$H = (b - PW/BP_2)P_2/P. \tag{46}$$

For the relation of a and b, or the projection of the curve f′f″ on a horizontal a/b plane,

$$(a - P_2/A)(b - PW/BP_2) = P. \tag{47}$$

The curve crosses the diagonal when, with $a = b = c$,

$$c = (P_2/A + PW/BP_2)/2 + \sqrt{(\quad)^2 + P(1 - W/AB)}. \tag{48}$$

If the base is strong,

$$(a - P_2/A)b = P, \qquad c = P_2/2A + \sqrt{(\quad)^2 + P}, \tag{49}$$

while if the acid is strong,

$$a(b - PW/BP_2) = P, \qquad c = (PW/2BP_2) + \sqrt{(\quad)^2 + P}. \tag{50}$$

With both strong, $ab = P$ and $c = \sqrt{P}$; in this case the curve f′f″ is vertically directly above curve e′e″ in Fig. 19–2; their projections on the a/b plane would be identical.

Again, as for the curve x′x″ in Chapter XVII for saturation with base and acid, the curves e′e″ and f′f″, as projections on the a/b plane, change only in position, not in their shapes, with respect to a change in the strength of either the acid or the base or both. From Eqs. (36–39) and (47–50) it is seen that the curve e′e″, for saturation with base and salt, is displaced to the left by the constant amount PW/AP_1 if the acid is strong and downward by the constant amount P_1/B if the base is strong. The curve f′f″, for saturation with salt and acid, is displaced to the left by the amount P_2/A if the acid is strong and downward by the amount PW/BP_2 if the base is strong.

For the dependence of the composition of the liquid on the curve f′f″ upon g,

$$a = -gP_2/2(P_2 + W) + \sqrt{(\ \ )^2 + P_2(P + P_2)/(P_2 + W)} + P_2/A, \quad (51)$$

$$a_{(g=0)} = \sqrt{P_2(P + P_2)/(P_2 + W)} + P_2/A, \quad (52)$$

$$b = gP/2(P + P_2) + \sqrt{(\ \ )^2 + P^2(P_2 + W)/P_2(P + P_2)} + PW/BP_2, \quad (53)$$

$$b_{(g=0)} = P\sqrt{(P_2 + W)/P_2(P + P_2)} + PW/BP_2. \quad (54)$$

For the relation between H and g, $H - OH = P_2/H - HP/P_2 + g$, and

$$H = \frac{gP_2}{2(P + P_2)} + \sqrt{(\ \ )^2 + P_2\left(\frac{P_2 + W}{P + P_2}\right)} \underset{(g=0)}{=} \sqrt{P_2\left(\frac{P_2 + W}{P + P_2}\right)}. \quad (55)$$

Finally, with $a = b = c_{48}$, either of the Eqs. (51) and (53) gives the value of g at point f_c (Fig. 19–2), the critical value of g below which the salt is congruently soluble and above which it is incongruently soluble, with respect to the acid; now if $g_{f_c} < 0$, the salt is incongruently soluble in pure water with respect to the acid. The critical relation of the constants is $P_2(P + P_2)(P_2 + W)(BP_2^2 - APW)^2 = (ABP_2)^2(PW - P_2^2)^2$.

b. Transition between single and twofold saturation. The equations of this section are used in the subsequent discussion of reactions involving precipitation of two solids, MOH + MX or MX + HX. These are reactions brought about by experimental variations such as the change of the ratio **a**/**b** for fixed g, the change of g for fixed values of **a** and **b**, etc. The relations at constant g have already been covered in the preliminary discussion of Fig. 19–1, which becomes quantitative by combination with the equations already developed. In reactions in which g is not constant and which should therefore be referred to Fig. 19–2, the composition of the solution saturated with its first precipitate travels on the saturation surface of that solid on a course determined by the intersection of the surface with some plane fixed by the particular experimental variation involved. Saturation with two solids then occurs when the solution reaches one of the curves e′e″, f′f″. As the experimental variation is continued, the solution, saturated with two solids, travels on that curve and may again leave it if its first precipitate comes to be completely redissolved in the reaction. The solution then begins to move across the surface of saturation with respect to the residual solid, eventually possibly also reaching, if this single solid is the salt, the second curve of twofold saturation.

It will be necessary therefore to express the conditions for the points at which the solution either reaches or leaves a curve of twofold saturation. If a solution reaches or leaves the curve e′e″ with MOH as its precipitate, the value of a at the point of contact is **a** since all the acid HX is in solution.

Hence this occurs when

$$H = WP/P_1(\mathbf{a} - PW/AP_1), \tag{56}$$

or in general when $a = \mathbf{a}$ in Eqs. (35), (36), and (40) so that no new expression is required. If the solution reaches or leaves e′e″ with MX as its precipitate, the condition, with MX as sole precipitate, is $a - b = \mathbf{a} - \mathbf{b} = \mathbf{h}$. Since the solution, however, is at the point of contact also just saturated with MOH, the required expression, obtained by combination of $P_1 = b\alpha' OH$ with Eq. (8), is

$$H = -(W/2P_1)(\mathbf{h} + P_1/B - PW/AP_1) + \sqrt{(\quad)^2 + PW^2/P_1^2}. \tag{57}$$

It is important to note that this expression holds only when the value of $a - b$ for the solution on the curve e′e″ is equal to $\mathbf{a} - \mathbf{b}$, or $\mathbf{h}$; although the solution is saturated with both solids, the quantity of solid MOH must be zero while the quantity of solid MX is unrestricted. The equation gives the value of H on the curve e′e″ for its intersection with a vertical plane of specified $\mathbf{h}$ in Fig. 19–2 such as that of Fig. 19–3. We may also express g, a, or b as a simple quadratic function of $\mathbf{h}$ and constants at the point defined by Eq. (57). For g we equate the expression for H in Eq. (45) with that in Eq. (57). The resulting equation will be called Eq. $H(45 = 57)$. With $\mathbf{h} = 0$, the Eq. $H(45 = 57)$ therefore gives the value of g at e_c, the crossing of the curve e′e″ with the diagonal plane $a = b = c$, the significance of which was discussed under Eqs. (37), (40), and (42). For a, H from Eq. (35) is equated with H from Eq. (57), giving Eq. $H(35 = 57)$; for b we use Eq. XVII(51) with Eq. (57), obtaining Eq. $H(57 = \text{XVII } 51)$.

For curve f′f″, the condition for a solution reaching it or leaving it with HX as sole precipitate is $b = \mathbf{b}$ in Eqs. (46), (47), and (53) so that

$$H = (\mathbf{b} - PW/BP_2)P_2/P; \tag{58}$$

and that with MX as sole precipitate is

$$H = -(P_2/2P)(\mathbf{h} - P_2/A + PW/BP_2) + \sqrt{(\quad)^2 + P_2^2/P}. \tag{59}$$

The corresponding value of g is then given by the Eq. $H(55 = 59)$; with $\mathbf{h} = 0$ in Eq. (59) this gives the value of g at the point f_c where the curve f′f″ crosses the diagonal $a = b = c$. For a we use the Eq. $H(59 = \text{XVII } 57)$ and for b the Eq. $H(46 = 59)$.

C. Effect of Change of g Upon Unsaturated Solution

1. General Relations. For a given unsaturated solution, one falling in the space in Fig. 19–2 closed off by the three solubility surfaces, precipitation of one of the solids may be brought about by change in g, which means a vertical motion in Fig. 19–2. The first solid to precipitate depends on the surface of saturation first reached by the change in g. If the acid and base

are both strong, when in ideal relations the solubility surface of the salt is vertical, the only possibilities are the precipitation of the acid at high g and precipitation of the base at low g. With both acid and base weak there is a bulge on the solubility surface of the salt, seen in section in Fig. 19–3. For some compositions then, the solubility surface of the salt could be crossed twice with change of g. Depending on the composition of the unsaturated solution, an increase in g may lead to either salt or acid as first precipitate and a decrease in g may lead to either salt or base as first precipitate. If only the acid is weak, the concentration increases with g on the salt surface, or the slope of the section, Fig. 19–3, is always positive; if only the base is weak the surface has the opposite slope. This may be seen from the derivative $d(ab)/dH$, and hence also of dc/dH when $a = b = c$, from Eqs. (4) and (5).

It follows then that in the system HX_s–MOH–H_2O, with HX_s a strong acid, an unsaturated solution may precipitate either MX_s or HX_s when g is raised, but it can precipitate only MOH when g is lowered. With increasing g, the salt precipitates first if the values of a and b are insufficient to satisfy the requirement for the curve of saturation with both salt and acid, defined by Eq. (50). If the acid precipitates first the salt can never be precipitated, but if the salt is first it is followed by precipitation of the acid as a second solid phase mixed with the salt. In the system M_sOH–HX–H_2O, with M_sOH a strong base, increase of g upon an unsaturated solution can cause only precipitation of HX, but if g is lowered either M_sOH or M_sX may precipitate, the salt precipitating first if a and b are large enough to satisfy Eq. (38). Again if the base is the first precipitate the salt is never precipitated, but if the salt precipitates first the base appears later as a second precipitate mixed with the salt.

2. Saturation With Both MOH and MX. We shall now consider in detail the relations involved in saturation with respect to both MOH and MX, in illustration of the possibilities. Fig. 19–4 shows, in schematic projection on a horizontal a/b plane and viewed from high g, the two saturation surfaces and their intersection curve, e′e″. This curve meets tangentially the hyperbola $ab = (ab)_{\min}$ at the point k. For the composition of point k: Eq. (35), with $H = \sqrt{WA/B}$, gives $a_k = (PW/AP_1)(1 + \sqrt{AB/W})$; from Eq. (7) or (36), $b_k = (P_1/B)(1 + \sqrt{AB/W})$; and from Eq. (44), $g_k = \sqrt{WA/B}(1 + P_1/W) - \sqrt{WB/A}(1 + P/P_1)$ so that g_k may be positive or negative. Looking down from high g we see the MX saturation surface twice, in the region between the curve ke″ and the curve $(ab)_{\min}$. The shapes of the surfaces are brought out by contours of constant g, or solubility curves at constant g. The curve $(ab)_{\min}$ on the right of point k is merely the envelope of these contours for the MX surface; it is the projection of the bulge on this surface along which $H = H_{ie}$ and $ab = (ab)_{\min}$.

If the composition of an unsaturated solution, **P**, with known values of **a** and **b**, falls on the left of curve e′k and the envelope $(ab)_{\min}$, the salt never precipitates either with increase or with decrease of g. Decrease of g causes MOH to appear as sole and final precipitate when its solubility surface is reached or when, from Eq. (1), $g_{req} = \mathbf{b}\alpha' - \mathbf{a}\alpha + H - OH$, with $H_{\mathbf{b}}$ from Eq. XVII(51). While MOH is the sole precipitate, the system is described by

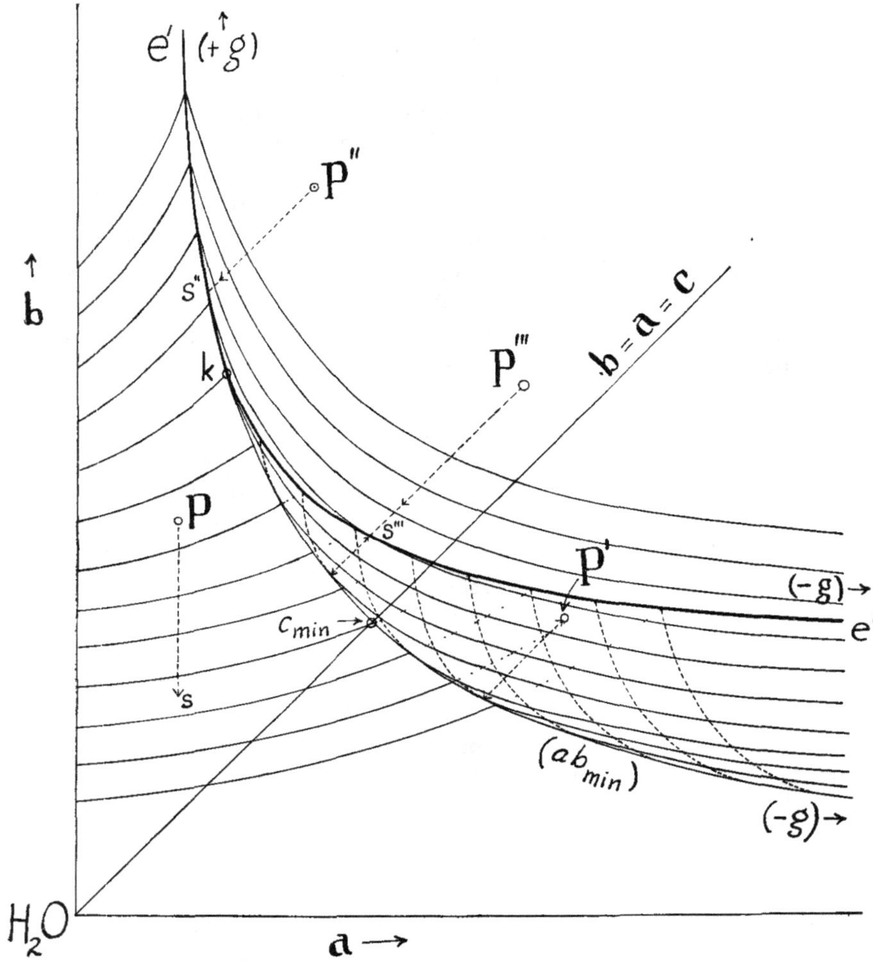

Fig. 19–4. Saturation with weak base MOH and its salt of weak acid MX.

Eqs. XVII(51–55). During this precipitation the concentration of the acid HX is constant (= **a**) and the solution travels downward on the solubility surface of MOH on a course fixed by its intersection with a plane of "constant **a**." In projection on Fig. 19–4, the solution, precipitating MOH, is therefore traveling vertically downward on the course **P** → s. Since it is traveling away from the curve e′e″, with decrease of b and constant a, it never becomes saturated with both MOH and MX. At any given value of g, b is given by Eq. XVII(53), and the number of moles of solid MOH precipitated is $\mathbf{b} - b$.

If the composition of the unsaturated solution, **P**′, is in the region between the curve ke″ and the $(ab)_{min}$ envelope, decrease of g causes MX to precipitate when the MX surface is reached; g_{req} follows from Eq. (1) with H_+ from Eq. (3). Precipitation of the salt causes the composition of the saturated solution to move to the left and downward along a line of constant **h** until the $(ab)_{min}$ envelope is reached, when the salt begins to redissolve and the solution begins to change its composition, along the same straight line, back toward **P**′. But the solution is then lying on the lower face of the MX saturation surface. For saturation with MX as sole precipitate, the composition at any specified value of g may be calculated through Eq. (32). When the composition returns to point **P**′, the solution is again unsaturated; g_{req} again comes from Eq. (1), now with H_- from Eq. (3). Further decrease of g then causes MOH to precipitate, as in the case of point **P**.

The curve e′e″ is reached, with decrease of g, only if the original composition lies above it in the projection of Fig. 19–4, as for points **P**″ and **P**‴; the first precipitate forming with decrease of g is again MX, and MOH appears as second precipitate, mixed with MX, when the solution reaches curve e′e″ as at points s″ and s‴ respectively, fixed by lines of constant **h** through the original or total compositions. In the first case (**P**″), with $(\mathbf{a}'' - \mathbf{b}'')$, or $(\mathbf{h})''$, less than $(a_k - b_k)$, precipitation of MX, caused by decrease of g, leads directly to the curve e′e″ at point s″; the solution does not first reach the hyperbola $ab = (ab)_{min}$. In the second case (**P**‴) the curve e′e″ is reached only after the solubility of the salt passes through its minimum, on this hyperbola. In either case the value of H when the curve e′e″ is reached is given both by Eq. (45) and by Eq. (57), so that g_{req} comes from Eq. $H(45 = 57)$, with $\mathbf{h} = \mathbf{a} - \mathbf{b}$. The full composition of the solution then follows from the various equations describing the curve e′e″, Eqs. (35–45),or from $P_1 = b\alpha' OH$ for b, etc. The number of moles of MX precipitated at this point is $\mathbf{a} - a = \mathbf{b} - b$.

With further decrease of g, the solution, saturated with both solids, follows the curve toward e″, and the composition for specified g comes from Eqs. (35–45). During this process the concentration a increases while b decreases; some of the solid salt therefore is being consumed while more of the base is precipitated, but the salt never redissolves completely.

With decreasing g the envelope $(ab)_{min}$ meets tangentially the curve f′f″ for saturation with salt and acid at a point k′ similar in significance to the point k here explained. This means that the curves e′e″ and f′f″, although always separated by the salt saturation surface in the three dimensional relations of Fig. 19–2, cross each other as projections on the a/b plane of Fig. 19–4. Such crossing in the projection is possible only if both acid and base are weak, as may be verified from Eqs. (36) and (47), but it is only in this general case that there is the $(ab)_{\min}$ bulge on the salt saturation surface. With $B = \infty$ for example, or for the salt of strong base and weak acid, the f′f″ curve always lies above and to the right of the curve e′e″ on the a/b projection if, according to Eqs. (38) and (49) combined, $PW < P_1P_2$, which is always true if the salt is a stable solid phase, as explained under Eq. xvii(82).

For $g < g_k$ and greater than $g_{k'}$ all the contours of the saturation surface of the salt of weak acid and weak base touch the $(ab)_{\min}$ envelope in their course between the curves e′e″ and f′f″. In the diagram at constant g for such values of g, as in Fig. 19–1, there will therefore be a point on the solubility curve of the salt at which $H = H_{ie}$ and the product $ab = (ab)_{\min} = c^2_{\min}$. The value of a at such a point, for given g, may be calculated from Eq. (11) with $H = H_{ie}$.

If the acid and the base are not both weak, there is no bulge on the solubility surface of the salt and no envelope of minimum solubility below the curve e′e″ in the a/b projection. The solubility contours for the salt, at constant g, then all leave the curve e′e″ as they do in Fig. 19–4 above point k if for example the acid is weak and the base is strong. The calculations are then made through the simplified equations, with either A or B or both equal to infinity.

3. Suspension of Salt M_sX (**c**) Treated with Foreign Strong Acid (a_s). The quantity of the salt is assumed to be such as to give the concentration **c** when dissolved. There will be two possibilities to consider, the salt dissolving before or after the precipitation of the weak acid HX; M_sOH is assumed to be a strong base. In either case the process consists of three periods.

(a) If the salt dissolves before the acid precipitates, there will be first a period of saturation with respect to the salt alone, described by Eqs. (20–22) with $g = a_s$, and with $a = b = c$ in Eq. (20), then a period with no solid phase present, during which the equations of Chapter x hold, and finally a period of saturation with respect to the acid alone, when the composition of the solution (values of a and H) are given by Eqs. xvii(59) and xvii(62) with **c** for "b" and a_s for g. The equivalent point, $a_s = g =$ **c**, may occur in any period. The salt dissolves completely when g has the value in Eq. (20) with $a = b =$ **c**, and the acid precipitates when $g_{\mathrm{HX}\downarrow} = P_2/A + D$, with H from Eq. xvii(57). This case occurs when $g_{20} < g_{\mathrm{HX}\downarrow}$.

(b) If the acid precipitates before the salt is completely dissolved, the first period, of saturation with respect to salt alone, is followed by one of saturation with respect to two solids simultaneously; and the third period is again one of saturation with respect to the acid alone. The second precipitate, the acid, appears when the system reaches point f_c in Fig. 19–2, with a value of g given, as already stated, by Eq. $H(55 = 59)$. While both precipitates are present the composition of the solution is on the curve f′f″ and its composition is therefore given by the equations for that curve or

Eqs. (46–55), for specified g ($= a_s$). The first precipitate, the salt, dissolves, leaving HX alone, when g reaches the value in Eq. (53) with $b = \mathbf{c}$.

D. Reactions Involving Salts

1. Special Sections of Fig. 19–2. "NaX" and "MNO_3" will represent soluble salts of the components HX and MOH with foreign strong base and strong acid respectively. A change in the quantity of such a salt involves both a change in the proportions of HX and MOH (i.e. of **a** and **b**) in the system, and a change in g. If a solution contains NaX at the concentration **a** and MNO_3 at the concentration **b**, it contains "HX" at **a** and "MOH" at **b**, with $g = \mathbf{b} - \mathbf{a}$.

There are three combinations to be considered:

Case I: reaction of HX and MNO_3;
Case II: reaction of NaX and MOH;
Case III: reaction of NaX and MNO_3.

With reference to Fig. 19–2, NaX is represented by a straight line on the right hand face, the HX face, originating at the point H_2O and falling at a 45° angle, since $g = -\mathbf{a}$. MNO_3 is represented by a similar line on the MOH face, rising at a 45° angle from the point H_2O, since $g = +\mathbf{b}$. Mixtures of HX and MNO_3 fall on the plane (plane I) defined by the lines HX and MNO_3; "the line HX" means the horizontal from H_2O at $g = 0$ on the HX face. Addition of HX to such a mixture causes the composition to move on this plane parallel to the HX face, while addition of MNO_3 causes it to move on the same plane parallel to the MOH face. Mixtures of NaX and MOH fall on the corresponding plane (plane II) defined by the lines NaX and MOH; this plane is perpendicular to the plane HX–MNO_3, but the only point in the three dimensional figure common to both planes is the point H_2O. Mixtures of the two salts, NaX and MNO_3, fall on the plane (plane III) defined by the lines NaX and MNO_3. This plane lies between the other two planes in the three dimensional figure. Equimolar mixtures of NaX and MNO_3 fall on the diagonal $\mathbf{b} = \mathbf{a}$ of Fig. 19–1; this line is the intersection of plane III with the horizontal section at $g = 0$. Again the addition of NaX to the mixture NaX + MNO_3 changes the composition on the inclined plane III in a motion parallel to the HX face, while addition of MNO_3 causes a change parallel to the MOH face.

When projected orthogonally on a horizontal section of Fig. 19–2, changes in the quantities of "NaX" and of "MNO_3" are identical with changes in the quantities of "HX" and "MOH" respectively; they are merely accompanied by corresponding changes in "g."

The equations already developed in this Chapter and in Chapter XVII for the surfaces and curves of Fig. 19–2 continue to apply, therefore, to processes involving these "salts" as reactants: I, with $g = +\mathbf{b}$ for HX (at **a**) and MNO_3 (at **b**); II, with $g = -\mathbf{a}$ for NaX (**a**) and MOH (**b**);

III, with $g = \mathbf{b} - \mathbf{a}$ for NaX (**a**) and MNO_3 (**b**). In these reactions therefore, g is no longer an independent variable, being itself determined by the quantities of the "reactants."

For orientation in each of the three cases, it is necessary to have the coordinates (a and b) of the intersections of the particular plane with the curves e′e″ and f′f″. The determination of the intersections will be explained in Case I, and since the process is similar in every case, the expressions for the others will therefore be written without discussion.

2. The Three General Cases.

a. Case I: Reaction of HX (**a**) *and* MNO_3 (**b**). Mixtures with *total* compositions made up from H_2O, HX, and MNO_3 fall on plane I, which may be defined as the plane $g = b$. If the system is heterogeneous, the composition of the solution phase is on this plane only if g (here **b**) $= b$ and therefore only if it is either unsaturated or saturated with HX as sole precipitate. If $\mathbf{b} > b$ as the result of precipitation of either MX or MOH, then the solution composition is above plane I.

For the intersection of any of the three planes with the curve e′e″, the equation of electroneutrality for the unsaturated solution is combined with $P_1 = \mathbf{b}\alpha' OH$ and with H from Eq. (56). For the unsaturated solution in plane I we have

$$H - OH = \mathbf{a}\alpha + \mathbf{b}\rho', \tag{60}$$

so that the intersection of the curve e′e″ with this plane occurs at

$$a_{e_I} = -PP_1/2B(P + P_1) + \sqrt{(\quad)^2 + P^2W/P_1(P + P_1)} + PW/AP_1, \tag{61}$$

$$b_{e_I} = P/(a_{61} - PW/AP_1) + P_1/B. \tag{62}$$

For the intersection with curve f′f″, the equation of electroneutrality of the unsaturated solution is combined with $P_2 = \mathbf{a}\alpha H$ and with H from Eq. (58). The result for plane I is

$$b_{f_I} = P^2W/2BP_2^2 + \sqrt{(\quad)^2 + P^2(P_2 + W)/P_2^2} + PW/BP_2, \tag{63}$$

$$a_{f_I} = P/(b_{63} - PW/BP_2) + P_2/A. \tag{64}$$

These four expressions therefore give the coordinates of the points e_I and f_I in Fig. 19–5.

Fig. 19–5 shows schematically the intersections of the three special planes I, II, III in addition to that of the plane at $g = 0$, with the surfaces and curves of Fig. 19–2, all projected on the a/b plane. For the sake of clarity the projections of the curves e′e″ and f′f″ are shown as not crossing each other in the region covered.

The points S_0 and S_0' are the solubilities of MOH and HX respectively in pure water, and the light curves, corresponding to Fig. 19–1, represent the solubility relations in pure water, or the plane for the system HX–MOH–H_2O at $g = 0$. The intersection S, of the salt solubility curve ef, on this plane, with the diagonal $\mathbf{b} = \mathbf{a}$, is therefore the solubility of the salt in pure water.

The curves joining the points $S_I e_I f_I S'_I$ constitute the solubility diagram on the plane I. S'_I, the solubility of HX with $[MNO_3] = 0$, is still S'_0. The curve $S'_I f_I$, for the solubility of HX in aq. MNO_3, has a negative slope if MOH is weak. Its limiting position is ideally vertical if MOH is strong.

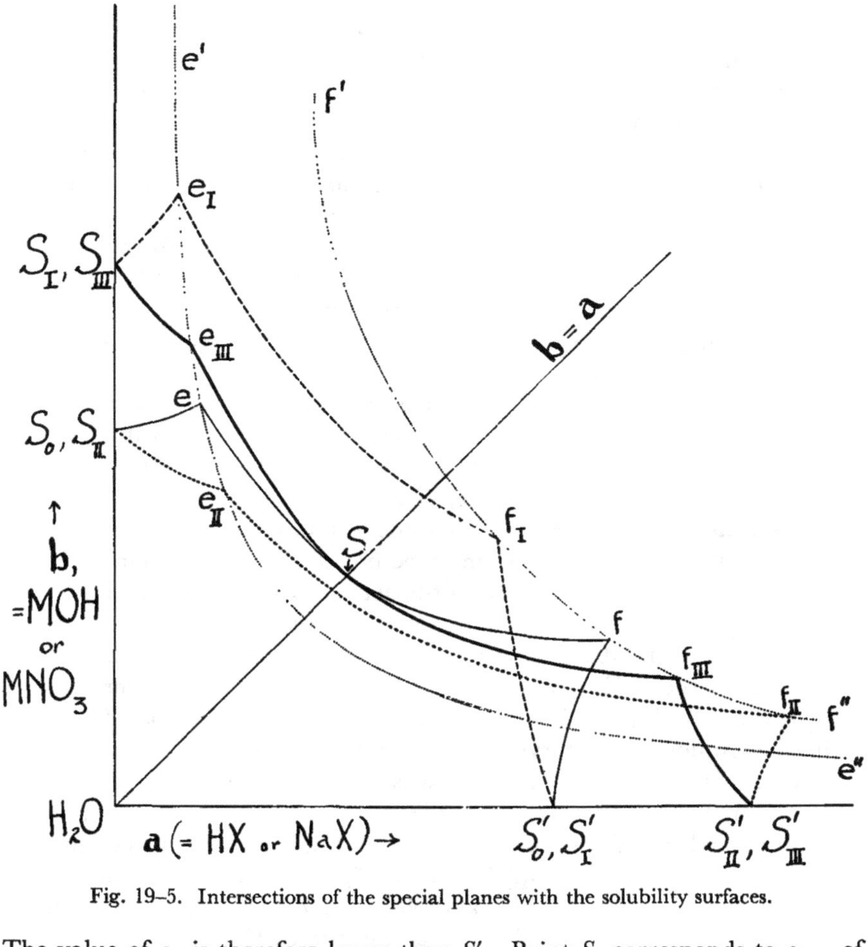

Fig. 19–5. Intersections of the special planes with the solubility surfaces.

The value of a_{f_I} is therefore lower than S'_I. Point S_I corresponds to c_{max} of MNO_3, the maximum concentration of MNO_3 in pure water without precipitation of MOH, and it is therefore higher than S_0, the solubility of MOH in pure water. The curve $S_I e_I$ represents the effect of the acid HX on this concentration so that $b_{e_I} > S_I$.

*a*1. *Addition of* MNO_3 *to aq.* HX (**a** *known*). If MNO_3 is added to an unsaturated solution of HX, with concentration **a** (less than S'_I), the first precipitate may be MOH, MX, or HX, depending on the value of **a** relative to e_I and f_I.

(a) For $\mathbf{a} < a_{e_I}$ the first precipitate is MOH. It appears when the composition of the solution reaches the curve $S_I e_I$; $\mathbf{b}_{req}$, for known $\mathbf{a}$, is given by Eq. XVII(52) with $a = \mathbf{a}$ and $b = g = \mathbf{b}$. While MOH is the sole precipitate, the composition (i.e. the value of b in the solution) for specified $\mathbf{b}$ is calculated from Eq. XVII(52) with $a = \mathbf{a}$ and $g = \mathbf{b}$. As MNO_3 is added the quantity of the precipitate, MOH, increases, but at the same time the concentration b, of MOH in solution, also increases, and the composition of the solution rises vertically, with $a = \mathbf{a}$, above the curve $S_I e_I$.

If $\mathbf{a}$ is smaller than PW/AP_1, the asymptotic limit of the curve e′e″, then the salt MX does not precipitate with continued addition of MNO_3. With $\mathbf{a} > PW/AP_1$, MX appears as second precipitate, mixed with MOH, when the solution, traveling on the surface of saturation for MOH, reaches the curve e′e″ at a point with $a = \mathbf{a}$ and hence above point e_I. The value of $\mathbf{b}_{req}$ follows from Eq. (40) with $a = \mathbf{a}$ and $g = \mathbf{b}$. With still further addition of MNO_3 the solution continues upward on the curve e′e″, and its composition for specified $\mathbf{b}$ and hence of g may be calculated directly through Eqs. (40) and (42) or with H from Eq. (45) through Eq. (35) for a and then Eq. (36) for b.

(b) For $\mathbf{a}$ between a_{e_I} and a_{f_I} the first precipitate is MX, appearing when the composition reaches the curve $e_I f_I$, and when H has the value given by Eq. (60) with $\mathbf{b} = P/\mathbf{a}\alpha\alpha'$; then with H calculated, $\mathbf{b}_{req} = P/\mathbf{a}\alpha\alpha'$. While MX is the sole precipitate, H for specified $\mathbf{b}$ is given by Eq. (32) with $\mathbf{h} = \mathbf{a} - \mathbf{b}$ and $g = \mathbf{b}$; then the composition, a and b, may be calculated through Eq. (9). It must be noted that the solution, saturated with MX alone, does not travel on the actual curve $f_I e_I$ of Fig. 19–5, since with precipitation of MX, g (or $\mathbf{b}$) is no longer equal to b, and the solution rises above plane I. MOH appears as a second precipitate when the solution, traveling on the MX surface, meets the curve e′e″ above point e_I, when $\mathbf{b}_{req}$ is given by Eq. $H(45 = 57)$, with $g = \mathbf{b}$ and $\mathbf{h} = \mathbf{a} - \mathbf{b}$. Thereafter, with the solution on the curve e′e″, the composition is given by Eqs. (35–45).

(c) For $\mathbf{a} > a_{f_I}$ the first precipitate is HX, appearing when $\mathbf{b}$, for known $\mathbf{a}$, is given by Eq. XVII(58) with $b = g = \mathbf{b}$ and $a = \mathbf{a}$. Since $\mathbf{b}$ (or g) remains equal to b during precipitation of HX, the solution, remaining on plane I, follows the curve $S_I' f_I$ in Fig. 19–5, its composition (value of a in solution) being given by Eq. XVII(58) with $b = g = \mathbf{b}$. MX therefore appears as the second precipitate when the curve f′f″ is reached at point f_I, with $\mathbf{b}_{req} = b_{f_I}$. With further addition of MNO_3 the solution, saturated with HX and MX, leaves plane I and travels on the curve f′f″ toward point f′, its composition for specified $\mathbf{b}$ being given by Eqs. (46–55) with $g = \mathbf{b}$. During this process the HX redissolves and the solution comes to leave the curve f′f″, with MX as sole precipitate, when $\mathbf{b}$ reaches the value given by the Eq. $H(55 = 59)$ with $\mathbf{h} = \mathbf{a} - \mathbf{b}$ and $g = \mathbf{b}$. With continued addition of MNO_3 the composition of the solution, saturated with MX alone, follows through Eq. (32) with $\mathbf{h} = \mathbf{a} - \mathbf{b}$ and $g = \mathbf{b}$. Eventually the curve e′e″

may be reached above point e_I, when MOH appears mixed with the MX; $\mathbf{b}_{req}$ comes from Eq. $H(45 - 57)$.

a2. Addition of HX *to aq.* MNO_3 (**b** *known*). If HX is added to a solution of MNO_3 with $\mathbf{b} < S_I$, the first precipitate is either MX or HX itself, while the precipitation of MOH is out of the question. If $\mathbf{b} < b_{f_I}$, saturation with HX occurs when $\mathbf{a}_{req}$ is given by Eq. XVII(58) with $a = \mathbf{a}$ and $b = g = \mathbf{b}$; in this case no second precipitate appears, further addition of HX merely adding to the solid. With $\mathbf{b} > b_{f_I}$, the first precipitate is MX, appearing when H has the value given by Eq. (60) with $\mathbf{a} = P/\mathbf{b}\alpha\alpha'$, whereupon $\mathbf{a}_{req}$ follows as $P/\mathbf{b}\alpha\alpha'$. Then for specified $\mathbf{a}$ the composition of the solution saturated with MX alone may be obtained through Eq. (32) with $\mathbf{h} = \mathbf{a} - \mathbf{b}$ and $g = \mathbf{b}$. HX appears as a second precipitate when the curve f′f″ is reached at some point above plane I and therefore above point f_I; $\mathbf{a}_{req}$ for given $\mathbf{b}$ may be calculated through Eq. $H(55 = 59)$, with $\mathbf{h} = \mathbf{a} - \mathbf{b}$ and $g = \mathbf{b}$. Further addition of HX again merely adds to the solid HX, the solution remaining invariant at some point on the curve f′f″, above point f_I.

a3. Reaction of solid HX *with aq.* MNO_3 (**b**). If solid HX, always in excess, is suspended in a solution of the salt MNO_3, then as long as HX is the sole precipitate the solution is on curve $S_I'f_I$; H is given by Eq. XVII(61) with $b = g = \mathbf{b}$, and the concentration of the dissolved HX follows from its solubility product as $a = P_2/\alpha H$, or directly from Eq. XVII(58) with $b = g = \mathbf{b}$. Above a certain value of **b** however, MX also precipitates upon the suspended HX, the critical value of **b** required being b_{f_I} of Eq. (63). With higher **b** the solution is saturated with both solids and its composition is determined by Eqs. (46–55) with $g = \mathbf{b}$. Thus the residual concentration of the base, or b, is calculated from Eq. (53) with $g = \mathbf{b}$. Then the number of moles of the cation M^+ (or of the salt MX) found in the precipitate per liter of solution treated with solid HX, is $\mathbf{q} = \mathbf{b} - b_{53}$. If the solid HX is suspended in a solution of MNO_3 of unknown concentration and if at equilibrium the number of moles of the cation M^+ found in the precipitate per liter of solution treated with the suspension is $\mathbf{q}$, then the original concentration **b** of the salt may be calculated through Eq. (53) with $b = \mathbf{b} - \mathbf{q}$ and $g = \mathbf{b}$; the actual concentration at equilibrium then follows as $b = \mathbf{b} - \mathbf{q}$.

b. Case II: Reaction of NaX (**a**) *and* MOH (**b**). The composition of the solution is now on the plane of the system, plane II, when $g\ (= -\mathbf{a}) = -a$ and hence only for the unsaturated solution or for the solution saturated with MOH as sole precipitate. The solution is below the plane when $a < \mathbf{a}$, as by precipitation of MX or HX.

For the unsaturated solution on plane II, $H - OH = -\mathbf{a}\rho - \mathbf{b}\alpha'$. For the intersections e_{II} and f_{II},

$$a_{e_{II}} = P^2W/2AP_1^2 + \sqrt{(\quad)^2 + P^2(P_1 + W)/P_1^2} + PW/AP_1, \qquad (65)$$

while $b_{e_{II}}$ is as in Eq. (62) with a_{65};

$$b_{f_{II}} = -PP_2/2A(P+P_2) + \sqrt{(\ \ )^2 + P^2W/P_2(P+P_2)} + PW/BP_2, \quad (66)$$

with $a_{f_{II}}$ as in Eq. (64) with b_{66}. Eqs. (65) and (66) may be written by symmetry from Eqs. (61) and (63) respectively.

For plane II, S_{II}, in Fig. 19–5, is the same as S_0, the solubility of MOH in pure water, and $S'_{II} = c_{max}$ for NaX, the maximum concentration of NaX without "hydrolytic" precipitation of HX. Now the curve $S_{II}e_{II}$ is horizontal if HX is a strong acid.

The detailed explanation for the reaction of MNO_3 with HX given under Case I applies by symmetry to the present case. The only change, in the use of the equations, is that g is now $-\mathbf{a}$.

*b*1. *Addition of* NaX *to aq.* MOH (**b** *known*). For $\mathbf{b} < b_{f_{II}}$, the first precipitate is HX, and MX appears as a second precipitate only if $\mathbf{b} > PW/BP_2$. It appears when the solution reaches the curve f′f″ with $b = \mathbf{b}$; $\mathbf{a}_{req}$ for given **b** is given by Eq. (53) with $b = \mathbf{b}$ and $g = -\mathbf{a}$. For **b** between $b_{e_{II}}$ and $b_{f_{II}}$ the first precipitate is MX, and HX appears as second precipitate with $\mathbf{a}_{req}$ given by the Eq. $H(55 = 59)$. For $\mathbf{b} > b_{e_{II}}$ MOH precipitates first followed by MX mixed with it when $\mathbf{a} = a_{e_{II}}$; the MOH redissolves, leaving MX as sole precipitate when $\mathbf{a}_{req}$ is given by the Eq. $H(45 = 57)$. Eventually HX may precipitate mixed with MX, with $\mathbf{a}_{req}$ from the Eq. $H(55 = 59)$.

*b*2. *Addition of* MOH *to aq.* NaX (**a** *known*). The only precipitate is MOH if $\mathbf{a} < a_{e_{II}}$. With a higher **a**, MX is first and MOH appears mixed with it when $\mathbf{b}_{req}$ is given by the Eq. $H(45 = 57)$.

*b*3. *Reaction of solid* MOH *with aq.* NaX (**a**). If solid MOH, always in excess, is suspended in aq. NaX, the solution is on curve $S_{II}e_{II}$ as long as MOH is the sole precipitate; H is given by Eq. XVII(55) with $a = -g = \mathbf{a}$, and the concentration of the dissolved MOH $= P_1/\alpha' OH$, or it may be calculated from Eq. XVII(52) with $a = -g = \mathbf{a}$. The value of $\mathbf{a}_{req}$ for precipitation of MX on the suspended solid MOH is $a_{e_{II}}$, Eq. (65). With higher **a** the value of H and the composition of the solution would be given by Eqs. (35–45) with $g = -\mathbf{a}$. Again the original concentration **a**, if unknown, may be calculated from the number of moles of X^- found in the precipitate per liter of solution at equilibrium through Eq. (40) with $a = \mathbf{a} - \mathbf{q}$ and $g = -\mathbf{a}$.

c. Case III: Reaction of NaX (**a**) *and* MNO_3 (**b**). The composition of the solution is now on the plane of the system if $g\ (= \mathbf{b} - \mathbf{a}) = b - a$ and hence when the solution is either unsaturated or saturated only with MX. With precipitation of HX the solution is below the plain and it is above it with precipitation of MOH.

For the unsaturated solution, $H - OH = -\mathbf{a}\rho + \mathbf{b}\rho'$. Then

$$a_{e_{III}} = (P/2P_1)(PW/AP_1 - P_1/B) + \sqrt{(\ \ )^2 + P^2W/P_1^2} + PW/AP_1, \quad (67)$$

with $b_{e_{III}}$ as in Eq. (62) with a_{67}; symmetrically,

$$b_{f_{III}} = (P/2P_2)(PW/BP_2 - P_2/A) + \sqrt{(\quad)^2 + P^2W/P_2^2} + PW/BP_2, \quad (68)$$

with $a_{f_{III}}$ as in Eq. (64) with b_{68}.

In Fig. 19–5, $S_I = S_{III}$ and $S'_{II} = S'_{III}$; $b_{e_{III}} < S_{III}$ since NaX (if HX is weak) decreases the solubility of MOH, and $a_{f_{III}} < S'_{III}$ since MNO_3 (if MOH is weak) decreases the solubility of HX. Plane III intersects the plane of $g = 0$ at the diagonal line $a = b = c$; hence the solubility curves of the salt for these two planes cross at the diagonal where $g = 0$ for plane III, at S, the solubility of the salt in water or ideally in aq. $NaNO_3$. On the left of the diagonal, $g > 0$ for plane III and hence $b_{e_{III}} > b_e$; similarly $a_{f_{III}} > a_f$ since $g < 0$ for plane III on the right of the diagonal. Again curve $S_{III}e_{III}$ is horizontal if HX is strong and curve $S'_{III}f_{III}$ is vertical if MOH is strong.

We shall consider the *addition of* MNO_3 *to* NaX (**a** known); the considerations for addition of NaX to MNO_3 are easily developed by symmetry.

(1) With $\mathbf{a} < a_{e_{III}}$, MOH is the first precipitate, followed by MX when the solution reaches the point $a = \mathbf{a}$ on curve e′e″; $\mathbf{b}_{req}$ comes from Eq. (40) with $a = \mathbf{a}$ and $g = \mathbf{b} - \mathbf{a}$.

(2) With **a** between $a_{e_{III}}$ and $a_{f_{III}}$, MX is the first precipitate, and while the solution is saturated with MX alone its composition follows the curve $e_{III}f_{III}$ in Fig. 19–5, up to point e_{III}, where MOH begins to precipitate, mixed with the MX. This occurs when $\mathbf{b} = b_{e_{III}} + (\mathbf{a} - a_{e_{III}})$ since $a - b = \mathbf{a} - \mathbf{b}$.

(3) With $\mathbf{a} > a_{f_{III}}$ the first precipitate is HX, appearing when the composition reaches the curve $S'_{III}f_{III}$. Then the solution, precipitating HX, travels on the HX surface below plane III, following a course on this surface which in projection on Fig. 19–5 is more vertical than the curve $S'_{III}f_{III}$ although still negative in slope. It therefore comes to meet the curve f′f″ on the right of point f_{III}, where MX appears as a second precipitate; at this point $\mathbf{b}_{req}$ is given by Eq. (53) with $b = \mathbf{b}$ and $g = \mathbf{b} - \mathbf{a}$. Then as the solution follows the curve f′f″ upward, HX redissolves with precipitation of MX, and when the point f_{III} is reached MX is left as sole precipitate; to reach this point $\mathbf{b}_{req} = b_{f_{III}} + (\mathbf{a} - a_{f_{III}})$. With MX as sole precipitate the solution then follows the curve $f_{III}e_{III}$ to point e_{III}, when MOH begins to precipitate mixed with MX; $\mathbf{b}_{req} = b_{e_{III}} + (\mathbf{a} - a_{e_{III}})$. Then, while the solution, on curve e′e″, is saturated with both MOH and MX, H is given by Eq. (45) with $g = \mathbf{b} - \mathbf{a}$. Eq. (45) thus gives the value of H in a suspension of MOH and MX formed by mixing the salts NaX and MNO_3 at concentrations **b** and **a** moles per liter respectively. But it is to be noted that this equation may not be used for arbitrary values of ($\mathbf{b} - \mathbf{a}$) unless the individual values of **a** and **b** satisfy the requirements just explained under Case III for saturation with both solids together.

Finally, although Eq. (17), showing that the solubility curve of the salt does not pass through a minimum of a with increasing b or through a minimum of b with increasing a, applies only for constant g, the solubility curve of the salt for the three inclined planes I, II, and III, which are not planes of constant g, still does not pass through such minima. This may be verified through the equations giving the values of H required for the appearance of MX as the first precipitate in every process considered. These equations, which have not here been written out explicitly, are cubic in every case but give always only a single value of H so that the salt does not redissolve as sole solid phase in the process. For Case III this condition is obtained by elimination of either $\mathbf{a}$ or $\mathbf{b}$ from $D = -\mathbf{a}\rho + \mathbf{b}\rho'$ and $P = \mathbf{ab}\alpha\alpha'$.

3. PRECIPITATION OF STRONG-STRONG SALTS.

a. General relations. The solubility curve of the salt is simply $ab = P$ and the value of H at saturation is given by Eq. (13) as $f(g,a)$ and by Eq. (16) as $f(g,b)$. Excess of either of the components of the salt decreases the solubility of the salt. Foreign acids and bases, whether strong or weak, have ideally no effect on the solubility of a strong-strong salt, M_sX_s. But the presence of salts such as M_sNO_3 and NH_4X_s, with "common ion," have ideally the same effect as M_sOH and HX_s themselves respectively, since again the foreign components of these salts, whether strong or weak, are without specific effect, adding only to the general ionic strength.

For reactions involving the salts NaX_s and M_sNO_3, forming the precipitates HX_s, M_sOH, and M_sX_s, the analytical relations are given by Eqs. (35–68), all simplified with $\alpha = \alpha' = 1$ and $A = B = \infty$. For example if M_sNO_3 is added to aq. NaX_s ($\mathbf{a}$ known), the precipitation of HX_s is out of the question, for the curve $S'_{III}f_{III}$ is vertical; the first precipitate is M_sX_s rather than M_sOH if $\mathbf{a}$ is greater (rather than less) than the critical value $a_{e_{III}} = P\sqrt{W/P_1}$, from Eq. (67). If M_sX_s is the first precipitate, the second, M_sOH, appears when $\mathbf{b} = b_{e_{III}} + (\mathbf{a} - a_{e_{III}}) = \mathbf{a} - P\sqrt{W/P_1} + P_1/\sqrt{W}$. The precipitate at the equivalent point ($\mathbf{b} = \mathbf{a}$) would therefore be contaminated with the base, M_sOH, only if $PW > P_1^2$, or failing this, if the concentration of excess of foreign strong base present is $b_s \geq OH - H$, or greater than the value $b_s = P_1/\sqrt{P} - W\sqrt{P}/P_1$, since $OH_{eq} = P_1/\sqrt{P}$.

b. Saturation with two related salts.

b1. Two precipitates, both 1 : 1 in type. We shall take as hypothetical examples "AgBr" (with P) and "AgCl" (with P'). For the present purpose of illustration these are assumed not to form solid solution(1). If $AgNO_3$ ($\mathbf{b}$) is added to a solution in which the concentration of NaBr is $\mathbf{a}$, AgBr precipitates when $\mathbf{ab} = P$; when $\mathbf{b} = \mathbf{a}$, for saturation with AgBr alone, $[Ag^+] = [Br^-] = \sqrt{P}$, while if $\mathbf{b} \neq \mathbf{a}$,

$$[Ag^+] = (\mathbf{b} - \mathbf{a})/2 + \sqrt{(\quad)^2 + P}, \tag{69}$$

$$[Br^-] = (\mathbf{a} - \mathbf{b})/2 + \sqrt{(\quad)^2 + P}. \tag{70}$$

(1) These salts actually form continuous solid solution. See for example H. C. Yutzy and I. M. Kolthoff, *J. Am. Chem. Soc.*, **59**, 916 (1937).

If the solution also contains NaCl at the concentration $\mathbf{a}'$, AgBr precipitates first if $\mathbf{a} > \mathbf{a}'P/P'$ and AgCl begins to precipitate when $[Br^-] = \mathbf{a}'P/P'$, and $\mathbf{b} = \mathbf{a} - \mathbf{a}'P/P' + P'/\mathbf{a}'$. When $\mathbf{b} = \mathbf{a} + \mathbf{a}'$ we have a suspension of AgBr and AgCl (in $NaNO_3$ solution); at this point, $[Ag^+] = \sqrt{P + P'}$, $[Br^-] = P/\sqrt{P + P'}$, and $[Cl^-] = P'/\sqrt{P + P'}$.

b2. Two precipitates, both 2 : 1 *in type.* With "PbI_2" (P) and "$PbCl_2$" (P') as hypothetical examples, we assume again that solid solution is not formed and that the acids and bases are strong. With $\mathbf{a}$ as the original concentration of NaI and $\mathbf{a}'$ that of NaCl and with $\mathbf{b}$ as the number of moles of $Pb(NO_3)_2$ added per liter, PbI_2 precipitates when $\mathbf{ba}^2 = P$; when $\mathbf{b} = \mathbf{a}/2$, for saturation with PbI_2 alone, $[Pb^{++}] = [I^-]/2 = \sqrt[3]{P/4}$, while if $\mathbf{b} \neq \mathbf{a}/2$, $[Pb^{++}]^3 + [Pb^{++}]^2(\mathbf{a} - 2\mathbf{b}) + [Pb^{++}](\mathbf{a} - 2\mathbf{b})^2/4 - P/4 = 0$. With NaI and NaCl both present, PbI_2 precipitates first if $\mathbf{a} > \mathbf{a}'\sqrt{P/P'}$ and $PbCl_2$ begins to precipitate when $[I^-] = \mathbf{a}'\sqrt{P/P'}$ and $\mathbf{b} = \mathbf{a}/2 + P'/(\mathbf{a}')^2 - (\mathbf{a}'/2)\sqrt{P/P'}$. When $\mathbf{b} = (\mathbf{a} + \mathbf{a}')/2$ we have a suspension of PbI_2 and $PbCl_2$ in aq. $NaNO_3$; at this point $[Pb^{++}] = \sqrt[3]{(P + P' + 2\sqrt{PP'})/4}$, with $[I^-]/[Cl^-] = \sqrt{P/P'}$.

b3. Two precipitates, type AgCl (P) *and* $PbCl_2$ (P'). Upon addition of NaCl ($\mathbf{a}$) to $AgNO_3$ ($\mathbf{b}$) and $Pb(NO_3)_2$ ($\mathbf{b}'$), AgCl precipitates first if $\mathbf{b} > P\sqrt{\mathbf{b}'/P'}$. $PbCl_2$ begins to precipitate when $[Cl^-] = P\sqrt{\mathbf{b}'/P'}$ and $\mathbf{a} = \mathbf{b} - P\sqrt{\mathbf{b}'/P'} + \sqrt{P'/\mathbf{b}'}$. When $\mathbf{a} = \mathbf{b} + 2\mathbf{b}'$, $[Cl^-]^3 - [Cl^-]P + 2P' = 0$ while $[Ag^+] = P/[Cl^-]$ and $[Pb^{++}] = P'/[Cl^-]^2$.

c. Titration of alkali halide with $AgNO_3$. Eqs. (69) and (70), for the reaction between $AgNO_3$ and alkali halide, are similar to Eq. IX(6) for the titration of strong acid against strong base. Eq. (69) may be rewritten as follows:

$$[Ag^+] = \sqrt{P}(\mathbf{Y} + \sqrt{\mathbf{Y}^2 + 1}), \tag{71}$$

with $\mathbf{Y} = (\mathbf{b} - \mathbf{a})/2\sqrt{P}$. Hence, for the change in $p[Ag^+]$, with $p[Ag^+] = -\log[Ag^+]$, during the titration of bromide from $(p/2)$ per cent before to $(p/2)$ per cent beyond the equivalent point, we have $\Delta_p = (2/2.3)\sinh^{-1}(p\mathbf{a}/400\sqrt{P})$, which is identical in form with Eq. IX(12) for the feasibility of the titration of strong acid with strong base, with P in place of W. This simple expression is exact on the assumption that the p per cent of $AgNO_3$ is being added with negligible change of volume of the solution and with the understanding that $\mathbf{a}$ is the concentration of the "NaBr" at the equivalent point in absence of precipitation.

For the effect of volume: if the alkali halide solution, with concentration c_a and V_a, is titrated with $AgNO_3$ solution of concentration c_b and if the volume of $AgNO_3$ solution added is V_b, then $\mathbf{Y}$ for Eq. (71) becomes $\mathbf{Y} = (c_bV_b - c_aV_a)/2(V_a + V_b)\sqrt{P}$. Hence, at $(p/2)$ per cent before and at $(p/2)$ per cent beyond the equivalent point, $V_{b(\mp)} = c_aV_a(1 \mp p/200)/c_b$.

Then if the total volume is $V_a + V_b$, $\mathbf{Y}_{(\mp)} = \mp pc_ac_b/400\sqrt{P}(c_a + c_b \mp pc_a/200)$, so that for small p, $\Delta_p \cong (2/2.3)\sinh^{-1}[pc_ac_b/400\sqrt{P}(c_a + c_b)]$. These equations are parallel to those in Section A5 of Chapter IX.

E. Fractional Precipitation of Salts of Monobasic Acid

General Relations; Cases to be Discussed. We shall here consider the general problem of the separation of two bases, at the initial concentrations $\mathbf{b}_M$ and $\mathbf{b}_N$, by precipitation of one of them as the salt of the monobasic acid HX (at $\mathbf{a}$ moles per liter of system); the solubility products will be P_M for the salt of the first base and P_N for the salt of the second base. The corresponding equations for the reverse problem, that of the separation of two acids by precipitation of one of them as the salt of a monoacid base, may easily be written by analogy from the equations here developed.

Several methods of precipitation will be considered:

(1) Addition of either HX or some soluble salt of HX, to the solution buffered at a known constant value of H.

(2) Addition of foreign base or acid, strong or weak, to the solution of known composition, including $\mathbf{b}_M$, $\mathbf{b}_N$, $\mathbf{a}$ (i.e. HX) and $\mathbf{g}$ (original value of g.)

(3) Addition of HX at constant g ($= \mathbf{g}$).

(4) Addition of NaX, which changes g, to a solution with known initial value of $g = \mathbf{g}$.

(5) Addition of "NH_4X," as in Method (4).

Of these the first, involving a solution presumably perfectly buffered at a suitable value of H, is mathematically the simplest problem since all the ionization fractions involved will be constants. Its application however depends on the calculation of the value or range of H required for the separation. This is a question considered under Method (2). Concerning Method (2), we note that this procedure is not possible if all the substances involved are strong since there will then be no effect of H and hence of foreign acids and bases on the solubility of either salt. All four other methods are always possible. We shall consider only two cases, that for weak acid (HX) and two strong bases, and that for strong acid (HX) and two weak bases.

The general equation of electroneutrality, if the solution is unsaturated, is $H - OH = a\alpha - (\mathbf{b}\alpha')_M - (\mathbf{b}\alpha')_N + g + \Sigma c\beta$ if the two bases to be separated are both univalent. In all the discussion g will represent the net equivalent concentration, $a_s - b_s$, of foreign strong acid. If the solution contains MNO_3 at the concentration $\mathbf{b}_M$, $N(NO_3)$ at $\mathbf{b}_N$, NaX at $\mathbf{a}$, and nitric acid at a', then $g = a' + \mathbf{b}_M + \mathbf{b}_N - \mathbf{a}$. The term $\Sigma c\beta$ is introduced if there are present in addition foreign weak acids and bases.

If the base is univalent, the condition for saturation is given by Eq. (4) for the salt of weak base and strong acid and by Eq. (5) for the salt of strong base and weak acid. The symbol $H_{4\mathbf{b}_M}$ for example will mean the expression for H in Eq. (4), with $b = \mathbf{b}_M$, etc. If the base is strong and divalent, $N(OH)_2$, then the condition for saturation with respect to its salt of the weak acid HX, or NX_2, with $P_N = [N^{++}][X^-]^2 = b_N(a\alpha)^2$, is

$$H = A(a\sqrt{b_N/P_N} - 1). \tag{72}$$

The relation between the concentrations b_M and b_N (to be distinguished from the original values $\mathbf{b}_M$ and $\mathbf{b}_N$) in a solution saturated with both salts will be independent of the method used in the precipitation. The relation is obtainable by combination of the two solubility product equations involved.

Case Ia, two strong univalent bases, MOH *and* NOH, *and weak* HX*:*

$$b_M = b_N P_M/P_N, \tag{73}$$

regardless of the values of H, a, and A (the ionization constant of HX).

Case Ib, MOH *and* $N(OH)_2$, *both strong, and weak* HX*:* $b_M = P_M\sqrt{b_N/P_N}$.

Case Ic, $M(OH)_2$ *and* $N(OH)_2$ *both strong, and weak* HX*:* Eq. (73). These expressions determine in any method which salt precipitates first; in Case I, MX is the first to precipitate if $\mathbf{b}_M > \mathbf{b}_N P_M/P_N$.

Case II, two univalent weak bases, MOH *and* NOH, *and strong* HX*:* by elimination of a from the two solubility products,

$$b_M = (b_N B_N P_M/B_M P_N)(B_M + OH)/(B_N + OH), \tag{74}$$

a relation directly useful when OH (or H) is known, as in Method (1); by elimination of H,

$$b_M = b_N B_N P_M/B_M P_N + P_M/a - B_N P_M/aB_M, \tag{75}$$

a relation useful if a is known or useful for the evaluation of a. From Eq. (75) it is seen that $b_M = P_M/a$ if MOH is strong regardless of H, b_N, P_N, and B_N.

Finally, we shall here define for subsequent use the function ϕ_{MX} of the salt MX:

$$\begin{aligned} \phi_{MX} &= P_{MX}/\alpha'_{MOH}\alpha_{HX} = b_{MOH}a_{HX}; \\ \phi_{M_2X} &= P_{M_2X}/(\alpha'_{MOH})^2(\alpha_2)_{H_2X} = b^2{}_{MOH}a_{H_2X}; \\ \phi_{MX_2} &= P_{MX_2}/(\alpha'_2)_{M(OH)_2}(\alpha_{HX})^2 = b_{M(OH)_2}a^2{}_{HX}. \end{aligned} \tag{76}$$

METHOD (1): ADDITION OF HX, OR ITS SALT, AT CONSTANT H. Since each degree of ionization, being a function of ionization constants and H, is now a constant, the ϕ functions defined in Eq. (76) are therefore also constants for this method. With ϕ in place of P, the problem may therefore be treated as the general case of strong-strong salts.

For the separation then of MOH and NOH by means of HX with MX as first precipitate, the value of $\mathbf{a}$ required to cause saturation is simply $\mathbf{a}_{req} = \phi_M/\mathbf{b}_M$. The value required to leave p per cent of the first base, MOH, unprecipitated, with b'_M defined as $p\mathbf{b}_M/100$, is

$$\mathbf{a}_{req} = \mathbf{b}_M - b'_M + \phi_M/b'_M. \tag{77}$$

Finally, the value required to cause the first appearance of the second salt, NX, mixed with the first precipitate, is

$$\mathbf{a}_{req} = \mathbf{b}_M - \mathbf{b}_N P_M/P_N + \phi_N/\mathbf{b}_N. \tag{78}$$

The range of $\mathbf{a}$ for the separation is therefore that between the values $\mathbf{a}_{77}$ and $\mathbf{a}_{78}$. Identical equations apply for $M(OH)_2$ and $N(OH)_2 + H_2X$ (all dibasic). For MOH and $N(OH)_2 + HX$, with MX as first precipitate, the only change is in the third term of Eq. (78), which becomes $\sqrt{\phi_N/\mathbf{b}_N}$. If NX_2 is the first precipitate we have $\mathbf{a}_{req} = \sqrt{\phi_N/\mathbf{b}_N}$ for the first appearance of NX_2, while Eqs. (77) and (78) become $\mathbf{a}_{req} = 2(\mathbf{b}_N - b'_N) + \sqrt{\phi_N/b'_N}$, and $\mathbf{a}_{req} = 2(\mathbf{b}_N - \mathbf{b}_M^2 P_N/P_M^2) + \phi_M/\mathbf{b}_M$. Similar equations are easily written for other combinations of two bases and acid.

METHOD 2: ADDITION OF FOREIGN ACID OR BASE.

Case Ia: two strong univalent bases, MOH *and* NOH, *and weak* HX. In this case a certain positive value of g (we ignore for the moment foreign weak solutes) is in general necessary for the solution to be unsaturated, and precipitation is then brought about by decrease of H, or experimentally by decrease of g. If MX is the first salt to be precipitated, the value of g required for saturation is given by the equation

$$H - OH = P_M/\mathbf{b}_M - \mathbf{b}_M - \mathbf{b}_N + g, \tag{79}$$

with $H_{5\mathbf{b}_M}$. When all but p per cent of the first base has been precipitated, as sole precipitate, the value of a is reduced to $a = \mathbf{a} - (\mathbf{b}_M - b'_M)$. With this value of a, $H_{5b'_M}$ is introduced into Eq. (79), now used with b'_M in place of $\mathbf{b}_M$, to find g_{req} to leave b'_M unprecipitated. When NX first appears as a second precipitate mixed with MX, and b_M is reduced to $b_M = \mathbf{b}_N P_M/P_N$, then $a = \mathbf{a} - (\mathbf{b}_M - \mathbf{b}_N P_M/P_N)$. Now $H_{5\mathbf{b}_M}$, with this value of a, is introduced into

$$H - OH = P_N/\mathbf{b}_N - \mathbf{b}_N P_M/P_N - \mathbf{b}_N + g, \tag{80}$$

to find g_{req} for the first appearance of NX as second precipitate. We thus find both the range of H and the range of g, for a solution with given initial values of $\mathbf{a}$, $\mathbf{b}_M$, and $\mathbf{b}_N$, required for a separation leaving not more than p per cent of MOH in solution, without precipitating any NOH as NX.

If the separation is controlled not with a foreign strong base but with a weak one, possibly in the presence of both strong and weak foreign acids, the calculation is still completely explicit, as explained under Eqs. XVII(37–39). We shall therefore make no further reference to this point.

The value of H in a solution saturated with both salts varies with g. The conditions, in addition to the equation of electroneutrality, are the solubility product $P_N = b_N a\alpha$, the relation in Eq. (73) and $\mathbf{a} - (\mathbf{b}_M + \mathbf{b}_N) = \mathbf{h}' = a - (b_M + b_N)$. These three conditions combined give

$$b_N = -\mathbf{h}'P_N/2(P_M + P_N) + \sqrt{(\quad)^2 + P_N^2/\alpha(P_M + P_N)}. \tag{81}$$

When this is introduced for $\mathbf{b}_N$ in the electroneutrality equation, Eq. (80), we could obtain an expression somewhat similar to Eq. (31) for H as $f(\mathbf{h}',g)$ for the solution saturated with both salts. With H so determined, b_N would follow from Eq. (81), b_M from Eq. (73), and a from P_N.

Case Ib: MOH *and* $N(OH)_2$, *both strong, and weak* HX. If MX is the first precipitate, it appears when g_{req} has the value defined by Eq. (79) with $2\mathbf{b}_N$ for $\mathbf{b}_N$ and with $H_{5\mathbf{b}_M}$ (used with $\mathbf{a}$). The upper limit of g (or "g_1") for the separation is given by the same equation, with b'_M for $\mathbf{b}_M$, $2\mathbf{b}_N$ for $\mathbf{b}_N$, and $H_{5b'_M}$, Eq. (5) being used with $a = \mathbf{a} - (\mathbf{b}_M - b'_M)$. The lower limit, "$g_2$," is given by $H - OH = \sqrt{P_N/\mathbf{b}_N} - P_M\sqrt{\mathbf{b}_N/P_N} - 2\mathbf{b}_N + g$, in which $H = H_{72\mathbf{b}_N}$, with $a = \mathbf{a} - (\mathbf{b}_M - P_M\sqrt{\mathbf{b}_N/P_N})$. *Note:* this means that H is given the value calculated from Eq. (72) used with $b_N = \mathbf{b}_N$ and with $a = \mathbf{a} - (\mathbf{b}_M - P_M\sqrt{\mathbf{b}_N/P_N})$.

If NX_2 is the first precipitate, it appears when g is defined by

$$H - OH = \sqrt{P_N/\mathbf{b}_N} - \mathbf{b}_M - 2\mathbf{b}_N + g, \tag{82}$$

in which $H = H_{72\mathbf{b}_N}$, with $a = \mathbf{a}$. The same Eq. (82), with b'_N in place of $\mathbf{b}_N$ and in which $H = H_{72b'_N}$, with $a = \mathbf{a} - 2(\mathbf{b}_N - b'_N)$, gives g_1 for the separation. For g_2,

$$H - OH = P_M/\mathbf{b}_M - \mathbf{b}_M - 2\mathbf{b}_M^2P_N/P_M^2 + g, \tag{83}$$

with $H_{5\mathbf{b}_M}$, in which $a = \mathbf{a} - 2(\mathbf{b}_N - \mathbf{b}_M^2P_N/P_M^2)$.

Case Ic: $M(OH)_2$ *and* $N(OH)_2$, *both strong, and weak* HX. The preceding operations should make clear how the range of H and of g for the separation may be calculated from the combination of Eq. (72) and the condition

$$H - OH = a\alpha - 2b_M - 2b_N + g. \tag{84}$$

Case II: MOH *and* NOH, *both weak, and strong* HX. In this case precipitation from an unsaturated solution is brought about by increase of g. If MX is the first precipitate, g_{req} for saturation is given by

$$H - OH = a - P_M/a - (\mathbf{b}\alpha')_N + g, \tag{85}$$

with $H_{4\mathbf{b}_M}$ and $a = \mathbf{a}$. With $H_{4b'_M}$ and with $a = \mathbf{a} - (\mathbf{b}_M - b'_M)$, this also gives the value of g_{req} to leave p per cent of the base MOH unprecipitated. For the first appearance of NX as second precipitate, g_{req} is calculated from

$$H - OH = a - P_M/a - P_N/a + g, \tag{86}$$

with $H_{4\mathbf{b}_N}$ and $a = \mathbf{a} - (\mathbf{b}_M - \mathbf{b}_NP_M/P_N)$.

For the dependence of H upon g, in a solution saturated with both salts, the conditions are $P_N = (b\alpha')_N a$, the relation in Eq. (74), and $\mathbf{h}' = a - (b_M + b_N)$, whence we obtain

$$b_N = -\mathbf{h}'/2(1 + \mathbf{J}) + \sqrt{(\quad)^2 + P_N/\alpha'_N(1 + \mathbf{J})}, \tag{87}$$

in which $\mathbf{J}$ is the function $B_N P_M(B_M + OH)/B_M P_N(B_N + OH)$, from Eq. (74). With this expression for b_N in

$$H - OH = P_N/b_N - b_N P_M/b_N - b_N\alpha'_N + g, \tag{88}$$

the desired relation is possible.

METHOD (3): ADDITION OF HX TO UNBUFFERED SOLUTION.

Case Ia: MOH *and* NOH, *both strong, and weak* HX. If HX is added to a solution containing the two bases with known and constant g ($= \mathbf{g}$), the value of $\mathbf{a}_{req}$ to cause saturation with the first salt, MX, is given by Eq. (19) with $b = \mathbf{b}_M$ and "g" $= \mathbf{g} - \mathbf{b}_N$. In order to find the value of $\mathbf{a}_{req}$ to leave b'_M of the first base in solution, we first calculate a, the concentration of unprecipitated acid, HX, from Eq. (79), with b'_M in place of $\mathbf{b}_M$ and with $H_{5b'_M}$; the expression is quadratic in a. Then $\mathbf{a}_{req} = a + \mathbf{b}_M - b'_M$. For the value of $\mathbf{a}_{req}$ when NX appears as second precipitate, we calculate a from Eq. (80) with $H_{5\mathbf{b}_N}$; then $\mathbf{a}_{req} = a + \mathbf{b}_M - \mathbf{b}_N P_M/P_N$.

For the value of H at a specified value of $\mathbf{a}$, when the solution is saturated with both salts, the relation would be the same as that discussed under Eq. (81), with the same definition of $\mathbf{h}'$.

Case Ib: MOH *and* $N(OH)_2$, *both strong, and weak* HX. If MX is the first precipitate, $\mathbf{a}_{req}$ for saturation is given by Eq. (19) with $b = \mathbf{b}_M$ and "g" $= \mathbf{g} - 2\mathbf{b}_N$. For $\mathbf{a}_{req}$ to leave b'_M in solution, we proceed as in Case Ia, but again with $2\mathbf{b}_N$ for $\mathbf{b}_N$ in Eq. (79). For the appearance of NX_2 as second precipitate, we calculate a from Eq. (82) with $\mathbf{b}_M = P_M\sqrt{\mathbf{b}_N/P_N}$ and with $H_{72\mathbf{b}_N}$; then $\mathbf{a}_{req} = a + \mathbf{b}_M - P_M\sqrt{\mathbf{b}_N/P_N}$.

If NX_2 is the first precipitate, $\mathbf{a}_{req}$ is calculated from Eq. (82) with $H_{72\mathbf{b}_N}$ used with $\mathbf{a}$ in place of a. For the value of $\mathbf{a}_{req}$ to leave b'_N unprecipitated, we use $H_{72b'_N}$ in Eq. (82), with b'_N in place of $\mathbf{b}_N$, to calculate a; then $\mathbf{a}_{req} = a + 2(\mathbf{b}_N - b'_N)$. For the appearance of MX as second precipitate, a is calculated from Eq. (83) with $H_{5\mathbf{b}_M}$; then $\mathbf{a}_{req} = a + 2(\mathbf{b}_N - \mathbf{b}_M^2 P_N/P_M^2)$.

Case Ic: $M(OH)_2$ *and* $N(OH)_2$, *both strong, and weak* HX. Similar procedures are used, through Eqs. (72) and (84), to find the range of $\mathbf{a}$ for the separation.

Case II: MOH *and* NOH, *both weak, and strong* HX. The value of $\mathbf{a}_{req}$ for saturation with the first salt, MX, is found from Eq. (85) with $H_{4\mathbf{b}_M}$, both Eqs. (85) and (4) being used with $a = \mathbf{a}$. The result is an equation of the fourth degree in $\mathbf{a}$ if both bases are weak (cubic if only MOH is weak). For $\mathbf{a}_{req}$ to leave b'_M unprecipitated, a is calculated from Eq. (85) with $H_{4b'_M}$;

then $\mathbf{a}_{req} = a + \mathbf{b}_M - b'_M$. *For the appearance of* NX *as second precipitate,* a is calculated from Eq. (86) with $H_{4\mathbf{b}_N}$, and $\mathbf{a}_{req} = a + \mathbf{b}_M - \mathbf{b}_N P_M/P_N$.

The dependence of H on $\mathbf{a}$ (and $\mathbf{g}$) for the solution saturated with both salts would be the same as that discussed under Eqs. (87) and (88).

METHOD (4): ADDITION OF NaX TO UNBUFFERED SOLUTION. Now the general electroneutrality equation is $H - OH = a\alpha - (b\alpha')_M - (b\alpha')_N + (\mathbf{g} - \mathbf{a})$, in which $\mathbf{g}$ is known and constant and $\mathbf{a}$ is variable, being the number of moles of NaX added per liter.

Case I: MOH *and* NOH, *both strong,* HX *weak.* If the first precipitate is MX, $\mathbf{a}_{req}$ for saturation is calculated from

$$H - OH = P_M/\mathbf{b}_M - \mathbf{b}_M - \mathbf{b}_N + (\mathbf{g} - \mathbf{a}), \tag{89}$$

with H from Eq. (5), which is used with $a = \mathbf{a}$ and $b = \mathbf{b}_M$. For the value of $\mathbf{a}_{req}$ to leave b'_M unprecipitated, Eq. (89) is used with b'_M for $\mathbf{b}_M$ and with H from Eq. (5), with $b = b'_M$ and $a = \mathbf{a} - \mathbf{b}_M + b'_M$, whereupon we may solve for $\mathbf{a}_{req}$. For the appearance of NX as second precipitate,

$$H - OH = P_N/\mathbf{b}_N - \mathbf{b}_N P_M/P_N - \mathbf{b}_N + (\mathbf{g} - \mathbf{a}); \tag{90}$$

in this we substitute H from Eq. (5) with $b = \mathbf{b}_N$ and $a = \mathbf{a} - \mathbf{b}_M + \mathbf{b}_N P_M/P_N$, to find $\mathbf{a}_{req}$. The equation for H as $f(\mathbf{h}'$ or $\mathbf{a}$, and $\mathbf{g})$, in the solution saturated with both salts, is obtained by combination of Eqs. (81) and (90).

Case II: MOH *and* NOH, *both weak,* HX *strong.* For saturation with MX as first precipitate, $\mathbf{a}_{req}$ is defined by

$$H - OH = -P_M/\mathbf{a} - (\mathbf{b}\alpha')_N + \mathbf{g}, \tag{91}$$

with H from Eq. (4), used with $b = \mathbf{b}_M$ and $a = \mathbf{a}$. For $\mathbf{a}_{req}$ to leave b'_M unprecipitated,

$$H - OH = a - P_M/a - (\mathbf{b}\alpha')_N + (\mathbf{g} - \mathbf{a}); \tag{92}$$

in this we substitute H from Eq. (4), with $b = b'_M$ and $a = \mathbf{a} - \mathbf{b}_M + b'_M$, to calculate $\mathbf{a}_{req}$. For the appearance of NX as second precipitate,

$$H - OH = a - P_M/a - P_N/a + (\mathbf{g} - \mathbf{a}); \tag{93}$$

now with H from Eq. (4) used with $b = \mathbf{b}_N$, and with a in terms of $\mathbf{a}$ and known quantities, Eq. (93) could be used to calculate $\mathbf{a}$. For the value of a it is necessary to use the relation $a = \mathbf{a} - \mathbf{b}_M + (b_M)_{75}$, whence

$$a = (\mathbf{a} - \mathbf{b}_M + \mathbf{b}_N B_N P_M/B_M P_N)/2 + \sqrt{(\quad)^2 + P_M(1 - B_N/B_M)}. \tag{94}$$

For H as $f(\mathbf{a},\mathbf{g})$ in a solution saturated with both solids, Eq. (87) would be substituted for $\mathbf{b}_N$, and Eq. (94) for a, in Eq. (93).

METHOD (5): ADDITION OF NH_4X TO UNBUFFERED SOLUTION.

Case I: MOH *and* NOH *both strong,* HX *weak.* We simply repeat the procedure for Method (4) but with $(\mathbf{g} - \mathbf{a}\alpha')$ in place of $(\mathbf{g} - \mathbf{a})$; α' is the degree of ionization of the weak base of the salt "NH_4X."

Case II: MOH *and* NOH *both weak,* HX *strong.* We proceed as in Method (4), with $(\mathbf{g} + \mathbf{a}\rho')$ for $\mathbf{g}$ in Eq. (91), and with $(\mathbf{g} - \mathbf{a}\alpha')$ for $(\mathbf{g} - \mathbf{a})$, in the rest.

F. Reaction Between Suspended Salt and Solution

1. System Involving Reciprocal Salt Pairs. If the solid salt MX, always present in excess, is suspended in a solution containing the base NOH at concentration $\mathbf{b}_N$ and the acid HY at concentration $\mathbf{a}_Y$, the precipitation of other salts, namely NX, MY, and NY, becomes possible, depending on the concentrations and on the value of H. If no solid solution is involved, the solid precipitate may change from one to two and even to three phases. If the quantity of new material precipitated is small relative to the quantity of MX, the solid will still *appear* as a single phase, and the phenomena involved are then related to those in so-called "ion-exchange" on the suspended solid. In general it is possible to have exchange of both cation and anion. If there is exchange of both ions, it must be caused by precipitation of the salt NY.

For the general problem, we shall consider a solution containing MOH, NOH, HX, and HY at the initial concentrations $\mathbf{b}_M$, $\mathbf{b}_N$, $\mathbf{a}_X$, and $\mathbf{a}_Y$ respectively, the solution being unsaturated as the result of our control of H through *foreign* soluble acid or base. Solid MX, in unlimited excess, is now suspended in this solution, and the concentrations and the value of H are assumed to be such that no phase reaction, in the sense of new precipitation, occurs. Such reaction or reactions will then require a decrease or increase of H, controlled by addition of foreign acid or base as for example by change of g. If we count MX as the "first precipitate," it will then be possible, with change of H, to reach saturation with a second precipitate and still later with a third. The remaining fourth salt can never be precipitated at equilibrium since the four salts can not coexist at equilibrium at arbitrary temperature whatever the value of H.

When three precipitates are present they will include either the combination MX + NY or the combination MY + NX; only one of these combinations is possible as a stable pair of coexisting solid phases for the system at equilibrium at a particular temperature for all possible values of H in the solution. If the pair MX + NY can exist together at one value of H, the reciprocal pair MY + NX can not coexist at equilibrium at any value of H. Only one is the stable pair regardless of the manner in which the four solutes are mixed and of the presence of other substances in the solution. For the pair MX + NY, the ion-product constant of the mixture of the solids is $P_{mix} = [M^+][N^+][X^-][Y^-] = P_{MX}P_{NY}$, while the ion-product constant for saturation with the reciprocal pair is $P_{mix} = [M^+][N^+][X^-][Y^-] = P_{MY}P_{NX}$.

As pointed out by van't Hoff[(2)], that pair is stable which has the smaller product of solubility products. The solution saturated with the unstable pair would be supersaturated with respect to the stable pair. (As the solubility products change with temperature, there may be a certain transition temperature at which the pair stability is reversed.) We shall suppose for particular discussion that MX + NY is the stable pair, with $P_{MX}P_{NY} < P_{MY}P_{NX}$. Therefore if two precipitates finally form on the suspended MX, they cannot be MY and NX at equilibrium; or if both cation and anion are "exchanged," one of the precipitates must be NY.

If all four solutes are weak, as acids and bases, so that each salt has an iso-electric point, then it is possible that some salts may require increase of H and others decrease of H for saturation. For simplicity of discussion we shall assume that the iso-electric points are either all low or all high so that within the experimental range the four solubilities all increase or all decrease with increase of H. We shall also assume that the concentrations are sufficient to give saturation with any of the possible salts.

If all the precipitations are expected to require decrease of H, then of the three possibilities, NX, MY, NY, that salt will appear as the second precipitate (MX being always the "first") which requires the highest value of H for saturation with the given initial values of $\mathbf{b}_M$; $\mathbf{b}_N$, $\mathbf{a}_X$, and $\mathbf{a}_Y$. A fixed condition is always that of P_{MX}, or $b_M a_X = P_{MX}/\alpha'_M\alpha_X = \phi_{MX}$. In this section ϕ is defined, in other words, as in Eq. (76); for specified and known H, ϕ is a constant since a quantity such as α'_M is then simply a constant coefficient, here $B_M/(B_M + OH)$. Another fixed condition is that the number of moles of cations (M^+ and N^+) leaving (or entering) the solid precipitate equals the number of moles of anions (X^- and Y^-) leaving (or entering) it. In general, this means

$$(a_X - \mathbf{a}_X) + (a_Y - \mathbf{a}_Y) = (b_M - \mathbf{b}_M) + (b_N - \mathbf{b}_N). \tag{95}$$

For saturation with each additional salt, we add the condition of the pertinent solubility function $\phi_{NX} = b_N a_X = P_{NX}/\alpha'_N\alpha_X$, $\phi_{MY} = b_M a_Y = P_{MY}/\alpha'_M\alpha_Y$, and $\phi_{NY} = b_N a_Y = P_{NY}/\alpha_N\alpha_Y$. It is to be noted also that if the solid MX is suspended in a solution containing neither MOH nor HX at the start, then both $\mathbf{b}_M$ and $\mathbf{a}_X$ are zero. If the original solution is one of the pure salt NY then, in addition, $\mathbf{b}_N = \mathbf{a}_Y$.

As long as MX is the sole precipitate, the conditions are $a_Y = \mathbf{a}_Y$ and $b_N = \mathbf{b}_N$ so that Eq. (95) and ϕ_{MX} give

$$a_X = (\mathbf{a}_X - \mathbf{b}_M)/2 + \sqrt{(\quad)^2 + \phi_{MX}}. \tag{96}$$

With b_M then fixed through ϕ_{MX}, this gives explicitly the composition of the solution as $f(H)$.

For NX to begin to precipitate upon the MX, the requirement is $a_Y = \mathbf{a}_Y$ and $b_N = \mathbf{b}_N$ together with ϕ_{NX}; hence $\phi_{NX}/\mathbf{b}_N - \mathbf{a}_X = \mathbf{b}_N\phi_{MX}/\phi_{NX} - \mathbf{b}_M$,

(2) J. H. van't Hoff and L. T. Reicher, *Z. phys. Chem.*, **3**, 482 (1889).

which defines H_{req} in terms of $\mathbf{b}_M$, $\mathbf{b}_N$, and $\mathbf{a}_X$. The expression is cubic for H. If $\alpha'_M = 1$, the expression is still cubic unless one of the other α's is also 1; but if either α_X or $\alpha'_N = 1$, it is quadratic.

For MY to precipitate upon MX, the requirement is $b_N = \mathbf{b}_N$ and $a_Y = \mathbf{a}_Y$ together with ϕ_{MY}. Hence H_{req} is given by $\mathbf{a}_Y\phi_{MX}/\phi_{MY} - \mathbf{a}_X = \phi_{MY}/\mathbf{a}_Y - \mathbf{b}_M$.

For NY, the required value of H is H_+ of Eq. (3) with $a = \mathbf{a}_Y$, $b = \mathbf{b}_N$, and A_Y, B_N and P_{NY} as the constants.

From the three values of H_{req} so calculated, we may determine which of the three salts appears as second precipitate mixed with the MX.

(a) If this second precipitate is NX, the third if any must be NY.

After the first appearance of NX, there will be a range of H for the system saturated with MX + NX. In this range the condition is simply $a_Y = \mathbf{a}_Y$ together with ϕ_{MX}, ϕ_{NX}, and Eq. (95) so that

$$a_X^2 - a_X(\mathbf{a}_X - \mathbf{b}_M - \mathbf{b}_N) - (\phi_{MX} + \phi_{NX}) = 0. \tag{97}$$

This gives explicitly a_X for any specified value of H; then b_M and b_N follow from the solubility products. The quantity of NX precipitated per liter of solution at specified H is $\mathbf{b}_N - b_N$.

With continued change of H, NY appears as an additional precipitate. The condition of ϕ_{NY} is now added to those leading to Eq. (97) so that

$$\mathbf{a}_Y\phi_{NX} = (a_X)_{97}\phi_{NY}, \tag{98}$$

in which $(a_X)_{97}$ represents the solution of Eq. (97); Eq. (98) may then be used to calculate H_{req} for the appearance of NY as third precipitate. Finally, when the solution is saturated with MX, NX, and NY, we simply combine ϕ_{MX}, ϕ_{NX}, and ϕ_{NY} with Eq. (95), to obtain

$$a_X^2(1 + \phi_{NY}/\phi_{NX}) - a_X(\mathbf{a}_X + \mathbf{a}_Y - \mathbf{b}_M - \mathbf{b}_N) - (\phi_{MX} + \phi_{NX}) = 0, \tag{99}$$

which defines a_X, and hence all the concentrations through the solubility products, as $f(H)$.

An entirely analogous procedure is followed if MY is the first precipitate to appear upon the MX, followed by NY; if the solution becomes saturated with MX, MY, and NY, the combination of Eq. (95) with ϕ_{MX}, ϕ_{MY}, and ϕ_{NY} gives the relation

$$a_X^2(1 + \phi_{MY}/\phi_{MX}) - a_X(\mathbf{a}_X + \mathbf{a}_Y - \mathbf{b}_M - \mathbf{b}_N) - (\phi_{MX} + \phi_{MX}\phi_{NY}/\phi_{MY}) = 0. \tag{100}$$

(b) If NY is deposited on the MX, there will be a range of H before the appearance of an additional precipitate, within which the conditions are $a_X - \mathbf{a}_X = b_M - \mathbf{b}_M$ and $a_Y - \mathbf{a}_Y = b_N - \mathbf{b}_N$ together with ϕ_{NY} and ϕ_{MX}. Hence a_X is given by Eq. (96), while

$$a_Y = (\mathbf{a}_Y - \mathbf{b}_N)/2 + \sqrt{(\quad)^2 + \phi_{NY}}; \tag{101}$$

with b_M and b_N fixed by the solubility products, these Eqs. (96) and (101) give the composition of the solution and hence the quantity of NY precipitated as $f(H)$.

NY may be followed either by NX or by MY as the next and final precipitate. The requirement for NX as the next precipitate is

$$(a_X)_{96}\phi_{NY} = (a_Y)_{101}\phi_{NX}, \tag{102}$$

and that for MY is

$$(a_X)_{96}\phi_{MY} = (a_Y)_{101}\phi_{MX}. \tag{103}$$

If H_{req} for Eq. (102) is greater than H_{req} for Eq. (103), NX rather than MY precipitates. If the solution comes to be saturated with three solids, then the composition is given either by Eq. (99) or (100), depending on the combination of solids present.

If the salt MX is suspended in a solution with known value of $\mathbf{b}_N$, if the constants P_{MX}, A_{HX}, B_{MOH}, and B_{NOH} are known, and if the reaction is such that only NX is deposited on the MX, then the solubility product of NX may be determined by simultaneous measurement, at equilibrium, of the value of H and either of the quantity, $\mathbf{q}$, of NX precipitated per liter of solution or of the change in the concentration of NOH, $\mathbf{b}_N - b_N$, with $b_N = \mathbf{b}_N - \mathbf{q}$. The calculation would be direct and exact through Eq. (97) rewritten with the substitution $a_X = \phi_{NX}/b_N$; it will be recalled that the ϕ's are functions of constants and H. In reverse, if P_{NX} is known, then the measurement of H and of the quantity of NX precipitated would allow the calculation of the original concentration $\mathbf{b}_N$ through the same equation.

2. Series of Salts With Common Ion.

a. Reaction at specified H. The solution is now assumed to contain a series of bases at the concentrations $\mathbf{b}_1, \mathbf{b}_2, \mathbf{b}_3 \cdots$, all of the type NOH, univalent, with solubility products $P_1, P_2, P_3 \cdots$; the concentrations and the constants are assumed to be such that the salts of these bases will precipitate on a suspension of the salt MX in the order $(NX)_1$, $(NX)_2$, $(NX)_3 \cdots$, as H is changed. There will be first a range of H in which the only foreign cation found in the solid at equilibrium is $(N^+)_1$, then a range with only $(N^+)_1$ and $(N^+)_2$, followed by one with $(N^+)_1$, $(N^+)_2$, and $(N^+)_3$, etc. The general condition in the j'th range is, from Eq. (95),

$$b_M + \Sigma_j b_N - a - \Sigma_j \mathbf{b}_N = 0, \tag{104}$$

in addition to the conditions set by the solubility products $P_{(NX)_1} \cdots P_{(NX)_j}$ and P_{MX}, abbreviated $P_1 \cdots P_j$ and P_M. We may therefore write the following relation between b_j and H in the j'th range:

$$b_j[1 + (\theta_M + \Sigma_{j-1}\theta_N)/\theta_j] - \theta_j/b_j\alpha - \Sigma_j \mathbf{b}_N = 0. \tag{105}$$

Here the function θ_M means P_M/α'_{MOH}, b_j is the concentration of the j'th base, and α is the degree of ionization of HX, or $A/(H + A)$. If a base is strong, $\theta = P$.

If all the constants and all the $\mathbf{b}_N$'s from $\mathbf{b}_1$ to $\mathbf{b}_j$ are known, the quantity of the j'th base precipitated in the j'th range, when the j'th salt constitutes the newest solid precipitating, may thus be calculated explicitly for specified H since the quantity $\mathbf{q}_j$ will be $\mathbf{b}_j - b_j$. In reverse, if all the constants and all the $\mathbf{b}_N$'s up to $\mathbf{b}_{j-1}$ are known, then the original concentration of the j'th base, or $\mathbf{b}_j$, may be calculated explicitly from the measurement in the j'th range either of b_j or of $\mathbf{q}_j (= \mathbf{b}_j - b_j)$ together with the value of H. Finally, if all the $\mathbf{b}_N$'s up to $\mathbf{b}_j$ and all the constants except P_j are known, P_j may be calculated from the measurement either of b_j or of $\mathbf{q}_j$ $(= \mathbf{b}_j - b_j)$, together with H, in the j'th range. In all three cases the calculation involves a simple quadratic equation.

By a stepwise procedure, it is possible at least theoretically to determine the whole series of solubility products of the j bases if all the $\mathbf{b}_N$'s are known and if P_M and all the ionization constants are known; P_1 for the first of the bases in the solution, determined as just explained in the first range, is used to determine P_2 in the second range, etc. Similarly it would be possible to determine the whole distribution of original concentrations, $\mathbf{b}_1 \cdots \mathbf{b}_j$ if all the constants are known since $\mathbf{b}_1$ determined in the first range is then used to determine $\mathbf{b}_2$ in the second.

When we calculate b_j in the j'th range for specified H, through Eq. (105), we may also thereby find explicitly all the other concentrations in the solution since for example $b_{j-1} = b_j\theta_{j-1}/\theta_j$. Hence the composition of the suspended solid is thus determined, in respect to $\mathbf{q}_1, \mathbf{q}_2 \cdots \mathbf{q}_j$, the quantities of all the bases precipitated at the specified value of H.

In reverse it is also possible to calculate through Eq. (105) the value of H required to cause the precipitation of a specified quantity $\mathbf{q}_j$ $(= \mathbf{b}_j - b_j)$ of the j'th base in the j'th range and hence also of a specified quantity of any of the bases up to j, in that range. With H as the dependent variable or as the unknown however, the equation is in general of high degree unless some of the ionization fractions equal 1.

If the suspended solid is the normal salt, MX, of a divalent base and a dibasic acid and if the solution contains a series of divalent bases, $N(OH)_2$, the relations in the j'th range are again given by Eq. (105), with θ defined as P/α_2', involving the second ionization fraction of the base, and with $\alpha = \alpha_2$, the second ionization fraction of the acid.

For a suspension of M_2X, salt of univalent base and dibasic acid, in a solution containing a series of univalent bases, type NOH, Eq. (104) involves $2a$ in place of a, and Eq. (105) becomes

$$b_j[1 + (\sqrt{\theta_M'} + \Sigma_{j-1}\sqrt{\theta_N'})/\sqrt{\theta_j'}] - 2\theta_j'/b_j^2\alpha_2 - \Sigma_j\mathbf{b}_N = 0. \quad (106)$$

Here α_2 is the second ionization fraction of H_2X and θ' is defined as $P/(\alpha')^2$, α' being the degree of ionization of a base.

If MX_2, salt of divalent base and monobasic acid, is suspended in a solution containing a series of divalent bases, type $N(OH)_2$, Eq. (104) involves $a/2$ in place of a, and Eq. (105) becomes

$$b_j[1 + (\theta''_M + \Sigma_{j-1}\theta''_N)/\theta''_j] - \sqrt{\theta''_j}/2\alpha\sqrt{b_j} - \Sigma_j \mathbf{b}_N = 0. \qquad (107)$$

Here α is the degree of ionization of HX and $\theta'' = P/\alpha'_2$, α'_2 being the second ionization fraction of a base. For the calculation of b_j, P_j, or $\mathbf{b}_j$ from various data at specified H, Eqs. (106) and (107), unlike Eq. (105), are cubic.

Finally, the equations of this section may easily be transformed by symmetry into the corresponding relations involving a series of acids forming insoluble salts with the cation of the suspended salt.

b. Reaction controlled by change of g. The preceding relations are simple because they are expressed in terms of H as the independent variable, which directly fixes all the degrees of ionization. We shall now illustrate for two simple cases the relations based upon the value of g as the actual experimental variable. The additional condition to be considered is the equation of electroneutrality.

Case I: MOH *and* NOH *both strong*, HX *weak*. The salt MX is suspended, always in excess, in a solution containing the base NOH at concentration $\mathbf{b}_N$ and with a certain value of g (net foreign strong acid or base). As long as MX is the sole precipitate the concentration of the salt, with $c = b_M = a$ (the analytical concentration of HX in the solution) is given by Eq. (21), with $P = P_M$ and $A = A_{HX}$, and with "g" $= g - \mathbf{b}_N$. The value of H in the solution is then given by Eq. (5) with $a = b = c$. As $\mathbf{b}_N$ is increased at constant g or as g is decreased at constant $\mathbf{b}_N$, precipitation of NX on the suspended MX becomes possible. With $P_N = \mathbf{b}_N a\alpha$ and $a = \sqrt{P_M/\alpha}$, the condition for the beginning of saturation with NX as second solid phase is $H = A(\mathbf{b}_N^2 P_M - P_N^2)/P_N^2$. With this expression for H in Eq. (80), we may calculate either the value of $\mathbf{b}_N$ required at specified g or the value of g required for given $\mathbf{b}_N$. If the MX is suspended in a pure solution of $N(NO_3)$ at the concentration $\mathbf{b}_N$, the same equation is used to find $(\mathbf{b}_N)_{req}$, with $g = \mathbf{b}_N$.

If the minimum requirement for precipitation of NX upon MX is satisfied, then for the solution saturated with both salts H varies with both $\mathbf{b}_N$ and g as explained under Eqs. (80) and (81); Eq. (81), with $\mathbf{h}' = -\mathbf{b}_N$, would be used for b_N in

$$H - OH = P_N/b_N - b_N P_M/P_N - b_N + g, \qquad (108)$$

to obtain the desired relation. For the case of the pure salt $N(NO_3)$ at $\mathbf{b}_N$, again $g = \mathbf{b}_N$. With H so determined, b_N follows from Eq. (81), and the number of moles of NX precipitated per liter of solution is then $\mathbf{b}_N - b_N$. It would thus be possible to calculate the quantity of NX precipitated for a given value of $\mathbf{b}_N$ as $f(g)$ or for given g as $f(\mathbf{b}_N)$.

According to Eq. (108) the solubility product P_N may be determined from

P_M and the equilibrium values of b_N, H, and g without knowledge of the original concentration $\mathbf{b}_N$ of the base NOH. If $\mathbf{b}_N$ is known, b_N may also be determined for the purpose from the quantity, $\mathbf{q}$, of NX found in the solid, as $\mathbf{b}_N - \mathbf{q}$. But with the three analytical data of $\mathbf{b}_N$, b_N, and g it is now possible to calculate P_N without measurement of H, if P_M and A are known. From Eq. (81), with $\mathbf{h}' = -\mathbf{b}_N$,

$$H = (Ab_N/P_N)[b_N(P_M + P_N)/P_N - \mathbf{b}_N] - A; \tag{109}$$

with this expression for H the only unknown in Eq. (108) is P_N. Finally if P_N is known, the combination of Eqs. (108) and (109) gives an expression which may be used for the determination of the original concentration of the base, $\mathbf{b}_N$, from the measurement of the quantity of NX precipitated upon MX at a measured value of g. This is possible by substitution of $\mathbf{b}_N - \mathbf{q}$ for b_N, leaving $\mathbf{b}_N$ as the only unknown.

Case II: MOH *and* NOH *both weak,* HX *strong.* With MX as the sole solid phase,

$$H - OH = a\rho'_M - (\mathbf{b}\alpha')_N + g. \tag{110}$$

With $a = \sqrt{P_M/\alpha'_M}$ this equation may be used to calculate H for the suspension of MX in a solution containing the base NOH at the concentration $\mathbf{b}_N$. For the precipitation of NX on the suspended MX, the requirement is, from $P_N = \mathbf{b}_N\alpha'_N a$ and $P_M = a^2\alpha'_M$,

$$(OH)^2 - (OH)B_N(\mathbf{b}_N^2 P_M B_N/P_N^2 B_M - 2) - B_N^2(\mathbf{b}_N^2 P_M/P_N^2 - 1) = 0. \tag{111}$$

With the value of OH fixed by Eq. (111) and with $a = \sqrt{P_M/\alpha'_M}$ in Eq. (110), the critical value of $\mathbf{b}_N$ at given g or of g at given $\mathbf{b}_N$ may be calculated. Again if MX is suspended in pure $N(NO_3)$ at $\mathbf{b}_N$, then the same equation holds, with $g = \mathbf{b}_N$.

For the variation of the concentrations with respect to $\mathbf{b}_N$ and g when the solution is saturated with both salts, b_N is given by Eq. (87) with $\mathbf{h}' = -\mathbf{b}_N$, and the expression would then be used in Eq. (88).

XX

Saturation with Salts of Dibasic Acid

We shall consider only relatively simple examples involving salts of dibasic acid with strong bases, both monoacid and diacid. But the dependence of the solubility upon pH, already discussed in general terms in Chapter IV, is readily extended to any type of salt, at least in the form of implicit equations. For the salt $M_2(H_4X)_3$, for example, the secondary salt of a hexabasic acid and a triacid base, both weak, we have $P = [M^{+3}]^2[H_4X^{-2}]^3$, and $S = [P/108(\alpha_2)^3(\alpha_3')^2]^{1/5}$, S being the formality of the saturating salt. Also,

$$\log S = (1/5)(\log P - \log 108 - 3 \log \alpha_2 - 2 \log \alpha_3'). \qquad (1)$$

For the limits at high and low H, we shall first note the limiting values approached by the pertinent ionization fractions for use in Eq. (1). For α_2, the second ionization fraction of the acid,

$$\alpha_2 = H^4K_{12}/(H^6 + H^5K_1 + H^4K_{12} + H^3K_{123} + H^2K_{1234} + HK_{12345} + K_{123456}). \qquad (2)$$

With increasing H, this approaches K_1K_2/H^2; with decreasing H, H^4/K_{3456}. For α_3', the third ionization fraction of the base,

$$\alpha_3' = K_{123}'/[(OH)^3 + (OH)^2K_1' + (OH)K_{12}' + K_{123}']. \qquad (3)$$

This approaches $K_{123}'/(OH)^3$ with increasing OH, and 1 with decreasing OH. Hence Eq. (1) becomes or approaches $\log S \rightleftharpoons (1/5)(\log P - \log 108 - 3 \log K_{12} - 6\text{p}H)$ in high H, and $\log S \rightleftharpoons (1/5)(\log P - \log 108 + 3 \log K_{3456} - 2 \log K_{123}' + 6 \log W + 18\text{p}H)$ in high OH.

A. Normal Salt of Dibasic Acid and Divalent Strong Base

1. Aqueous $Ca(OH)_2 + CO_2$. This is a specific example of the general case of the reaction of a strong divalent base, $M(OH)_2$, with a dibasic acid, "H_2X," forming an insoluble normal salt, MX. With a as the concentration of the acid and b as that of the base, the equation of electroneutrality is $D = [HX^-] + 2[X^=] - 2[M^{++}] + g$, or

$$H - OH = a(\alpha_1 + 2\alpha_2) - 2b + g, \qquad (4)$$

while the condition for saturation with respect to MX is

$$P = [M^{++}][X^=] = ba\alpha_2. \qquad (5)$$

For a suspension of the pure salt, with $a = b = c$, in presence of foreign strong (monobasic) acid or base at the net concentration $g = a_s - b_s$,

$$H - OH = c(\alpha_1 + 2\alpha_2 - 2) + g, \tag{6}$$

while $P = c^2\alpha_2$.

The solubility of this type of salt, which has no finite iso-electric point, does not pass through a minimum with simple change of pH, or as function of g. The solubility is increased by foreign acid and decreased by foreign base; this is seen directly from $c^2 = P/\alpha_2$, since α_2 rises continuously with pH. There is therefore a single value of H for saturation at a specified value of c, namely

$$H_{\text{req}} = -K_1/2 + \sqrt{(\quad)^2 + K_1K_2(c^2/P - 1)}, \tag{7}$$

the minimum value of c required for saturation being $c = \sqrt{P}$. With H_7, the concentration of foreign strong acid or base corresponding to a specified solubility is then calculated explicitly from Eq. (6), whereby it is possible to plot a curve for the effect of g upon the solubility.

The same single effect of pH and hence of g upon the solubility holds even when $b \neq a$, or when the salt is accompanied by excess either of its acid or of its base. This is seen from Eq. (5); for fixed a (or for fixed b) the value of b (or that of a) will fall continuously with increasing pH since α_2 increases continuously with pH. For the general case, $a \neq b$, Eq. (7) becomes

$$H = -K_1/2 + \sqrt{(\quad)^2 + K_1K_2(ab/P - 1)}. \tag{8}$$

With reference, however, to the effect of the specific acid or of the specific base upon the solubility of the salt or to the effect of a upon b and of b upon a at saturation (all for a fixed value of g in the solution) the relation is not simple. If both the base and the acid were weak in respect to their second ionization constants, the solubility would pass through a minimum both in excess of the base and in excess of the acid. For the case under consideration, in which the base $M(OH)_2$ is assumed strong, excess of the base decreases the solubility, or a continually decreases as b is increased above $b = a$. Excess of the acid, however, causes the solubility, now measured as "b," to pass through a minimum.

Thus, by elimination of a from Eqs. (4) and (5) we obtain as the condition of saturation for specified values of b and g

$$H_{\text{req}} = (b^2 - P - gb/2)/(P/K_2 - b) \pm \sqrt{(\quad)^2 - bW/(P/K_2 - b)}. \tag{9}$$

If CO_2 is passed into a solution of $Ca(OH)_2$ at the molarity b, let us say with $g = 0$, Eq. (9) gives two values of H for saturation with respect to the normal salt $CaCO_3$, one for its precipitation and one for its redissolving. Eqs. (4–9) therefore allow the calculation of the solubility curve of $CaCO_3$ in the system $Ca(OH)_2$–CO_2–H_2O: curve ef of Fig. 20–1. For a given value of b, H_9 is used in Eq. (5) to find a; for given a, $P/a\alpha_2$ could be substituted for b

in Eq. (4) to calculate H, and then b as $P/a\alpha_2$. If $g \neq 0$, the same calculations give the relations for saturation in the presence of foreign strong acid or strong base: the relation between a and b for specified g, or the value of g for saturation for specified values of a and b.

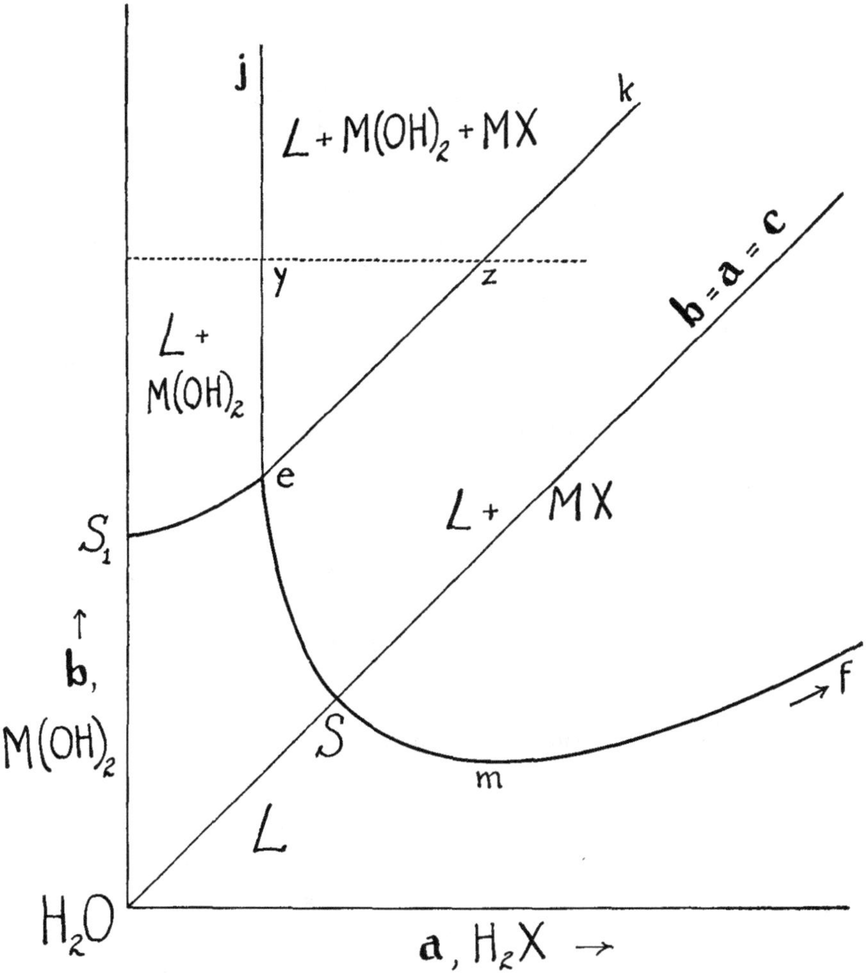

Fig. 20–1. Normal salt of dibasic acid with congruent solubility.

If $b > P/K_2$, Eq. (9) gives one and only one value of H, marking the appearance of the precipitate, which then never redissolves. On the other hand, the equation has no real positive root, if b is below a certain minimum value, to be called b_{min}. In the intermediate range, with b between b_{min} and $b = P/K_2$, there are two values of H for each value of b. The lower value, H_- of Eq. (9), determines the appearance of the precipitate, and the higher value, or H_+, its redissolving upon continued increase of a or addition of CO_2.

The corresponding value of a for given b follows in each case from Eq. (5).

The value of b_{min} is determined, according to Eq. (9), by the two conditions:

$$(b^2 - P - gb/2)^2 + b^2W - bPW/K_2 = 0, \tag{10}$$

and

$$b^2 - P - gb/2 > 0, \tag{11}$$

with b_{min} increasing with increasing g. The value of H at b_{min} is $\sqrt{bW/(P/K_2 - b)}$. For the actual case of $CaCO_3$, with $P = 5 \times 10^{-9}$, $K_1 = 4 \times 10^{-7}$, and $K_2 = 5 \times 10^{-11}$, Eq. (10) gives $b = 1.28 \times 10^{-4}$ and $b = 2.1 \times 10^{-5}$, for $g = 0$; since b_{min} must be greater than $\sqrt{P}$, the required value is $b_{min} = 1.28 \times 10^{-4}$. If a solution of $Ca(OH)_2$ of this concentration is treated with CO_2, saturation with $CaCO_3$ is just touched when, from Eq. (9), $H = 1.13 \times 10^{-10}$ and, from Eq. (5), $a = 1.28 \times 10^{-4}$; this would schematically be point m in Fig. 20–1. [The numerical identity of a and b at this point for the system $Ca(OH)_2$–CO_2–H_2O is not an exact or necessary relation. In general b_{min} will be smaller than the solubility of the normal salt in pure water, or point S, with the corresponding value of a larger than b_{min}. In the present case b_{min} is in its numerical value hardly distinguishable from the solubility of $CaCO_3$ (see Chapter XVIII, Section B2) and hence the ratio b/a at point m is very close to 1.] As b increases above b_{min}, the two values of H given by Eq. (9) diverge on either side of $H = 1.13 \times 10^{-10}$. For example: if $\mathbf{b} = 1.4 \times 10^{-4}$, the precipitate appears at $H = 0.605 \times 10^{-10}$ ($a = 7.89 \times 10^{-5}$) and redissolves at $H = 2.315 \times 10^{-10}$ ($a = 2.01 \times 10^{-4}$); if $\mathbf{b} = 10^{-3}$, $CaCO_3$ appears when $H = 5 \times 10^{-12}$ ($a = 5.5 \times 10^{-6}$) and it redissolves when $H = 1.99 \times 10^{-8}$ ($a = 2.09 \times 10^{-3}$). During the course of the addition of CO_2 the solubility of the salt, measured as "b," passes through a minimum, which is of course the value fixed by Eqs. (10) and (11).

For $MgCO_3$, with $P = 5 \times 10^{-5}$, the minimum value of b for Eq. (9) is 7.67×10^{-3}, which happens again to be numerically indistinguishable from the solubility of the carbonate as calculated from its solubility product. But such a concentration of $Mg(OH)_2$ in pure water is not possible at equilibrium since its solubility is only 1.36×10^{-4}. The significance of these relations will be brought out later in connection with a schematic phase diagram (Section B2).

While the solution is saturated with $CaCO_3$ as sole precipitate, or while the total composition of the system, with coordinates $\mathbf{a}$ and $\mathbf{b}$, falls in the region $L + MX$ of Fig. 20–1, the condition, again with $\mathbf{h} = \mathbf{a} - \mathbf{b} = a - b$, is $P = a(a - \mathbf{h})\alpha_2 = b(b + \mathbf{h})\alpha_2$. This may be used with Eq. (4) to eliminate a and b, for an expression defining H in terms of $\mathbf{h}$, g, and constants, which we shall here simply refer to as

$$H_{MX} = f(\mathbf{h}, g, \text{constants}). \tag{12}$$

2. Reaction of MCl_2 (at **b**) With H_2X (at **a**). The numerical values considered in the preceding section apply at $g = 0$. The value of $b_{\min}$ increases with the algebraic value of g, as seen in Eqs. (10) and (11). Moreover, for any given value of **b** there will be a maximum (algebraic) value of g above which precipitation is impossible unless $\mathbf{b} > P/K_2$. For $\mathbf{b} > P/K_2$ precipitation is possible for any value of g; but otherwise g must be less than $2[\mathbf{b} - P/\mathbf{b} - \sqrt{W(P/K_2 - \mathbf{b})/\mathbf{b}}]$. This means that the salt MX can not be precipitated from pure MCl_2 solution, in which $g = 2\mathbf{b}$, by addition of H_2X, unless $\mathbf{b} > P/K_2$. Combination of Eqs. (4) and (5), with $b = \mathbf{b}$ and $g = 2\mathbf{b}$, shows that the precipitation of MCO_3 requires a value of H defined by $H^2(\mathbf{b} - P/K_2) - 2HP - \mathbf{b}W = 0$. This expression requires $\mathbf{b} > P/K_2$ and then gives only one positive root. For the actual case of $CaCl_2$, it would require the impossible concentration of $\mathbf{b} = 5 \times 10^{-9}/5 \times 10^{-11} = 100$. At any rate, if a precipitate forms it never redissolves on continued addition of H_2X. This follows also according to Eq. III(64), where it was seen that the concentration of the $X^{=}$ ion rises continuously with increase of a, unless there is present in the system an excess of base over acid, with which the H_2X may react. In MCl_2 the number of equivalents of strong acid equals that of strong base.

For $\mathbf{b} < P/K_2$, then, the "$CaCl_2$" solution must not be pure, for precipitation of $CaCO_3$ with CO_2. The salt may be precipitated (and redissolved) only if the normality of the $Ca(OH)_2$ exceeds that of the HCl by the quantity $2P/\mathbf{b} + 2\sqrt{W(P/K_2 - \mathbf{b})/\mathbf{b}}$, and then the relations are defined by Eqs. (9), (4), and (5). Also, if the $CaCl_2$ solution contains some foreign base at concentration b', so that $D = \mathbf{a}(\alpha_1 + 2\alpha_2) - b'\beta'$ for electroneutrality, then $H^2(\mathbf{b} - P/K_2) + H(\mathbf{b}b'\beta' - 2P) - \mathbf{b}W = 0$. Now two positive roots are possible if $b'\beta' > 2P/\mathbf{b}$ together with $\mathbf{b}K_2 < P$; and the precipitate will redissolve on continued addition of CO_2. When the values of H fixed by this expression (a simple quadratic if the added base is strong and β' is an integer) are substituted in the equation of electroneutrality, we may calculate the values of **a** required for the appearance and disappearance of the precipitate.

Finally, since only K_2 is involved, these various relations hold whether or not the dibasic acid is "strong in respect to K_1." Whether or not sulfuric acid is strong in K_1, therefore, and whatever the value of g in the solution, $BaSO_4$, according to these ideal equations, can not redissolve in concentrated H_2SO_4 if $\mathbf{b} > P/K_2$, in which **b** is the concentration of Ba^{++} to be kept in solution. "Ideally," this would be a very small concentration indeed, if $P = 10^{-10}$ and $K_2 = 10^{-2}$. The fact that $BaSO_4$ has a considerable solubility in concentrated H_2SO_4 shows to what extent such "constants" fail to apply when we leave the domain of truly dilute aqueous solutions. While the usual "explanation" of this dissolving of $BaSO_4$ in concentrated H_2SO_4 is that $[SO_4^{=}]$ decreases at the expense of $[HSO_4^{-}]$ as the concentration of the acid increases, it is clear that the effect is rather due essentially to great changes in the values of the "constants."

3. WINKLER'S METHOD[1] FOR TITRATIONS INVOLVING ALKALI CARBONATES OR CARBONIC ACID. Strong base at concentration b_s, in presence of alkali carbonate, or Na_2CO_3, at concentration **a**, may be titrated with strong acid, a_s, after addition of **b** moles per liter of $BaCl_2$, with $\mathbf{b} > \mathbf{a}$. The titration is made in the presence of the precipitated $BaCO_3$ with the use of an indicator like phenolphthalein. The equivalent point, $a_s = b_s$, is a suspension of $BaCO_3$ in a solution of NaCl and $BaCl_2$, and it is slightly alkaline since $Ba(OH)_2$ is a strong base. The error involved in titrating to the point $H_{\text{end}} = 10^{-9}$ will be negative but extremely small, while the sharpness or feasibility of the titration is fairly good.

For the estimation of the error when the titration is stopped at some specified value of H such as $H_{\text{end}} = 10^{-9}$, we write first the general condition of electroneutrality for saturation with respect to $BaCO_3$, or

$$D = a_s - b_s + 2(\mathbf{b} - \mathbf{a}) - 2[Ba^{++}] + [HCO_3^-] + 2[CO_3^=] \tag{13}$$

$$= a_s - b_s + 2(\mathbf{b} - \mathbf{a}) - 2P/[CO_3^=] + [CO_3^=](2 + H/K_2), \tag{14}$$

in which P is the solubility product of $BaCO_3$. For the exact expression for $[CO_3^=]$ we note that with $BaCO_3$ as the sole precipitate,

$$\mathbf{b} - \mathbf{a} = [Ba^{++}] - [CO_3^=]/\alpha_2 = [Ba^{++}] - [HCO_3^-] - [CO_3^=] - [CO_2], \tag{15}$$

and that

$$[CO_3^=] = (\mathbf{a} - \mathbf{b})\alpha_2/2 + \sqrt{(\quad)^2 + P\alpha_2}. \tag{16}$$

If the value of H_{end} is somewhere between K_1 and K_2, $\alpha_2 \cong K_2/H$, and $[CO_3^=] \cong P/(\mathbf{b} - \mathbf{a})$; i.e. $[Ba^{++}] \cong (\mathbf{b} - \mathbf{a})$. Hence if the concentration of strong acid required in the titration is $(a_s)_{\text{end}}$, then, from Eq. (14),

$$(b_s)_{\text{end}} \cong (a_s)_{\text{end}} + P(H_{\text{end}} + 2K_2)/K_2(\mathbf{b} - \mathbf{a}) + W/H_{\text{end}}. \tag{17}$$

For $\mathbf{a} = 0.1$, $\mathbf{b} = 0.2$, $P = 5 \times 10^{-9}$ and $K_2 = 5 \times 10^{-11}$, the error is only -0.011 per cent if the titration is stopped at $H_{\text{end}} = 10^{-9}$ and if b_s at the end-point is 0.1.

An approximate evaluation of the feasibility of the titration may be made as follows. If $+p$ is the percentage of titration beyond the equivalent point or if $a_s = b_s + pb_s/100$, Eq. (13) combined with Eq. (15), becomes $D = -[HCO_3^-] - 2[CO_2] + pb_s/100$. Combined with the definitions of K_1, K_2, and P, this gives $D = -HP/K_2[Ba^{++}] - 2H^2P/K_1K_2[Ba^{++}] + pb_s/100$, or, with $[Ba^{++}] \cong \mathbf{b} - \mathbf{a}$,

$$2H^3P/K_1K_2(\mathbf{b} - \mathbf{a}) + H^2[1 + P/K_2(\mathbf{b} - \mathbf{a})] - Hpb_s/100 - W \cong 0. \tag{18}$$

(1) C. Winkler, *Practische Uebungen in der Maassanalyse*, Leipzig, 3rd ed., 1902; see F. P. Treadwell and W. T. Hall, *Analytical Chemistry*, Wiley, N.Y., 1942; vol. II, p. 498.

For $b_s = 0.1$ this gives the following points for the titration curve through the equivalent point (with $K_1 = 4 \times 10^{-7}$ and with $\mathbf{b} - \mathbf{a} = 0.1$):

p	H	pH
-0.1	10^{-10}	10.00
0	3.14×10^{-9}	8.50
$+0.1$	7.32×10^{-8}	7.14

Similarly, CO_2 at concentration **a** may be determined by treatment with $Ba(OH)_2$, at **b** moles per liter, followed by titration of the excess of $Ba(OH)_2$ in presence of the precipitated $BaCO_3$ with phenolphthalein as indicator (we assume $H_{\text{end}} = 10^{-9}$). The equivalent point, at $a_s = 2(\mathbf{b} - \mathbf{a})$, is again a suspension of $BaCO_3$ in aq. $BaCl_2$. Since $[CO_3^{=}]$ is still given, through Eq. (16), as approximately $P/(\mathbf{b} - \mathbf{a})$, the relation between **a** and the value of a_s required to reach the end-point is now

$$\mathbf{a} \cong \mathbf{b} - (a_s)_{\text{end}}/2 - P(H_{\text{end}} + 2K_2)/2K_2(\mathbf{b} - \mathbf{a}) - W/2H_{\text{end}}, \quad (19)$$

so that the error becomes

$$\% \text{ Error} = + (100/2\mathbf{a})[P(H_{\text{end}} + 2K_2)/K_2(\mathbf{b} - \mathbf{a}) + W/H_{\text{end}}]. \quad (20)$$

Relatively this is a larger error than that in the preceding case since the concentrations, **a** and **b**, are usually small. If 0.3 milli-mole of CO_2 is being determined, with 0.6 milli-mole of $Ba(OH)_2$ used for precipitation and with a final volume of 200ml, the error is $+ 2.8$ per cent, for $H_{\text{end}} = 10^{-9}$.

For the feasibility, the equation for H as function of the percentage of titration is the same as Eq. (18), with $2(\mathbf{b} - \mathbf{a})$ in place of b_s. The values of H as $f(p)$ there calculated therefore apply in the present case for $(\mathbf{b} - \mathbf{a}) = 0.05$.

Note: the type of error here calculated is that called Type 1 in Chapter VII, in which either the concentration of the indicator is considered negligible or the end-point is established potentiometrically at the specified value of H.

4. Separation by Precipitation. We shall consider the separation of two cations, M^{++} and N^{++}, by precipitation with the anion of H_2X. Given a solution containing the two chlorides, MCl_2 and NCl_2, at concentration c_1 and c_2 respectively, to be treated with H_2X (or with its salt Na_2X) at concentration a, the problem is to find the range of H for the separation. With P_1 as the solubility product of MX and P_2 for NX, and with $P_1 < P_2$, we find H to leave p per cent of M^{++} in solution, or "H_{pc_1}," from the relation $P_1 = a\alpha_2 pc_1/100$ and hence from Eq. (8) with $P = P_1$ and $b = pc_1/100$; we also find H to saturate with NX, or H_{c_2}, from $P_2 = c_2 a\alpha_2$ or from Eq. (8) with $P = P_2$ and $b = c_2$. If $H_{pc_1} > H_{c_2}$ and if the range is practical, the separation is practical.

a. Separation of cadmium from manganese. In the familiar separation of Cd^{++} from Mn^{++} with H_2S a value of H in the desired range is commonly maintained by regulation of the concentration of "free" strong acid. To leave

p per cent of the Cd^{++} in solution, in precipitation with H_2X, a_s must be smaller than the value in

$$D = a_s + a(\alpha_1 + 2\alpha_2) + 2c_1(1 - p/100), \tag{21}$$

used with H_{pc_1}. If a, the total concentration of H_2X in solution, is assumed constant despite the change in the concentration of the hydrogen ion (which is approximately true when an acid solution is kept saturated with H_2S), then the MnX begins to precipitate when $a\alpha_2 = P_2/c_2$ and $[Cd^{++}] = P_1/a\alpha_2 = c_2P_1/P_2$ or

$$D = a_s + HP_2/c_2K_2 + 2P_2/c_2 + 2(c_1 - c_2P_1/P_2), \tag{22}$$

since $\alpha_1 = \alpha_2H/K_2$. With H_{c_2}, this gives the value of a_s for the beginning of saturation with MnX. The separation therefore requires a value of a_s between that of Eq. (21) and that of Eq. (22).

b. Separation of barium from strontium. In the case of the usual separation of Ba^{++} from Sr^{++} by precipitation of $BaCrO_4$ with K_2CrO_4, a value of H in the desired range is fixed and maintained by a buffered solution (acetic acid and ammonium acetate). Here P_1 ($= 2 \times 10^{-10}$) is the solubility product of $BaCrO_4$ and P_2 ($= 3.6 \times 10^{-5}$) is that of $SrCrO_4$; K_1, K_2, and K are the equilibrium constants for chromic acid discussed in Chapter XV. The value of H required to leave p per cent of the barium in solution is found when the value of $P_1(100)/pc_1$, for $[CrO_4^=]$, is equated with the expression for $[CrO_4^=]$ obtainable from Eq. XV(40), through the relation $[CrO_4^=] = [HCrO_4^-]K_2/H$. The result is

$$H = -K_2/(4xK + 2K_2/K_1) + \sqrt{(\quad)^2 + K_2^2(a - x)/(2x^2K + xK_2/K_1)}, \tag{23}$$

in which $x = 100P_1/pc_1$. This same equation, with $x = P_2/c_2$, gives the value of H at which $SrCrO_4$ begins to precipitate. We thereby calculate that for a solution in which $[Ba^{++}] = [Sr^{++}] = 0.01$ and in which the analytical concentration of K_2CrO_4 is 0.1, one per cent of the barium remains in solution at $H = 4.43 \times 10^{-3}$ and one tenth per cent at $H = 4.47 \times 10^{-4}$; $SrCrO_4$ does not begin to precipitate until (now with $[K_2CrO_4] = 0.09$) H is decreased to 2.27×10^{-6} [(2)].

5. Separation by Dissolving. We now consider the separation by dissolving of one of two precipitates (MX and NX) sufficient to give respectively the concentrations c_1 and c_2 when dissolved; we assume again that MX is the less soluble. If the mixture of the two solids is suspended in water and treated with strong acid, NX dissolves completely when $[M^{++}] = c_2P_1/P_2$ and $c_2^2(1 + P_1/P_2)\alpha_2 = P_2$, or when H is given by Eq. (8) with $P = P_2$ and $ab = c_2^2(1 + P_1/P_2)$. With this value of H, a_s required to dissolve the NX is given by

$$D = a_s + c_2(1 + P_1/P_2)(\alpha_1 + 2\alpha_2 - 2). \tag{24}$$

(2) In this chapter, the equilibrium constants of chromic acid are used with the rounded values $K_1 = 0.2$, $K_2 = 3 \times 10^{-7}$ and $K = 40$.

When p per cent of the solid MX has dissolved, $(pc_1/100)(c_2 + pc_1/100)\alpha_2 = P_1$, with H therefore from Eq. (8) used with $ab = (pc_1/100)(c_2 + pc_1/100)$ and $P = P_1$. The required value of a_s is then given, with this value of H, by

$$D = a_s + (c_2 + pc_1/100)(\alpha_1 + 2\alpha_2 - 2). \qquad (25)$$

If the value of $[M^{++}]$ when all the NX has just dissolved, or c_2P_1/P_2, is very small and if the difference between $(a_s)_{24}$ and $(a_s)_{25}$ for very small p is large and practical, the separation is feasible. The second precipitate, MX, dissolves completely when $c_1(c_1 + c_2)\alpha_2 = P_1$ or when H is given by Eq. (8) with $P = P_1$ and $ab = c_1(c_1 + c_2)$, and when $D = a_s + (c_1 + c_2)(\alpha_1 + 2\alpha_2 - 2)$.

6. Hinman's Volumetric Method for Sulfate.[(3)] A solution of alkali sulfate, at $c_1 \simeq 10^{-3}$, is treated with a solution of pure $BaCrO_4$ in aq. HCl; after boiling it is "neutralized" with excess of $CaCO_3$ (Hinman used ammonia originally), cooled, and filtered. Iodometric determination of total dissolved chromate (at concentration c_2) should then measure the original sulfate, mole per mole. For zero error the analytical concentration of chromate should be found to be c_1 moles per liter if the process is assumed to involve no change in volume.

When the mixture is filtered the solution is saturated with both solids so that $[Ba^{++}][SO_4^{=}] = P_1 = 10^{-10}$ for $BaSO_4$, and $[Ba^{++}][CrO_4^{=}] = P_2 = 2 \times 10^{-10}$ for $BaCrO_4$. For the low final concentration of sulfate it will be assumed that none of it is present as HSO_4^-. For chromic acid the effect of K_1 we have seen is small at $H \ll K_1$, so that we shall take $[CrO_4^{=}]$ to be given, from Eq. xv(73) with c_2 in place of a, as

$$[CrO_4^{=}] = [HCrO_4^-]K_2/H = (K_2/H)[-(H + K_2)/4HK + \sqrt{(\quad)^2 + c_2/2K}]. \qquad (26)$$

The number of moles of sulfate entering the precipitate, or $c_1 - [SO_4^{=}]$, equals the number of moles of dissolved chromate in excess of dissolved barium, or $c_2 - [Ba^{++}]$. The error therefore is

$$c_2 - c_1 = [Ba^{++}] - [SO_4^{=}] = P_2/[CrO_4^{=}] - P_1[CrO_4^{=}]/P_2. \qquad (27)$$

The error is zero when $[CrO_4^{=}] = P_2/\sqrt{P_1}$. On equating this with Eq. (26) for $[CrO_4^{=}]$ and setting $c_2 = c_1$, we obtain the required value of H for zero error as

$$H = -K_2\sqrt{P_1}/4KP_2 + \sqrt{(\quad)^2 + K_2^2P_1(c_1 - P_2/\sqrt{P_1})/2KP_2^2}, \qquad (28)$$

which is possible only if $c_1 > P_2/\sqrt{P_1}$. If $c_1 = 10^{-3}$, this gives $H_{req} = 1.37 \times 10^{-5}$. If the solution is actually "neutralized" then to $H = 10^{-7}$ before filtration, for this value of c_1, there will be a negative error. For

(3) C. W. Hinman, *Am. J. Sci. and Arts*, **114,** 478 (1877). See Treadwell and Hall, *op. cit.*, Ref. xx–1, Vol. II, p. 657.

the calculation of the error for a value of H other than that fixed by Eq. (28), we introduce Eq. (26) into Eq. (27) and solve for c_2 in terms of c_1. We first obtain

$$c_2 - c_1 = HP_2/K_2[\mathrm{HCrO_4^-}] - P_1K_2[\mathrm{HCrO_4^-}]/HP_2. \tag{29}$$

With $H \cong 10^{-7}$ the first term is clearly negligible compared to the second (which essentially constitutes the error) so that it will be dropped. The result is a quadratic in c_2, or

$$c_2^2 - c_2\left[2c_1 + \frac{P_1K_2(HP_2 + K_2P_2 + K_2P_1)}{2H^2KP_2^2}\right] + \left[c_1^2 + \frac{c_1P_1K_2(H + K_2)}{2H^2KP_2}\right] = 0. \tag{30}$$

With $c_1 = 0.001$ and with $H = 10^{-7}$, this gives $c_2 = 0.000718$, which represents a considerable error. This assumes of course that equilibrium is established upon neutralization, before filtration.

It might appear then as though there were something wrong with the values or with the interpretation of the constants quoted or perhaps with the assumption of pure solid phases rather than solid solutions, for the method is reported to be satisfactory even when the filtration is performed in presence of excess of ammonia to overcome the interference on the part of salts of iron, nickel, and zinc. The alternative is that in the actual analysis complete solubility equilibrium is not established upon neutralization of the acid solution. In the original acid solution, from which no $BaCrO_4$ has yet precipitated, the sulfate is practically completely precipitated. This is so as long as the original concentration of total chromate, "c_2," is high compared to c_1 so that $[Ba^{++}]$ is also high, or $[Ba^{++}] \cong$ "c_2" $- c_1$. The establishment of equilibrium upon neutralization then requires the precipitation of the excess of chromate and simultaneously the redissolving of some of the barium sulfate. It is this process that finally leads to $c_2 < c_1$ in the neutralized solution, at equilibrium. But if the redissolving of barium sulfate is too slow or if the barium sulfate becomes coated with the precipitating barium chromate, a nonequilibrium relation may persist, with a final value of c_2 almost equal to c_1 and theoretically possibly slightly larger. If in other words the solid $BaSO_4$ is now ignored as a reactant, we have merely $[Ba^{++}][CrO_4^=] = (c_2 - c_1)[CrO_4^=] = P_2$, which combined with Eq. (26) for $[CrO_4^=]$ gives

$$c_2^3 - 2c_2^2c_1 + c_2[c_1^2 - P_2(1 + H/K_2)] + [c_1P_2(1 + H/K_2) - 2KH^2P_2^2/K_2^2] = 0. \tag{31}$$

Here the last term is certainly negligible at $H = 10^{-7}$; hence since $c_1^2 \gg P_2$, c_2 is almost exactly equal to c_1.

These conditions, leading to Eq. (31), would hold therefore if the $BaSO_4$ is filtered off while the solution is acid and if the excess of $BaCrO_4$ is then filtered off after neutralization of the filtrate. If the two solids are removed together then, the success of the procedure depends on the nonattainment of theoretical equilibrium in the neutral solution. Presumably the error increases with time and stirring before filtration, with consequent approach toward equilibrium. The situation is similar to that in the Volhard method for the determination of chloride. The original procedure of Volhard[(4)] was the titration of excess of $AgNO_3$ with KCNS in the presence of the precipitated AgCl. If the titration is made very rapidly, only a negligible amount of the solid AgCl redissolves as compared to that demanded by the equilibrium relations. The error then increases with time and stirring during the titration. The procedure was therefore first modified by filtration of the AgCl before titration[(5)], and later by the use of nitrobenzene to coat the precipitated AgCl before titration[(6)]. Without such precautions the equilibrium error depends both as to sign and as to magnitude upon the concentration of ferric ion used as indicator, and conditions for minimizing the error have been elaborated[(7)].

B. Saturation With Both $M(OH)_2$ and MX

1. Possible Solids and Invariant Solutions in System $M(OH)_2$–H_2X–H_2O. If both the base and the acid are dibasic and weak, we may expect as possible solid phases in addition to the acid and base themselves not only the normal salt but also basic and acid salts. The solubility product of the basic salt, $M_2(OH)_2X$ or $(MOH)_2X$, may be written in two ways:

$$[MOH^+]^2[X^=] = P_B, \tag{32a}$$

$$[M^{++}]^2[OH^-]^2[X^=] = P'_B = P_B/(K'_2)^2, \tag{32b}$$

in which K'_2 is the second ionization constant of the base. For the acid salt,

$$[M^{++}][HX^-]^2 = P_A, \tag{33a}$$

$$[M^{++}][H^+]^2[X^=]^2 = P'_A = P_A/K_2, \tag{33b}$$

in which K_2 is the second ionization constant of the acid. The two forms of each constant are interrelated then so that the equilibrium measurements on the saturated aqueous solution can not distinguish between $(MOH)_2X$, $M_2(OH)_2X$, and even $M(OH)_2{\cdot}MX$, as "formulas" for the basic salt, or between $M(HX)_2$, MH_2X_2, and $MX{\cdot}H_2X$ as "formulas" for the acid salt.

Since the solubility product of the acid salt involves at least implicitly the

(4) J. Volhard, *J. prakt. Chem.*, **117,** 217 (1874).
(5) M. A. Rosanoff and A. E. Hill, *J. Am. Chem. Soc.*, **29**, 269 (1907).
(6) J. P. Caldwell and H. V. Moyer, *Anal. Chem.*, **7,** 38 (1935).
(7) E. H. Swift, G. M. Arcand, R. Lutwack, and D. J. Meier, *Anal. Chem.*, **22,** 306 (1950).

species HX^- and the ionization constant K_2, such a salt is not expected to exist unless the acid is weak in respect to this constant. The same applies for a basic salt; such a salt implies the species MOH^+ and the finite constant K_2', and it is therefore not expected to form if the base is completely strong. A compound such as $Mg_2(OH)_2CO_3$, then, would not be a "basic salt" in this strictly defined sense, if $Mg(OH)_2$ were a strong diacid base with both K_1' and $K_2' = \infty$. Its existence on the other hand does not necessarily imply the weakness of the base. Somewhat analogously, the existence of $KIO_3{\cdot}HIO_3$ and $Ag_2SO_4{\cdot}2H_2SO_4$ does not imply the dibasicity of iodic acid or the tribasicity of sulfuric acid. These compounds together with "$Mg_2(OH)_2CO_3$" (if $Mg(OH)_2$ is "strong") would be classified as "addition compounds" or "double compounds," possibly complex. The solubility product of $Mg_2(OH)_2CO_3$ if $Mg(OH)_2$ is strong could not be written as in Eq. (32a) but only as in Eq. (32b).

In the form of P_B', or Eq. (32b), moreover, the ion-product resembles in form that of a mixture of $M(OH)_2$ and MX but differs from it in value. For the mixture, $P_{\text{mix}} = [M^{++}]^2[OH^-]^2[X^=] = PP_1$, in which P is the solubility product of the normal salt and P_1 that of the base. As pointed out for the analogous but simpler situation in Fig. 17–5, the ion-product P_B' must be smaller than P_{mix} or PP_1 for the intermediate compound, whether or not it is strictly a basic salt, to exist as a stable solid phase in contact with aqueous solution. Similar remarks apply for the ion-product of the salt $M(HX)_2$ as compared to that of the mixture $MX + H_2X$.

With five possible solid phases in the system acid–base–water, there will be four possibilities of solutions saturated with two adjacent solids: (1) base and basic salt; (2) basic salt and normal salt; (3) normal salt and acid salt; (4) acid salt and acid. For any such solution there will be an ion-product constant for saturation with the mixture of two solids expressible in various ways by multiplying or dividing the two individual solubility products involved. For a solution for example saturated with the normal salt and the acid salt: $[M^{++}]^2[HX^-]^2[X^=] = P_AP$, $[M^{++}]^2[H^+]^2[X^=]^3 = P_APK_2^2$, $[HX^-]^2/[X^=] = P_A/P$, $[H^+][HX^-] = P_AK_2/P$, $[H^+]^2[X^=] = P_AK_2^2/P$, $[M^{++}]/[H^+]^2 = P^2/P_AK_2^2$. If the system contains only the acid, base, and water, it is ternary, and this solution saturated with two solids is isothermally invariant as a condensed system, with $H = f(\text{constants})$ and with fixed values of the concentrations, a and b. In the presence of the salt of foreign strong acid and strong base such as $NaNO_3$, H continues to have the same invariant value if the solution is ideal; the $NaNO_3$ merely affects the activity coefficients. Hence the composition of the solution is changed only little by such foreign salt. If the solution contains *net* foreign strong acid or any foreign weak acid or base, the various ion-products are still constant for saturation, but H now depends on the concentration of the foreign solutes and hence the concentrations of a and b (of H_2X and $M(OH)_2$ respectively) in the solution are variable.

As a second example we note that in the absence of a basic salt the solution saturated with solid base, $M(OH)_2$, and solid normal salt, MX, would involve the following ion-product constants, one of which has already been mentioned:

$$\text{(a)}\ [M^{++}]^2[OH^-]^2[X^=] = PP_1; \qquad \text{(b)}\ [X^=]/[OH^-]^2 = P/P_1. \qquad (34)$$

2. The Equilibrium Liquid, Solid Base and Solid Normal Salt.

a. Congruent solubility of normal salt; $CaCO_3$. This case, presumably exemplified by the system $Ca(OH)_2$–CO_2–H_2O, is represented schematically in Fig. 20–1, which has already been discussed in part. Solution e is the solution saturated with both solid $M(OH)_2$ and solid MX. For a given value of b along the curve S_1e, which represents saturation with $M(OH)_2$, we have $OH = \sqrt{P_1/b}$, and the corresponding value of a is obtained from Eq. (4); for a given value of a, H is first calculated through Eq. (4) with $b = P_1/(OH)^2$ and then b as $P_1/(OH)^2$. The curve for saturation with salt, curve eSmf, was discussed under Eqs. (5–12).

For the composition of solution e, the combination of the two solubility products, $P = [M^{++}][X^=] = ba\alpha_2$ and $P_1 = [M^{++}][OH^-]^2 = b(OH)^2$, with the general Eq. (4) gives

$$2H^4P_1/W^2 + H^3 - H^2g - HW(1 + PW/P_1K_2) - 2PW^2/P_1 = 0. \qquad (35)$$

If $g = 0$, this refers either to point e of Fig. 20–1 for the pure ternary system or to a suspension of $M(OH)_2$ + MX in the presence of some foreign strong-strong salt such as $NaNO_3$. With H fixed by Eq. (35), $b = P_1H^2/W^2$ and $a = P/b\alpha_2 = PW^2(H^2 + HK_1 + K_1K_2)/H^2K_1K_2P_1$; again these give the coordinates of point e in Fig. 20–1 if $g = 0$ in Eq. (35).

For the actual case of the calcium compounds, point e is practically indistinguishable numerically from S_1. At any rate, if a suspension of $Ca(OH)_2$, with $\mathbf{b} < b_e$, is treated with CO_2, the solid would dissolve when the curve S_1e is reached and the carbonate would then precipitate on the curve eSm. But for higher values of **b**, above b_e, the solid $Ca(OH)_2$ becomes mixed with the carbonate when the vertical line ej is reached, as at point y, with $\mathbf{a}_{req} = a_e$; the further considerations would be the same as those explained under Fig. 19–1. The pertinent numerical data for the calcium system, with $K_1 = 4 \times 10^{-7}$, $K_2 = 5 \times 10^{-11}$, $P_1 = 8 \times 10^{-6}$, and $P = 5 \times 10^{-9}$, are given in Table 20–1.

b. Incongruent solubility of the normal salt; $MgCO_3$. These relations would apply to the system $Mg(OH)_2$–CO_2–H_2O if there were no precipitation of basic carbonate; Fig. 20–2. Here point e, the intersection of the two solubility curves, is on the right of the diagonal, with $a > b$ in the solution; the solubility of the carbonate in pure water, or S_{inc}, is metastable, lying in the field for saturation with respect to the hydroxide. The minimum m on the solubility curve of the carbonate may lie on either side of the invariant solution e. Addition of CO_2 to a suspension of $Mg(OH)_2$ merely causes the dissolving of the hydroxide, on curve S_1e, if $\mathbf{b} < b_e$, and the carbonate never subsequently precipitates unless point m should be on the right of point e. If $\mathbf{b} > b_e$, the carbonate appears, mixed with the hydroxide, when the vertical line ej is reached, and the considerations regarding the points y and z are the same as for Fig. 20–1 (and Fig. 19–1). The solid hydroxide is all consumed at point z, and the carbonate would redissolve if the curve ef is reached, on the right of point m. The numerical values pertaining to the magnesium system, with $P_1 = 10^{-11}$ and $P = 5 \times 10^{-5}$, and with $Mg(OH)_2$

Table 20–1. Numerical Coordinates for Figs. 20–1 and 20–2.

		$Ca(OH)_2$–CO_2–H_2O	$Mg(OH)_2$–CO_2–H_2O With $K_2' = \infty$	$Mg(OH)_2$–CO_2–H_2O With $K_2' = 2.6 \times 10^{-3}$
S_1	H	3.97×10^{-13}	3.69×10^{-11}	3.62×10^{-11}
	b	1.26×10^{-2}	1.36×10^{-4}	1.45×10^{-4}
point e	H	3.97×10^{-13}	3.97×10^{-10}	3.96×10^{-10}
	b	1.26×10^{-2}	0.0158	0.0157
		~ 0	0.0284	0.0285
S (or S_{inc})	H	1.13×10^{-10}	8.76×10^{-12}	1.00×10^{-11}
	b $(= a = c)$	1.28×10^{-4}	7.67×10^{-3}	9.12×10^{-3}
point m		practically same as S	practically same as S_{inc}	

considered as a strong divalent base, are shown in the second column of Table 20–1. With $K_2' = 2.6 \times 10^{-3}$ for $Mg(OH)_2$ however, one calculates the slightly different values in column 3. Here the solubility, S_1, of $Mg(OH)_2$ is calculated through the base analog of Eq. XVII(14) for the value of OH_0, with then $b = P_1/\alpha_2'(OH_0)^2$. The solubility, S_{inc}, of $MgCO_3$ was already calculated in Chapter XVIII. For point e, Eq. (35) is changed to the extent that the coefficient of the term in H^3 becomes $(1 + P_1/K_2'W)$.

3. Reaction of Solid $Ca(OH)_2$, in Excess, With Aq. Na_2CO_3; Production of Caustic Soda. When for an initial Na_2CO_3 concentration **a** the solution is saturated with both $Ca(OH)_2$ and $CaCO_3$, H is defined by Eq. (35) with $g = -2\mathbf{a}$ since the concentration of the foreign strong base NaOH is 2**a**. For an appreciable value of **a** the first two terms of the equation are negligible in the actual calcium system, leaving $OH \cong -P_1/4P + \sqrt{(\ \ )^2 + \mathbf{a}P_1/P}$ or less accurately $OH \cong 2\mathbf{a}[1 - 2\mathbf{a}(2\mathbf{P}/\mathbf{P}_1)/\gamma^2]$

$\cong 2\mathbf{a}[1 - 2\mathbf{a}(0.00125)/\gamma^2]$; here γ is as usual the activity coefficient of a univalent ion. With excess of solid $Ca(OH)_2$ always present therefore, the final concentration of "NaOH" (strictly that of the hydroxyl ion) is equivalent very nearly to the starting concentration of Na_2CO_3 and there is no limit, except that fixed by the solubility of either Na_2CO_3 or NaOH, to the concentration attainable. With a as unprecipitated carbonate, the efficiency is $1 - a/\mathbf{a} = 1 - [CO_3^=](1 + H/K_2 + H^2/K_1K_2)/\mathbf{a} \cong 1 - [CO_3^=]/\mathbf{a} \cong 1 - P(OH)^2/P_1\mathbf{a} \cong 1 - 4\mathbf{a}P/P_1 \cong 1 - \mathbf{a}(0.0025)/\gamma^2$. For constant γ the efficiency thus increases with dilution.

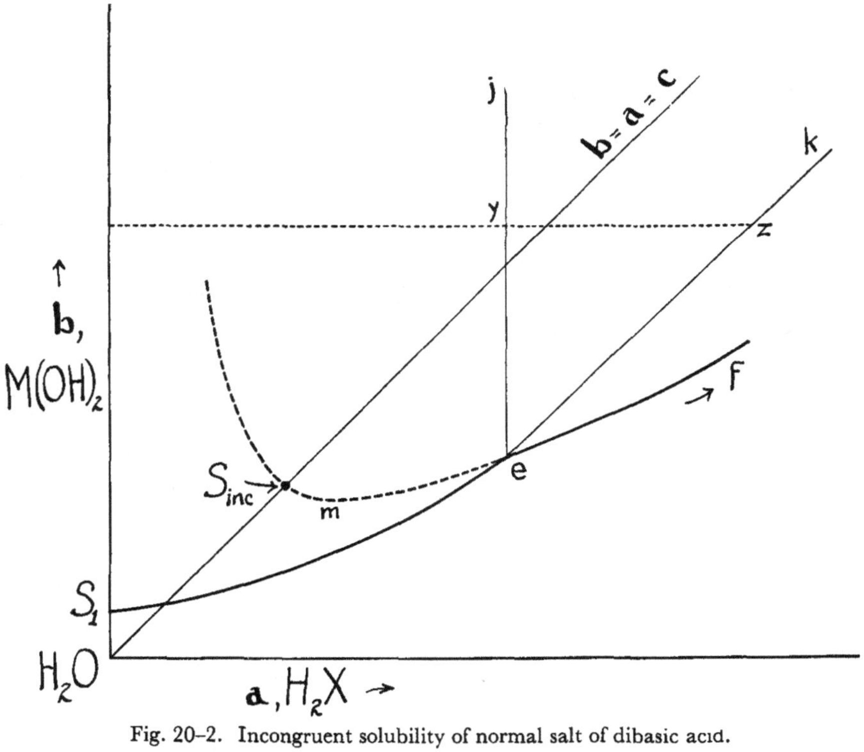

Fig. 20–2. Incongruent solubility of normal salt of dibasic acid.

The composition, including the hydroxyl ion concentration, of the solution in twofold saturation is invariant only for the simple ternary system $Ca(OH)_2$–CO_2–H_2O, namely point e of Fig. 20–1. But the system $Ca(OH)_2$–Na_2CO_3–H_2O, with solid $CaCO_3$ present, is quaternary. The composition *of the solution saturated with the two solids is then no longer invariant* although the ion-products of Eqs. (34a,b) are still constant; and from Eq. (34b) it is seen that the concentration of hydroxyl ion at equilibrium increases with that of the carbonate. If $\mathbf{a} = 1$ (industrially[8] a solution of

(8) E. R. Riegel, *Industrial Chemistry*, Reinhold, N.Y., 1942; pp. 94–95.

10–20 per cent Na_2CO_3 is used, with **a** from 1 to 2) and if the final concentration of NaOH is approximately 2, then $\gamma \cong 0.70$ (Ref. v–3, p. 560). Hence $(\mathbf{a} - a)/\mathbf{a} \cong 1 - 0.005$, at $\mathbf{a} = 1$, representing 99.5 per cent conversion. This is at any rate the theoretical equilibrium figure for the values of the constants used. Industrially lower conversions are reported but the temperatures are variable and equilibrium may not be fully attained.

The minimum value of **a**, the concentration of the salt Na_2X, required for the precipitation of the salt CaX upon a suspension of the base $Ca(OH)_2$ may be found as follows. From the conditions $P_1 = [Ca^{++}][OH^-]^2$ and $P = [Ca^{++}][X^=] = [Ca^{++}]\mathbf{a}\alpha_2$, we have

$$\mathbf{a}_{req} = PW^2(1 + K_1/H + K_1K_2/H^2)/K_1K_2P_1. \tag{36}$$

When this expression for **a** is equated with that for **a** from Eq. (35), in which $g = -2\mathbf{a}$, the critical value of H is given by the equation

$$2H^3P_1/W^2 + H^2 + 2HPW^2/K_1K_2P_1 - W(1 - PW/P_1K_2) = 0, \tag{37}$$

which shows that it is necessary for P_1K_2 to be greater than PW. With H calculated from Eq. (37), $\mathbf{a}_{req}$ then follows from Eq. (36). For the system $Ca(OH)_2 + Na_2CO_3$, Eq. (37) gives $H \cong W/\sqrt[3]{2P_1}$, whereupon Eq. (36) gives $\mathbf{a}_{req} \cong P\sqrt[3]{4/P_1} = 4.0 \times 10^{-7}$. The reaction of $Ca(OH)_2$ with Na_2CO_3 corresponds to the reaction of MOH and NaX discussed for plane II in Chapter XIX, Section D–2*b*3, and $\mathbf{a}_{req}$ therefore corresponds to point $a_{e_{II}}$ or the intersection of the curve e′e″ for saturation with base and normal salt, with the special plane II, Fig. 19–5.

4. Reaction of the Salts, $Na_2X(\mathbf{a})$ and $MCl_2(\mathbf{b})$. We again consider as possible solids only MX, with solubility product P, and $M(OH)_2$, with solubility product P_1. $M(OH)_2$ is a strong base and H_2X is a soluble dibasic acid with ionization constants K_1 and K_2. This system is graphically similar to "case III," the reaction "NaX + M_sNO_3," considered in connection with Figs. 19–2 and 19–5. Considering only the precipitation of $M(OH)_2$ and MX, we are concerned only with the position of the point e_{III} of Fig. 19–5. S_{III} is now "c_{max}" of MCl_2, with $b = P_1/W$ as in Section D–2*c* of Chapter XIX. The general equation of electroneutrality for the system is

$$H - OH = a(\alpha_1 + 2\alpha_2) - 2\mathbf{a} + 2\mathbf{b} - 2b, \tag{38}$$

and if the solution is unsaturated,

$$H - OH = \mathbf{a}(\alpha_1 + 2\alpha_2 - 2). \tag{39}$$

With $P_1 = b(OH)^2$ and $P = ab\alpha_2$ in Eq. (39), the composition at point e_{III} is

$$\sqrt{\mathbf{b}_{crit}} = -PW/K_1K_2\sqrt{P_1} + \sqrt{(\quad)^2 + (P_1/W - P/K_2)}, \tag{40}$$

$$\mathbf{a}_{crit} = P/\mathbf{b}_{crit} + PW/K_2\sqrt{\mathbf{b}_{crit}P_1} + PW^2/P_1K_1K_2. \tag{41}$$

From Eq. (40) we see that unless $P_1K_2 > PW$, the plane defined by $Na_2X + MCl_2$, or "plane III," does not intersect the curve of twofold saturation for the solids $M(OH)_2$ and MX.

a. Addition of MCl_2 *to aq.* Na_2X (**a** *known*).

(1) If $\mathbf{a} < \mathbf{a}_{crit}$, the first precipitate is $M(OH)_2$, which appears when, with $P_1 = \mathbf{b}(OH)^2$ in Eq. (39),

$$x^4W^3 + x^3P_1^{1/2}W^2(2\mathbf{a} + K_1) + x^2P_1W(\mathbf{a}K_1 + K_1K_2 - W) - xP_1^{3/2}K_1W - P_1^2K_1K_2 = 0, \quad (42)$$

in which $x = \sqrt{\mathbf{b}_{req}}$. While $M(OH)_2$ is the sole precipitate, the concentration b may be calculated from Eq. (38) with $a = \mathbf{a}$ and $OH = \sqrt{P_1/b}$.

For MX to appear as a second precipitate the conditions are $\mathbf{a} = P/b\alpha_2$ and $P_1 = b(OH)^2$ so that the required value of H is defined by

$$H^2(\mathbf{a}P_1K_1K_2/PW^2 - 1) - HK_1 - K_1K_2 = 0. \quad (43)$$

Now with $a = \mathbf{a} = P/b\alpha_2$ and $b = P_1/(OH)^2$ and with H_{43}, Eq. (38) gives $\mathbf{b}_{req}$; but according to Eq. (43) the salt MX never precipitates as second solid phase unless $\mathbf{a} > PW^2/P_1K_1K_2$. When the solution is saturated with both solids, H is given by Eq. (35) with $g = 2(\mathbf{b} - \mathbf{a})$; then $b = P_1/(OH)^2$ and $a = P/b\alpha_2$.

As pointed out in connection with Eq. XIX(45) in Section D2c of Chapter XIX, the present Eq. (35) may not be used for the calculation of H for arbitrary values of g ($= 2\mathbf{b} - 2\mathbf{a}$) unless the individual values of **a** and **b** are such as to give saturation with both $M(OH)_2$ and MX together, according to the conditions indicated in connection with Eqs. (38–43).

(2) If $\mathbf{a} > \mathbf{a}_{crit}$, MX appears as the first precipitate when $\mathbf{b} = P/\mathbf{a}\alpha_2$ with the value of H in pure Na_2X solution at concentration **a**. With MX as sole solid phase the composition of the solution travels on the curve $e_{III}f_{III}$ toward point e_{III}, with $H = f(\mathbf{h},g)$, as in "Eq. (12)." At point e_{III}, $M(OH)_2$ appears as a second precipitate, and this requires $P_1K_2 > PW$, as seen under Eq. (40). The value of $\mathbf{b}_{req}$ for $M(OH)_2$ to appear as second solid phase is then $\mathbf{a} + (b_{40} - a_{41})$ since these are the coordinates of the point e_{III}.

b. Addition of Na_2X *to aq.* MCl_2 (**b** *known*). If $P_1K_2 < PW$, the first precipitate must be MX.

(1) If $P_1K_2 > PW$ and $\mathbf{b} > \mathbf{b}_{crit}$, then the first precipitate is $M(OH)_2$, appearing when $\mathbf{a}_{req}$ has the value in Eq. (42) with **b** known. For MX to appear as a second precipitate, the conditions are $a = \mathbf{a} = P/b\alpha_2$ and $P_1 = b(OH)^2$ in Eq. (38) so that H_{req} is given by

$$2H^3P_1/W^2 + H^2 - 2H(\mathbf{b} - PW^2/P_1K_1K_2) - W(1 - PW/P_1K_2) = 0. \quad (44)$$

(This is seen to be the same as Eq. (37) except for the quantity $-\mathbf{b}$ in the third term.)

(2) If $\mathbf{b} < \mathbf{b}_{\text{crit}}$, the first precipitate is MX, appearing when, with $\mathbf{a} = P/\mathbf{b}\alpha_2$ in Eq. (39),

$$2H^3P/\mathbf{b}K_1K_2 + H^2(1 + P/\mathbf{b}K_2) - W = 0. \qquad (45)$$

With continued addition of Na_2X, $H = f(\mathbf{h},g)$ and the composition of the solution travels on the curve $e_{III}f_{III}$ away from point e_{III}. No further precipitate appears unless saturation with an acid salt or with the acid itself is reached.

5. SEPARATION OF FERROUS IRON FROM MAGNESIUM, BY MEANS OF $(NH_4)_2S$. We shall consider a system in which $[FeCl_2] = c_1$, $[MgCl_2] = c_2$, and $[(NH_4)_2S] = a$; the solubility products for FeS, $Mg(OH)_2$, and $Fe(OH)_2$ are P, P_1, and P_1' respectively, and both hydroxides are assumed strong. The separation depends on the range of H between the value required to leave p per cent of the iron in solution and the value required to precipitate any $Mg(OH)_2$. The separation is practical if the solution can easily be buffered within this range. We shall assume that a is maintained constant during the precipitation as by saturation with H_2S. To leave p per cent of the iron unprecipitated, H must satisfy the relation $a\alpha_2pc_1/100 = P$ while to cause $Mg(OH)_2$ to precipitate the requirement is $H = W\sqrt{c_2/P_1}$. $Fe(OH)_2$ can appear as a second precipitate, together with FeS, only when $P/a\alpha_2 = P_1'/(OH)^2$, or when

$$OH = -W/2K_2 + \sqrt{(\quad)^2 + aP_1'/P - W^2/K_1K_2}, \qquad (46)$$

which is possible only if $a > PW^2/K_1K_2P_1'$. With $P = 10^{-19}$, $P_1' = 1.7 \times 10^{-15}$, and $a = 0.1$ or larger, this value of OH is very large. Therefore the significant range for the separation is that between $H = W\sqrt{c_2/P_1}$ and the value of H given by Eq. (8) with $b = pc_1/100$. With $a = 0.1$, $c_1 = 10^{-3}$, $c_2 = 0.1$ and $p = 0.1$, this range, with $P_1 = 10^{-11}$ for $Mg(OH)_2$, lies between pH = 5.00 and pH = 9.00. The requirement is usually met by buffering with NH_3 and NH_4Cl.

C. NORMAL SALT OF DIBASIC ACID AND UNIVALENT STRONG BASE

1. GENERAL RELATIONS. If a is the concentration of the acid H_2X and b that of the specific strong base MOH, then

$$H - OH = a(\alpha_1 + 2\alpha_2) - b + g. \qquad (47)$$

The condition for saturation with respect to the salt M_2X is $P = [M^+]^2[X^=] = b^2a\alpha_2$; and the combination of these expressions gives

$$H = (b^3 - 2P - gb^2)/2(P/K_2 - b^2) \pm \sqrt{(\quad)^2 - b^2W/(P/K_2 - b^2)}. \qquad (48)$$

These are the analogues of Eqs. (4), (5), and (9) for saturation with the normal salt of divalent strong base. Again there is a minimum value of b

below which there is no precipitation upon addition of the acid H_2X; the minimum value of b is determined by the conditions for a real positive root in Eq. (48):

$$(b^3 - 2P - gb^2)^2 + 4b^4W - 4b^2PW/K_2 = 0, \quad (49)$$

together with $b^3 - 2P - gb^2 > 0$, corresponding to Eqs. (10) and (11). If $b > \sqrt{P/K_2}$, the precipitate does not redissolve in excess of the acid. With b in the intermediate range, the lower value of H from Eq. (48) gives the condition for precipitation and the higher value the condition for the redissolving of the precipitate for specified values of b and g. With H_{48}, the value of a_{req} is in each case then obtained from $P = b^2a\alpha_2$.

Finally, for the effect of foreign strong acid or base on the solubility c of the pure salt, when $c = a = b/2$, the equation for electroneutrality is again Eq. (6). Then with the condition for saturation fixed by $P = 4c^3\alpha_2$, or

$$H = -K_1/2 + \sqrt{(\quad)^2 + K_1K_2(4c^3/P - 1)}, \quad (50)$$

Eq. (6) can be used to calculate g for specified values of c, for plotting the curve of the effect.

The analog of Eq. (8), for saturation with $a \neq b/2$, is simply Eq. (50) with ab^2 in place of $4c^3$.

Finally, if a solution is saturated with both the base, MOH, with solubility product P_1, and the normal salt, M_2X, the ion-product constants, corresponding to those of Eq. (34), are $[M^+]^3[OH^-][X^=] = PP_1$ and $[X^=]/[OH^-]^2 = P/P_1^2$. For the composition of the solution we obtain as the analog of Eq. (35)

$$H^3(1 + P_1/W) - H^2g - HW(1 + PW/P_1^2K_2) - 2PW^2/P_1^2 = 0, \quad (51)$$

whereupon $b = P_1H/W$ and $a = P/b^2\alpha_2 = PW^2/H^2P_1^2\alpha_2 = PW^2(H^2 + HK_1 + K_1K_2)/H^2K_1K_2P_1^2$. If $g = 0$, these expressions give the coordinates of point e in Fig. 20–1, in which however the salt composition would be on the line **b** = 2**a**. For a suspension of AgOH and Ag_2CO_3 in pure water, with $P = 8.2 \times 10^{-12}$ and $P_1 = 2 \times 10^{-8}$, Eq. (51) gives $H = 1.90 \times 10^{-10}$.

2. Mohr's Method for Chloride.[9] A solution containing NaCl (concentration c) and $_2KCrO_4$ (concentration a) is titrated with $AgNO_3$; b is the number of moles of $AgNO_3$ added per liter of solution. P (= 2×10^{-10}) is the solubility product of AgCl, and P' (= 10^{-12}) is that of Ag_2CrO_4. For chromic acid, rounded values of the equilibrium constants are used, as already mentioned.

For electroneutrality, while the solution is unsaturated or with AgCl as sole precipitate,

$$H - OH = [HCrO_4^-] + 2[CrO_4^=] + 2[Cr_2O_7^=] - 2a. \quad (52)$$

(9) F. Mohr, *Ann. der Chemie*, **97,** 335 (1856).

As long as $H < K_1$, we may neglect $[H_2CrO_4]$ and assume

$$a = [HCrO_4^-] + [CrO_4^=] + 2[Cr_2O_7^=]; \tag{53}$$

and since $[HCrO_4^-] = (H/K_2)[CrO_4^=]$ and $[Cr_2O_7^=] = K[HCrO_4^-]^2$, Eq. (52) becomes

$$H - OH = -(H/K_2)[CrO_4^=] - 2K(H/K_2)^2[CrO_4^=]^2. \tag{54}$$

Furthermore, $[CrO_4^=]$ in Eq. (54) will be taken as in Eq. (26), with $c_2 = a$.

AgCl precipitates first if $P/c < \sqrt{P'/[CrO_4^=]}$. While AgCl is the sole precipitate, H is of course constant for a given value of a and $[Ag^+]$ is given by Eq. XIX(69), with c for **a**. With H still fixed by Eq. (54), Ag_2CrO_4 appears as the second precipitate, marking the end-point of the titration, when $[Cl^-] = P\sqrt{[CrO_4^=]/P'}$. At this point the theoretical (ideal) titration error, or $b - c$, is

$$b - c = [Ag^+] - [Cl^-] = \sqrt{P'/[CrO_4^=]} - P\sqrt{[CrO_4^=]/P'}, \tag{55}$$

which is also obtainable by setting $[Ag^+] = \sqrt{P'/[CrO_4^=]}$ in Eq. XIX(69).

The titration error then is zero when $[CrO_4^=] = P'/P$ or when, by Eq. (54), $2H^3K(P'/PK_2)^2 + H^2(P'/PK_2 + 1) - W = 0$. For the numerical values involved, this gives $H \cong \sqrt{K_2PW/P'} = 7.75 \times 10^{-10}$. With this value of H in Eq. (53), which with the condition $[CrO_4^=] = P'/P$ becomes $a = HP'/PK_2 + P'/P + 2K(HP'/PK_2)^2$, we calculate $a_{req} = 5.01 \times 10^{-3}$. This is the concentration of K_2CrO_4 therefore required in the titration of pure aqueous alkali chloride for the end-point or precipitation of Ag_2CrO_4 to coincide with the theoretical equivalent point. Although the solution is slightly alkaline, AgOH does not precipitate; since $[Ag^+] = \sqrt{P}$ at the equivalent point and since H has the approximate value $\sqrt{K_2PW/P'}$, the product $[Ag^+][OH^-] \cong \sqrt{P'W/K_2}$, which is smaller than the solubility product ($P_1 = 2 \times 10^{-8}$) of AgOH.

For any arbitrary value of a, the concentration of K_2CrO_4 used as indicator, the error depends on the value of H; $[CrO_4^=]$ is calculated through Eq. (26) for use in Eq. (55). For the usual concentration used, $a = 2.5 \times 10^{-3}$,(10) the error has the values shown on page 419.

The theoretical error with this concentration of the indicator is therefore always positive, approaching a minimum value of 10^{-5} with increasing pH, the limit approached being $(\sqrt{P'/a} - P\sqrt{a/P'})$, from Eq. (55). If the value of H is allowed to be simply that caused by the K_2CrO_4, we calculate it through Eq. XV(78) as $H = 9.8 \times 10^{-10}$ for $[K_2CrO_4] = 2.5 \times 10^{-3}$ so that the error is in that case almost the minimum.

A value of pH higher than 7 moreover does not incur danger of precipitation of AgOH. The value of H required for AgOH to precipitate before

(10) I. M. Kolthoff and E. B. Sandell, *Textbook of Quantitative Analysis*, Macmillan, N.Y., 1943; p. 569.

Ag_2CrO_4 for a given concentration of K_2CrO_4 in solution may be calculated by equating $\sqrt{P'/[CrO_4^=]}$ with P_1/OH, in which P_1 is the solubility product of AgOH. With $[CrO_4^=]$ expressed as in Eq. (26) the result is

$$H = P'W^2/2aP_1^2K_2 + \sqrt{(\quad)^2 + (P'W^2/aP_1^2)(1 + 2KP'W^2/K_2^2P_1^2)}. \qquad (56)$$

pH	Error ($= b - a$)
4	3.95×10^{-4}
5	1.24×10^{-4}
6	3.92×10^{-5}
7	1.47×10^{-5}
8	1.05×10^{-5}
9	1.005×10^{-5}
10	1.0005×10^{-5}
11	1×10^{-5}

With $a = 2.5 \times 10^{-3}$, this gives H as almost exactly equal to $\sqrt{P'W^2/aP_1^2}$, or $H = 10^{-11}$. Hence any value of pH between 7 and 11 would give a negligible error theoretically without precipitation of AgOH.

Finally, we note that the error can be negative only if $a > P'/P$ since a is the limiting value of $[CrO_4^=]$ approached with increasing pH.(11)

3. Product K_1K_2 from Solubility Measurements. With $c = a = b/2$, $[X^=] = c\alpha_2 = P/4c^2$, and $\alpha_1 + 2\alpha_2 = 1 + \alpha_2 - \rho$ in Eq. (47),

$$K_1K_2 = H^2P/4c^2\,(g - c + P/4c^2 - D). \qquad (56a)$$

Hence the product K_1K_2 may be obtained, without even an approximate estimate of either separate constant, from simultaneous measurements of the solubility c of the salt and of the pH in a suspension in a solution of varying g, if P is known. For the salt MX, similarly,

$$K_1K_2 = H^2P/c\,(g - c + P/c - D). \qquad (56b)$$

It is significant that these equations hold even in the complex case of chromic acid, regardless of the value of the complex constant K, provided that P properly represents the product $[M^+]^2[CrO_4^=]$ or $[M^{++}][CrO_4^=]$ at saturation.

D. Saturation With Acid Salt and Normal Salt

1. Divalent Strong Base.

a. General. In Fig. 20–3 the curve ef is the solubility curve of the normal salt MX with the composition $b = a$, and the curve fn is the solubility curve

(11) For a similar discussion of the more complicated relations involved in the argentometric titration of cyanide, see J. E. Ricci, *J. Phys. and Coll. Chem.*, **51,** 1375 (1947).

of the acid salt $M(HX)_2$ with composition $2b = a$. The solubility of the acid salt in pure water is S_A, the crossing of the solubility curve with the line $2\mathbf{b} = \mathbf{a}$. If the acid salt is congruently soluble, the solution saturated with both normal salt and acid salt is on the left of S_A as at point f; if it is incongruently soluble with respect to the normal salt, the two solubility curves cross at point f′, on the right of point S_A, which is then metastable. In the first case the solid acid salt is stable in contact with its pure aqueous solution; in the second case the process of concentration of the pure aqueous solution of

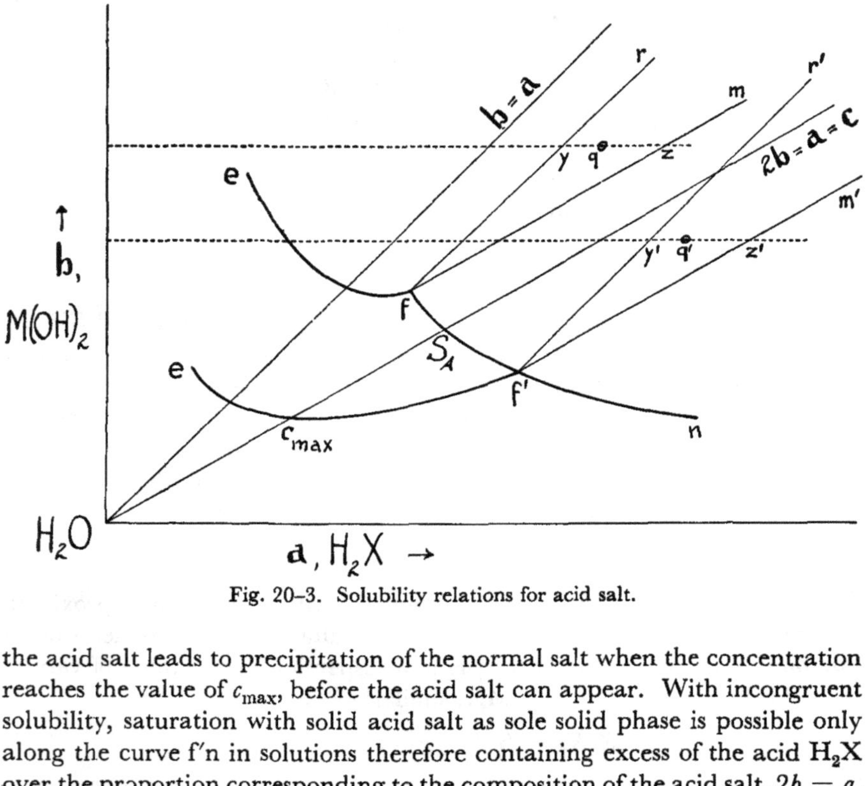

Fig. 20–3. Solubility relations for acid salt.

the acid salt leads to precipitation of the normal salt when the concentration reaches the value of c_{max}, before the acid salt can appear. With incongruent solubility, saturation with solid acid salt as sole solid phase is possible only along the curve f′n in solutions therefore containing excess of the acid H_2X over the proportion corresponding to the composition of the acid salt, $2b = a$.

If the solubility product of the acid salt is defined as $P_A = [M^{++}][HX^-]^2 = b(a\alpha_1)^2$, as in Eq. (33a), the substitution of $\sqrt{P_A/b}$ for $a\alpha_1$ in Eq. (4) gives

$$H = -(2b - \sqrt{P_A/b} - g)/2 + \sqrt{(\quad)^2 + (2K_2\sqrt{P_A/b} + W)}, \quad (57)$$

as the condition for saturation for specified values of b and g. With this value of H the corresponding value of a is calculated as $a = (\sqrt{P_A/b})/\alpha_1$. With $g = 0$ this describes the curve fn of the ternary system of Fig. 20–3. The solubility curve of the acid salt does not pass through a minimum of b as a is increased; there would be a minimum of a on its upper branch, above S_A, with increasing b, but only if the base $M(OH)_2$ were weak at least in respect to its second ionization constant K_2'.

If a suspension of the pure acid salt in water (assumed to be congruently soluble), with $c = b = a/2$, is treated with foreign strong acid or base, we have

$$H - OH = 2c(\alpha_1 + 2\alpha_2 - 1) + g. \tag{58}$$

The value of g corresponding to a specified solubility c may be calculated from Eq. (58) with H from $P_A = 4c^3\alpha_1^2$, or

$$H = K_1(\sqrt{4c^3/P_A} - 1)/2 \pm \sqrt{(\quad)^2 - K_1K_2}. \tag{59}$$

There is no precipitation and no positive value of H for Eq. (59) unless c is greater than the minimum solubility of the salt with respect to pH. This occurs at the iso-electric point of the salt when $H = \sqrt{K_1K_2}$ so that $c_{\min} = [P_A(1 + 2\sqrt{K_2/K_1})^2/4]^{1/3}$. This minimum occurs in acid solution if $K_1K_2 > W$ and in alkaline solution if $K_1K_2 < W$. The value of g required to bring the solubility to the minimum is simply the value of g required for the iso-electric point, or $g_{\text{ie}} = \sqrt{K_1K_2} - W/\sqrt{K_1K_2}$, which is the same whether or not the solution is saturated with the salt. With $c > c_{\min}$, Eq. (59) gives two values of H; and the two corresponding values of g, one of which will be positive, the other negative, since $g = a_s - b_s$, may then be calculated from Eq. (58). As in the case of the 1 : 1 salt of weak acid and weak base, both univalent, the two values of H given by Eq. (59) are equidistant in pH from the pH of the iso-electric point since their product equals K_1K_2 and ρ/α_2 at H_- equals α_2/ρ at H_+. If $c > c_{\min}$, the solution is supersaturated if H is between the limits of the two values fixed by Eq. (59).

With a^2b in place of $4c^3$, Eq. (59) gives H_{req} for saturation at given values of a and b as independent variables, with $a \neq 2b$; g_{req} then comes from Eq. (4). The minimum value of a^2b for saturation is $4c_{\min}^3$.

b. Saturation with both acid salt and normal salt. For the composition of the solution saturated with both normal salt and acid salt, the combination of Eq. (4) with the two solubility products gives

$$2H^4P^2/P_AK_2^2 + H^3 - H^2g - H(P_AK_2/P + W) - 2P_AK_2^2/P = 0. \tag{60}$$

With $g = 0$ this refers to the suspension of the two solids either in pure water, which would be point f (or f′) of Fig. 20–3, or in the solution of a salt like $NaNO_3$. Using this value of H in the combination of the solubility products, $P = ba\alpha_2$ and $P_A = b(a\alpha_1)^2$, we may then calculate the values of a and b in the solution as $a = K_2P_A/HP\alpha_1$ and $b = H^2P^2/K_2^2P_A$ (since $\alpha_2 = \alpha_1K_2/H$). If this calculation gives $a > 2b$, the acid salt is incongruently soluble with respect to the normal salt. The value of $c_{\max}$ could be calculated through the combination of Eq. (58) for the pure salt solution at the concentration c

and the condition for saturation with the normal salt, $P = ba\alpha_2 = 2c^2\alpha_2$. Substitution of $c = \sqrt{P/2\alpha_2}$ in Eq. (58) allows calculation of H_{req}, and c_{max} then comes from $P = 2c^2\alpha_2$.

If a suspension of the normal salt, with $\mathbf{b} > b_f$, is treated with the acid H_2X, the acid salt appears as a solid phase, mixed with the normal salt, when the total composition reaches the line fr, parallel to the diagonal $\mathbf{b} = \mathbf{a}$. The value of $\mathbf{a}$ required to reach this line is $a_f + (\mathbf{b} - b_f)$. As addition of the acid is continued, the normal salt is consumed and more solid acid salt is formed, while the solution remains fixed at point f. When the line fm is reached, parallel to the line $2\mathbf{b} = \mathbf{a}$, the normal salt has vanished as solid phase. The value of $\mathbf{a}$ required to reach this line is $a_f + 2(\mathbf{b} - b_f)$. When the total composition is at some intermediate point such as point q, the ratio of normal salt to acid salt in the solids is qz/yq. The number of moles of acid salt present at point q is $\mathbf{a}_q - \mathbf{a}_y = \mathbf{a}_q - a_f - (\mathbf{b} - b_f)$ while the number of moles of normal salt left at point q is $\mathbf{a}_z - \mathbf{a}_q = 2(\mathbf{b} - b_f) - (\mathbf{a}_q - a_f)$.

2. Univalent Strong Base. In this case the normal salt M_2X has the composition $b = 2a$ and the acid salt MHX has the composition $b = a$, but the general relations of the solubility curves shown in Fig. 20–3 otherwise remain the same. With $P_A = [M^+][HX^-] = ba\alpha_1$, Eq. (47) gives

$$H = -(b - P_A/b - g)/2 + \sqrt{(\quad)^2 + (2K_2P_A/b + W)}, \qquad (61)$$

as the condition for saturation for specified values of b and g. For the composition of the solution saturated with both normal salt and acid salt, with $P = [M^+]^2[X^=] = b^2a\alpha_2$, we have

$$H^3(1 + P/P_AK_2) - H^2g - H(P_A^2K_2/P + W) - 2P_A^2K_2^2/P = 0. \qquad (62)$$

For the effect of foreign strong acid or base on a suspension of the pure salt in water, we have first the equation of electroneutrality with $c = b = a$, or $D = c(\alpha_1 + 2\alpha_2 - 1) + g$, while from $P_A = c^2\alpha_1$ we have

$$H = K_1(c^2/P_A - 1)/2 \pm \sqrt{(\quad)^2 - K_1K_2}. \qquad (63)$$

(If $a \neq b$, this equation applies with ab in place of c^2.) At the iso-electric point, when $g = \sqrt{K_1K_2} - W/\sqrt{K_1K_2}$, the solubility is at its minimum,

$$c_{min} = \sqrt{(ab)_{min}} = \sqrt{P_A(1 + 2\sqrt{K_2/K_1})}. \qquad (64)$$

If $c > c_{min}$, Eq. (63) gives two values of H, and the corresponding values of g for saturation are obtained from the equation of electroneutrality. Again the two values of H given by Eq. (63) are equidistant in pH from the iso-electric point, and ρ/α_2 at H_- equals α_2/ρ at H_+.

XXI

Saturation with Ampholytes and Their Salts

A. General Relations

Saturation with respect to an ampholyte means constancy of the activity (or of the concentration if its activity coefficient equals 1) of the uncharged species of the ampholyte, U^0 or $U^{\pm}$, or of the sum of $[U^0]$ and $[U^{\pm}]$, if both forms exist; see Eqs. VI(43) and VI(45). Hence if S is the solubility and α_0 the fraction of this solute in the forms U^0 and $U^{\pm}$ combined, $S\alpha_0 = \mathbf{k}$. Because of the interrelation of α_0 and the various ionization fractions of an ampholyte, through H and the over-all apparent ionization constants, given in Eqs. I(33–35), I(73), and I(74), the solubility product may be written in several related forms, the most important of which are those as a base, P_b, as an acid, P_a, and as an incompletely dissociated "salt," P_S. With $\mathbf{k} = P_b/K_{b_1} \cdots K_{b_{z+}} = P_a/K_{a_1} \cdots K_{a_{z-}}$, we have

$$P_b = [U^{+z_+}][OH^-]^{z_+} = S\alpha_{+z_+}[OH^-]^{z_+}; \tag{1}$$

$$P_a = [H^+]^{z_-}[U^{-z_-}] = S\alpha_{-z_-}[H^+]^{z_-} = P_b K_{a_1} \cdots K_{a_{z-}}/K_{b_1} \cdots K_{b_{z+}}; \tag{2}$$

$$P_S = [U^{+z_+}]^{z_-}[U^{-z_-}]^{z_+} = (S\alpha_{+z_+})^{z_-}(S\alpha_{-z_-})^{z_+}$$

$$= \frac{P_b^{z_-} P_a^{z_+}}{W^{z_+z_-}} = \frac{P_b^{(z_+ + z_-)}}{W^{z_+z_-}} \left(\frac{K_{a_1} \cdots K_{a_{z-}}}{K_{b_1} \cdots K_{b_{z+}}}\right)^{z_+}. \tag{3}$$

The derivative of S with respect to pH has already been discussed (Chapter IV), showing that S is a minimum at the iso-electric point. From Eqs. (1) and (2), $S = P_b/(OH)^{z_+}\alpha_{+z_+} = P_a/H^{z_-}\alpha_{-z_-}$. Increasing on both sides of H_{ie}, S tends to vary linearly with $(OH)^{z_-}$ on the alkaline side and with H^{z_+} on the acid side. With increasing OH, α_{-z_-} increases toward a limiting value of 1 so that in highly alkaline solution $S \stackrel{\rightarrow}{=} (OH)^{z_-}P_a/W^{z_-}$, or $-\log S \stackrel{\rightarrow}{=} \text{p}P_a - z_-\text{p}H$; in highly acid solution, the limit approached is a value of 1 for α_{+z_+}, giving $S \stackrel{\rightarrow}{=} H^{z_+}P_b/W^{z_+}$, or $-\log S \stackrel{\rightarrow}{=} \text{p}P_b - z_+\text{p}W + z_+\text{p}H$. When the solubility then is found to be proportional to the z_-'th power of OH, in highly alkaline solution, we may deduce that the maximum negative charge of the ampholyte is z_-; and similarly if S approaches proportionality to the z_+'th power of H in acid solution, the highest positive charge of the ampholyte is z_+. From these limiting proportionalities, then, the "basicities," (z_- and z_+) of the ampholyte are established.

Such relations, however, follow purely from mathematics and thermodynamics. The fact that the solubility of zinc oxide is proportional to $[OH^-]^2$ in highly alkaline solution tells us merely that the ampholyte has a maximum negative valence of 2 but it has nothing to do with deciding between possible "mechanisms" of the process or possible constitutional formulas of the ion; it no more supports or corroborates the formula $Zn(OH)_4^=$ than the formula $ZnO_2^=$. The mathematical relations do not, in short, depend on any theory of the mechanism of the interaction between these ampholytes and water, acids, and bases or, that is, on whether we assume the "ampholyte behavior" to be ionization (even dissociation) of a simple molecule in two ways, "dipolar" or "zwitter-ion" relations (as for amino acids) or complex ion formation such as combination of cation and OH^- to form "hydroxyl complexes," as suggested for "amphoteric metallic hydroxides."

B. Saturation With Simple Ampholyte

A 1 : 1 ampholyte (MOH, or simple amino acid) with $z_- = z_+ = 1$ has three aqueous equilibrium constants characterizing it: two ionization constants, K_a and K_b, and a solubility product constant, which may be written in the three forms

$$P_b = S\alpha_+(OH), \tag{4a}$$

$$P_a = S\alpha_- H = P_b K_a/K_b, \tag{4b}$$

$$P_s = S^2\alpha_+\alpha_- = P_a P_b/W = P_b^2 K_a/K_b W. \tag{4c}$$

We may therefore write

$$S = P_b/K_b + P_b/OH + P_a/H \tag{5a}$$

$$= (P_b/K_b W)\mathbf{Y} = (P_a/K_a W)\mathbf{Y} = \mathbf{Y}\sqrt{P_s/K_a K_b W}, \tag{5b}$$

in which $\mathbf{Y} = K_b H + K_a OH + W$. The minimum of S, occurring at $H_{ie} = \sqrt{WK_a/K_b}$, may also be expressed in various ways, such as

$$S_{min} = (P_b/K_b)\mathbf{J} = (P_a/K_a)\mathbf{J} = \mathbf{J}\sqrt{P_s W/K_a K_b}, \tag{6}$$

with $\mathbf{J} = (1 + 2\sqrt{K_a K_b/W})$.

1. The Pure Aqueous Solution. With $D_0 = S_0\beta_0 = S_0(\alpha_- - \alpha_+)_0$, H_0 and S_0 may be calculated from the three equilibrium constants. By combining this expression with Eq. (4) we obtain

$$H_0 = \sqrt{W(P_a + W)/(P_b + W)}; \quad OH_0 = \sqrt{W(P_b + W)/(P_a + W)}. \tag{7}$$

S_0 then follows from Eq. (5). These formulas should be compared with $H_0 = \sqrt{P_2 + W}$, Eq. XVII(19), for simple acid, and with $OH_0 = \sqrt{P_1 + W}$, Eq. XVII(20), for saturation with simple base, to which they reduce of course if either P_b or P_a is zero. On the other hand, given H_0, S_0, and one of the ionization constants such as K_a, we may calculate P_b and the other ionization constant, K_b: with K_b eliminated from $D_0 = S_0\beta_0$ and Eq. (4a), $P_b = W[K_a(S_0 - D_0) - H_0D_0]/H_0(2K_a + H_0)$ and then $K_b = W[K_a(S_0 - D_0) - H_0D_0]/H_0^2(S_0 + D_0)$. From S_0 and the ionization constants, H_0 is found directly from $D_0 = S_0\beta_0$ and P_b then from Eq. (4a).

Finally, the concentration of the uncharged form at saturation, ($\mathbf{k} = S\alpha_0$) may be calculated from these constants as $\mathbf{k} = \mathbf{P}_b/\mathbf{K}_b = \mathbf{P}_a/\mathbf{K}_a$.

2. TREATMENT OF SATURATED SOLUTION WITH STRONG ACID OR BASE. With strong acid, the solubility increases from the start if $H_0 < \sqrt{W}$, or $K_a < K_b$; otherwise it first falls to a minimum at H_{ie} and then rises again. With strong base, the solubility increases from the start if $K_a > K_b$; but if the pure aqueous solution is alkaline, S passes through a minimum at H_{ie} before rising again. The condition for saturation at the concentration S is, from Eq. (5),

$$H_{req} = (W/2P_b)(S - P_b/K_b) \pm \sqrt{(\quad)^2 - WK_a/K_b}. \tag{8}$$

Provided that $S > S_{min}$, this equation gives two positive values of H. The two values of H (H_- and H_+) are equidistant in pH from the pH of the isoelectric point since $H_+H_- = WK_a/K_b = (H_{ie})^2$; and α_-/α_+ at the higher value of H_{req} equals α_+/α_- at the lower value. With $S > S_{min}$, the ampholyte can not be completely dissolved if H is between the limits of the two values fixed by Eq. (8). The value of a_s or b_s then required for saturation at the concentration S, or the value of g to dissolve or to precipitate the ampholyte at the concentration S, is calculated, with H_s, from

$$H - OH = S(\alpha_- - \alpha_+) + g. \tag{9}$$

Note: On the analogy between the 1 : 1 ampholyte and the alkali acid salt of a dibasic acid, or M_sHX. Through the transformations of constants given in Chapter I, the equation defining H in a *solution* of the ampholyte is the same as that for the salt M_sHX at the same concentration. But the correspondence does not hold for the relations of the *saturated* solutions. The formula for S_{min}, or Eq. (6), is not interchangeable with that for the acid salt, Eq. XX(64). Eq. (8), for the condition of saturation at a specified concentration, involves the first power of the concentration as compared to the square in the corresponding equation for the acid salt, Eq. XX(63). Finally, while H_0 for the pure saturated solution of the ampholyte is given by the simple formula, Eq. (7), in terms of constants, the corresponding expression defining H_0 in the pure saturated solution of the salt M_sHX in terms of constants, which would be obtained by elimination of S from $H - OH = S(\alpha_1 + 2\alpha_2 - 1)$ and $P = S^2\alpha_1$, is an equation of the sixth degree in H_0. In the same way the combination of the condition of saturation with the equation of electroneutrality for a solution of a pure monobasic acid, whether weak or strong, results in a simple expression for H_0 in terms of constants, Eq. XVII(19); but for a simple 1 : 1 salt of monobasic acid and monoacid base the equation is of the fifth degree, Eq. XVIII(5), even if the base is strong. In its solubility relations an ampholyte may be said to behave therefore at least partly as something which is both an acid and a base rather than a salt, while in its buffering properties, those shown in its unsaturated solution, it is mathematically a salt. The ampholyte remains a separate category of solute, despite all analogies.

3. DETERMINATION OF CONSTANTS FROM SOLUBILITY MEASUREMENTS. It is possible to determine the equilibrium constants from data on the solubility and H in presence of strong acid or strong base. From Eq. (9), with $g = a_s - b_s$,

$$K_a = H(1 + K_bH/W)[S\alpha_+ - g + D]/[g + S(1 - \alpha_+) - D], \tag{10}$$

$$K_b = OH(1 + K_aOH/W)[S\alpha_- + g - D]/[- g + S(1 - \alpha_-) + D]. \tag{11}$$

These should be compared with Eqs. XIX(23) and XIX(25). If g is a_s and H is high enough, Eq. (10) simplifies as α_+ approaches 1; then a temporary approximate value of K_a is obtainable by neglect of the quantity K_bH/W. If g is $- b_s$ and OH is high enough, Eq. (11) similarly becomes simplified as α_- approaches 1, and then K_b may be evaluated either through the neglect of the quantity K_aOH/W or through the use of the approximate value of K_a. It is clear then that with data on the solubility both in acid and in alkaline solution, these two equations may be used in combination until the constants are known with the desired precision.

The question of the introduction of activity coefficients was discussed in connection with Eqs. V(117–120) and Eqs. VI(52) and VI(53). If the unionized species is assumed to consist entirely of molecular U^0, then the mass constants of these equations are replaced by $\mathbf{K}_a/\gamma^2$ and by $\mathbf{K}_b/\gamma^2$ respectively, γ being, according to the convention of Chapter V, the activity coefficient of a univalent ion. If the "unionized species" is the zwitter-ion $U^{\pm}$ exclusively, then the mass constants are $\mathbf{K}_a\gamma_{U^{\pm}}/\gamma^2$ and $\mathbf{K}_b\gamma_{U^{\pm}}/\gamma^2$.

If K_a and K_b are both known, the solubility product follows through its definition from the measurement of S and H under any conditions at all. From Eq. (4a) for example for either assumption concerning U^0 and $U^{\pm}$, $\mathrm{p}\mathbf{P}_b = - \log S + (\mathrm{p}\mathbf{W} - \mathrm{p}\mathbf{H}) + \log \alpha_+ - \log \gamma$. The fraction α_+ may be calculated from the ionization constants and $\mathbf{H}$ and the whole function may be either extrapolated to $\mu = 0$ or evaluated through calculation of γ from μ.

Furthermore, substitution of Eq. (4a) in Eq. (9) may be used to write $P_b = OH(S\alpha_- + g - D)$. With an approximate value of K_a available or with g as $- b_s$ and OH high enough to give $\alpha_- \cong 1$, this may be used directly to determine P_b. Similarly, Eqs. (4b) and (9) give $P_a = H(S\alpha_+ - g + D)$

for the determination of P_a from S (and H) in presence of a_s when $\alpha_+ \rightleftharpoons 1$, or through an approximate value of K_b. These solubility products then help in determining the ionization constants since $P_b/P_a = K_b/K_a$. Finally, P_S from Eq. (4c) may be used with Eq. (9) so that $P_S = S^2\alpha_+^2[1 - (g - D)/S\alpha_+]$ and $P_S = S^2\alpha_-^2[1 + (g - D)/S\alpha_-]$. Again, the first of these would be useful in acid solution when α_+ approaches 1 and the second in alkaline solution when α_- approaches 1; *these formulas are to be compared with* Eqs. XIX(26) and XIX(27) respectively.

4. PRECIPITATION OF AMPHOLYTE FROM ITS SOLUBLE SALT. For a solution of the soluble salt (the "hydrochloride"), MCl, at concentration c, of a simple ampholyte MOH, $D = c(1 + \alpha_- - \alpha_+) + g$, in which g is the net concentration, $a_s - b_s$, of foreign strong acid and base, other than that contributed by the salt itself. The ampholyte precipitates when $c\alpha_+OH = P_b$, which is possible only if $c > S_{min}$ for the ampholyte, Eq. (6). By combining P_b with the equation for the pure salt solution, or $D = c(1 + \alpha_- - \alpha_+)$, we may find the condition for the maximum concentration of the pure salt, MCl, without "hydrolytic" precipitation of the amphoteric MOH. The result is

$$H = P_b/2K_b + \sqrt{(\quad)^2 + (2P_a + W)}, \tag{12}$$

and $c_{\max}$ is then calculated from Eq. (5) with this value of H; $c_{\max}$ is of course always greater than $S_{\min}$ of the ampholyte.

If c is smaller than $c_{\max}$ for MCl but greater than $S_{\min}$ for MOH, then the two values of H_{req} from Eq. (8) may be used in $D = c(1 + \alpha_- - \alpha_+) + g$ to find g required to precipitate and then to redissolve the ampholyte. This precipitation and redissolving would in some cases be brought about by acid, in others by base; the two values of g will be either both positive or both negative. If $c > c_{\max}$, such calculation gives a positive and a negative value of g; the positive value represents a_s and the negative value b_s required to prevent the precipitation of MOH in the salt solution.

The value of g required to bring the solution of the hydrochloride salt, at concentration c, to the iso-electric point of the ampholyte is $g = H_{ie} - OH_{ie} - c = \sqrt{WK_a/K_b} - \sqrt{WK_b/K_a} - c$, which may be positive or negative. It is also possible for the *pure* salt solution to be at the iso-electric point of the ampholyte. The required concentration of the pure salt is $c = \sqrt{WK_a/K_b} - \sqrt{WK_b/K_a}$, which is therefore possible only if $K_a > K_b$. If the acid of the salt is weak, the salt being MX rather than MCl, then $D = c(\alpha + \alpha_- - \alpha_+) + g$. Now the value of g required for the iso-electric point of the ampholyte in salt solution of concentration c is $g = H_{ie} - OH_{ie} - c(\alpha)_{ie}$, the value of H in α_{ie} being H_{ie}. The concentration of the pure salt itself giving this iso-electric point is then $c = (H_{ie} - OH_{ie})/(\alpha)_{ie}$.

The maximum concentration for the alkali salt of the ampholyte, or NaM, in pure water, without precipitation of the ampholyte MOH, is given by Eq. (5) with $OH = P_a/2K_a + \sqrt{(\quad)^2 + (2P_b + W)}$.

The following are two numerical examples of c_{max} for such salts of simple ampholytes:

(1) $K_a = 10^{-10}$, $K_b = 10^{-6}$, $P_b = 10^{-12}$ ($P_a = 10^{-16}$)

(a) For MCl, $H_{\text{max}} = 1.01 \times 10^{-6}$, $c_{\text{max}} = 1.02 \times 10^{-4}$.

(b) For NaM, $OH_{\text{max}} = 2.00 \times 10^{-6}$, $c_{\text{max}} = 1.52 \times 10^{-6}$.

(2) $K_a = 10^{-7}$, $K_b = 10^{-10}$, $P_b = 10^{-13}$ ($P_a = 10^{-10}$)

(a) For MCl, $H_{\text{max}} = 10^{-3}$, $c_{\text{max}} = 0.011$.

(b) For NaM, $OH_{\text{max}} = 10^{-3}$, $c_{\text{max}} = 10.001$.

C. Solubility Relations of Simple Ampholyte and Monobasic Acid

1. Analytical equations. We here consider the system HX–MOH–H_2O ($\pm g$), which is assumed to involve no solid salt of the acid HX and the ampholyte MOH. The relations are therefore similar to the hypothetical case considered in Chapter xvii, Section E, Figs. 17–1,2. For the general electroneutrality, with a as the concentration of HX and b as that of MOH, the equation of electroneutrality is $D = a\alpha + b(\alpha_- - \alpha_+) + g$.

For saturation with respect to HX, the condition is $P_2 = a\alpha H$, so that H is given by Eq. xvii(57) for specified a. For saturation at specified values of b and g, substitution of $a\alpha = P_2/H$ in the electroneutrality equation gives

$$H^4K_b + H^3[K_b(b-g) + W] - H^2[W(g - K_a + K_b) + K_bP_2] - HW[K_a(b+g) + P_2 + W] - K_aW(P_2 + W) = 0. \quad (13)$$

For saturation with respect to MOH at a specified value of b, we have Eq. (8). For H at specified values of a and g, we have

$$H - OH = a\alpha + P_a/H - P_b/OH + g, \quad (14)$$

$$H^3(1 + P_b/W) + H^2[A(1 + P_b/W) - g] - H[A(a+g) + P_a + W] - A(P_a + W) = 0. \quad (15)$$

For saturation with both MOH and HX, we have

$$H - OH = P_2/H + P_a/H - P_b/OH + g, \quad (16)$$

whence

$$H = gW/2(P_b + W) + \sqrt{(\quad)^2 + W(P_a + P_2 + W)/(P_b + W)}. \quad (17)$$

Combination of Eq. (14) with Eq. xvii(57) gives

$$a = -gP_2/2(P_a + P_2 + W) + \sqrt{(\quad)^2 + P_2^2(P_b + W)/W(P_a + P_2 + W)} + P_2/A, \quad (18)$$

which permits calculation of either a or g from the other for solutions saturated with both MOH and HX, describing a three-dimensional curve which, being similar to that of Fig. 17–2, will be called the curve x′x″. For the relation between b and g we may calculate g directly from b through Eq. (16), using H as given by Eq. (8) for specified b $(= S)$; for the value of b at specified g, Eq. (18) would be used in combination with the direct relation between a and b.

This direct relation between a and b allowing the plotting of the projection of the curve x′x″ on the a/b plane is obtained by combination of the two solubility products, as

$$a = (b - P_b/K_b)P_2/2P_a \pm \sqrt{(\quad)^2 - P_2^2 P_b/P_a W} + P_2/A, \qquad (19)$$

or

$$b = (a - P_2/A)P_a/P_2 + P_2 P_b/W(a - P_2/A) + P_b/K_b. \qquad (20)$$

Again the asymptotic limit for low a at $b = \infty$ is $a = P_2/A$. In this case, however, the curve passes through a minimum of b, point m, which occurs at $b_m = S_{min}$ or Eq. (6), with $H_m = H_{ie} = \sqrt{WK_a/K_b}$, and with $a_m = P_2/A + P_2\sqrt{K_b/K_a W}$. With $b > b_m$ there are two values of a, as seen in Eq. (19). The minimum may occur on either side of the diagonal $a = b$, depending on the relative magnitudes of the constants. The value of g at this minimum in the curve is given by Eq. (16), with H_{ie}, as $g_m = (\sqrt{WK_a/K_b} - \sqrt{WK_b/K_a}) - P_2\sqrt{K_b/K_a W}$. The curve crosses the diagonal when c $(= a = b)$ is given by the expression

$$c^2 A(1 - P_a/P_2) - c(P_2 + AP_b/K_b - 2P_a) + P_2(P_b/K_b - AP_b/W - P_a/A) = 0. \qquad (21)$$

2. Graphical Relations, System MOH–HX–H_2O. Fig. 21–1 shows the relations at the curve x′x″, for saturation with MOH and HX, in schematic projection on a horizontal a/b plane; the figure is to be compared with Fig. 17–3. This is an orthogonal view from high g. The two saturation surfaces, shown by a number of constant g contours, meet along the curve x′x″. On the convex side of the curve, the MOH surface lies below the HX surface. Part of the MOH surface is seen, unobstructed by the HX surface, on the concave side of the curve.

The solubility of MOH in the absence of HX is S_{min} at $g = g_{ie} = (H - OH)_{ie}$. At this value of g therefore, HX increases the solubility of MOH as it does also at $g > g_{ie}$. If $g < g_{ie}$ however, the addition of HX first decreases the solubility to $b = S_{min}$ and then increases it again. This effect is brought out by the shape of the contours for the MOH surface for $g < g_{ie}$. They all fall to the value of S_{min} and then rise again, as long as g is still greater than g_m, to end at the curve x′x″. For $g < g_m$, the contours reach the curve x′x″ without passing through the minimum. The line S_{min}m is horizontal and it represents the envelope of these contours or the projection of the bulge on the MOH solubility surface. Along this bulge, projected as the line S_{min}m, H and b are constant, being equal to H_{ie} and S_{min} respectively, but a varies from zero to a_m and g varies from $(H - OH)_{ie}$ to g_m.

The contours on the HX solubility surface change their inclination when g has the value g_m. This particular contour (or solubility curve at $g = g_m$) is vertical since the addition of the ampholyte to a solution in which $H = H_{ie}$, which is the value of H on this contour, causes no change in H, so that at this value of g the solubility of HX is not affected by MOH. At higher g

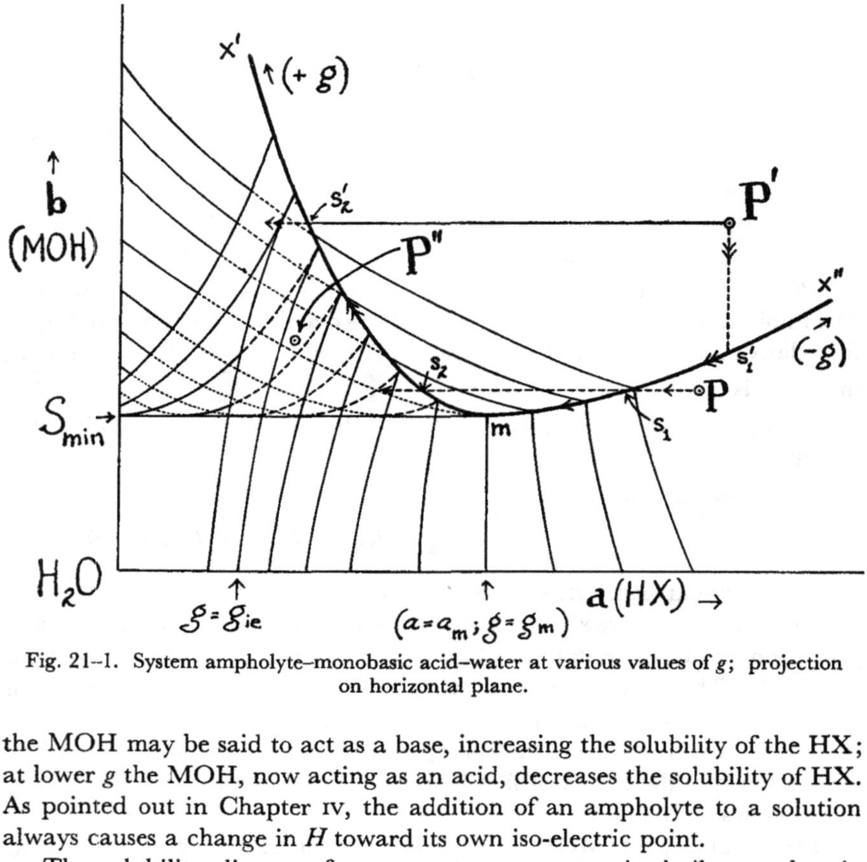

Fig. 21–1. System ampholyte–monobasic acid–water at various values of g; projection on horizontal plane.

the MOH may be said to act as a base, increasing the solubility of the HX; at lower g the MOH, now acting as an acid, decreases the solubility of HX. As pointed out in Chapter IV, the addition of an ampholyte to a solution always causes a change in H toward its own iso-electric point.

The solubility diagram for constant g at $g > g_{ie}$ is similar to that in Fig. 17–1, for base (MOH) and acid (HX); each increases the solubility of the other. The diagram at $g < g_m$ is essentially the same except that the solubility of each component is decreased by the other; this is schematically the typical diagram of a system of two acids, MOH + HX. For g between g_{ie} and g_m the effect of MOH on the solubility of HX is the effect of a base upon it, increasing the solubility; but the solubility curve of MOH first falls and then rises again with the addition of HX before reaching saturation with

HX at a point on the curve x′x″. Except for these differences the solubility relations at constant g are similar to those discussed under Figs. 17–1,2,3.

3. Effect of Change of g on Unsaturated Solution. We consider next the effect of change of g upon an unsaturated solution containing HX at concentration **a** and MOH at concentration **b**. If the composition of the unsaturated solution is below the line S_mm and the curve mx″, no precipitation is possible by decrease of g; increase of g causes HX to precipitate when the HX surface is reached. The value of g required to reach this point is given by Eq. (18) with $a = \mathbf{a}$. Furthermore, MOH does not appear as a temporary second precipitate mixed with the HX unless the initial composition, as for point **P** in the diagram, lies to the right of the curve mx″ with $\mathbf{b} > S_{\min}$. For such a point however, the precipitation of HX, causing the composition of the saturated solution to move to the left, horizontally at constant b, brings the solution to the curve mx″ at point s_1, when MOH appears as a second precipitate. H_{req} is given as H_- of Eq. (8) with $S = \mathbf{b}$, and g_{req} then follows from Eq. (16). As g is increased further, the solution follows the curve, passing through point m, and the composition of the solution while it is saturated with both solids is calculated for specified g through Eqs. (17) and (18) followed by Eq. (20) or through other sets of Eqs. (17), (18), (20), and (5) and Eq. xvii(57). During this stage the quantity of solid MOH mixed with HX first increases and then decreases; the MOH redissolves completely when H reaches the value H_+ of Eq. (8), g_{req} following again from Eq. (16). This redissolving of MOH, leaving HX again as sole precipitate, occurs at point s_2 on the horizontal line fixed by point **P**. The course of the composition of the saturated solution is indicated by the single arrows.

If the composition, now **P**′, falls above the curve x′x″, decrease of g causes no precipitation, but increase of g causes MOH to precipitate when the MOH surface is reached at H_- of Eq. (8). With precipitation of MOH the composition of the saturated solution moves vertically downward at constant a to the curve x′x″, where, at point s_1', HX begins to precipitate; g_{req} to reach point s_1' comes directly from eq. (18) for the given value of $a = \mathbf{a}$. The solution then follows the curve, passing through point m, and the MOH redissolves at point s_2', on the horizontal line through **P**′. At this point $H = H_+$ of Eq. (8) with $S = \mathbf{b}$, and g_{req} comes from Eq. (16). Thereafter the sole precipitate is HX. The course of the saturated solution in this case is shown by the double arrows. If **P**′ is on the left of point m, but still above the curve x′x″, the precipitation of MOH causes the composition of the saturated solution to travel down to the line S_mm to a minimum of b before it begins to rise again, in the projection, to a point s_1' on the curve mx′. The phenomena are otherwise the same as those already discussed for the point **P**′. The value of g required to reach the line S_mm is $g = (H - OH - \mathbf{a}\alpha)_{ie}$, with $H = H_{ie}$.

Finally, if the composition, **P**″, is above the line S_mm but below the curve x′m, it may lie, as an unsaturated solution, either between the MOH surface and the HX surface or below the bulge of the MOH surface. For a given pair of values of **a** and **b** this is a question of the value of H and hence of g in the unsaturated solution. If **P**″ lies below the MOH surface MOH precipitates when g is increased; Eq. (8) gives the two values of H_{req} for its precipitation and redissolving, and g_{req} comes in each case from Eq. (14). Further increase of g causes HX to precipitate when $g = D - \mathbf{b}(\alpha_- - \alpha_+) - P_2/H$ with $H = P_2/(\mathbf{a} - P_2/A)$ from Eq. XVII(57). If **P**″ lies between the two surfaces, increase of g causes HX to precipitate and decrease of g causes MOH to precipitate and then redissolve.

Table 21–1. Systems involved in Fig. 21–2.

System		Curve	
I	**II**	A	B
base	base	x′m_a	m_bx″
ampholyte	base	x′$m_a m_b$	
acid	base	$m_a m_b$	
acid	ampholyte	$m_a m_b$x″	
acid	acid	m_bx″	x′m_a
ampholyte	acid		x′$m_a m_b$
base	acid		$m_a m_b$
base	ampholyte		$m_a m_b$x″

If the acid HX is strong the various equations are simplified with $A = \infty$ or $\alpha = 1$. Eq. (15) for example becomes $H^2(1 + P_b/W) - H(a + g) - (P_a + W) = 0$. Eq. (20) for the relation between b and a on the curve x′x″ becomes $b = aP_a/P_2 - P_2P_b/aW + P_b/K_b$; but in Eq. (19), as also in Eq. (18), we merely drop the term P_2/A. The curve crosses the diagonal only once, when c ($= a = b$) is

$$c = P_bP_2/2K_b(P_2 - P_a) + \sqrt{(\quad)^2 + P_bP_2^2/W(P_2 - P_a)}. \tag{22}$$

Eqs. (16, 17) remain the same. The limiting value of a at $b = \infty$ shifts from P_2/A to zero, and the value of a_m similarly shifts to the left by P_2/A, to $a_m = P_2\sqrt{K_b/K_aW}$, while H_m and b_m remain the same. In fact the whole curve x′x″ is merely displaced to the left by the constant quantity P_2/A, according to Eq. (19). This is similar to the effect in the case of curve x′x″ in Chapter XVII for saturation with respect to acid and simple base.

4. General System of Two Ampholytes and Water. For general orientation, we shall consider saturation with respect to two ampholytes, the first, or solute **I**, with the concentration a and the second, or solute **II**, with

the concentration b. The first will furthermore be distinguished as the substance with primed constants in this special discussion. The relation between the concentrations a and b in the solution saturated with both solids may be obtained by substitution of H_8 for substance **I** in Eq. (5a) as applying to substance **II**. The result is a quadratic for a in terms of b and constants, giving two values of a for $b > b_{min}$, or a quadratic for b in terms of a giving two values of b for $a > a_{min}$. Projected on the a/b plane the curve appears schematically therefore as in Fig. 21–2. There are two possibilities in respect to the direction of change of g, depending on whether substance **I**

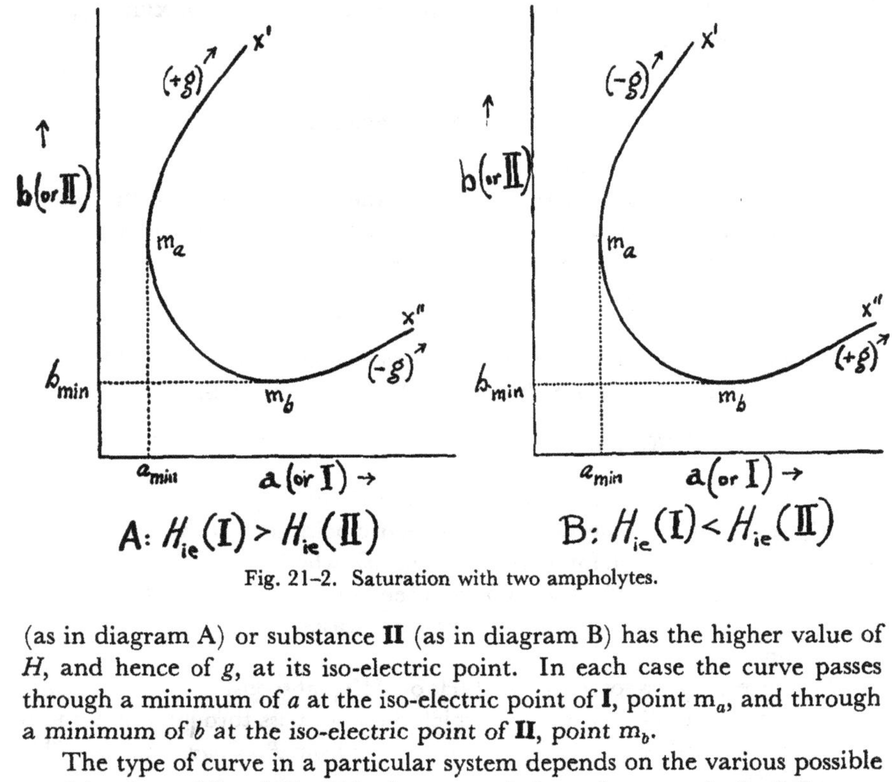

Fig. 21–2. Saturation with two ampholytes.

(as in diagram A) or substance **II** (as in diagram B) has the higher value of H, and hence of g, at its iso-electric point. In each case the curve passes through a minimum of a at the iso-electric point of **I**, point m_a, and through a minimum of b at the iso-electric point of **II**, point m_b.

The type of curve in a particular system depends on the various possible combinations of **I** and **II**. If both are ampholytes the curve is the full curve of either part A or part B of Fig. 21–2. Otherwise only a section of the full curve applies, as shown in Table 21–1. The iso-electric point of a simple acid may be considered mathematically to be at $H = \infty$ or $g = \infty$; that for a base at $OH = \infty$ and $g = -\infty$. The curve for saturation for simple acid and base therefore lies entirely between the (infinite) iso-electric points, m_a and m_b, and is hyperbolic in shape (see Chapter XVII). For two acids it is on the low g side of both and for two bases on the high g side of both; in these two cases it becomes a straight line of positive slope as seen in Eq. XVII(41) for saturation with two bases.

Finally, the composition of the solution saturated with the two ampholytes, as $f(g)$, may be calculated through the value of H as $f(g)$; from $D = P_a/H - P_b/OH + P'_a/H - P'_b/OH + g$, we have

$$H = gW/2(P_b + P'_b + W) + \sqrt{(\quad)^2 + W(P_a + P'_a + W)/(P_b + P'_b + W)}. \quad (23)$$

With H so determined the concentration b $(= S)$ follows from Eq. (5a), while a is given by the same equation used with the primed constants. If the substance **I** is a simple acid, with $P'_b = 0$ and $P'_a = P_2$, while **II** is an ampholyte, Eq. (23) becomes Eq. (17). For two bases we have Eq. XVII(40), and for acid and base Eq. XVII(78).

D. Salt of Ampholyte and Strong Monobasic Acid

1. Analytical Equations.

a. Saturation with salt alone. With b as the concentration of the simple ampholyte MOH and a as that of the specific monobasic acid HX_s, the equation of electroneutrality, with g as *foreign* strong acid or base, is

$$H - OH = a + b(\alpha_- - \alpha_+) + g. \quad (24)$$

The condition for saturation with respect to the salt MX_s is $P = [M^+][X_s^-] = ab\alpha_+$, so that for a given value of ab, H required for saturation is

$$H = W/2K_b(ab/P - 1) + \sqrt{(\quad)^2 + K_aW/K_b(ab/P - 1)}; \quad (25)$$

ab (or c^2 if $a = b = c$) must be greater than P for saturation.

This salt, however, has no finite iso-electric point so that there is only one value of H for saturation for a given value of ab, whether $a \neq b$ or $a = b = c$. The value of g required for saturation then comes from Eq. (24) with H_{25}. The solubility of the salt decreases as g increases, all along its surface of saturation.

Moreover, excess of either MOH or HX_s at constant g continually decreases the solubility of the salt, which does not pass through a minimum with respect to either a or b. If we eliminate b from $P = ab\alpha_+$ and Eq. (24) the result is

$$H^3 - H^2(a - P/a + g) - HW - PK_aW/aK_b = 0, \quad (26)$$

which gives a single value of H for saturation at specified a for any specified value of g; then the corresponding value of b comes from $b = P/a\alpha_+$. If we eliminate a from the same equations we obtain

$$H^5K_b + H^4K_b(b - g - P/b + W/K_b) - H^3W(g + K_b + 2P/b - K_a) - H^2K_aW(b + g + 2P/b + PW/bK_aK_b + W/K_a) - HK_aW^2(1 + 2P/bK_b) - K_a^2PW^2/bK_b = 0. \quad (27)$$

This gives H for saturation at specified b (and g); a then follows from $a = P/b\alpha_+$. That Eq. (27) must give a single value of H for specified b (and g) is not evident by inspection unless we recall that for any actual ampholyte the product $K_a K_b$ is smaller than W; but from $P = ab\alpha_+$, we have $db/da = -P[a(d\alpha_+/da) + \alpha_+]/(a\alpha_+)^2$, which is always negative since $d\alpha_+/da$ is positive at constant g.

If the solution is saturated with MX_s alone, and if **a** and **b** are the number of moles of HX_s and MOH respectively per liter of system, then $\mathbf{h} = \mathbf{a} - \mathbf{b} = a - b$, and

$$P = a(a - \mathbf{h})\alpha_+ = b(b + \mathbf{h})\alpha_+. \tag{28}$$

Combination of this with Eq. (24) would give H as $f(\mathbf{h},g,\text{constants})$.

b. Saturation with salt and ampholyte. In considering the relations for saturation with both MOH and MX_s, or the intersection of the two solubility surfaces, we note first that the surface of saturation with respect to MOH is the same as that already discussed under Fig. 21–1 for the system ampholyte and strong acid. Since however it is now intersecting a different surface, that of MX_s rather than that of HX_s, the intersection curve, or the curve of twofold saturation, now called curve $e'e''$, is different.

For the general composition of the solution on the curve $e'e''$, the combination of Eq. (24) with the solubility products gives

$$H - OH = P(OH)/P_b + P_a/H - P_b/OH + g, \tag{29}$$

$$H = gW/2(P_b + W) + \sqrt{()^2 + W^2(P + P_b + P_s)/P_b(P_b + W)}, \tag{30}$$

and

$$a = -gP/2(P + P_b + P_s) + \sqrt{()^2 + P^2(P_b + W)/P_b(P + P_b + P_s)}. \tag{31}$$

The solubility products alone give

$$H = PW/aP_b, \tag{32}$$

and when this is combined with Eq. (8), the direct relation between a and b, describing the projection of the curve on the a/b plane, is

$$a = (b - P_b/K_b)P/2P_s \pm \sqrt{()^2 - P^2/P_s}, \tag{33}$$

or

$$b = P/a + aP_s/P + P_b/K_b. \tag{34}$$

Since $db/da = -P/a^2 + P_s/P$, the curve passes through a minimum of b at a point m where $b_m = S_{\min}$ for MOH, and $H = H_{ie}$. Also,

$$a_m = P/\sqrt{P_s}, \tag{35}$$

and

$$g_m = \sqrt{WK_a/K_b} - \sqrt{WK_b/K_a} - P/\sqrt{P_s}. \tag{36}$$

If the salt is a stable solid phase the solubility product $[M^+][X_s^-]$ for the salt will be smaller than the product $[M^+][X_s^-]$ for the mixture of the solids $MOH + HX_s$. Hence if the salt is a stable solid phase, a_m will be smaller and g_m higher as plotted below in the projection of Fig. 21–3, than in Fig. 21–1.

If we set $a = b = c$ in Eq. (34) we have $c^2(1 - P_s/P) - cP_b/K_b - P = 0$. The curve crosses the diagonal therefore only if $P_s < P$ and then it crosses it only once. If $P_s > P$ however, the curve is not asymptotic to the diagonal with increasing a and b but bends back upon itself when its slope reaches the value 1 at a point which will be called point n, where, from $db/da = 1$,

$$a_n = P/\sqrt{P_s - P}. \tag{37}$$

Hence from Eqs. (32) and (34) respectively

$$H_n = (W/P_b)\sqrt{P_s - P}, \tag{38}$$

$$b_n = P_b/K_b + \sqrt{P_s - P} + P_s/\sqrt{P_s - P}. \tag{39}$$

The value of g_n follows from Eq. (29) as

$$g_n = (W/P_b)\sqrt{P_s - P} - (2P + P_b)/\sqrt{P_s - P}. \tag{40}$$

It will also be necessary, as in connection with Eq. XIX(57), to have an expression for the composition on the curve $e'e''$ for a specified value of $\mathbf{h}$, applying when, with MX_s as sole precipitate, the solution is either just reaching or just leaving the curve. This will represent the intersection of curve $e'e''$ with a vertical plane of specified value of $\mathbf{h}$. Combination of Eq. (28) with $P_b = b\alpha_+OH$ gives

$$H = -W(\mathbf{h} + P_b/K_b)/2P_b \pm \sqrt{(\quad)^2 - W^2(P_s - P)/P_b^2}. \tag{41}$$

If this expression for H moreover is equated with H_{30}, the resulting quadratic equation, called Eq. $H(30{=}41)$, gives g in terms of $\mathbf{h}$ and constants; similarly the equation $H(32{=}41)$ gives a in terms of $\mathbf{h}$ and constants. These expressions give only one value of H, g, and a if $P_s < P$ whatever the value of $\mathbf{h}$, for as explained in connection with point n the curve $e'e''$ is then cut only once by any line of constant $\mathbf{h}$. But if $P_s > P$, $\mathbf{h}$ must be negative and the expressions give two values of H, g, and a for a specified value of $\mathbf{h}$ $(< \mathbf{h}_n)$; as explained in connection with point n, and as may be seen from Eqs. (37) and (39), $\mathbf{h}_n$ (or $a_n - b_n$) is negative if $P_s > P$.

The corresponding condition for a solution reaching or leaving the curve with MOH as sole precipitate is given by either Eq. (31) or (32) with $a = \mathbf{a}$.

2. GRAPHICAL CONSIDERATIONS. The case with $P_s > P$ is shown in schematic projection on the a/b plane, in Fig. 21–3, which may then be used in connection with the consideration of the effect of change of g upon an unsaturated solution with known values of $\mathbf{a}$ and $\mathbf{b}$ and hence of $\mathbf{h}$. If the composition, $\mathbf{P}$, lies below the line S_mm and the curve me″, no precipitation occurs with decreasing g. With increasing g, MX_s precipitates when its

solubility surface is reached. H_{req} is given by Eq. (25) and g_{req} then by Eq. (24). Moreover, MOH never appears as a second precipitate in this case unless the point **P** lies between the curve ne″ and the line $\mathbf{h}_n$ tangent to the curve at point n. Now the precipitation of MX_s, causing the composition

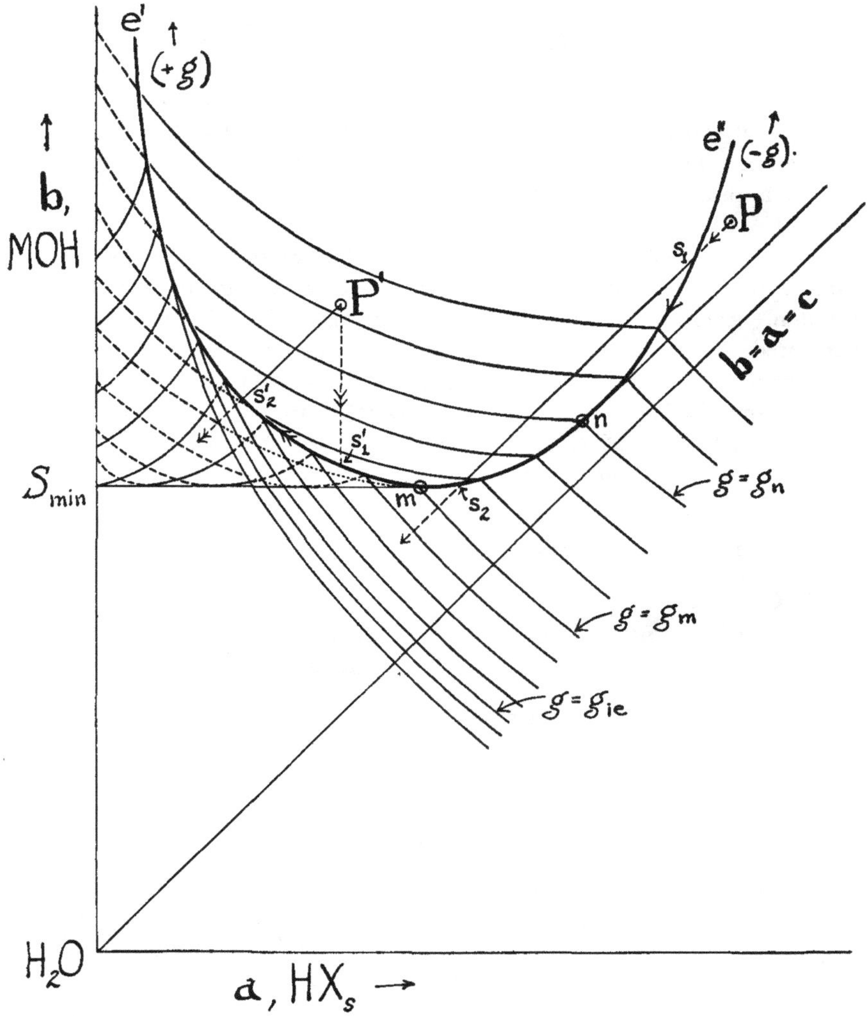

Fig. 21–3. Saturation with ampholyte MOH and its salt of strong acid, MX_s.

of the solution to move toward the H_2O corner along a line of constant **h**, brings the solution to the curve ne″ at point s_1. At this point $H = H_-$ from Eq. (41), and g_{req} comes from Eq. (29) with this value of H, or as the lower value of g from Eq. $H(30=41)$. With continued increase of g the solution follows the curve e′e″, and in the process the quantity of solid MOH first increases and then decreases. While the solution is saturated with both solids, its composition for specified g may be calculated through Eqs. (30–34). Depending on the value of **h** for **P**, the solution, saturated with the two solids, may or may not traverse the point m, but it always traverses the point n before the MOH redissolves. MOH redissolves when the point s_2 is reached, again fixed by the line of constant **h** through **P**. At this point the value of H is H_+ from Eq. (41) and g_{req} is the higher value of g from Eq. H(30=41).

If the composition of the unsaturated solution, originally at low g, is between the line S_mm and the curve e′m, then increase of g causes MOH to precipitate and redissolve before MX_s precipitates. Finally, if the original composition lies above the curve e′e″, as at **P′**, the first precipitate is again MOH, appearing when H has the value H_- of Eq. (8), with $S = \mathbf{b}$; g_{req} then comes from Eq. (24). Precipitation of MOH, causing the composition of the solution to move vertically downward from **P′**, finally brings the solution to point s_1' on the curve e′e″; this point may be on either side of point m and of point n. At point s_1' with $a = \mathbf{a}$, g_{req} is given by Eq. (31). The solution, saturated with both solids, now follows the curve e′e″ toward point e′, its composition being given through Eqs. (30–34). At point s_2', fixed by the line of constant **h** through **P′**, a line parallel to the diagonal $\mathbf{b} = \mathbf{a}$, MOH redissolves, leaving MX_s as sole precipitate. At this point $H = H_+$ from Eq. (41) and g_{req} is the higher value of g from Eq. H(30=41).

E. Salt of Ampholyte and Weak Monobasic Acid

1. General.

a. Saturation with salt alone. The general conditions for electroneutrality and for saturation with respect to the salt MX are

$$H - OH = a\alpha + b(\alpha_- - \alpha_+) + g, \tag{42}$$

and $P = [M^+][X^-] = b\alpha_+ a\alpha$. For the pure salt at concentration c ($= a = b$), $P = c^2\alpha_+\alpha$, and

$$H - OH = c\beta + g = c(\alpha + \alpha_- - \alpha_+) + g. \tag{43}$$

Since the charge coefficient β for this salt can have the value zero, there will be an iso-electric point for the salt besides one for the ampholyte. For the salt the value of H_{ie}, when $\beta = 0$, is given by the equation

$$H^3K_b - HW(K_a + A) - 2AK_aW = 0; \tag{44}$$

the minimum solubility of the salt, c_{min}, is then determined from $P = c^2\alpha_+\alpha$ with H_{ie}. As in the case of the simple weak-weak salt in Chapter XIX, there will therefore be a bulge on the solubility surface of this salt at constant $H = H_{ie}$ and with variable g; and along this bulge the product ab is a minimum, with $(ab)_{min} = c^2_{min}$. The projection of this bulge on a horizontal a/b plane is the hyperbola $ab = c^2_{min}$. Since H_{ie} of the salt is not the same as that of the ampholyte, c_{min} of the salt must be greater than S_{min} of the ampholyte. According to Eq. (44) moreover it is seen that H_{ie} of the salt is higher than H_{ie} of the ampholyte, which is at $H = \sqrt{WK_a/K_b}$.

The condition of saturation for a specified value of ab is, from $P = ab\alpha_+\alpha$,

$$H^3K_b - H^2[AK_b(ab/P - 1) - W] + HW(A + K_a) + AK_aW = 0, \quad (45)$$

which applies with c^2 for ab if $a = b = c$. This gives a positive value of H only if ab (or c^2) $> c^2_{min}$. This minimum must therefore be large enough for the coefficient of the second term of this equation to be negative; c^2, or ab, must be at least greater than $P(1 + W/AK_b)$, but the actual value of c_{min} requires the calculation of H_{ie} for use in $P = c^2\alpha_+\alpha$, as already pointed out. If this minimum requirement for c or ab is satisfied, then Eq. (45) gives two values of H required for saturation with respect to the salt MX for specified values of a and b, or of c, one on each side of the iso-electric point of the salt. Since the charge coefficient of this salt is not symmetrical however, the two values of H given by Eq. (45) will not be equidistant in pH from the iso-electric point. In each case the value of g required for saturation follows from Eq. (42) or (43).

For saturation with respect to MX at specified a and g, Eq. (42) gives, with $P = ab\alpha_+\alpha$,

$$\begin{aligned} &H^4(1 + P/aA) + H^3(A + 2P/a - g) - H^2[A(a + g - P/a) \\ &\quad + W(1 + PK_a/aAK_b)] - HW(A + 2PK_a/aK_b) - APK_aW/aK_b = 0; \quad (46) \end{aligned}$$

then the corresponding value of b may be calculated from $P = ab\alpha_+\alpha$, with H_{46}. Eq. (46) with $A = \infty$ reduces to Eq. (26). For H at specified b and g, we have in fact Eq. (27), which does not involve explicitly the ionization constant of the acid. In this way the solubility curve (the variation of a with respect to b) at constant g may be plotted.

Finally, neither a nor b passes through a minimum in the solubility curve at constant g. This curve passes through a point at which $H = H_{ie}$ of the salt and $ab = (ab)_{min}$. As noted under Eq. (44), H_{ie} of the salt is higher than H_{ie} of the ampholyte. Increase of a relative to the point where $ab = (ab)_{min}$ must cause further increase of H. Increase of b relative to the same point decreases H, since H is greater than H_{ie} of the ampholyte, but it can never bring H down to H_{ie} of the ampholyte (see Chapter IV, Section D3). Or, for saturation with the salt H is always higher than H_{ie} of the ampholyte and the ampholyte always acts as a base in respect to its effect upon H. (Since this holds for any given value of g, then the value of H anywhere on the surface of saturation with respect to MX is always higher than H_{ie} of the ampholyte.) Therefore the relation between a and b is similar to that for the salt of simple weak base and weak acid, discussed in Chapter XIX, Section B2a.

From $P = ab\alpha_+\alpha$ and Eq. (42) at constant g,

$$d(b\alpha_+)/db = [\alpha_- + b(d\alpha_-/db) - dD/db]/[1 + P/(b\alpha_+)^2]. \quad (46a)$$

If the ampholyte acts as a base, decreasing H, then $d\alpha_-/db$ is positive and dD/db negative, so that $d(b\alpha_+)/db$ is positive. At the same time,

$$da/db = - [P/(b\alpha_+\alpha)^2][\alpha d(b\alpha_+)/db + b\alpha_+(d\alpha/db)]. \tag{46b}$$

Hence, since $d\alpha/db$ is positive when the ampholyte acts as a base, da/db is always negative, and there is neither a minimum of a nor one of b in the solubility curve of the salt.

b. Saturation with salt and ampholyte. For the composition of the solution on the curve $e'e''$, for saturation with both MOH and MX, we have, proceeding as in Section D, Eqs. (29) and (30) applying unchanged, while

$$a = a_{31} + PW/AP_b. \tag{47}$$

In place of Eq. (32), we have

$$H = PW/P_b(a - PW/AP_b); \tag{48}$$

and the direct relation between a and b is

$$a = a_{33} + PW/AP_b, \tag{49}$$

$$b = P/(a - PW/AP_b) + aP_s/P - P_a/A + P_b/K_b. \tag{50}$$

According to Eqs. (47) and (49) therefore, the curve $e'e''$ of the present case is the same as that for Fig. 21-3, displaced to the right by the amount PW/AP_b. Thus the asymptotic value for low a at $b = \infty$ is zero in Fig. 21-3 and PW/AP_b in the present case. The curve passes through a minimum of b (point m) where $b_m = S_{min}$ for MOH, $H = H_{ie}$, and

$$a_m = P/\sqrt{P_s} + PW/AP_b, \tag{51}$$

a value between that in Fig. 21-1 and that in Fig. 21-3; g_m is still given by Eq. (36).

As in Fig. 21-3, the curve passes through a point n, where its slope is 1, only if $P_s > P$; but because of the displacement of the curve to the right by the constant amount PW/AP_b, point n may be on either side of the diagonal $\mathbf{a} = \mathbf{b}$ (or $\mathbf{h}_n$ may be either positive or negative), so that the curve may cross the diagonal twice. If $P_s < P$ then, the curve crosses the diagonal only once; if $P_s > P$, it crosses it twice if $\mathbf{h}_n > 0$ and not at all if $\mathbf{h}_n < 0$. H_n is given by Eq. (38), g_n by Eq. (40), and b_n by Eq. (39), while

$$a_n = P/\sqrt{P_s - P} + PW/AP_b, \tag{52}$$

so that $\mathbf{h}_n = PW/AP_b - 2\sqrt{P_s - P} - P_b/K_b$.

Finally, the intersection of the curve with a plane of specified $\mathbf{h}$ is given by

$$H = -W(\mathbf{h} + P_b/K_b - PW/AP_b)/2P_b \pm \sqrt{(\quad)^2 - W^2(P_s - P)/P_b^2}, \tag{53}$$

to be compared with Eq. (41) and with Eqs. xix(57) and xix(59). Again this equation, giving two values of H if $P_s > P$, gives the value of H for a solution reaching or leaving the curve e′e″ with MX as sole precipitate. For the value of a as $f(\mathbf{h})$ we would equate H_{53} with H_{48} as Eq. $H(48{=}53)$; for g we use Eq. $H(30{=}53)$.

The corresponding equations for a solution reaching or leaving the curve with MOH as sole precipitate are simply Eqs. (47) and (48) with $a = \mathbf{a}$.

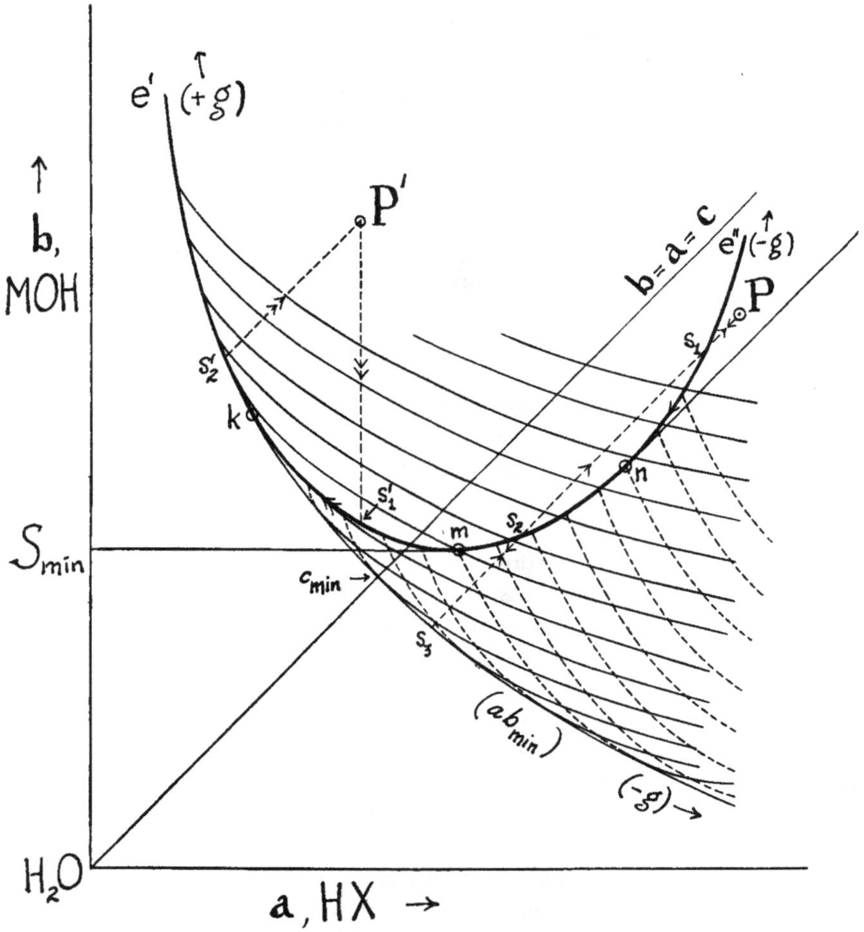

Fig. 21–4. Saturation with ampholyte MOH and its salt of weak acid, MX.

The projection of the relations involving saturation with both MOH and MX may appear schematically as in Fig. 21–4, combining some features of Fig. 19–4 with some of Fig. 21–3. Again the MOH saturation surface is identical with that in Figs. 21–1 and 21–3 up to its intersection with a second saturation surface. The contours of the MOH surface have therefore been omitted in Fig. 21–4 for the sake of clarity; all the contours shown pertain to the surface for saturation with MX, which, like that in Fig. 19–4, rolls back on itself along a bulge of minimum solubility, the projection of which is the hyperbola $ab = (ab)_{min} = c^2_{min}$. In the projection the hyperbola is merely the envelope of the solubility contours of the salt at constant g as they pass through the point of minimum ab. The envelope $(ab)_{min}$ meets the curve of twofold saturation, e′e″, tangentially at the point k. The MX saturation surface is doubled, in this projection, between the curve e′e″ and the envelope $(ab)_{min}$; above curve e′e″ it is a single surface. The position of point k is that point on the curve e′e″ where the product ab is a minimum. The values of a_k, b_k, and g_k could be obtained from Eqs. (48), (50), and (30) respectively after calculation of H for the iso-electric point of the salt, from Eq. (44).

The value of H along the envelope S_{min}m is H_{ie} of the ampholyte; along the envelope $(ab)_{min}$, $H = H_{ie}$ of the salt. On the curve e′e″, H increases above H_{ie} of the ampholyte from point m to point e′ and decreases from point m to point e″; at point k, $H = H_{ie}$ of the salt. All the salt contours, or salt solubility curves at constant g, are hyperboloid in shape, with neither a minimum of a nor one of b.

2. System HX–MOH–H_2O; Effect of g.

a. Graphical considerations. Fig. 21–4 may be used in following the effect of change of g upon an unsaturated solution with composition **P**, fixed by its original values of **a** and **b**. Certain simple principles used in the graphical interpretation of the effect will here be recapitulated. If the solution is saturated with MOH as sole precipitate, its composition, traveling on the MOH surface, moves vertically up or down in the projection on the line of constant **a** through **P**, according to whether the solid is being dissolved or precipitated. If the solution is saturated with MX as sole precipitate, its composition moves on a line of constant **h** through **P**, a line with unit slope or parallel to the diagonal, toward or away from the H_2O corner, according to whether the solid is being precipitated or dissolved. If **P** is below the curve e′e″ (with the exception of the special region between the section ne″ and the line $\mathbf{h}_n$, which is tangent to the curve at point n), the composition of the solution, which refers to its coordinates a and b, returns to point **P** without ever reaching the condition of twofold saturation. When the solution returns to **P** the solid first precipitated will have redissolved and the solution is again unsaturated. If the solution, however, saturated with its first precipitate, reaches the curve e′e″, the second solid there appears, mixed with the first. The solution saturated with both solids then follows the curve in the direction of point e′ with increasing g and of point e″ with decreasing g, until one of the solids redissolves. This happens when the composition of the solution reaches a point on the curve e′e″ lying again on either a line of constant **h** or a line of constant **a** passing through the fixed point **P**. Then the solution, saturated with only one solid, leaves the curve to travel on the saturation surface of the residual solid, until the solid redissolves. This occurs when the composition of the saturated solution, following a line of constant **h** or constant **a** back toward **P**, finally reaches the composition **P** again to become an unsaturated solution. Since both the ampholyte and the salt have isoelectric points, in other words, the solution, whatever its composition in respect to **a** and **b**, is always unsaturated either at sufficiently high g or at sufficiently low g; this disregards the possible precipitation of solids other than MOH and MX.

If the composition of the unsaturated solution, then, falls between the H_2O corner and the two envelopes, S_mm and $(ab)_{min}$, change of g causes no precipitation. If it lies above S_mm and to the left of the curve e′k and the $(ab)_{min}$ envelope, change of g causes merely the precipitation and redissolving of MOH. If it lies above $(ab)_{min}$ and below S_mm, mn, and the line $\mathbf{h}_n$, then change of g causes merely the precipitation and redissolving of MX. In the small region between the two envelopes and the curve km, increase of g from low g causes precipitation and redissolving of MOH and then precipitation and redissolving of MX, but again the solution never becomes saturated with both solids. In all these cases the value of H_{req} for precipitation and redissolving of each solid as *sole* precipitate, which we shall distinguish as $H_{MOH\downarrow}$, $H_{MOH\uparrow}$, $H_{MX\downarrow}$, $H_{MX\uparrow}$, may be calculated through the pertinent solubility product for the known values of **a** and **b**, or from Eq. (8) for MOH, with $S = \mathbf{b}$, and from Eq. (45) for MX; g_{req} follows in each case from Eq. (42). The value of g required to reach the envelope S_mm, or the value of g for maximum precipitation of MOH, is $g = (H - OH - \mathbf{a}\alpha)_{ie}$, with $H = H_{ie}$ of the ampholyte. The value of g required to reach the envelope $(ab)_{min}$, or the value of g for maximum precipitation of MX, is $g = (H - OH - \mathbf{h}\alpha)_{ie} = [H - OH + \mathbf{h}(\alpha_- - \alpha_+)]_{ie}$, with $H = H_{ie}$ of the salt.

For a solution with the composition **P**, shown in Fig. 21–4, between the curve ne″ and the line $\mathbf{h}_n$, the first precipitate appearing with increasing g is MX. When the solution composition reaches the curve ne″ at point s_1, MOH appears as a second precipitate mixed with the MX. This occurs when H has the value H_- of Eq. (53), and g_{req} follows as g_- (the lower value) from Eq. $H(30{=}53)$. The solution then follows the curve, passing through the point n; while it is saturated with both solids, the composition of the solution, for specified g, may be calculated through Eqs. (30) and (47–50). During the course of the solution along the curve, the quantity of MOH first increases and then decreases. When the solution reaches a point, s_2, on the left of point n, and fixed again by the line of constant **h** through **P**, MOH will have redissolved, leaving MX again as sole solid. This occurs with H_+ of Eq. (53), while g_{req} is g_+ of Eq. $H(30{=}53)$. Then the solution leaves the curve and travels on the MX surface, always on a course fixed by the **h** line through **P**. It therefore travels around the bulge of the MX surface, touching the $(ab)_{min}$ curve at point s_3, when MX begins to redissolve. Then the solution travels back to **P** (along the line s_3**P**, a straight line in projection) on the upper face of the MX surface, which is of course above the intersection curve e′e″ in respect to g, and returns to point **P**, becoming again unsaturated at

$H_{MX\uparrow}$. Point s_2 may be on the left of point m and even of point k, depending on the value of **h** at **P**. If it is on the left of point k, the only difference is that the solution remaining when MOH redissolves begins at once to return to **P** since the MX surface above point k does not extend to the hyperbola $(ab)_{min}$.

Finally, for a composition **P**′ above the curve e′e″, increase of g causes MOH to appear as first precipitate. MX appears mixed with MOH when the curve e′e″ is reached at some point s_1', which is vertically below **P**′, and which may be anywhere on the curve. This occurs when H has the value given by Eq. (48) and g has the value in Eq. (47), both with $a = \mathbf{a}$. While saturated with both solids, the solution travels on the curve toward point e′, with its composition determined by Eqs. (30) and (47–50) for specified g. At point s_2', fixed by the line of constant **h** through **P**′, the original precipitate of MOH will have redissolved, leaving MX as sole solid. This occurs at H_+ of Eq. (53), and at g_+ from Eq. H(30=53). The solution, then redissolving the MX, travels back to **P**′ where it becomes unsaturated again at $H_{MX\uparrow}$. If point s_2' is on the right of point k, then, after the redissolving of MOH, the solution first travels to the bulge of the MX surface, at $ab = (ab)_{min}$, before returning to point **P**′ on the upper face of the MX surface.

b. Algebraic considerations. In practical application the foregoing considerations require that the curves e′e″ and $(ab)_{min}$ have been plotted. The necessary calculations for a single given solution, however, may be made without the plotting of the two curves. For the effect of g upon a specific unsaturated solution with the composition **a** and **b** and with known values of the constants K_a, K_b, P_b, A, and P, we fix point m, for which $b = S_{min}$ of Eq. (6) and $a = a_m$ of Eq. (51), and point n, also given explicitly through Eqs. (39) and (52). Subsequent considerations then depend on the principle that the sequence of phase changes (appearance and disappearance of any precipitate) must occur in the order of increasing H for increasing g, and of decreasing H for decreasing g. (This principle was used in Section F1 of Chapter XIX.)

If $\mathbf{b} < b_m$, the only effect if any of increase of g will be the precipitation and redissolving of MX. Whether or not the salt ever precipitates, or whether or not, in other words, the original solution lies above the unknown curve $(ab)_{min}$, is determined through Eq. (45), which gives a real and positive value of $H_{MX\downarrow}$ only if $\mathbf{ab} > (ab)_{min}$.

If $\mathbf{b} > b_m$ and $\mathbf{a} < a_m$, the first precipitate must be MOH, appearing at $H_{MOH\downarrow}$. Now if $\mathbf{a} < PW/AP_b$, the asymptotic value of a at point e′, MX never appears as a second precipitate and the process ends with the mere redissolving of MOH at $H_{MOH\uparrow}$. Whether or not, with larger **a**, MX does appear as a second precipitate mixed with MOH (or whether or not the original solution is above the unknown curve e′e″) is determined by the value of H_{48} with $a = \mathbf{a}$. If this is greater than $H_{MOH\uparrow}$, the solution never becomes saturated with both solids. (Whether or not, in this case, the composition of the solution is in the special region between the curve km and the envelope

$(ab)_{min}$, depends again simply on whether Eq. (45) gives positive values of H for the precipitation and redissolving of MX after the redissolving of MOH.) If $H_{48} < H_{MOH\uparrow}$, MX precipitates before MOH redissolves, as at point s_1' in Fig. 21–4. Then MOH redissolves, leaving MX as sole precipitate, when $H = H_+$ of Eq. (53), as at point s_2'.

If $\mathbf{b} > b_m$ and $\mathbf{a} > a_m$, the first precipitate is MOH, or the original composition is above the curve e′e″, if $H_{MOH\downarrow} < H_{MX\downarrow}$. In this case MX later appears as second precipitate mixed with MOH (point s_1') and the original MOH redissolves (point s_2') to leave MX alone, all as in the quadrant just discussed. If, however, $H_{MOH\downarrow} > H_{MX\downarrow}$, MX is the first precipitate, and saturation with two solids is not reached unless $\mathbf{h} < \mathbf{h}_n$ together with $\mathbf{a} > a_n$ (or $\mathbf{b} > b_n$); for such special composition MOH appears and redissolves, as a temporary precipitate mixed with the MX, when H has the successive values given by Eq. (53).

3. Reactions Involving the Salts NaX and MNO_3. The considerations given in connection with Fig. 21–4 may also be applied for the effect of "g" upon solutions containing combinations such as : I, HX($\mathbf{a}$) and MNO_3($\mathbf{b}$), the salt of the ampholyte and foreign strong acid; II, NaX($\mathbf{a}$) and MOH($\mathbf{b}$); III, NaX($\mathbf{a}$) and MNO_3($\mathbf{b}$). The diagram is the same and all the calculations are the same; but the quantity "g" in the equations must be replaced by $g + \mathbf{b}$ in I, $g - \mathbf{a}$ in II, and $g - \mathbf{a} + \mathbf{b}$ in III, if g is the net concentration of foreign strong acid and base other than that contributed by the salts NaX and MNO_3.

a. Reaction between NaX *and* MNO_3. We shall here consider some questions regarding the reaction between the salts NaX($\mathbf{a}$) and MNO_3($\mathbf{b}$), with $\mathbf{a}$ and $\mathbf{b}$ as the experimental variables. This reaction is the analog of that studied as "Case III" in Section D2*c* in Chapter XIX, with MOH as a simple base rather than an ampholyte. We shall concern ourselves only with that region of the solubility diagram on "plane III" involving saturation with MOH and MX, on the curves $S_{III}e_{III}$ and $e_{III}f_{III}$ respectively. S_{III}, as in Fig. 19–5, corresponds to c_{max} of the salt MNO_3; its value, determined by H_{12}, is greater than S_{min} of the ampholyte. For the coordinates of the intersection of the plane with the curve e′e″, or point e_{III}, we combine the equation of electroneutrality for the unsaturated solution, $D = \mathbf{b}(1 + \alpha_- - \alpha_+) - \mathbf{a}\rho$, with $P_b = \mathbf{b}\alpha_+ OH$ and with H from Eq. (48) used with $a = \mathbf{a}$, obtaining

$$a_{e_{III}} = \frac{P}{2(P_b + 2P_S)}\left(\frac{PW}{AP_b} - \frac{P_b}{K_b}\right) + \sqrt{(\quad)^2 + \frac{P^2W}{P_b(P_b + 2P_S)} + \frac{PW}{AP_b}}. \qquad (54)$$

This is to be compared with Eq. XIX(67). This value of a in Eq. (50) then gives $b_{e_{III}}$.

Fig. 21–5, which is to be compared with Fig. 19–5, shows a possible schematic arrangement for the solubility curves for the part of plane III under consideration. If $a_{e_{III}}$ in Eq. (54) is less than a_m in Eq. (51), as assumed in the particular arrangement in this figure, the solubility curve of MOH in the plane of the system meets the curve e′e″ on the left of its minimum point, m, and otherwise on the right of it. Unless the solubility of the salt is too low, the curve $S_{III}e_{III}$ passes through a minimum, as shown in the Figure. If HX is a strong acid the curve $S_{III}e_{III}$ is horizontal since the salt NaX_s would have no effect on the solubility of MOH. Since the ampholyte acts as a base in respect to saturation with the salt MX, the solubility curve

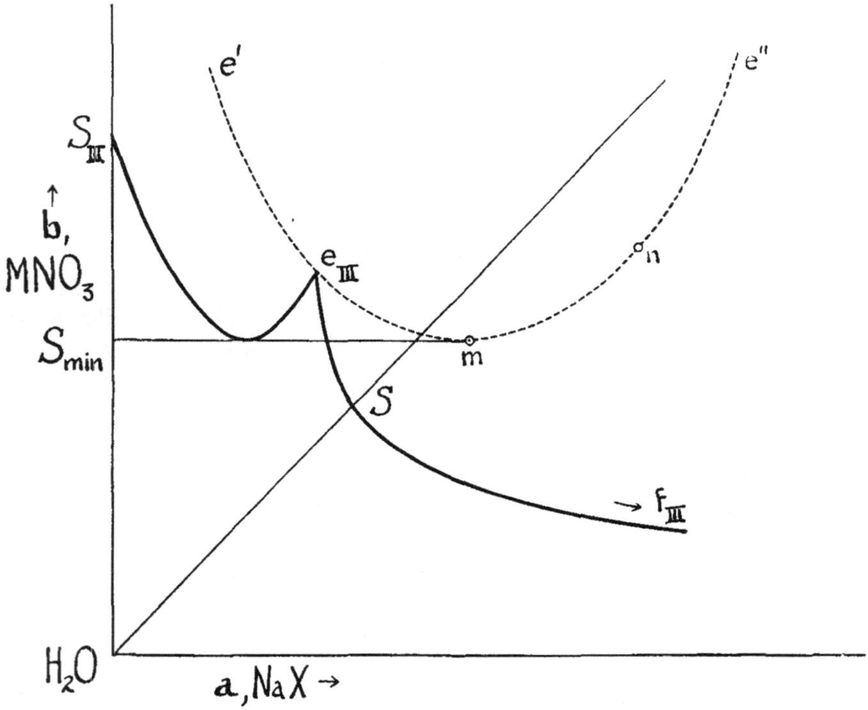

Fig. 21–5. Part of the system MNO_3–NaX–H_2O.

$e_{III}f_{III}$, like that in Fig. 19–5, is expected not to pass through either a minimum of a or one of b.

*a*1. *Addition of* NaX *to aq.* MNO_3 (**b** *known*). If $\mathbf{b} < S_{min}$ of the ampholyte, MOH can never precipitate, and MX precipitates when the composition reaches the curve $e_{III}f_{III}$. For saturation with MX as sole precipitate at specified **b**, $H_{MX\downarrow}$ and $H_{MX\uparrow}$ would be calculated from

$$H - OH = \mathbf{b}(1 + \alpha_- - \alpha_+) - HP/\mathbf{b}A\alpha_+, \tag{55}$$

and the values of $\mathbf{a}_{req}$ then follow from $\mathbf{a} = P/\mathbf{b}\alpha\alpha_+$.

While the solution is saturated with MX alone in this system, the addition of either NaX or MNO_3, it will be recalled, causes its composition to follow the actual curve $e_{III}f_{III}$ on plane III.

If **b** is greater than S_{min} of the ampholyte and less than $b_{e_{III}}$, MOH precipitates and redissolves before the precipitation of MX; the two values of $\mathbf{a}_{req}$ then follow from $D = \mathbf{b} + P_a/H - P_b/OH - \mathbf{a}\rho$, with the values of H from Eq. (8) for specified **b**. If $S_{III} < b_{e_{III}}$ therefore, saturation with both solids can not occur upon the addition of NaX to aq. MNO_3. The curve e′e″ can be reached only if $S_{III} > b_{e_{III}}$, and then if $\mathbf{b} > b_{e_{III}}$. In this case solid MOH appears when **a** has the value required by Eq. (55) with H_+ of Eq. (8); the solution then travels on the MOH saturation surface on a course which is above the plane of Fig. 21–5 and which is horizontal in projection on the figure. Its composition for specified **a** is determined by Eq. (15) with $a = \mathbf{a}$ and $g = \mathbf{b} - \mathbf{a}$; and with H_{15} the concentration b follows from P_a or P_b. MX appears as a second precipitate mixed with MOH when the curve e′e″ is reached at some point above point e_{III}, with $a = \mathbf{a}$. Hence $\mathbf{a}_{req}$ may be calculated from Eq. (47) with $a = \mathbf{a}$ and $g = \mathbf{b} - \mathbf{a}$. With continued addition of NaX the solution, saturated with both solids, follows the curve e′e″ in the direction of point e″, with composition for specified **a** determined by Eqs. (30) and (47–50). When the solution reaches point e_{III}, on the plane of the figure, MOH will have redissolved, leaving MX as sole precipitate. At this point therefore, $\mathbf{a}_{req} = a_{e_{III}}$. Continued addition of NaX now causes the composition of the solution, saturated with MX alone, to follow the salt solubility curve $e_{III}f_{III}$ toward point f_{III}.

a2. Addition of MNO_3 *to aq.* NaX (**a** *known*). If the first curve encountered with increasing **b** is $S_{III}e_{III}$, the first precipitate is MOH, appearing when H has the value given by

$$H - OH = P_a/H\alpha_- + P_a/H - P_b/OH - \mathbf{a}\rho; \qquad (56)$$

then $\mathbf{b}_{req} = P_a/H\alpha_-$. The solution then travels on the MOH saturation surface on a course of constant a ($= \mathbf{a}$), its composition being determined again by Eq. (15) for specified **b**, with $a = \mathbf{a}$ and $g = \mathbf{b} - \mathbf{a}$. If $\mathbf{a} < PW/AP_b$, the asymptotic limit of the curve e′e″, MX never appears as second precipitate; nor does the MOH redissolve, since Eq. (56), a cubic in H, gives only one value of H for any values of the parameters involved. With $\mathbf{a} > PW/AP_b$, MX appears as second precipitate when the curve e′e″ is reached, above point e_{III}, at $a = \mathbf{a}$; now $\mathbf{b}_{req}$ comes from Eq. (47) with $a = \mathbf{a}$ and $g = \mathbf{b} - \mathbf{a}$. With continued addition of MNO_3 the solution, saturated with both solids, follows the curve e′e″ toward point e′, with composition determined by Eqs. (30) and (47–50) for specified **b**. Neither solid now ever redissolves.

If $\mathbf{a} > a_{e_{III}}$ the first curve encountered is $e_{III}f_{III}$ and the first precipitate is MX, appearing when H is defined by $D = P(1 + \alpha_- - \alpha_+)/\mathbf{a}\alpha\alpha_+ - \mathbf{a}\rho$; then $\mathbf{b}_{req} = P/\mathbf{a}\alpha\alpha_+$. The solution saturated with MX as its first precipitate follows the curve $f_{III}e_{III}$ up to point e_{III}, where MOH appears mixed with the MX. At this point $\mathbf{b}_{req} = b_{e_{III}} + (\mathbf{a} - a_{e_{III}})$. With continued increase of $\mathbf{b}$ the solution, saturated with both solids, follows the curve e′e″ toward point e′.

If the original unsaturated solution of one salt to which the other is being added also contains a net concentration $\mathbf{g}$ of foreign strong acid or base other than that contributed by the two salts themselves, the various electroneutrality equations of this Section (E3*a*) still apply, with $+\mathbf{g}$ as an additional term, while Eqs. (15), (30), and (47–50) are used with $g = \mathbf{g} + \mathbf{b} - \mathbf{a}$.

b. Reaction of solid MOH *with aq.* NaX. If the solid ampholyte, always present in excess, is suspended in a solution of NaX at concentration $\mathbf{a}$ (with or without foreign strong acid at the constant net concentration $\mathbf{g}$), the value of H of the saturated solution is given by Eq. (15) with $a = \mathbf{a}$ and $g = \mathbf{g} - \mathbf{a}$, as long as MOH is the only solid phase. The concentration b, of dissolved MOH, is then $P_a/\alpha_- H$. With $\mathbf{g} = 0$, this is the curve $S_{II}e_{II}$ on "plane II" for the system NaX–MOH–H_2O; see Fig. 19–5 for orientation. From $D = b(\alpha_- - \alpha_+) - \mathbf{a}\rho + \mathbf{g}$, with $P_b = b\alpha_+ OH$ and with H from Eq. (48) used with $a = \mathbf{a}$, the value of $\mathbf{a}$ required for precipitation of MX on the suspended MOH is

$$\mathbf{a}_{req} = \frac{P}{2(P_b + P_s)}\left[\frac{PW}{AP_b} - \mathbf{g}\right] + \sqrt{(\quad)^2 + \frac{P^2}{P_b}\left(\frac{P_b + W}{P_b + P_s}\right)} + \frac{PW}{AP_b}. \quad (57)$$

With $\mathbf{g} = 0$, this is $a_{e_{II}}$ on plane II, to be compared with Eq. XIX(65) for MOH as a simple base. If the acid HX is strong, the expression, with $\mathbf{g} = 0$, reduces to $\mathbf{a} = P\sqrt{(P_b + W)/P_b(P_b + P_s)}$. When the solution is saturated with both MOH and MX, the composition is given by Eqs. (30) and (47–50) with $g = \mathbf{g} - \mathbf{a}$. With the concentration a thus given by Eq. (47), the number of moles of MX in the precipitate per liter of solution, or $\mathbf{q}$, is $\mathbf{a} - a$.

In reverse, if the suspension, with MX in the precipitate, is allowed to come to equilibrium, the single measurement of H, a, or $\mathbf{q}$ at equilibrium suffices theoretically for the calculation of one of the six quantities $\mathbf{a}$, A, P, K_a, K_b, P_b if the five other quantities and $\mathbf{g}$ are known. From the value of H we may calculate $\mathbf{a}$, P, or one of the ampholyte constants directly through Eq. (30) with $g = \mathbf{g} - \mathbf{a}$; for A as the unknown we would express a as $f(H, \text{constants})$ from the equation $H - OH = P_a/H - P_b/OH + a\alpha - \mathbf{a} + \mathbf{g}$ and substitute the result in Eq. (48) to solve for A. From the value of a, or of $\mathbf{q}$ $(= \mathbf{a} - a)$, the unknown quantity is given directly by Eq. (47). If both $\mathbf{q}$ (or a) and H are measured and if $\mathbf{a}$ is known, certain combinations of two unknown constants may be determined. For example we may calculate P from Eq. (29) with $g = \mathbf{g} - \mathbf{a}$ and then A from Eq. (48), or P_b from Eq. (48) with $a = \mathbf{a} - \mathbf{q}$ and then one of the ionization constants of the ampholyte from one of the Eqs. (29), (30), and (47).

F. Higher Order Ampholytes

1. Precipitation of $Zn(OH)_2$. With b as the concentration of the 2 : 2 ampholyte of type $Zn(OH)_2$,

$$P_b = [Zn^{++}][OH^-]^2 = b\alpha_{+2}(OH)^2. \tag{58}$$

With the transformation of constants explained in Eq. 1(73), or $W/K_{b_2} = K_1^*$, $W/K_{b_1} = K_2^*$, $K_{a_1} = K_3^*$, $K_{a_2} = K_4^*$, then α_{+2} of the ampholyte is taken as ρ of a tetrabasic acid, Eq. 1(74); hence from Eqs. 1(21) and 1(24),

$$\alpha_{+2} = \frac{HK_{b_2}}{W} \bigg/ \left(1 + \frac{HK_{b_2}}{W} + \frac{W}{HK_{b_1}} + \frac{WK_{a_1}}{H^2K_{b_1}} + \frac{WK_{a_1}K_{a_2}}{H^3K_{b_1}}\right). \tag{59}$$

For c_{max} of pure $ZnCl_2$ solution, without "hydrolytic" precipitation of $Zn(OH)_2$, the expression for b from Eq. (58) is substituted for c in

$$H - OH = c(\alpha_{-1} + 2\alpha_{-2} - \alpha_{+1} - 2\alpha_{+2}) + 2c + g, \tag{60}$$

with $g = 0$; with H so found, c_{max} is calculated from Eq. (58).

Eq. (58), if all the constants are known, gives two values of H for saturation at the concentration b: the higher value for precipitation of $Zn(OH)_2$ by addition of b_s to $ZnCl_2$ solution and the lower value for its redissolving in excess of b_s; or the lower value for precipitation of $Zn(OH)_2$ by addition of a_s to sodium zincate solution and the higher value for its redissolving in excess of a_s. The corresponding values of g_{req} or $(a_s - b_s)_{req}$ may be found from Eq. (60) through these critical values of H for the case of $ZnCl_2$ solution; for sodium zincate solution, at concentration c, the electroneutrality equation is still Eq. (60), but with reversed sign of the term $2c$. In either case, whether the solution is one of zinc chloride or of sodium zincate, if $c < c_{max}$ but greater than S_{min} for $Zn(OH)_2$, these values of g_{req} are for the precipitation and redissolving of the precipitate; if $c > c_{max}$, the values are for the prevention of the precipitate.

2. Precipitation of ZnS. With $ZnCl_2$ (b known and fixed) and H_2S (at fixed a), ZnS appears as the first precipitate, rather than $Zn(OH)_2$, when $b\alpha_{+2}a\alpha_2 = P$, in which P is the solubility product of ZnS. As H is decreased, by decrease of g, the first precipitate may be either ZnS or $Zn(OH)_2$, depending on H_{req} for $Zn(OH)_2$ to precipitate, from Eq. (58), and H_{req} for ZnS, from P as just defined, with b and a known. For g, or $(a_s - b_s)$, required, the required value of H is substituted in

$$H - OH = b(\alpha_{-1} + 2\alpha_{-2} - \alpha_{+1} - 2\alpha_{+2}) + 2b + a(\alpha_1 + 2\alpha_2) + g. \tag{61}$$

If $ZnCl_2$ (b known) is treated with $Na_2S(a)$, $Zn(OH)_2$ appears first if H_{req} from Eq. (58) is greater than H_{req} for ZnS, which may be calculated by substitution of $a = P/b\alpha_{+2}\alpha_2$ in

$$H - OH = b(\alpha_{-1} + 2\alpha_{-2} - \alpha_{+1} - 2\alpha_{+2} + 2) + a(\alpha_1 + 2\alpha_2 - 2). \tag{62}$$

Then a_{req} may be calculated from P for the appearance of ZnS as the first precipitate, and from Eq. (62) for the appearance of $Zn(OH)_2$ as the first precipitate.

3. Suspension of an Ampholyte (Example, $Al(OH)_3$) in a Solution of a Foreign Salt. If a salt solution (concentration **a**) is treated with excess of solid, insoluble metallic hydroxide, which may be an ampholyte, the pH of the resulting suspension will be a function of the salt concentration and of the various equilibrium constants involved. Such observations are used in the study of the "exchange" between the anions of the salt solution and the hydroxyls upon the surface of the solid precipitate(1). In this connection, it may be important to know for comparison purposes the pH change to be expected on the basis of the solubility and ionization equilibria possible in the system. In the case of a suspension of "hydrous alumina," or aluminum hydroxide, not all the possible equilibrium constants are known. The simple case for a suspension of a 1 : 1 ampholyte in a solution of the salt NaX has already been treated in Section E3*b* of this chapter, which may therefore serve as reference.

For an approximation in the important case of $Al(OH)_3$, we shall treat it as a 1 : 3 ampholyte having one acid constant, K_a, and three basic constants, the first of which, K_{b_1}, may be assumed to be very large. Then, while $Al(OH)_3$ is the only precipitate in a suspension of excess of the hydroxide in a solution of the salt NaX, such as sodium benzoate, at concentration **a**, we have, with $P_b = [M^{+++}][OH^-]^3$ and $P_a = [H^+][M^-]$, $D = [X^-] - \mathbf{a} + [M^-] - 3[M^{+++}] - 2[M^{++}] - [M^+]$, or

$$D = -\frac{\mathbf{a}H}{H + A} + \frac{P_a}{H} - \frac{P_b}{(OH)^3}\left[3 + \frac{2(OH)}{K_{b_3}} + \frac{(OH)^2}{K_{b_2}K_{b_3}}\right], \qquad (63)$$

$$3H^5P_b + H^4P_b(3A + 2W/K_{b_3}) + H^3W[2AP_b/K_{b_3} + W(P_b/K_{b_2}K_{b_3} + W)] + H^2W^2[W(a + A) + AP_b/K_{b_2}K_{b_3}] - HW^3(P_a + W) - AW^3(P_a + W) = 0. \qquad (64)$$

If AlX_3, such as the benzoate, with solubility product P, is also precipitated when **a** is sufficiently high, the value of H for the solution saturated with both $Al(OH)_3$ and AlX_3 is given, with $(\mathbf{a} - OH\sqrt[3]{P/P_b})$ in place of $(\mathbf{a} - [X^-])$ or of $\mathbf{a}H/(H + A)$, in Eq. (63), by the expression

$$3H^4P_b + 2H^3P_bW/K_{b_3} + H^2W^2(W + P_b/K_{b_2}K_{b_3}) + H\mathbf{a}W^3 - W^4(1 + \sqrt[3]{P/P_b} + P_a/W) = 0. \qquad (65)$$

For the minimum value of **a** required for AlX_3 to be precipitated by the suspension of $Al(OH)_3$, we combine Eq. (65) with the condition

$$\mathbf{a} = [X^-]/\alpha = OH\sqrt[3]{P/P_b}(1 + H/A), \qquad (66)$$

(1) Thus: R. P. Graham and A. E. Horning, *J. Am. Chem. Soc.*, **69**, 1214 (1947).

obtaining

$$3H^4P_b + 2H^3P_bW/K_{b_3} + H^2W^2(W + P_b/K_{b_2}K_{b_3}) + HW^4\sqrt[3]{P/P_b}/A - W^3(P_a + W) = 0. \qquad (67)$$

With H_{67}, $\mathbf{a}_{req}$ follows from Eq. (66).

The known constants for $Al(OH)_3$ are $P_b = 1.9 \times 10^{-33}$, $K_a = 4 \times 10^{-13}$, $K_{b_3} = 7.1 \times 10^{-10}$ (Latimer, Ref. III–1, p. 214).

We shall then assume $K_{b_2} = 10^5 K_{b_3}$ and $K_{b_1} = 10^5 K_{b_2}$, whereupon $P_a = 2.1 \times 10^{-33}$. For benzoic acid, $A = 6.3 \times 10^{-5}$ (Harned and Owen, Ref. v–3, p. 210). For the possible precipitate of aluminum benzoate, reported as "very slightly soluble," we shall assume three different values of P, as 10^{-23}, 10^{-20}, and 10^{-17}, for purposes of illustration. The calculated relations are as follows, for the value of $\mathbf{a}$ required for precipitation of AlX_3 on solid $Al(OH)_3$ suspended in a solution of NaX at the concentration $\mathbf{a}$.

P for AlX_3	H, Eq. (67)	$\mathbf{a}_{req}$, Eq. (66)
10^{-23}	3.2×10^{-8}	0.00054
10^{-20}	3.6×10^{-9}	0.048
10^{-17}	3.6×10^{-10}	4.8

If $\mathbf{a} < \mathbf{a}_{req}$ the pH of the suspension as function of $\mathbf{a}$ is given by Eq. (64), and by Eq. (65) if $\mathbf{a} > \mathbf{a}_{req}$. Both of these equations, for actual use, may be simplifiable to only two or three significant terms, depending upon the values of the constants involved. The three calculations given in the numerical example required only the last three terms of Eq. (67).

Index

GPSR Authorized Representative: Easy Access System Europe - Mustamäe tee 50, 10621 Tallinn, Estonia, gpsr.requests@easproject.com

www.ingramcontent.com/pod-product-compliance
Lightning Source LLC
Chambersburg PA
CBHW060423220326
41598CB00021BA/2273
9780691627243